The
$100
Hamburger

About the Author

John Purner is an avid pilot and freelance writer. His $100 Hamburger website was awarded the top rating from *Plane & Pilot* magazine.

The
$100
Hamburger

A Guide to Pilots' Favorite Fly-In Restaurants

Third Edition

John Purner

New York   Chicago   San Francisco   Lisbon   London   Madrid
Mexico City   Milan   New Delhi   San Juan   Seoul
Singapore   Sydney   Toronto

The **McGraw·Hill** Companies

Cataloging-in-Publication Data is on file with the Library of Congress

Copyright © 2007 by The McGraw-Hill Companies, Inc. All rights reserved. Printed in the United States of America. Except as permitted under the United States Copyright Act of 1976, no part of this publication may be reproduced or distributed in any form or by any means, or stored in a data base or retrieval system, without the prior written permission of the publisher.

1 2 3 4 5 6 7 8 9 0   DOC/DOC   0 9 8 7 6

ISBN-13:   978-0-07-147925-7
ISBN-10:   0-07-147925-2

The sponsoring editor for this book was Steve Chapman, the editing supervisor was Patty Mon, production supervisor was George Anderson, and the project manager was Samik Roy Chowdhury (Sam). It was set in Perpetua by International Typesetting and Composition. The art director for the cover was Margaret Webster-Shapiro.

RR Donnelley was printer and binder.

McGraw-Hill books are available at special quantity discounts to use as premiums and sales promotions, or for use in corporate training programs. For more information, please write to the Director of Special Sales, McGraw-Hill Professional, Two Penn Plaza, New York, NY 10121-2298. Or contact your local bookstore.

This book was printed on acid-free paper.

Contents

Acknowledgments

No man is an island. Certainly not a man with a family. Flying and writing take a lot of time. The income from the latter doesn't always offset the requirements of the former. Flying is expensive in both money and time. Not just the time when you're at the airport or flying the plane, but also the time that you're thinking about and dreaming of it, the time you're planning the next flight and reliving the last one. Time and money. That's all any of us have, really. Of the two, time is the more precious. It is finite; you can't earn more of it or borrow a couple of hours from the corner bank. Once it is gone, it is gone. A memory is made at the expense of an opportunity. Time in the air is time not spent training my daughter's new puppy. Time writing a book is time that can't be spent loving my wife.

I am passionate about flying and writing. I like to communicate. I like to exchange ideas. I like crafting a good sentence and using it as the foundation for a winning paragraph and then hammering more sentences and more paragraphs together to stir you to action to share our passion for flight and our excitement for all of the hardware that makes it possible. I like to debate the rules and the changes and try to sort it all out. I like to weigh in.

I didn't know that I liked writing. After five books and many magazine articles, I now know that I love it. Writing is what I was born to do. It is why God made me. I am grateful to know that, and I am grateful to have the chance to live His plan.

I have the chance to do it because I have a wife who is a saint and a daughter who is an angel. Renée is a photographer. If she wasn't, I don't know if she could possibly understand why I need all of this time to write and why I am driven to do it. She is built of the same stuff. Each of her photographs is like a movie in a single frame. They tell the story of a life well lived or reveal an episode of one about to change. Renée has an understanding heart and possesses the gift of encouragement. "You can do it," "That's great," "Keep at it," and "Try again" are just words. It is the expression on her face and the softness of her voice that makes me believe them.

Jillian's body is just 6 years old; her mind is 35 and her soul is eternal. Jillian is, in a word, good. She is the little girl that her mother and I dreamed of and prayed for. Now she is here and better in every way than our best dream. Just the other day as we worked together on Jillian and Dad's first collaboration a book titled *Little Tear*, she asked me if we could get a yellow airplane. Sounds perfect to me!

Time and money. That's all we have. Money can't buy time, not one second, so of the two it is time that is the more important. Thanks for the time, girls. The time you give me to write and fly, and the time you spend with me when I'm not.

Introduction

Good news!

Since the publication of the second edition of *The $100 Hamburger*, the general aviation landscape has gotten better than we could have imagined. Not since World War II has so much progress been made in aviation in so little time.

In 2003 the least expensive brand-new-from-the-factory "ready-to-fly" bird went for just over $150,000. Today there are several *Burger Flight*–ready ships that can be yours for less than half of that. I'm not considering kit planes that you will spend years assembling in your *spare-time*. I am talking about ready to go, fully certified airplanes.

One of them really has my heart pounding. It's the *Mermaid*, manufactured by **Czech Aircraft Works** (CAW). The *Mermaid* is the first certified amphibian to be offered for sale since the Lake. There is one major difference: the *Mermaid* can be yours or mine for only $87,500, ready to fly. It is an all-aluminum, two-place, 120 mph amphibian. You may operate from land or water or both as you see fit. The gear retracts hydraulically, just as you would expect. The *Mermaid* is the most interesting of the new LSA (**Light Sport Aircraft**) fleet. CAW's U.S. affiliate, **Sport Aircraft Works**, is headquartered in Indiantown, FL at the Indiantown Airport (X58). The *Mermaid* and their other offerings, the *Parrot* and the newly certified *Sport Cruiser*, are all there and available for demo flights. The *Sport Cruiser* was designed in America using American components. It cruises at 125 mph while burning just 4.5 gph of auto fuel. The price? Bring $65,000 and you'll fly it home. $12,000 more will get you a complete glass panel and an autopilot. Amazing!

So far, about 20 wonderful aircraft have been certified under the new LSA rule. Oddly, most of them come from companies based outside of the United States—our rule, our country, our market, and their airplanes. Why? I don't know, but that is currently the way it is. Two U.S. companies are building look-alike J3 Cubs and offering them as LSAs. Those yellow tube and fabric wannabe's are being offered for almost $90,000! That's an outrageous amount. A perfectly restored **REAL** .J3 Cub can be purchased for $40,000 or less. I don't know about you, but I'd rather eat worms than have a knockoff anything. It reminds me of the guys who sport fake Rolex watches. Everybody knows it isn't real. Why embarrass yourself? Currently, RANS is the only American company to offer a new design in this new market segment.

I mention the LSA segment as the first part of our good news story because it promises to bring a ton of new products and interest to the recreational segment of the general aviation market. The best news of all is that you don't have to take a medical every two years to fly one. You must have a pilot's license and a driver's license. That's about it.

Every major aviation convention will have an LSA display area during the 2006 and 2007 flying seasons. Get to Oshkosh or a show near you and check them out. You will be amazed!

There are more new certificated aircraft being offered for sale now than at anytime in my memory. If you haven't taken a demo flight in a Cirrus, Columbia, or Diamond, do so at once. Technology has moved well beyond the Spam can that we all grew up with. The Cessna 172 just celebrated its fiftieth birthday. It is still a very good airplane but is out classed by today's designs. Interesting to note that both the Cirrus SR22 and its counterpart, the Columbia 350, cruise at nearly 190 knots. That is remarkable because neither of them sports retractable landing gear. They do it with straight legs dangling in the breeze. The world's fastest production reciprocating engined, propeller-driven aircraft is the Columbia 400, which cruises in the flight levels at speeds up to 235 knots. It also is a nonretract. The Diamond DA40, also a straight-legged ship, is able to clock 147 knots with a 180 hp engine. All of these modern ships have one thing in common: they are constructed of composite material, not aluminum. They are rivetless. The most interesting fact of all is that the Cirrus SR22 is, today, the world's bestselling private airplane and has been for the past four years.

While these new designs have amazing performance, that is not what has brought them to market prominence. Interest was ignited by their magical panels. Cirrus started the trend by providing GA pilots with cockpit equipment rivaling jetliners. We are all now becoming familiar with the terms **PFD** (**P**rimary **F**light **D**isplay) and **MFD** (**M**ulti-**F**unction **D**isplay). It is wonderful to fly behind those two—or sometimes three—large pieces of glass. The day of the steam gauge is ended. It is difficult to market an airplane today that doesn't have a "glass panel." Every major GA manufacturer has switched over to either the Garmin G1000 or an Avidyne designed panel. All of the information that pilots have always needed to keep their plane going right side up and headed where they want to go are included. Additionally, you can have terrain and traffic avoidance information and real-time NexRAD weather that makes airborne radar systems cry uncle. All of that info and more is presented right on top of your GPS-driven moving map being displayed on the MFD side of the panel. Situational awareness has been raised to a level previously unimagined. The G1000 is totally integrated, including an autopilot. One system from one manufacturer, doing it all.

That brings me to the next area of good news. The Very Light Jets, just a twinkle in our eyes three years ago, are now a reality. Eclipse the Big Daddy in this segment, will make the first customer delivery of its Eclipse 500 VLJ just as this book comes to market. The Eclipse 500 truly is a replacement for any airliner. It is a reliable, high-altitude, high-speed traveling machine. It is a revolutionary product in many, many ways. That is why Vern Raburn, whom I believe is a national hero, deserves all of the success that is coming his way. Already Eclipse has a backlog of over 2,000 orders. It currently sells for $1.4 million. That is approximately 25 percent of the cost of the cheapest business jet and does exactly the same job. The E500 cruises at 375 knots at an altitude of 41,000 feet.

Working with United Airlines and others, Eclipse has put together an impressive transition training program for new/owner pilots. The emphasis is on safety with a clear expectation that not everyone will be able to qualify to fly this aircraft. Those that can't make the grade **MUST** hire a pro pilot to assist them with the flying chores or forfeit the right to buy one. The very light jet (VLJ) segment will definitely change the personal transportation landscape. Some of us will choose to own one; others will use them as short haul charters or air taxis. Many, many will become fractional share owners. **VLJs** deliver on the promise of personal air transportation in ways that no previous product has been able too. They truly are mini-airliners.

Can a **VLJ** be used for Burger Hunting? Absolutely! The burger you seek with one of these is likely to be further away and a tad more expensive. Can you think of a better way to impress a client? I can't. It adds adventures to the invitation, "got time for lunch?" It would be hard to refuse a jump to the Keys for stone crab or Kansas City for barbeque.

The news keeps getting better, doesn't it? What about not buying a new ship and just upgrading your present ride? This is a great time to do it. The FAA has just approved WAAS-enabled instrument approaches. That adds the vertical to the horizontal and allows ILS-style approaches. It means that soon many airports that didn't have precision approaches will have them at virtually NO cost to the FAA or the airport. Adding ILS to either end of an existing runway cost $1.5 million. The cost of adding a WAAS-enabled GPS approach, which provides the same function, is $0. It gets even better when you consider that a WAAS-enabled GPS approach requires NO maintenance.

Adding a WAAS-enabled GPS system to your panel is a VERY good idea. Currently, only one is available: the Garmin 480, the one they picked up from II Morrow. For whatever reason, no other Garmin instrument approach certified panel–mounted GPS system, including the G1000, is WAAS enabled. The good news for Garmin is that nobody else is offering one, either. This is a great time to pick up a USED 430. The pricing is generally attractive, and Garmin has long committed to upgrade the 430, 530, and G1000 for very little money. Someday they will. In the meantime, it is approach certified for nonprecision operations. The 430, like its larger cousin the 530, can display NexRad weather, terrain depiction, and traffic information overlaid on its color display. The cost gets pretty high when you add those other features. Each requires another expensive piece of equipment in addition to your GPS. The GDL 69, which is required for NEXRAD depiction on a 430, runs $7,000.

Many pilots are getting situational awareness on the cheap by picking up Garmin's miracle portable box, the 396GPS. At $2,600 it is a great bargain. The 396GPS comes with NEXRAD capability built in. So buy a 430 to be approach certified and pick up a 396GPS for weather and terrain at a low cost. If you don't see yourself flying GPS approaches anytime soon, pick up the 396GPS and skip the 430. You'll thank me. The 396GPS is worth every cent. It is the only MUST have box that I am aware of. Some others are nice but it is truly a MUST have. By the way, it is WAAS enabled! Go figure.

If I were flying an older Cherokee or a 172, there are only two upgrades that I would consider, and I would add them both: the 396GPS, because it is wonderful and will likely save your bacon, and S-TEC's System 20 or System 30 autopilot. Flying gets so much better and safer with an autopilot. It really does. The System 20 is the best single axis add-on product for a certified aircraft out there, period. The System 30 adds altitude hold. The System 20 can be added to your ship for about $6,000 installed, money very well spent. Single pilot flying becomes relaxing with a good autopilot. Set your course and let "George" help you fly it. The GPS396 and the S-Tec 20 will extend the useful life of any aging ship while you wait to see what boxes will finally trickle down to the aftermarket.

It is a bad time to invest BIG BUCKS in terrain awareness or traffic information. The FAA is pushing hard to get us all to move to ADS-B. I think we will one day, once we know who is going to build the boxes that support it and what they'll cost. Today, no one is! When they do, utter amazement will enter a panel near you. Maybe it will actually be here by the time I write the fourth edition of The *$100 Hamburger*. For today, stick with upgrades that will help you out no matter what the FAA does. Remember, just a few years ago they were telling us that we *must* upgrade to Mode S transponders so we could receive TIS broadcast. Today? Well, they have canceled construction of new Mode S radar sites and are actually uninstalling the ones they already have. What if you had invested in an expensive Mode S transponder? The FAA answers, "Sorry, too bad." That's the same song they're singing to the tax payers who paid for those fancy and expensive Mode S–capable radar sites that will not become rust balls.

For the Burger Hunter, the news is even better. When I wrote the second edition of The *$100 Hamburger* in 2003, many aviation friendly restaurants were not able to make a go of it and were closing at an alarming rate. Today, the opening rate exceeds the closure rate. More new airport restaurants are opening than those that are closing. In 2006, I asked the $100 Hamburger Website Subscribers (**www.100dollarhamburger.com**) to tell us who they believed the best aviation friendly restaurants in the Untied States were. Hundreds of votes and suggestions came in. From all of the chatter, it became clear that 26 restaurants stood head and shoulders above all the others that I cover. I call them the "Best of the Best." They were given a special page of honor on the $100 Hamburger Website (**www.100dollarhamburger.com**) so that all of our Subscribers could make plans to visit them as soon as possible. Next, I commissioned the design of a special plaque to commemorate the achievement of each of the Best of the Best, and the 26 **Best of the Best** are now being honored with a special chapter in this book.

Here's some really good news. After operating my website, **www.100dollarhamburger.com**, for more than a decade as an open and FREE site, I made a decision to improve it greatly. The changes I planned and the ones I want to make in the future require a large investment. To raise the money, I asked my website readers to become paying Subscribers. On Labor Day 2005 I made it official, and the $100 Hamburger website (**www.100dollarhamburger .com**) has since then been accessible only to subscribers, and thousands of pilots have signed on. I really do appreciate their financial support and the pireps they continue to send in. In fact, I receive more pireps today than I did when The Burger (**www.100dollarhamburger.com**) was a free site!

The website (**www.100dollarhamburger.com**) has really improved. Today it is 100 percent database driven. This allows the pireps our subscribers input to go online instantly! In addition, the Burger Flight Planner feature has been added. It allows a user to input his or her departure airport and the maximum distance he or she wishes to fly. The system returns *all* of the aviation-friendly restaurants within that radius. Airport and runway information has been added for every airport. Best of all, phone numbers are now provided for *every* restaurant I cover. Please make a "know before you go" telephone call before making any burger flight. A restaurant's hours can change without warning.

Subscribers also have **FREE** access to our companion websites:

1. Pilot GetAways (great places to spend a weekend or longer)
2. Best Aviation Attractions (flight museums, air shows, adventures, and so on)
3. FlyIn CampOut (aviation-friendly campsites)
4. Terminal Faire (restaurants at airline terminals)
5. FlyIn Golf (aviation-friendly golf courses)
6. CrewCar (free ground transportation)
7. 100 LL (fuel reports)
8. Burger Hunters (individual burger hunters, their stories, and planes)
9. Editorials (our view on important aviation issues)
10. Surveys (give your view on key questions)
11. eNewsletter (the world of burger hunting published monthly)

I am often asked why the website exists, since I have written this book. Great question—here's the answer: it is hard to curl up in bed with a computer and dream about where you'll fly next. Sometimes you get the hungrys when you're flying cross-country on business. It is hard to throw an Internet-connected computer into your flight bag. Books are user friendly no matter where you are. But books are static; they can't be updated once they're printed. The website is dynamic. It can be instantly updated if a new restaurant opens or an old one closes or a five-burger spot goes downhill.

What does a Burger Hunter look like these days and why do they spend so much for a hamburger? I've long wondered the same thing. Are they all like me? The monthly website (**www.100dollarhamburger.com**) Subscriber Surveys have really filled in the blanks. Here's what we know. Burger hunters fly almost as frequently for breakfast as they do for lunch and hardly ever for dinner. Eighty percent of all Burger Flights are made to restaurants less than 100 nautical miles away. Ninety percent of our subscribers make a burger run at least once per month. When they go, 86 percent of them go alone or with just one passenger. Those figures are all derived from averages. When the averages are disregarded and we look only at the responses made by 50 percent or more of our participants here's what we learned: *Burger people fly with one buddy twice a month to a restaurant between 50nm's and 100nm's away for lunch.* If that sounds like you then, my friend, you are normal!

Here's the most important thing we learned: The pilots who make burger runs do so to stay current and to train. If they just wanted to get in the air and poke around aimlessly, they'd shoot touch-and-goes at their home 'port and cruise the local countryside. Our guys want to fly to a field they have never been to before. This approach causes them to practice good flight planning and takeoff, en route, approach, and landing skills. It causes them to work within the system and become sharper. It causes them to communicate with ATC, tower, and other aircraft. It makes them more comfortable. Airplanes are about travel—high speed travel! It takes work to obtain the skills to do it and practice to keep them.

So there is a lot of good news out there for burger pilots. Is there any bad news, any danger on the horizon? Yes, lots of it. Fuel prices are up and going higher. They can easily be offset by cruising at higher altitudes, practicing precision leaning, and sailing along using 55 percent power not 75 percent.

So if fuel cost really isn't a danger, what is? *User fees!* The Bush administration is set on going to user fees. The airlines want it, so do major airport operators. It will likely happen in some form.

I actually think the biggest danger we face is *lifestyle marketing*. What's that all about?

I love the new composite airplanes, all of them, but I am very opposed to the way they are being marketed. If left unchecked it will cause general aviation to go through another **bust** cycle. The following words headlined an ad for a well-known GA manufacturer of single engine, composite, nonpressurized, nonturbo charged aircraft in the November 2005 issue of ***AOPA Pilot*** magazine: "And to think, some frustrated guy is still waiting barefoot to get through airport security." These words were super-imposed on an ad picturing one of this manufacturer's aircraft flying over a snow dusted mountain range. The implication is clear: *our product will replace airline travel for anyone who owns it*. Really?

The problem is that similar ads are being placed in nonpilot magazines, like the *Robb Report*. Lifestyle marketing is a valid approach, but some general aviation marketing acolytes have glossed over the training and currency requirements and have soft pedaled the fact that all of these ships, other than the tall climbing Columbia 400, **must** navigate their way through clouds, not over them. Clouds—that's where turbulence and lighting and ice dwell. Clouds—that's where many pilots get into serious trouble. In the past, lifestyle marketing nearly killed the GA industry. I uncovered an Ercoupe advertisement from 1947:

> To the man who never expected to fly his own airplane...the doctor, executive, food-grower, lawyer... Ercoupe is a revelation of ease and safety. New pleasures and profit open up before you and more than make up the cost. Thousands of successful people fly Ercoupe, the leader in sales to individuals.

Can you imagine relying on an Ercoupe for corporate travel? And Ercoupe certainly wasn't the only company guilty of this kind of marketing. Ercoupe crashed and the industry along with it. By the way, there were 3,000 Ercoupes shipped that year. No airframe manufacturer is anywhere close to that number today, even though our population has risen from 100 million to 300 million. The problem with **lifestyle marketing** is that it *over commits*. It suggests an **unrealistic** outcome. Eventually, the market discovers the truth and abandons the category. What follows is an ugly crash. Not the metal kind, but the green folding money kind. Sales stop almost overnight and soon factories padlock their doors. It turned out badly in 1947, and it will turn out badly today, not just for the companies engaged in it but for the entire industry. That is exactly what happened in 1947. Fortunately, there are only a couple of companies involved in this ill-advised practice. Already their sales have begun to trickle off as the market learns that the facts of aircraft ownership are different than the fiction of aircraft advertising.

Train, fly, enjoy!

I look forward to hearing from you. E-mail me anytime you have something on your mind: jpurner@100dollarhamburger.com.

The Best of the Best
🏆 2006 🏆
from
1996-The $100 Hamburger-2006

PELL CITY, AL (ST CLAIR COUNTY - PLR)	Sammie's Touch & Go
LAKEVIEW, AR (GASTONS - 3M0)	Gaston's Restaurant
SEDONA, AZ (SEDONA - SEZ)	Sedona Airport Restaurant
NAPA, CA (NAPA COUNTY - APC)	Jonesy's Famous Steak House
PASO ROBLES, CA (PASO ROBLES MUNI - PRB)	Matthew's at the Airport
GEORGETOWN, DE (SUSSEX COUNTY - GED)	Jimmy's Fly-In Restaurant
VERO BEACH, FL (VERO BEACH MUNI - VRB)	CJ Cannon's
JEKYLL ISLAND, GA (JEKYLL ISLAND - 09J)	Jekyll Island Club
CHICAGO/SCHAUMBURG, IL (SCHAUMBURG REGIONAL - 06C)	Pilot Pete's
MUNCIE, IN (DELAWARE COUNTY - JOHNSON FIELD - MIE)	Vince's at the Airport
HUTCHINSON, KS (HUTCHINSON MUNI - HUT)	The Airport Steak House
LOUISVILLE, KY (BOWMAN FIELD - LOU)	Le Relais
STOW, MA (MINUTE MAN AIR FIELD - 6B6)	Nancy's Air Field Cafe
MARSHALL, MI (BROOKS FIELD - RMY)	Schuler's Restaurant
LITCHFIELD, MN (LITCHFIELD MUNI - LJF)	Peter's On Lake Ripley
MILLVILLE, NJ (MILLVILLE MUNI - MIV)	Antino's Cornerstone Grill
WESTHAMPTON BEACH, NY (FRANCIS S GABRESKI - FOK)	Belle's Cafe
PORT CLINTON, OH (CARL R KELLER FIELD - PCW)	Mon Ami

The
$100
Hamburger

Alexander City, AL (Thomas C Russell Field—ALX)

Burke's Lake Hill Restaurant 🍔🍔🍔

Rest. Phone #: (256) 215-5400
Location: Short drive from the airport
Open: Daily: Lunch and Dinner

PIREP FBO Comments: I called ALX and told them we wanted to go out to the Lake Hill Restaurant and would like to borrow their courtesy car. They said they would be closed when we arrived but they would leave the keys in the car. I asked them how much would it cost for me to use the car, and they refused to take any money and told me not to worry about putting any gas in it. What more could you ask for?

PIREP Burke's Lake Hill Restaurant is wonderful and it overlooks a beautiful lake. On Friday and Saturday night they have a seafood buffet that is reasonably priced, or you can order off the menu. ALX has a courtesy car and I was told by the restaurant owner's wife that if they are not too busy they will come pick you up from the airport.

The restaurant is just a short distance from the airport, but you will need a car to get to it. You may need to call ahead to reserve the courtesy car at ALX, but they will do that for you. Give this place a try—you won't be disappointed.

Cecil's Public House 🍔🍔🍔

Rest. Phone #: (256) 329-0732
Location: Short drive from the airport
Open: Daily: Lunch and Dinner

PIREP FBO Comments: This quiet, small town airport was a delight to fly into on a weekday morning. The gentleman working at the FBO was courteous and helpful, offering us the courtesy car for two hours and helping with directions to Cecil's. Everyone I talked to for help with directions, restaurants, and transportation was extremely friendly and informative. We were even tied down right in front of the FBO for a mere 20-foot walk in.

PIREP   Cecil's is a nice, moderately priced restaurant with a surprising menu and a lot of ambiance. It is housed in an attractive restored house with a very helpful, attentive staff. Cecil's is in downtown Alexander City and directions from the airport are a MUST.

There are a few older shops in the area for jewelry, antiques, and some great clothing buys! A small, friendly town; great for a morning getaway.

PIREP   This is an excellent place to fly just to be going somewhere. Excellent motels within walking distance, plenty of food places, and friendly people. The airport is attended on Sunday and they also have a courtesy car for your convenience.

Aside from the regular food places, Cecil's is an excellent upscale restaurant, but you will need directions to get there. It is well worth the effort. It is an old, original restored home and is very good with modest prices.

Dauphin Island, AL (Dauphin Island—4R9)

Seafood Galley 🍔🍔🍔🍔

Rest. Phone #: (251) 861-8000
Location: Close by: call for FREE pickup
Open: 7 days a week, 7 AM to 7 PM

PIREP   FBO Comments: Airfield is unmanned, no fuel, no nothing. The runway is newly surfaced, a little bumpy. Paint markings very good.

PIREP   My dad and I went there for a day flight for Redstone Arsenal, AL. We landed without incident (meaning there were no seashells on the runway) in which other aircraft were already there. Make sure you do a low fly-by to check for the shells dropped by the seagulls. Once we landed, I called the Seafood Galley and the owner was there in ten minutes in a nice Caddy to take us to her restaurant. BEST GROUPER I HAVE EVER EATEN. Once we finished, she took us back to the runway. Excellent service and food.

Price: avg: $15 per meal.

PIREP   FBO Comments: Pilot controlled lighting on CTAF (122.8) has been installed and is operational. However, key the mic. slowly to activate the lighting. The pay phone is on the field. Frequent birds are present on the runway and in the area—the "crunching" you might hear while taxiing is shells from hermit crabs dropped by the area birds. Be aware that helicopters operate at the field to shuttle workers to and from the many oil/gas rigs in the Mobile Bay area—some aren't so courteous with their rotor wash and land at the southern end of the tie-down area. You'd be smart to park close to the windsock and walk the extra distance to the gate.

PIREP   The Seafood Galley continues to be a favorite. Try the stuffed grouper supreme—you won't be disappointed! They will pick you up if you call, but the walk from the airport to the restaurant is pleasant and less than a mile.

PIREP   An excellent experience from beginning to end! The airport is almost completely surrounded by water. The north and west sides of the runway are right on the waterfront, and half the south side is on the water. There is a small tie-down area that accommodates about six or seven planes, if I remember correctly. Every parking space has tie-down ropes. 4R9 has no fuel, or an FBO for that matter, so be sure to plan accordingly. I've been told there is a pay phone there, but unfortunately I forgot to confirm that when I was there. The flight into 4R9 is gorgeous—very scenic both on the way and at the island. There is also a large wildlife watch area right there. Lots of signs that illustrate the birds around there. A great destination for nature lovers!

We called the Seafood Galley (251) 861-8000, and they had a car there to pick us up in five minutes! It's only a two- or three-minute drive to the restaurant. The meal was terrific—a large serving for a very reasonable price—and very speedy as well. Great quality food, too! The restaurant is right across the street from Dauphin Island's main beach, so after going to eat, you can walk across the street and be a tourist/beach bum. By the way, if you call the Seafood Galley and don't get an answer, try again in a couple minutes. They do get busy and will answer the phone eventually. Definitely head on down and give them a try! Five stars!

PIREP   Just spent the night on Dauphin Island and ate four meals at the Seafood Galley.....all excellent! The seafood dinner we had (grouper stuffed with crab and shrimp covered in hollandaise sauce) was possibly the best seafood dinner I've ever had.

Recommend five stars!

PIREP   We went to the Seafood Galley and it was good food at a good price. We went on a Saturday for lunch, and I had a rib-eye sandwich in which the meat was twice as big as the bun, and it was a big bun.

If no answer, call back. They have call waiting and may have put the phone back...it happened to us. They are more than happy to give you a ride. Beach is right across the street.

PIREP   There is a good seafood restaurant, Seafood Galley, at Dauphin Island airport (4R9), Dauphin Island, AL. Next to the phone is a placard with their number. Call, and they will pick you up and return you after the meal.

Have a good meal.

Barnacle Bill's 🍔🍔🍔

Rest. Phone #: (334) 861-5255
Location:
Open:

PIREP   Barnacle Bill's is a great restaurant on the island, near the base of the causeway bridge from Mobile. I was told the last time I visited that they would come pick you up from the airport if you call.

Decatur, AL (Pryor Field Regional—DCU)
Greenbrier Bar-B-Que 🍔🍔

Rest. Phone #: (205) 353-9769
Location: Several miles. Airport car available.
Open: Daily: Lunch and Dinner

PIREP   Fly in to Decatur/Athens airport (DCU) and have lunch at Greenbrier Bar-B-Que. It's located several miles off field but Clay, Cass, Stan, or Nicole at Decatur/Athens Aero will let you use one of several airport cars.

Food is good and staff at DAAS very accommodating. They are the local Comanche specialists for those of you who fly REAL airplanes.

Eufaula, AL (Weedon Field—EUF)
Riverboat Restaurant 🍔🍔🍔

Rest. Phone #: (334) 687-3132
Location: On the airport
Open: Daily: Lunch and Dinner

PIREP   Fully renovated restaurant with new owners.

PIREP   Alligators Ally's has now changed names. They are now called Riverboat Restaurant and Lounge.

At this present time they are open for dinner from 5–9 PM, Monday–Saturday. They are also open for lunch, but the waitress wasn't exactly sure when. They do not have the same menu as Alligators Ally's.

I flew in there on 1/31/06 and was going to write a pirep for that visit. I called them today to find out their hours and found out they have changed names. I suggest you call them before you fly in.

Their phone number is still the same: (334) 687-3132.

PIREP   I flew to Alligators Ally's for dinner. The food was good and reasonably priced. The waitress was very friendly. The owners came around and spoke to everyone. It is a place you would not be ashamed to take your kids to. I don't believe they are open on Sundays and Mondays. At one time they were not open for lunch, but this may have changed with the new owners. I suggest you call ahead to be sure.

PIREP   Dennis will make you feel right at home. Good food at the airport and at his bar-b-que shack in town.

PIREP   The airport restaurant has now changed hands and is called Alligator Ally's at the Airport. It is also now open every day, along with a Saturday lunch menu. Near future improvements are going to include a huge covered deck off the back of the newly renovated bar that will overlook the ramp/runway, and they are planning on having bands on the weekends. Sounds great, and I can't wait for the opening.

Fairhope, AL (H L Sonny Callahan—4R4)

The Grand Hotel 🍔🍔🍔🍔🍔

Rest. Phone #: (251) 928-9201
Location: FREE transfer to the hotel. FBO will call.
Open: Always

PIREP   The Grand Hotel, run by Marriot, is a lovely ($150/night and up) old resort on the eastern shore of Mobile Bay at Point Clear. It does live up to its name. The elegantly casual architecture, magnificent grounds, and glorious view over Mobile Bay would be enough for most of us. But wait, there's more: the hotel offers every amenity you could possibly ask for—sailboats, two 18-hole golf courses, tennis courts, huge new health spa, kiddie program, miles of walking/jogging trails through moss-draped live oaks and azaleas, and about five acres of swimming pool, both indoor and out. There are exercise rooms, game rooms, horseback riding, bicycles, fishing trips... not sure if they still have the polo field or not. There are half a dozen restaurants right there at the hotel and lots more just up the road a few miles in charming and artsy Fairhope. If you came in your 120-foot oceangoing boat, you could dock it right there in the hotel's own yacht basin, but since you flew into the Fairhope airport (4R4), three miles east, the hotel will meet you there.

If you happen to be there on a particularly hot and still summer midnight, you might even luck into a very special event called a jubilee. Happens rarely and nowhere else in the world, I'm told, and you have to be exactly at the right place at the right time to see thousands of flounders, crabs, and eels flop right up on the beach, somewhere between Daphne and Point Clear, and wait to be picked up.

PIREP   If you are looking for a nice place to take your significant other for a fabulous brunch, the Grand Hotel is IT!!!

Definitely call ahead and get reservations. There are two seatings on Sunday, and they are always packed. Once you go there you will know why. The limo driver (courtesy driver from the GH) said that there are people who fly in every week to enjoy the brunch. It is without a doubt one of the best I have ever had. All the boxes were checked:

Variety: Check (buffet style with everything from quiche to beef to seafood to peel-n-eat shrimp)

Quality: Check (two words: Yum. Yum.)

Service: Check (close, but not always in your face)

Friendly: Check

Atmosphere: Check

View: Check

The FBO was also very nice...I don't remember the fuel prices, but the food "coma" made it so I did not care (couldn't have been that bad). They monitor Unicom and will call the limo from the GH for you.

Enjoy.

"Troll"

MICHAEL T. HOEPFNER, Major, USAF

Commander, F-16 Flight 85TES

PIREP The Grand Hotel, located on Mobile Bay, owned by the Marriott chain, is an excellent place to go. On Sundays they have a champagne brunch that is out of this world; the cost is about $15 plus gratuities per person. Of course, you pilots shouldn't participate in the champagne, but your passengers can. The grounds are beautiful and you can walk along Mobile Bay after dinner. Land at (4R4) and the FBO will call a courtesy car to pick you up. There is fuel available at the FBO, but I don't remember the cost.

Fayette, AL (Richard Arthur Field—M95)

LB's Bar-B-Q 🍔🍔🍔

Rest. Phone #: (205) 932-7100
Location: 50 yards from the FBO
Open: Daily: Lunch and Dinner

PIREP One of the best food stops in Alabama is at Arthur Airport (M95) near Fayette. Look on the Atlanta Sectional 50 miles west-northwest; of Birmingham. Check out LB's Bar-B-Q located next to the Kentucky Fried Chicken.

No auto needed here, both restaurants are located within 50 yards of the FBO. The FBO is not open on weekends, so bring enough gas to get home.

Florala, AL (Florala Muni—0J4)

FBO 🍔🍔🍔

Rest. Phone #: (334) 858-6173
Location: On the airport
Open: Daily: Lunch

PIREP FBO Comments: Friendly staff, but no 100LL. JetA only.

PIREP Good fly-in place to eat. Flew in today (Saturday) in my C-152 and was treated very well. Buffet-style food served hot right in the FBO. Nothing fancy...all you can eat barbecue (Boston Butt), hot dogs, nachos, dessert, and drink for $6. Served seven days a week.

Lots of military traffic, especially helios and trainers.

PIREP The FBO on the field has an honor system buffet. All kinds of goodies. Dave smokes the pork right at the airport. Be sure to drop some money in the kitty so we can all continue to enjoy!! Watch out for military choppers; these guys flock to this place for lunch and fuel. Quite fun.

Keep in mind no avgas is available at Florala.

It's been about two months since I was there. I used to live nearby and flew in there a lot.

It's open seven days a week, year around. They're really set up to cater to the military choppers (they hot-fuel onsite, fun to watch) and the Navy trainers out of Pensacola. It's a very friendly bunch and GA is always welcome.

Hope this helps.

PIREP   On the Florida/Alabama border northwest of Pensacola (note the name: Flor-Ala), this small airport mainly caters to military aviation; Navy from Pensacola and Army from Ft. Rucker. Great barbecue, hot dogs, and burgers, cheap. Pay on the honor system and eat up.

Watch out for student helicopter pilots!!

Foley, AL (Foley Muni—5R4)

Lambert's 🍔🍔🍔🍔

Rest. Phone #: (251) 943-7655
Location: Close by. Call and they'll come get ya!
Open: 10:30 AM to 9 PM

PIREP   Check out Foley, AL for the "throwed" rolls. We went there.
They picked us up at Foley...
Great food and delicious rolls...
We also have been to Lambert's in Sikeston.

PIREP   The home of the famous throwed rolls has opened in Foley, AL. Lambert's Restaurant will pick you up and return you to the Foley Airport. Be conservative on your order as they come around with many extras. Open 0600–2300, bring your appetite.

PIREP   Foley, AL is a great getaway destination. Lambert's Cafe has excellent food and service. They picked us up from the airport, brought us to the cafe for lunch and, after we shopped at the famous outlet malls, they took us back to the airport. Not to mention you are only eight miles from the beach.

Gulf Shores, AL (Jack Edwards—JKA)

Lulu's 🍔🍔🍔

Rest. Phone #: (251) 967-LULU
Location: Short walk from the airport
Open: Daily: Lunch and Dinner

PIREP   This is a great place to stop for a meal. Just a short drive/walk from the Jack Edwards Airport, Lulu's sits on the intracoastal waterway with large open-air dining overlooking the water. I only tried the seafood gumbo—it was some of the best I've had on the Gulf Coast. The rest of the menu items I saw go by looked good too!

Lulu is Jimmie Buffett's sister, so live music is the norm.

Lambert's 🍔🍔🍔🍔

Rest. Phone #: (251) 943-7655
Location: Close by. FREE shuttle to the restaurant.
Open: Daily: Lunch and Dinner

PIREP   FBO Comments: Great FBO! They let us take a rental car to lunch for free in exchange for bringing back some rolls.

PIREP   We got stranded at JKA due to thunderstorms in Pensacola and ended up heading to Lambert's Cafe to check out this whole "throwed rolls" thing. This place is widely known and with good reason. Great atmosphere, great food, lots of fun overall. Definitely worth checking out. One important note though: CASH ONLY!!! They DO NOT accept credit cards. There is an ATM at the front door. I think it charged $1.50. Also, the lines do tend to get very long, especially at peak dining hours, but they move fast because the place is huge.

PIREP   Fly into Gulf Shores, AL for a good chow experience at Lambert's Restaurant, home of the "throwed rolls." If you haven't been, you should go, but go with an appetite. Call the FBO about 20 minutes out and ask for the Lambert's van. They'll call them and the van should be waiting for you or be there quickly.

Kirk Kirkland's Hitchin' Post Restaurant 🍔🍔🍔

Rest. Phone #: (251) 968-5041
Location: Adjacent to the airport—walkable
Open: Daily: Lunch and Dinner

PIREP   My family and I just returned off a long weekend trip from northeastern Ohio to Gulf Shores, AL for some sun and fun in a Cessna Stationair. FBO at KJKA is great; since we were there for the weekend we picked up a rental car at the FBO but stopped for lunch on the way out at the approach end of runway 9. Kirk Kirkland's Hitchin' Post Restaurant has great seafood, steaks, and barbecue… at least a four-burger. Good kid's menu, and make sure you save room for dessert! Restaurant is off field but estimate walkable in good weather down the access road to get around the fence to the approach end of niner.

Hamilton, AL (Marion County-Rankin Fite—HAB)

Hamilton Holiday Hotel 🍔🍔

Rest. Phone #: (205) 921-2171
Location: Less than 1 mile. Crew car available.
Open: Daily: Breakfast, Lunch, and Dinner

PIREP   Looking for a great steak at an even better price…fly into Hamilton, AL and ask Steve for directions to the hotel less than one mile away. They had more than ten steaks on the menu and none over $10. Steve will even share his crew car—first come, first served. Great $100 destination.

Huntsville, AL (Huntsville Intl-Carl T Jones Field—HSV)

Sheraton Hotel Restaurant Grill 🍔🍔🍔

Rest. Phone #: (256) 772-9661
Location: On the airport. Will send van to FBO.
Open: Breakfast, Lunch, and Dinner

PIREP   The Sheraton Hotel in the main terminal has a nice restaurant with good food, reasonable prices, and a view of the runways. And it's only a very short hike from the FBO.

Greenbriar BBQ 🍔🍔

Rest. Phone #: (256) 353-9769
Location: A short drive
Open: Daily: Lunch and Dinner

PIREP   FBO Comments: Friendly staff!

PIREP   Some of the best barbeque around. During the weekdays, Signature takes vanloads of Air Force cadets over there when they take their lunch break from the T-1s.

Check at the desk and they'll give you directions.

Jasper, AL (Walker County-Bevill Field—JFX)

Hoyett's 🍔🍔

Rest. Phone #: (205) 384-1888
Location: 2 miles. Airport car available.
Open: Daily: Lunch and Dinner

PIREP Fly in to Jasper, AL's Walker County-Bevill Field (JFX) and try the burgers or the catfish at Hoyett's. It's about 2 miles from the airport, and there is usually a nice airport car available. The food is good and the staff attractive.

Mobile, AL (Mobile Regional—MOB)

Wentzell's Oyster House 🍔🍔🍔

Rest. Phone #: (251) 341-1111
Location: Off the airport. Crew car available.
Open: Daily: Lunch and Dinner

PIREP I went to Alabama last week. Stopped in at MillionAir in Mobile. What a beautiful facility with a great staff. Ashley met us at the airplane, we had the Turbo 206, they were quick to give us a crew car—a new car at that. We only took on about 20 gallons of fuel, but that did not seem to matter to them. The owner/manager of the place is Arve Henrikson, a "good ole boy" whom I have known for years who has been in the aviation business for at least 20 years. Anyone should stop in and say hi to him. His office is right in front with a window to the lobby.

We were given a map at the FBO that showed about 10 nearby restaurants on the road next to the airport, and all were very easy to find and within 10 minutes of the airport. We chose Wentzell's Oyster House—the food was great, the service was a little slow, but the atmosphere was very pleasant.

Upon our return we were greeted by another line assistant, Chrissie, who also was extremely pleasant and helpful. We were given a ride back to our plane by Ashley and Chrissie and shown good ole Southern Hospitality by all.

After a long flight it was a very pleasant relief to all of us to be treated so well, and we did not have to arrive in a jet or large twin to get that kind of service. Arve believes that all customers are good customers as long as you are coming to his place, but we certainly cannot keep great FBOs like this if we don't at least buy some fuel from them when we stop in—at least when they aren't charging any ramp fees!!

Happy flying.

Monroeville, AL (Monroe County—MVC)

David's Catfish 🍔🍔🍔

Rest. Phone #: (251) 575-3460
Location: Adjacent to the airport
Open: Daily: Lunch and Dinner

PIREP David's Catfish offers great (all you can eat) catfish and catfish fillets. They serve up cheese grits and the best coleslaw I have ever eaten. Well worth the flight.

For overnighters, there is a motel within walking distance of the landing strip and restaurant.

PIREP On a recent flight, we were diverted to Monroeville, AL. While there, we discovered a couple of restaurants off the approach end of runway 21. One restaurant serves Mexican food while the other (David's Catfish) offers catfish.

Both restaurants are within walking distance of the FBO. To get there, walk down the side of runway 21 (in the grass!) and cross the highway—you are there. However, we were fortunate enough to get a ride from the line guy.

Montgomery, AL (Montgomery Regional (Dannelly Field)—MGM)

DJ's Old Post Office BBQ 🍔🍔🍔

Rest. Phone #: (334) 281-4887
Location: 10 minutes away by crew car
Open: Daily: Lunch and Dinner

PIREP Fuel is expensive at MGM. They have a courtesy car or they can set you up a rent-a-car for the day. A taxi is the other option.

For the best BBQ I've ever had, visit DJs BBQ—and I've flown all over the U.S.A. You will have to use the courtesy car from Montgomery Aviation to get there. Takes about 10 minutes but worth it. A locally owned and operated restaurant with breakfast and lunch M–F. Dinner on Friday and Saturday until 8 PM CST.

To get there, depart Montgomery Aviation and exit the airport east (right turn) onto Highway 80 (Selma Hwy). Go through one traffic light (not including the one for the airline terminal), across the bridge and take your first right, Lamar Road (County Rd 15). Go approximately 3 miles (the road will end in a T) and take a left onto Wasden Road (County Rd 42). Cross the RR tracks, and DJ's will be the first place on your left, approximately .7 miles. Be there early for lunch or you'll have to wait for a table. I've never been there when the place was not packed. Great service, friendly, and great barbecue, hamburgers, and biscuits.

Pell City, AL (St Clair County—PLR)

Sammie's Touch & Go 🍔🍔🍔🍔🍔

Rest. Phone #: (205) 338-9500
Location: On the airport
Open: Tues.–Fri. 11 AM–2 PM, Tues.–Fri. 5 PM–9 PM, Sat. 10:30 AM–9 PM, Sun. 11 AM–2 PM

PIREP This one really deserves its five burgers. The flight to Pell City provides some beautiful scenery. The runway is just adequate, but the restaurant is terrific. Ample menu, fun aviation decor, excellent service, parking at the door, and fuel service to the plane. The only thing that would have improved it is curb service. You see a lot of locals who drive in, which is a positive comment on the quality of food. NOT just airport burgers. I'll be back.

PIREP Pell City, St. Clair County: a flying gourmet's dream! The food has a New Orleans lean to it. Great gumbo! The burgers deserve the $100 label. Friendly service and taxi up to the porch. Indoor and outdoor seating both give a great view of the runway. The log cabin structure has much aviation memorabilia.

A must-visit!

PIREP Deserves a 5 rating. Just the best for food, service, and convenience. The burger is a burger lover's dream. The gumbo is fantastic. The only drawback is that they need to open a tower on Saturdays!

PIREP My wife and I (kids in tow) flew down to Pell City (St. Clair Co.) airport to try the famed Sammie's Touch & Go. The trip to the airport was beautiful and, despite the somewhat heavy air traffic, the FBO was courteous and helpful. We taxied down and parked in front of Sammie's on the grass with several other aircraft. The service was friendly and fast, the prices were more than reasonable, and the outside seating gave us the perfect view of the runway and surrounding area while we ate. Sammie's is located in a large log cabin and has a friendly and comfortable airport atmosphere with lots of room for the kids to stretch their legs. From the restaurant the

FBO can be called and they will deliver fuel to your plane while you eat. Sammie's was a great lunch trip that we definitely will be making again.

PIREP   This place rules!
   A grass taxiway leads to a grass ramp where you essentially park on the front lawn of this log cabin/restaurant.
   The Cajun pot roast was average, but the steaks are phenomenal.
   The runway is situated between a picturesque lake and a golf course.
   This will always be one of my favorites.

PIREP   FBO Comments: Airport/runway in good condition. People are friendly. Airport has a restaurant, and there's a nice golf course adjacent to the runway.

PIREP   My wife and I flew another couple to Pell City for lunch. Sammie's Touch & Go has a nice enough atmosphere for a double date (although I wish the nonsmoking section had a big screen TV like the smoking section has, but don't tell my wife). Sammie's has a buffet, but we all ordered off the menu. Food was very good and their chicken and sausage gumbo was excellent! Service was good and friendly. Next time we'll try the buffet (which I think was only $4.95). Plenty of parking, although it's on the grass.

PIREP   We stopped at Sammie's for dinner. Excellent, excellent food and service and very reasonable prices! You are severely shortchanging them by only ranking them as 3 burgers. They easily earned 4 or 5 in my opinion. This is especially true since, by comparison, the Lambert's (Gulf Shores, Sikeston, and Ozark) are ranked as four, four, and three burgers respectively. I've been to two out of the three and, I'm sorry, but Lambert's is into quantity and NOT quality. Honestly, Lambert's food sucks but you get a lot of it and many people like that. Sammie's blows their doors off when it comes to quality and service! If Sammie's is a culinary citation, then Lambert's is a culinary ultralight!

PIREP   FBO Comments: Airport nice, FBO helpful, taxi right up to place to eat.

PIREP   Food is great—service is great—view is great—parking is great—prices are great too.

PIREP   Flew to PLR with my wife for Saturday lunch. First time there, and we will be regulars. Next time we will take the kids. Pull right up to the front door of the restaurant and enjoy your meal (great food and service) while watching planes come and go. Very busy weekends if the weather is nice. An unofficial fly-in, get to see all types of aircraft. Log cabin style with aviation decor that young and old can enjoy.

PIREP   Sammie's Touch & Go is a great place for a great meal. My wife and I enjoyed our first visit there. They served a fantastic lunch buffet with all the fixin's at a very reasonable price. We will visit often and encourage others to do the same!

PIREP   Wow, what a neat airport. Lakes around it, golf course beside it, and neat airplanes live there. The restaurant had an excellent menu, good food, friendly people, aviation artifacts all over, and a bar for the nonflyers.
   Highly recommended for the $100 Hamburger lover.

PIREP   We have a new restaurant on the airport:
   Sammie's Touch & Go, run by Sammie Moore and his wife Sheila. Featuring Cajun-style cooking along with the usual fantastic steaks or just plain cheeseburgers. On the second Saturday of every month, Sammie's hosts our local EAA Chapter #1320 fly-in breakfast. Fantastic food, and you can taxi right up to the restaurant. (205) 338-9500.
   St. Clair County Airport has a 5002' paved runway, uncontrolled, vasi lights, avgas and Jet A fuel. (205) 338-9456.

PIREP   Now, located right on the airport with taxiway access, is possibly one of the finest airport restaurants this pilot has ever seen. Owned by a local pilot, Sammie's Touch & Go is a beautiful log building with an emerald green roof, decorated with extensive aviation memorabilia. Dining inside and out is available with a wide variety of inviting menu choices, friendly people, and a beautiful setting. Nice runway, great FBO, affordable fuel, what more could you want? Closed Sunday night and Monday. Call ahead at (205) 338-9500. FIVE FORKS in this pilot's humble opinion!

Tuscaloosa, AL (Tuscaloosa Regional—TCL)

Dreamland Bar-B-Q 😋😋😋😋😋

Rest. Phone #: (205) 758-8135
Location: It's a drive. Reserve the crew car.
Open: 10 AM–10 PM, Mon.–Fri. 10 AM–11 PM, Sat. 11 AM–9 PM

PIREP It's about a 15-minute drive from the airport. Get specific directions because it's on a back road. And if you haven't been to a Dreamland before, please realize that it's rib slabs and white bread. That's it. No sides.

Honestly, this isn't one of my favorite barbecue places. I've been several times and the ribs can be tough sometimes. Maybe you have to experience it once because of its reputation, but it just doesn't make the cut for me. Certainly not worth a long fly-in.

PIREP Slabs are only $14. My wife and I flew down last weekend. We have made this trip a couple of times lately…

Yes, it's a 5 +++.

PIREP Dreamland BBQ has some of the best barbecue in the country, period. Slabs are $15.50. A quart of sauce will run you a mere $500. The restaurant phone number is (205) 758-8135. Park at Bama Air; call ahead and reserve the courtesy car (1-800-937-1716). Ask for directions to Dreamland once you get to the FBO as it is located off the beaten path. Hours of operation: 10 AM–10 PM, Mon.–Fri.: 10 AM–11 PM, Sat.: 11 AM–9 PM.

This place is worth the trip.

City Cafe 😋😋😋

Rest. Phone #: (205) 349-0114
Location: 1.5 miles away. Borrow the crew car.
Open: Daily: Breakfast, Lunch, and Dinner

PIREP Stopped for fuel in Tuscaloosa—there are two FBOs with fuel there; we chose Dixie Air. The people at Dixie were extremely nice. Very attentive to directing us to a parking area—offered to wash the windshield and check oil in addition to fueling. The courtesy van was gone, but the fuel attendant told us we could use his car if we wanted.

He recommended the City Cafe about 1.5 miles from the airport. Very wide variety of choices of entrees and vegetables available as a "meat and two" combo. The food was very good, as were the portions. Service was excellent as were the prices—my wife and I had an excellent meal for about $10 total, including tip! I highly recommend stopping at Dixie Air and eating at the City Cafe. Fuel at Dixie was up there.

Bettles, AK (Bettles—BTT)

Bettles Lodge 🍔🍔🍔

Rest. Phone #: (907) 692-5111
Location: On the field
Open: Daily: Breakfast, Lunch, and Dinner

PIREP   One of the best places to get a burger above the Arctic Circle is in the fly-in only community of Bettles, AK (50 residents) in the foothills of the Brooks Range (180 nm north of Fairbanks). The airstrip officially enjoys the most clear weather days out of anywhere in the state and is a well-maintained 5,200-foot gravel strip with a nearby 5,000-foot floatpond. A historic lodge acts as a pilot roadhouse and has a full-service kitchen with many hamburgers on the menu. Stay overnight for the Midnight Sun in the summer or spend an afternoon exploring the unique and friendly community where everything (mail, fuel, food) is flown in daily. Visiting pilots enjoy talking with bush pilots and seeing the diverse amount of aircraft (from Cubs on tundra tires to DC-4 fuel carriers) that arrive daily. Pilots from as far away as Florida make the trip to Bettles annually.

Big Lake, AK (Big Lake—BGQ)

Big Lake 🍔🍔🍔

Rest. Phone #: (907) 892-9230
Location: On the field
Open: Daily: Breakfast, Lunch, and Dinner

PIREP   FBO Comments: Big Lake Airport is located 4.9 nm on the 90-degree radial from Big Lake VOR. The 2,400-foot gravel runway is well maintained and has lighting at 122.8. There is no fuel for sale on the field, but it is available at Wasilla approximately 10 miles west of BGQ. Transient parting is available.

PIREP   "The Hangar," located across the street from Big Lake Airport, is a favorite for locals and visitors. The food is great and the motel is excellent. Fishing is great in the area lakes and rivers.

Mile 1.5 South Big Lake Road
HC 34 Box 2406
Wasilla, AK 99654

Tel: (907) 892-7976 Fax: 892-9229
Motel, restaurant, lounge
Cindy and Mark Riley, Owners

Circle Hot Springs, AK (Circle Hot Springs—CHP)
Circle Hot Springs Lodge 🍔🍔🍔

Rest. Phone #: (907) 520-5113
Location: On the field
Open:

PIREP   Circle Hot Springs: Located 95 nm northeast of Fairbanks on the 135-degree radial at 67.5 DME off the Fort Yukon VORTAC.

(GPS Coord. N65.29,13/W144.36,65), deep in the interior of Alaska.

Access is by wheels only.

This lodge has great prime rib and other specialties, not to mention the hot springs themselves. Lodging is also available (year round).

Manley Hot Springs, AK (Manley Hot Springs—MLY)
Manley Hot Springs Lodge 🍔🍔🍔

Rest. Phone #: (907) 672-3161
Location: On the field
Open: Daily: Breakfast, Lunch, and Dinner

PIREP   Manley Hot Springs: Located 70 nm west of Fairbanks on the 079-degree radial at 40.4 DME off the Tanana VOR/DME.

(GPS Coord. N64.59,85/W150.38,65).

Access by wheels and floats.

Great "grease burgers" and rustic atmosphere. It too boasts relaxing hot springs and lodging (year round).

Hope to see you there.

Northway, AK (Northway—ORT)
Airport Cafe 🍔🍔🍔

Rest. Phone #: (907) 778-2311
Location: On the field
Open: Daily: Breakfast and Lunch

PIREP   We have flown to Alaska twice in the last four years and plan another trip this summer. The restaurant at the FBO in Northway makes great pies! We haven't had any of the other menu items, but they look great! Of course to make this trip from the states costs considerably more than $100, but it is definitely worth it!

PIREP   Great pie! Great sub! Can't wait to go back to Alaska!

Sitka, AK (Sitka Rocky Gutierrez—SIT)

Nuggett Restaurant 🍔🍔🍔🍔

Rest. Phone #: (907) 966-2480
Location: In the terminal
Open: Daily: Lunch and Dinner

PIREP Believe it or not, the best peanut butter pie in the world is right here at the terminal cafe in Sitka. For good greasy burgers and other fare, a short trip to town from the manmade aircraft carrier–style runway is in order. As a note, there are always more things open when the cruise ships arrive. Things really crank into full swing! Four winged burgers are in order and that's just for the PIE.

Skagway, AK (Skagway—SGY)

Red Onion Saloon 🍔🍔🍔🍔

Rest. Phone #: (907) 983-2222
Location: Downtown
Open: Daily: Lunch and Dinner

PIREP Just 79 nautical miles southwest of Whitehorse, Yukon, Skagway is nestled amidst the coastal mountains of the Alaskan panhandle at the end of the Taiya Inlet. Skagway is famous as the trailhead of the fabled Chilkoot Pass, where gold seekers lugged their ton of supplies up over the mountain into Bennett Lake.

What makes this a great destination for the $100 Hamburger is:

1. It is a breathtakingly beautiful spot.
2. The strip is quite literally in town—your own two feet are all you need for transportation.
3. There are many things to see and do, especially during the summer.

I recommend Dee's Restaurant for fresh Alaskan halibut fish and chips. For a more frontier setting, try the Red Onion Saloon.

Other things to do and see:

1. Check out the historic cemetery where many graves date back to the Gold Rush.
2. Ride the train to the summit of the White Pass (very scenic but a little pricey).
3. Do the tourist thing and visit the many historic buildings and quaint shops along Main Street.

Flight Logistics:
Coming from Whitehorse, access to Skagway VFR will depend upon the condition of the pass, which can often be solidly blocked with cloud. Be sure you have a good idea what the weather is before going. Check the Skagway winds (they can be a bit excessive and gusty at times).

It is important to be on time for the customs officer because he/she will be driving from the border to meet with you. They also do the train passengers and will not wait for you if there is a train coming in.

Recommended route: fly direct Whitehorse to Carcross (FA4), then follow the highway from there over the pass. Alternately, fly over Bennett Lake following the railway until it intersects the highway.

The downwind for the strip (wind is usually coming off the water) is a right-hand one. This takes you right along a mountainside and does not allow for a good view of the strip. Turn base when you see the waterfall on the river. Traffic can be heavy during days when there are cruise ships in the dock. Return flight plans can be filed with Juneau FSS by dialing 1-800 WXBRIEF.

Tok, AK (Tok Junction—6K8)

Fast Eddies 🍔🍔🍔

Rest. Phone #: (907) 883-4411
Location: On the field
Open: Daily: Breakfast, Lunch, and Dinner

PIREP   Tok, AK has a good breakfast or dinner cafe right next to the airport called Fast Eddies. Try it, you will like it.

Bullhead City, AZ (Laughlin/Bullhead International—IFP)

The Gourmet Room 🍔🍔🍔

Rest. Phone #: (702) 298-2535
Location: Call for FREE pickup
Open: 24/7

PIREP   An update on the Bullhead/Laughlin Airport. You can no longer walk across the street to catch the river shuttle to the casinos. All the general aviation parking is now on the east side of the airport. Most of the casinos have shuttle service to and from the airport. The staff at the GA terminal is always accommodating and will call the hotel you wish to go to. Also the prices are a little higher than my earlier post.

PIREP   I have eaten at the Hickory Pit in the Edgewater Hotel on three occasions in the past two years and find that the food is excellent as well as the service.

PIREP   This is one of our favorite places to fly in. The airport is on the Arizona side of the Colorado River, just across the river is Nevada with about 10 casinos. Hotel rooms are reasonably priced and range from about $15–$20 during the week and higher on weekends. All the hotels have courtesy van service to and from the airport. You can walk directly across the highway and catch a water taxi over to the hotels. I recommend this if you are there just for the day. All the hotels have coffee shops that are open 24 hours. However, the best food is in the dinner houses. The Gourmet Room at the Riverside is expensive, but the food and service is outstanding. And the view from any of the tables is spectacular. They will take reservations and, if you plan on eating there, you better make them. Our other favorite place to eat is the Hickory Pit at the Edgewater. Steak and lobster for $22, and it's outstanding. There is a lot to do on the river too, especially in the summer.

Chandler, AZ (Chandler Muni—CHD)

Hangar Cafe 🍔🍔🍔

Rest. Phone #: (480) 899-6965
Location: On the field
Open: 7 AM to 2:30 PM, 7 days a week

PIREP   The Hangar Cafe was excellent. Located right next to the ramp, near Chandler Air Service, the aerobatic flight school with the Pitts mentioned elsewhere, it is easy to access once you have landed. The food was very good, although I ate alone and did not try much. I do know several locals who eat there often and love the food. The service was excellent: I was promptly seated, asked if I was ready to order, and given my food more quickly than any normal restaurant. The waitress was also very friendly. Great aviation ambiance, tons of aviation posters, signs, and so on, along with many windows giving a great view of the Pitts and Great Lakes tied down nearby along with a great view of the runway. There is a patio you may eat at outside with an even better view, rarely used in the summer but where the majority eat in the winter. The price was about the same you would see at a normal nonformal sit-down restaurant. My meal cost me about $8 including a drink, but I didn't order much. But then again, if you're renting an aircraft (or have bought one) to fly somewhere for a meal, you shouldn't be too concerned with the price.

PIREP   There is no security fence between the airplane parking area and the restaurant…very friendly atmosphere. You walk past several Pitts biplanes en route to the cafe.

Breakfast: all day, lunch weekdays: 11 AM to 2:30 PM, lunch–weekends: 11:30 AM to 2:30 PM.

PIREP   Location: ramp side. There are about five tie-downs near the front of the restaurant next to a couple of above-ground fuel tanks. When it's busy with airplanes, parking is a problem. We were there on Sunday at 10:30 AM and were the second plane to park. When we left at about 12:30 PM, we were the only airplane in the restaurant parking area.

Food: very good. The veggie omelet comes with green chili peppers along with all the rest of the veggies—a very spicy omelet! So if you do not care for spicy food, check with the waitress.

Service: very good, friendly. Earned a good tip!

Ambience: a bit noisy. Lots of windows, great view of the active runway. There is a shaded patio available.

Price: reasonable, about $8 per meal, including drinks.

The Hangar Cafe is part of a hangar used by an aerobatic school. Lot of Pitts biplanes to see.

This was a Sunday. Restaurant was crowded, although we got a table immediately. It appeared that most of the customers were locals who drove to the airport. Many families with young children.

This is a towered airport. Ask ground control for directions to the Hangar Cafe (or "the restaurant") parking.

Busy airport. We approached from the south along I-10 and called the tower from Sun Lakes (see Phoenix Term. Area Chart) with the ATIS. They gave us a straight in to runway 04 left. The restaurant is adjacent to taxiway M off runway 4/22. If you exit the runway at any other taxiway, take taxiway A to the restaurant parking.

PIREP   Remodeling is tasteful and comfortable. For about three and a half bucks had a great egg, sausage, and cheese sandwich. The A/C is fantastic in the building, which is key when the OAT was pushing 110 at about 11:30! Lots of folks—pilots and nonflyers alike—but still, not much downtime waiting for food. You can basically wheel your aircraft around and park almost in front of the building.

PIREP   GREAT breakfasts and burgers for very reasonable prices. It is a very small place so there can be a small wait. They also stop serving food midafternoon. Some interesting aircraft are kept at Chandler, and there is an FBO there that rents several aerobatic aircraft.

PIREP   The Hanger Cafe is a good restaurant worthy of at least a 3-hamburger rating! Great place to go for breakfast. I don't think they are open beyond about 3 PM. Try one of their slow roll breakfast burritos.

PIREP   The Hanger Cafe in Chandler has the BEST frozen yogurt in the world! I plan my trips from California to Texas to stop there before 3 PM just so I can get a big insulated cup of yogurt to go (after fueling at the self-serve pumps—some of the best prices in the area).

PIREP   The Hangar Cafe in Chandler has just expanded and has lots more seating. The food is still great including the breakfasts, and the price and service is always right.

Cottonwood, AZ (Cottonwood—P52)

Hobo Joe's 🍔🍔🍔

Rest. Phone #: (928) 634-2651
Location: Long walk
Open: Daily: Breakfast, Lunch, and Dinner

PIREP   Although this restaurant is quite a walk from the airport, the breakfast is probably one of the best that I've ever had. They have a large selection, and the service is good. Best of all, the prices are cheap and you get filled up. The one downside is that you have to walk through the smoking section to get to the nonsmoking section. Just hold your breath, I guess.

Eloy, AZ (Eloy Muni—E60)

Bent Prop Saloon & Cookery 🍔🍔🍔

Rest. Phone #: (520) 466-9268
Location: On the field
Open: Daily: Breakfast, Lunch and Dinner

PIREP   The Bent Prop does not open during the week until 10 AM. Don't know about weekends.
There is a coffee and donut place in an adjacent building that opens about 6 AM.
Eloy is a great place to watch parachute jumping. You can watch them repacking the chutes. Sometimes the landings are hilarious.
Listen carefully to CTAF for parachute activity and the jump plane returning—jump plane usually lands on runway 20 and takes off runway 2. Stay to the northwest of the field and make right traffic to runway 20.

PIREP   Eloy is halfway between Tucson and Phoenix. The restaurant, the Bent Prop, is situated in the middle of the skydiving area, a few hundred yards from the transient tie-downs. They serve breakfast, lunch, and dinner. The food is good, service and prices are great. Cheeseburger w/the usual trimmings, batter-dipped fries, and bottomless Coke is about $5. You can eat indoors or out on the porch overlooking the skydiving landing area.
Fuel is available with credit card or cash; you pump.
Skydiving is extremely active most of the year. Approach from the west and do not overfly. Skydivers land on the east. They request land 20, T/O 02 for calm conditions.

PIREP   This is a quaint find—an airfield frequented primarily by both young and old but always energetic skydivers from around the world—to be sure.
About as much in the middle of nowhere as exists in southeastern Arizona, the field is several miles NW of the relatively abandoned town of Eloy in the middle of a large farming valley.
We flew two planes in, dodging drop-planes in so much of a hurry to get another load up that they forget such niceties as crossovers to taxiways and just cut across the dirt, and the happy-go-lucky attitude about the whole place didn't end there. The tie-downs are nice and convenient and the short hike to the facilities pleasant. The service at the Bent Prop restaurant was as outstanding as possible, decent burger bordering on unique, and a menu with a wide variety and very reasonable prices. Namesakes were on display just outside the window.
The setting is not on the tarmac but overlooking the central business square of an Old West–styled skydiving community—eat inside (cool A/C!) or on the porch (later in the year)—and the place is large and clean. After an unhurried meal at a table with room to stretch, self-serve fuel (Jet A or 100LL) is convenient near the tie-downs. Quite impressed.

Page, AZ (Page Muni—PGA)
The Ranch House Grille 🍔🍔

Rest. Phone #: (928) 645-1420
Location: 1 mile
Open: Daily: Lunch and Dinner

PIREP The Ranch House Grille is a nice little restaurant about a mile from the FBO. The FBO courtesy shuttle will drop you off and pick you up after for free. I ordered the Green Chile Burger, which included fries for $5.95, and it was excellent! One of the best burgers I have ever had. The fries were great too. The prices were unbeatable, and the service was great.

Even though it was a 2 hour and 45 minute flight from Salt Lake City, I will definitely be back! Besides the great food, the scenery of Lake Powell is magnificent! I filled up the memory card in my digital camera in no time.

I would highly recommend this as a fly-in $100 Hamburger destination.

Parker, AZ (AVI Suquilla—P20)
Blue Water Casino 🍔🍔🍔🍔

Rest. Phone #: 888-243-3360
Location: FREE shuttle
Open: 24/7

FBO Comments: Five-cent discount on fuel if you use a Chevron credit card. CRIT Air. CRIT stands for Colorado River Indian Tribe. The Tribe owns the FBO.

PIREP The airport, the FBO, and the casino are all owned by the Colorado River Indian Tribe (CRIT). There is no landing or tie-down fee; however, overnight parking is $2/night. The airport runway (1/19), taxiways, and ramp are smooth and well maintained. The many tie-downs are newly painted with new chains. No chain link fences with barbed wire and no locks on the gate. John Ashcroft would not like it! The airport is east of and very close to the town of Parker. You are advised to avoid flying over the town at low altitude. The landing pattern for runway 01 is right hand to stay away from the town. East of the field is flat uninhabited desert. Some hills about six miles to the east.

There is no restaurant on field. The FBO manager is happy to chauffeur you over to the Blue Water Casino/Hotel in his Chevy Van. The casino is located on the bank of the Colorado River. It's less than a mile away but not a convenient walk. When you are ready to return to the airport, ask the casino hotel registration desk to call the "airport van." They are most friendly and happy to do so. If the airport telephone line is busy or there's no answer (the FBO manager works the place alone), they keep trying until the Chevy Van is on its way.

The casino has a breakfast buffet that ends at 11 AM. Lunch buffet thereafter. There is a coffee shop at the casino; however, we did not check it out.

A very pleasant experience. Very friendly and helpful people with standard Las Vegas/Laughlin type casino buffet food and an airport operation right out of the 1960s.

An excellent $100 omelet…'er, hamburger flight.

PIREP I flew in there on Saturday, April 20 with a gaggle of five other planes and 12 people. About 106 nm from Glendale, AZ (GEU). The airport is Avi Suquilla (P20) but everyone refers to it as Parker. Nice airport with plenty of tie-downs. The friendly FBO folks at CRIT-Air will drive you in their shuttle van to the Blue Water Casino, which is one mile away on the Colorado River. They have a nice restaurant there with a good breakfast buffet or a la carte at reasonable prices. Then you can play the slots. When you're ready to return, tell the concierge and they will call the shuttle van for you. Didn't buy gas so don't know what they charge. A nice flight en route and good food.

What more can you ask?

Payson, AZ (Payson—PAN)
The Crosswinds Cafe 🍔🍔🍔

Rest. Phone #: (928) 474-1613
Location: On the field
Open: Breakfast, Lunch, and Dinner

PIREP   Payson is one of the best places to go in Phoenix for breakfast! It is a short hop and is 10 to 15 degrees cooler, which can make a big difference in the summers. Great pancakes and lots of friendly people. They have a small airshow every April.

PIREP   Very reasonably priced. My friends swear that they have the best biscuits and gravy in AZ. Once again, inexpensive. The airport can be a little tricky to fly into. The country surrounding the airport is beautiful. It makes a very nice short breakfast flight from Phoenix.

Peach Springs, AZ (Grand Canyon Caverns—L37)
Grand Canyon Caverns and Inn 🍔🍔🍔

Rest. Phone #: (928) 422-3223
Location: Close by. Call for pickup.
Open: Daily

PIREP   It has a great restaurant on the property along with a fantastic underground cave tour. The restaurant is at the top of the hill and free rides from the motel area are provided.

Phoenix, AZ (Phoenix Deer Valley—DVT)
Airport Cafe 🍔🍔🍔

Rest. Phone #: (602) 582-5454
Location: On the field
Open: Breakfast, Lunch, and Dinner

FBO Comments: Spent a week tied down there. Found we had a very dead battery on Sunday, the day before we were to leave. Cutter Aviation maintenance worked with us for over an hour trying to get it charged (didn't work), had new battery installed and aircraft ready for us before 10 AM Monday morning. Nice people, nice facility.

PIREP   Airport Cafe gets a lot of local (nonflying) traffic, and it should. Food, service, and prices are first-rate. Took some family (Phoenix residents) there for Sunday breakfast, heard them making plans to make it a regular dining-out place.
   Walk-in from ramp, parking lot, FBOs.

PIREP   A great cafe at the base of the tower. I recently flew my 172 from Mesa Falcon Field just to get some lunch at DVT. Unknowingly, I forgot my credit cards and cash in my car back in Mesa. After my guest and I were seated, I related the problem to the server, a very nice lady named Barbara. She told me that we could order anything we wanted and call her when we got back to Mesa with a credit card number.
   Any place that treats pilots with such respect is number one on my list.
   Highly recommend it to all.

Phoenix, AZ (Williams Gateway—IWA)

Baci's Italian Bistro 🍔🍔🍔

Rest. Phone #: (480) 988-1302
Location: Short walk
Open:

PIREP The airport is great. Where else can you fly with about every aircraft ever made? From homemades to Russian commercial biplanes to fighters. But I have to say that the food available at the airport is not good at all. You can walk about 1/4 mile off the airport grounds and reach the dining hall for ASU East and get good diner-style food for the same price but unlimited portions (all you can eat). Or walk about 20 minutes farther and reach Baci's Italian Bistro, my favorite. Great lunch specials on about 10 dishes, and the food is great. It's not real Italian, but pretty close. (I'm used to listening to staff screaming in Italian at real Italian eateries; everyone here has a SW accent!) But all that matters is that they love the aviation industry, and they take great care of us when we go and spread maps all over the tables. Service is fast and my drink glass was always full. The portions are good sized and the bread is yummy, baked right there. Even after the lunch specials expire, the regular prices are great. A four-olive rating!

BP is generally quick with the fuel.

The Flight Deck Cafe 🍔🍔🍔

Rest. Phone #: (480) 988-9517
Location: On the field
Open: Daily: Breakfast, Lunch, and Dinner

PIREP Williams Gateway Airport located at the old Williams AFB in Mesa, AZ has a great little hamburger joint called The Flight Deck Cafe. Since I live in Mesa, I drive there to eat. But almost everyday at lunch there are several transient Air Force fighters parked out front while the jocks get a bite to eat.

Go Sun Devils!!

Phoenix, AZ (Phoenix Sky Harbor Intl—PHX)

The Left Seat 🍔🍔🍔

Rest. Phone #: (602) 220-9407
Location: On the field
Open: Daily

PIREP Hey, what about the food at The Left Seat at Phoenix Sky Harbor? You can fly in or, if in town, why not drive in? Great atmosphere on one of the busiest runways in the country, 8 left and 26 right. The walls are covered with some cool airplane pictures and some dangling airplanes over the tables. They have great specials and the staff is very down to earth and friendly (nice people). The burgers are good, as well as the turkey burger's taste if you are trying to cut down on the weight and balance of your airplane. The French dip is something to write the French about, and don't forget the salad bar and chili. When done eating, stroll up to the deck on top of the restaurant for a better view of Phoenix and airport surroundings. Oh, by the way, the wood bar is the first ticket counter that PHX airlines were able to sell flights behind. So history, views, food, and friendly people are all here. Don't take my word for this reasonably priced aviation burger frenzy place—check it out for yourselves!!!!!

Prescott, AZ (Ernest A. Love Field—PRC)

Antelope Hills Golf Course 🍔🍔🍔🍔

Rest. Phone #: (928) 445-8727
Location: Short walk
Open:

PIREP If you don't mind a 20-minute walk, the restaurant at the Antelope Hills Golf Course adjacent to the airport is quite good and relatively inexpensive. I think they open at 6 AM every day. You can get there by walking around the fence at the west end of the airport and following the road into the residential area. The restaurant is at the clubhouse and is a bit SW of the airport.

Sedona, AZ (Sedona—SEZ)

Sedona Airport Restaurant 🍔🍔🍔🍔🍔

Rest. Phone #: (928) 282-3576
Location: On the field
Open:

PIREP The Sedona airport is the most beautiful pilot getaway location in the whole world. Breathtaking views surround you. Make sure you pay attention to the approach. A number of pilots have been distracted by the scenery and ended up in the weeds. The winds can sometimes be tricky. There are five windsocks on the field and in the afternoon they're often pointing in five different directions at the same time. The greatest concern is a potential big sink off the approach end of runway 3 when the wind is right down the runway. I got caught in it once in a 172 and with full power, 10 degrees of flaps, and best rate of climb we were still coming down at 600 fpm. Just be prepared.

The Sedona Airport Restaurant is a delightful place to eat. The owners, Bob and Debbie, are wonderful folks and are happy to give advice to visitors about what to see. I have eaten breakfast there about 100 (no kidding) times. My favorites are eggs Florentine and the veggie croissant. I have never eaten a hamburger there.

Definitely sit outside if the winds are calm and watch the luckiest pilots in the world land on the USS Sedona. The airport manager will even certify your landing on a suitable-for-framing certificate with an image of the airport morphed to look like an aircraft carrier. For good reason. That's what it does look like.

PIREP As mentioned, the airport sits on a mesa at 4,830' msl in the center of Sedona. It is about 500 feet above the surrounding city. Both ends of runway 03/21 (5,132' × 75') terminate with very steep drop-offs (well, you might call them cliffs). There are 7,000' and 8,000' peaks surrounding Sedona—not recommended for night operations. There is a GPS approach to 03 (8,000' msl at the IAF). There is no tower, but there is an AWOS-2. So announce approaching the airport, crossing, entering downwind, turning base, and turning final. 03 is the landing runway (1.8 percent grade uphill). 21 is the takeoff runway. There can be a lot of traffic. Wacos to Citation Jets. Announce taking 21 for takeoff and be sure no one is turning base to final or is on final for 03. I have seen jets making a long final. The GPS approach is 5.7 nm straight in from the FAF to the runway. Radio communications and eyeballs are very important.

The views of the surrounding red rock mountains is fabulous—good for passengers...pilots, watch for airplanes, there are Wacos in the area taking tourists for flightseeing rides—along with transient flights viewing the sights.

The transient parking is at the terminal. The restaurant is about 50 yards from the terminal building. There is a security fence between the terminal and restaurant and the parking ramp. Be sure to note the security gate key code when going from the ramp side to the terminal side—there's a sign next to the gate. Good flight planning facilities at the terminal.

Very good food. I had a veggie omelet that was delicious. Excellent service, very friendly, many coffee refills without having to ask. Florida room area where you can watch the planes coming and going while enjoying the fine food.

Cost is about $12/person for breakfast.

Scottsdale, AZ (Scottsdale—SDL)
The Blue Fig ☕☕☕

Rest. Phone #: (480) 948-8585
Location: On the field
Open: Daily

PIREP   Had lunch, two couples.
The Blue Fig is a ramp-side cafe.
Many dishes on the menu are Italian.
Cheese soup, loaded with mushrooms—outstanding!! They call it "mushroom soup."

Seligman, AZ (Seligman—P23)
Lilo's ☕☕☕

Rest. Phone #: (520) 422-5456
Location: Short walk
Open:

PIREP   Seligman is still a great place to go to beat the heat in summer. A nice runway and ramp area with plenty of tie-downs. Airport is fenced and has good security. Clean bathrooms. A bulletin board by the ramp shows where all the eating establishments are in town, which is a short walk. Lilo's food was excellent with huge portions and reasonable prices.

PIREP   Have heard that Lilo's in Seligman is a great restaurant with really good German food. Decided to give it a try this weekend. Went there for breakfast. P23 has been recently repaved. Nice airstrip. Lilo's is about a half-mile walk from the airport (rumor is that if you call they might pick you up: (520) 422-5456). What a GREAT place!!!! Really friendly people and really good food. The cinnamon rolls are about 4 pounds each and 8–10" across.
Give this one a try.

Temple Bar, AZ (Temple Bar—U30)
The Temple Bar Marina ☕☕☕

Rest. Phone #: The Temple Bar M[M]
Location: Call on unicom for marina pickup
Open: Breakfast, Lunch, and Dinner

PIREP   Stopped by the Temple Bar Resort on some recreational flying out of Las Vegas. Breakfast at the cafe was great, people were friendly, and the weather was of course sunny. After breakfast, head over to Grand Canyon West for some spectacular views.

PIREP   The Temple Bar Marina is located on the southeast portion of Lake Mead, very near to the lower reaches of the Grand Canyon. The marina sports a paved airstrip that is located about one mile from the water. The folks at the marina monitor unicom and will come for a pickup. Bring your own tie-down ropes. The cables are in place, but there are no tie-downs.

Food is good at the restaurant.

Tucson, AZ (Ryan Field—RYN)

Todd's Restaurant 😋😋😋

Rest. Phone #: (520) 578-6337
Location: On the field
Open: Breakfast and Lunch

PIREP   FBO Comments: They have self-serve or pumper truck. Good fuel prices. Easy to find. Just follow the blue taxi line.

PIREP   The place is open again! Todd's at Ryan Field is in business and is great. Food is good. People are great. Prices are reasonable. Todd's also has a very successful restaurant in town (Tucson), so they have experience in the business. They will be around awhile!

You can park the airplane, with tie-downs, directly in front of the restaurant. The ramp is fenced off for security, and you have public access through the restaurant. While you are eating, you can watch the other airplanes and all the action at the field. Great place!

PIREP   There is a good fly-in restaurant at Ryan Field just west of Tucson, AZ. They serve lunch and dinner, and the meals are reasonably priced. A good place to stop in for dinner. I always consider the local folk eating there as a good sign of quality.

Winslow, AZ (Winslow-Lindbergh Regional—INW)

Famous Falcon Restaurant 😋😋😋

Rest. Phone #: (928) 289-2628
Location: 1-mile walk
Open: Daily: Lunch and Dinner

PIREP   The Famous Falcon Restaurant is a great place to eat when you fly into Winslow, AZ. It's on historic Route 66 and has been in business since the late 1940s. Pretty nice write-up in the *AZ Republic* a few years ago. It's a family-owned business and the food is homemade. Really good chicken-fried steak (made by hand); their burgers are handmade, too. The restaurant has been visited by many celebrities. I spotted a picture of Kate Moss and Johnny Depp in the restaurant.

The restaurant is a ways from the airport—I'd say over one mile. You have to just hoof it. It probably takes a good 15 minutes to walk to the restaurant. Ask any local who's standing on a corner in Winslow, AZ, and they all know the way to the Falcon. If I remember correctly, you take the airport road to Highway 87, then hang a left and go under the old Santa Fe Railroad tracks. The airport is south of the tracks and most of the town is north. The Falcon Restaurant is at the northeast part of town. Only two main east/west streets in town (Old Rte. 66). Lots of old shops, the old train depot, and hotel. You turn right at the first eastbound one-way street. There are large black billboards for the Falcon Restaurant that guide you there (you're less than halfway there by now). The Falcon will be on the left. The old Winslow Airport Cafe has been closed for a while, so I sought to find another restaurant. This is a good one if you're into a little hike. Besides, it helps digest the food better with a little exercise.

Coming in from the west (Flagstaff—FLG), you pass near the Meteor Crater, quite a spectacular view.

El Torito 🍔🍔🍔🍔

Rest. Phone #:
Location: On field
Open: Daily

PIREP   New diner open, or maybe an old one on the field. E&O Mexican restaurant, had the abondagas torta and a tamale there this week on a cross country from a flight school at DVT. Food was very reasonably priced, and the owner was very nice. Nice quick lunch flight from anywhere in Phoenix, plus you get to say you "stood on the corner of Winslow, AZ."

PIREP   Business at the airport is down this year because the fire tankers are grounded for safety inspections. Winslow was a prime refueling spot for them. New owners are expected to take over in the next few months where the Last Resort was. Until then El Torito Cafe is the closest to the airport. The courtesy car was in the shop but the FBO took us close on their golf cart. It isn't able to pass over the berm north of the field, but it was only a five-minute walk from there. They serve good, affordable Mexican food but obviously not an airport atmosphere. Typical small-town diner with friendly staff.

PIREP   El Torito Mexican restaurant is very close to the airport, and it serves some of the greatest food in the world. The people of Winslow and others across the nation enjoy their great cooking. It's north of the east end of the east-west runway. Take the airport road to Highway 87. Turn left and go north to the first left turn, go west, and it is the first restaurant on the right. It's a short distance by car, or you can walk to the east end of the east-west runway and jump the fence, and it is on the north side of the street.

Yuma, AZ (Yuma Mcas/Yuma Intl—YUM)

La Casa Deli 🍔🍔🍔

Rest. Phone #: (928) 341-0486
Location: On the field
Open: Lunch

PIREP   La Casa Deli—1 mile from BetKo Air. Should be able to catch a ride. Mexican cuisine and hamburgers. Free chips and salsa with lunch. Local pilots have private table reserved from 11:30 AM to 12:30 PM, Mon.–Fri. Seats eight, but get there early. It's usually full.

Eureka Springs, AR (Silver Wings Field—5A5)

Basin Park Springs Hotel 🍔🍔🍔

Rest. Phone #: (479) 253-5008
Location: Call for pickup
Open: Daily

PIREP   The airport is run by Mr. Errol Severe who lives right next door and also runs the Aviation Cadet Museum located on the field. He gave us a ride to town (three miles) and also picked us up. Telephone is (479) 253-5008 or (479) 253-9471. We ate at the Basin Park Springs Hotel, which I would rate as 3 stars, but there are numerous restaurants, shops, and attractions. The field is only 1,900', turf, but it is on top of a mountain with no obstructions and in good shape.

Fort Smith, AR (Fort Smith Regional—FSM)

Jerry Neel's BBQ 🍔🍔🍔🍔

Rest. Phone #: (501) 646-8085
Location: Crew car required
Open: Daily: Lunch and Dinner

PIREP   Jerry Neel's is a must if you are looking for great food in Fort Smith. The outstanding barbecue is outdone only by their extraordinary hospitality.

PIREP   One of my absolute favorite places to "lunch-and-back" for the day is Jerry Neel's BBQ. Lip-smackin' good barbecue pork, beef, or chicken with a generous menu of sides, including fried okra. I always get free cobbler (with ice cream, too!) after a meal. The FBO at Ft. Smith, TAC Air, has a couple of courtesy cars and will also give you a 30% off coupon to use at Jerry Neel's. What a deal! They are open everyday but Sunday. Another place a stone's throw from there is The Catfish Cove, which does an outstanding buffet on Sundays. They have more food than you would believe, including shrimp, barbecued beef, fried chicken, casseroles, fresh bread, and homemade cobblers (and catfish, of course!). MMMMM!

Both places are a very short drive from the FBO. I give both places five burgers, at least!

Heber Springs, AR (Heber Springs Muni—HBZ)
McPherson's Restaurant 🍔🍔🍔

Rest. Phone #: (501) 362-9702
Location: Call for pickup
Open: Mon.–Thurs. 6:30 AM–2:00 PM, Fri.–Sat. 6:30 AM–9:30 PM

PIREP   McPherson's Restaurant has an excellent weekday noonday buffet for $5. It may be a little higher on Sunday. The food was well prepared with lots of choices of salads, meats, vegetables, and desserts. They also say they have a seafood buffet on Friday and Saturday nights that we have not tried. They are located about one half to three quarters of a mile from the airport but will gladly send a courtesy car to pick you up and take you back to the airport. Actually, it was the restaurant owner that picked us up. They are open seven days a week. Their phone number is (501) 362-9702.

Horseshoe Bend, AR (Horseshoe Bend—6M2)
Karen's Kitchen 🍔🍔🍔

Rest. Phone #: (870) 670-5553
Location: Across the street
Open: Mon.–Thurs. 6 AM–8 PM, Fri.–Sat. 6 AM–9 PM, Sun. 6 AM–2 PM

PIREP   There is a real good restaurant across the street from the airport—I go there often.
Breakfast, lunch, dinner.
Lunch special everyday.

Hot Springs, AR (Memorial Field—HOT)
McClard's Bar-b-que 🍔🍔🍔🍔

Rest. Phone #: (501) 767-4601
Location: 2 miles
Open: Daily: Lunch and Dinner

PIREP   The secret to McClard's Bar-b-que is to get there early, 11:30, or wait until the noon rush is over. The barbecue is out of this world. In business since 1928 and only two miles from the airport, McClard's is not to be missed! It is rated as one of the best in the nation, so good that the Prez goes there when he's in town. It has served many other celebrities. Try the Tamale Spread, but be warned, it'll take a big appetite to consume this tasty dish. The ribs and fries are a plateful—it's a slab of tasty ribs and so many fries you need a front-end loader to shovel them in. The fries are very good. One note: bring cash; they do not accept credit cards or checks. Just ask the fellows around the airport to take you to McClard's, they'll be glad to do it.

There is one more restaurant near the HS airport, called the "Old Feed House," an all-you-can-eat fish and seafood restaurant, excellent catfish, seafood gumbo, chicken and dumpling, frog legs, barbecued ribs, etc. Opens at 1700 hours and closes at 2100 hours. Food price is a flat $9.95 plus tax. No credit cards but will accept checks, local only.

Hot Springs (HOT) has the following approaches: ILS5, VOR A&B NDB, GPS runway 23-5, and no approaches on 13-31. The airport is very well maintained, and there is a lot to see in Hot Springs. Overnight tie-downs, $3.

Lakeview, AR (Gastons—3M0)
Gaston's Restaurant 🍔🍔🍔🍔

Rest. Phone #: (870) 431-5202
Location: On the field
Open: 24/7

PIREP   I went to Gaston's last weekend after much talk and reading about it. I was a little nervous about landing from some of the other PIREPs. I am a new pilot with under 100 hours, and this was my first landing on a grass strip. The strip is in good shape, though a little bumpy on the approach end. The approach is not nearly as bad as I imagined. It is not hard at all to land there. The resort is very nice and the river is beautiful. The restaurant is great! The food was good, the service was quick and, best of all, the price was not bad. There is also a gift shop to look through on the way out. I will recommend the trip to anyone.

PIREP   My brother and I recently decided to take a weekend and meet at Gaston's. For anyone who wants absolute top quality in every respect, it's one of the nicest places you can go. The food is outstanding (albeit not cheap) and the people, all of them, are extremely helpful and well trained. The entire operation is an example of what "quality" really means.

I would like to share some interesting notes concerning the airport. It was built back in the days when most people that flew in there were in Tri-Pacers, 182's, Bonanza's, etc. and small conventional gear aircraft. The airport is exceptionally well maintained, but you need to really pay attention on the approach because the trees that line both sides of the runway are now very mature and reasonably close to the runway on both sides.

It sits in a valley along the White River between two relatively high ridges that curve around both ends of the runway. As a result, a carrier type approach is required to some degree and makes it more difficult than your average local airport. And because of those trees, as you come around on final you won't be able to see all of the runway until you're "in the slot." Since most all knowledgeable people land to the west, sometimes a light tailwind will result in a ground speed of, let's say, about 20 mph faster than you are used to if the wind is from the east or southeast at 10 kts or so. That can produce some surprising results and, if the grass is a little wet, braking action is reduced. Be careful.

The other interesting thing about the runway is the fact that it has some undulations. Not bad, but enough to produce a noticeable reaction in the aircraft. And in some types of aircraft, major reactions. My brother was in his Cessna 190 and I was in my Howard. On the first try, I got the Howard on the end perfectly (wheel landing and a 10 kt tailwind) but then came the whoop-dee-do's. An aircraft with a soft, springy gear is going to give you a ride in a wheel landing you may not like. And the go-around? Well, it's interesting.

I ended up having to go around the first time since the Howard just wouldn't stop galloping. The next time I was prepared, much slower, and nailed it down as much as I dared. The ride was still exciting but no problem.

Then, when taking off to the east, due to the undulations, you may possibly get launched prematurely. Once again, depending on the type of aircraft you're in. It's no problem if you expect it and have good forward visibility. The Howard was exciting. (Remember, in this case, poor forward visibility, tall trees, and a bowling alley effect.)

They have aircraft like King Airs come in, so the airport can accommodate some fairly good-size airplanes. But, depending on the nature of your airplane, and if you are not a little on the adventurous side, might I suggest going over to Baxter County Airport in Mountain Home? It's an exceptionally nice airport and Gaston's will gladly come pick you up. It's about a 15-minute drive. Don't forget, Gaston's is a private airport and, while it's very well maintained, you are still in charge. Land at your own risk.

PIREP   Well, guess who's been and chowed down at Gaston's? About four years ago, my wife and I made the trip on a Saturday afternoon (about a 2.5-hour trip from the DFW area). The runway was in pristine condition, as smooth as many paved ones. We taxied to park next to an old hangar and fuel pump (fuel is available here). The restaurant is about a 100-yard walk from the airplane parking area. Once inside we were seated in a big open dining room, built with large timbers (log cabin style), that hangs out just a bit over the White River. The popular tasty dish here is rainbow trout, caught right out of the White River, which flows from Bull Shoals Lake. You can watch fishermen as they float down the fast-paced river, dragging their fishing poles. This is a great getaway for the day, weekend, or week. There are small cabins available for rent, and you can also rent fishing boats and float fish all day. I know this isn't a hamburger place, but it sure is worth the trip.

Little Rock, AR (Adams Field—LIT)
Flight Deck 🍔🍔🍔

Rest. Phone #: (501) 975-9315
Location: On the field
Open: Mon.–Fri. 7:30 AM–5 PM, Sat. 10 AM–2 PM

PIREP   Went in last week to Central Flying Service and found the Flight Deck Restaurant. It was a great atmosphere overlooking the ramp, and the service was quick and the food superb, all for about 8 bucks—not bad. Central treated me in my 152 as good as the jet guys. Nice place!!

PIREP   The Flight Deck at Adams Field (LIT) in Little Rock makes a great food stop. The restaurant is located inside the Central Aviation FBO building on the GA ramp at Adams. If you are refueling at Adams, you can grab a good meal while the avgas is being pumped.

The Flight Deck has a full restaurant menu and daily lunch specials. Yesterday's special was a small rib-eye steak with sides for $6.95. Not bad pricing for an airport. On the burger topic, the burgers are really good. The fries are also excellent—crisp, fresh, skins on, just the way I like them.

The restaurant has plenty of seating space (20 tables or so) and good fast service and is located overlooking the GA ramp. Lots of local folks drive to the restaurant to eat, so I guess it rates as one of the better spots around. The local folks probably enjoy the plane-watching, since the GA ramp is a really busy place at LIT.

Morrilton, AR (Petit Jean Park—MPJ)
Mather Lodge Restaurant 🍔🍔🍔🍔

Rest. Phone #: (501) 727-5441
Location: On the field
Open: 24/7

PIREP   The restaurant has improved. The last three trips I got very good food. Last trip, I got the ham steak and it was very good. They have a nice salad bar, and the view is absolutely breathtaking. Keep it simple on the food and you'll do just fine. The FBO is a bit casual, but they have a loaner vehicle and they either take you to the restaurant or give you the keys.

There is an EXCELLENT picnic/camping area at the opposite end of the runway from the FBO that is just for pilots. Very clean restrooms, showers, camping spots, the works.

PIREP   It is a nice place to drop in. The airstrip is *huge* (6,000 × 75'), given its traffic volume and rural Ozark location, and that was a mystery until I chatted with the ranger who came to ferry us to the lodge. Thousands of acres of land around about are owned by some Rockefeller foundation or other. They built the field and donated it to the state at some point, hence the impressive dimensions. Also it might be worth noting the airfield sits high up on a bluff, 600' above and south of the meandering Arkansas River valley. Taking off to the north is interesting. Attention-grabbing (on a hot day!) rising terrain of solid forest, until you clear the ridge...and there suddenly are the Arkansas Mountains River. An effect reminiscent of the opening sequence from *The Sound of Music* minus Julie Andrews.

Some notes: You need to bring your own tie-down ropes. There are cables but no ropes or chains at the field. I could not find the number at the payphone to call for a ride, but I had copied the number from *$100 Hamburger*, so all was well. The lodge is a splendid old rustic log edifice whose entryway frames a gorgeous view to the west down a spectacular gorge. There are lodge rooms and cabins (some with kitchen; one with hot tub) available for overnighting.

That's the good news. The food at the lodge was totally disappointing, though. It reminds me of the depressing Fred Harvey catering at the Grand Canyon. However, if you avoid anything pretentious on the menu, you might be okay. I ordered some fried mushrooms whose centers were still frozen and that were very greasy and unpleasant. There's a fair salad bar, and, should you be staying overnight, be aware no wine or liquor is served (or allowed) in the park. If I give it three hamburgers, two would be for the view. Ask for a table by a window (which are the nonsmoking ones).

I bought fuel at Russellville (RUE) about ten miles NW of MPJ. I'd recommend it as a friendly field for a top-up when visiting MPJ.

Mountain Home, AR (Ozark Regional—BPK)

Airport Pizza and Subs 🍔🍔🍔

Rest. Phone #: (870) 481-5000
Location: In the terminal
Open: Dinner: Mon.–Sat.; Lunch: Mon.–Fri.; Closed: Sunday

PIREP   FBO Comments: Also known as Mountain Home Airport.

PIREP   Small and friendly airport pizza and sub place that makes its own pizza dough, has a variety of pizzas and calzones, as well as sandwiches and salads. A short walk or crew car ride from the FBO to the new terminal where the restaurant is located. Prices are very reasonable. Check ahead for hours—open for dinner Monday–Saturday and open for lunch Monday–Friday, but closed Sundays.

Mount IDA, AR (Bearce—7M3)

Crystal Inn Hotel and Restaurant 🍔🍔🍔🍔

Rest. Phone #: (870) 867-2643
Location: Across the street
Open: Daily

PIREP   Was flying over Central Arkansas a few weeks ago enjoying the fall scenery and decided to put a few extra gallons in the Cheetah before heading back to Bentonville. The service was excellent at the friendly FBO, and they directed me to walk over to the Crystal Inn Hotel and Restaurant to fill up the other empty tank!

Absolutely fantastic burgers and all the treats I could stuff before heading home. Everyone acted like they were genuinely glad to see me. All pilots are treated like royalty at this airport. Wish other airports could be run like this.

This restaurant deserves five flying hamburgers from this airport restaurant junkie!

PIREP   This is a real pleasant place to visit. The airport is managed by Dr. Fowler, who goes out of his way to make you feel right at home. The field is well kept, and both Jet A and 100LL are available. Easy place to get into and out of.

Adjacent to the field is The Crystal Inn Hotel and Restaurant. We had their breakfast buffet. You can eat enough to demand recalculations of your weight and balance.

Pine Bluff, AR (Grider Field—PBF)

Grider Field Restaurant 🍔🍔🍔

Rest. Phone #: (870) 536-4293
Location: On the field
Open: Mon.–Sat. 7 AM–3 PM

PIREP   FBO Comments: Linemen very helpful and friendly.

PIREP   Took a friend who had never flown before to lunch at the Grider Field Restaurant. It being Friday, we were treated to a great catfish buffet with all the trimmings. I quote my friend: "What a great treat, flying low and slow on a clear day, and a catfish buffet.....it doesn't get any better."

PIREP   Have just passed through Pine Bluff, AR, and find that the Grider Field Restaurant is under new management, who are serving very good lunch. Both the buffet lunch and the BLT's we ordered were fine, and service is mucho friendly. Breakfast, lunch.

Rogers, AR (Rogers Municipal-Carter Field—ROG)
Aviate Cafe 🍔🍔🍔

Rest. Phone #: (479) 621-2312
Location: On the field
Open: Mon.–Fri. 7 AM–1:30 PM

PIREP   FBO Comments: Very comfortable with friendly attentive staff.

PIREP   The Aviate Cafe is open for breakfast and lunch with daily specials, good, hot food. Not your usual airport fare. Lots of seating (indoor and outdoor) with a view of the runway.

Siloam Springs, AR (Smith Field—SLG)
Cathy's Corner 🍔🍔

Rest. Phone #: (479) 524-4475
Location: On the field
Open: Daily: Breakfast, Lunch, and Dinner

PIREP   FBO Comments: Very friendly! Airport cars available.

PIREP   Excellent big hamburgers. I could see what looked like lots of good home cooking, lots of daily specials. Ask at the FBO for directions to Cathy's.

Springdale, AR (Springdale Muni—ASG)
The Airport Cafe 🍔🍔🍔

Rest. Phone #: (479) 756-3339
Location: In the GA building
Open: Breakfast and Lunch, Mon.–Sat., closes at 2:30 PM

PIREP   Springdale, AR has a cafe located inside the GA bldg. It features okay food, but it has an outdoor "deck" where you can enjoy your meal and soak in the sights, sounds, and smells of all the airport's activities. The GA bldg is a two-story modern facility, and the cafe is located on the second level.

They also have a courtesy car that you can use to run into town if you have need of some shopping or a bigger selection of eateries.

Good place to stop, excellent facility with loads of room to park.

PIREP   The cafe in the modern GA building has excellent hamburgers and very good plate lunch specials. The tower is now open 7 days a week. Be careful not to wander into class delta airspace surrounding the new Northwest Arkansas Regional Airport (KXNA) to the northwest.

Walnut Ridge, AR (Walnut Ridge Regional—ARG)
Parachute Inn 😋😋😋😋

Rest. Phone #: (870) 886-5918
Location: On the airport
Open: Closed Sun. & Mon.; Tues.–Fri. 7 AM–2 PM; Fri. & Sat. night, open again at 5 PM.

PIREP   All-you-can-eat catfish on Fridays. Part of the restaurant is an old Southwest 737.

West Memphis, AR (West Memphis Muni—AWM)
Payne's Bar-B-Q 😋😋😋😋

Rest. Phone #: (870) 272-1523
Location: Downtown cab or crew car
Open: Daily: Lunch and Dinner

PIREP   Some travel is required from any of the Memphis area airports. Believe me, if you love barbecue, as I do, it might be worth the effort. Landed at West Memphis Airport (AWM), just under the Class B airspace, on the flat, flat, Mississippi floodplain, west of the river. Busy, but friendly place, with rental cars, courtesy car, and a splendid airport cat. The courtesy car was available for the quick drive east to Memphis, TN. And a quest for a Memphis barbecue fix.

Take I-40, then I-55 south (posted Jackson, MS) across the river; head for the international airport (MEM) via I-55S and I-240; hang a left on Airways Blvd. northbound, to US 78 (Lamar Ave.); there, take a left on Lamar to about McLean to Payne's Bar-B-Q. It is on the right as you head west on Lamar, just past the eye-catching Mel's Seafood restaurant, 1762 Lamar. Payne's is a hideous-looking white cinder-block place—a converted gas station, no less.

Inside, it's a different story. The atmosphere is like—how can I sum it up?—a temple or cathedral, devoted to the worship of meat; more specifically, pork. A hushed silence among the orderly crowd awaiting service. The interior is tasteful and clean, at least by most BBQ standards.

You can eat in or take out. The menu is short. Shoulder sandwiches (specify sliced or chopped meat); ribs (full or half slab); hot links; bologna; potato chips…and that's about it. However, no more is necessary. The food is great. And the barbecue sauce (you have a choice of hot or mild; I never did hear anyone order mild, and having tasted the hot, I can see why) is about the best I ever remember tasting. A hint of orange (?) fruitiness but wholesome and totally savory. It goes with the meat perfectly, and dunking a sandwich in it turns the meal into a deeply religious experience. As with all Memphis barbecue, coleslaw will be added to your sandwich unless you specify otherwise. In this case, it's no hardship. A piquant version of the coleslaw genre. No creaminess to mask the meat flavor. Perfect!

The staff consists of an older black couple. He fishes out hunks of shoulder or ribs from a vast pitch-black pit. She wields a mean cleaver, hacking away at the ribs and shoulder meat; and also assembles orders and collects your money. There was a constant stream of customers, patiently waiting as each order was painstakingly prepared to perfection. While the neighborhood is of the auto body shop–slightly grungy variety, it's certainly quite respectable. I would have no worries venturing there at any time. And even if I did, I would still go for some more of their ribs, and a "sliced shoulder" on the side. Owing to bad forecasts, we flew to AWM, ate at Payne's, and were in the air back home in less than three hours. Six hours flying in a day for barbecue may seem excessive. But then you'd have to try it before you pass judgment on my sanity.

Happy eating and safe flying!

Agua Dulce, CA (Agua Dulce Airpark—L70)

L70 Cafe and Lounge 🍔🍔🍔

Rest. Phone #: (661) 268-8835
Location: On the field
Open: Lunch 7 days a week

PIREP I visited this airpark twice recently, the latest in early October, and all I have to say is AWESOME! The new owner has done one heck of a job up there. The entire field has been repaved, and there is a fresh, brand-new grass strip that they are opening soon. The restaurant/pilot lounge is right on the ramp and is first rate with free pool tables, plasma TVs, horseshoe pits, two sand volleyball courts, a pool, and a spa. My buddies and I flew up there for my birthday just so we could bum around, enjoy the barbecue and the quiet environment. They have barbecue through the summer months; during the rest of the year, the restaurant inside the lounge is great. The waffle breakfast is one of the best.

All in all, a must see!

PIREP Today we flew in to Agua Dulce (please say "dull-say," not "dull-chay"—it is Spanish, not Italian!). It is a delightful little airport, nestled among the hills north of the L.A. basin—what a wonderful escape from the smog! The 4,600-foot runway is a pleasure to land on, and on Saturday and Sunday, they have a barbecue from 11 AM to 3 PM. Delicacies such as tri-tip sandwich, chicken breast sandwich, hot dogs, and burgers are on offer. We had a cheeseburger for $5 each, which included a soda and chips or barbecue beans. You add your own condiments, including lettuce, tomato, onion, relish, mayo, mustard, and ketchup. It was very good. I would give it four hamburgers; my husband loved his and voted for a five.

Amazing but true, today avgas is cheaper here than premium gasoline in San Diego. You have to support effort like this!!

One small comment: there is quite a lot of loose gravel in the parking area, so it can be worth clearing a path before you apply significant amounts of power.

PIREP I flew into Agua Dulce Airpark (L70) today because I heard that there was again food available there. There used to be a diner-type restaurant that closed some time ago. Seems Arizona investor Barry Kirshner has purchased L70 and refurbished the former restaurant building. It is no longer a restaurant (instead it has been transformed into a clubby atmosphere, replete with billiard tables). The "managers" are holding a fly-in barbecue (hamburgers only) on Saturdays and Sundays at lunch.

I purchased a cheeseburger for $5 that included a can of soda and a small bag of chips. Not exactly Harris Ranch, but it filled the void and I was back on my way. Some nice picnic tables have been set up under the trees, and some older airport flyers were comparing notes as they enjoyed their burgers.

Try it out next time you're in the area.

Angwin, CA (Angwin-Parrett Field—2O3)
Pacific Union College Cafe 😊😊😊

Rest. Phone #: (707) 965-6327
Location: Short walk
Open: Sun.–Thurs. 9 AM–5 PM, Fri. 9 AM–4 PM, Sat. closed

PIREP The College cafeteria has good, wholesome, inexpensive meals. Summer hours are reduced (lunch starts at 11:30). Even then there are friendly students around who can direct you to the cafeteria and tell you the hours. The campus is about half a mile due west and 160' elevation drop below the tie-down area. Just follow the paved road downhill, and you'll arrive at the campus.

Being Seventh Day Adventist, the cafeteria won't accept payment of any type (either cash or credit card) on Saturdays. Not to fear, they have worked out an acceptable system for visitors: just write your name and address on one of the forms at the checkout, then when you get home send your remuneration in the preaddressed envelope provided.

One of the unique fly-in attractions of Angwin is the miles of mountain biking trails that start right at the airport parking lot. Several mountain bike races have been run on these trails, including the Howell Mountain Challenge, the Napa Knobular, and the Napa Valley Dirt Classic. The trails seem well drained, but if your bikes get muddy while riding in winter, there is a hose at the airport office you may use to wash them off before loading them in the plane for the return trip.

PIREP Fly into this slightly challenging hilltop Napa County airport, then walk a few blocks down the hill to Pacific Union College for a genuine Seventh Day Adventist veggie burger in the cafeteria.

Atwater, CA (Castle—MER)
Flight of Fancy Grill and Banquet Room 😊😊😊😊

Rest. Phone #: (209) 723-2177
Location: On the field
Open:

PIREP FBO Comments: Gemini was great. They have a van available to take pilots and passengers to the air museum. When you want to come back, just have the museum gift shop call up the FBO and the van will come to bring you back. Sweet.

PIREP Went to the little grill that's part of the gift shop building. I didn't see the "banquet room" anywhere, but who knows, maybe they actually have one. Burgers, fries, o-rings, and a couple of different standard sandwiches here. Not a lot of selection, but at any rate, the Bomber Burger was a cut above the typical greasy spoon fare. Not bad, indeed. Great place to fly in. How often do you get to land on a nearly two-mile long runway without a landing fee?

The museum is awesome—fantastic history in those planes. We went during a weekend when the "Boomers" were having a reunion. We got to see two HUGE tankers landing there—one was a big pregnant guppy of a plane, had four turbo props, didn't catch the type. The other was a C135, which joined two more jet-powered tankers already on the tarmac. What a sight that was, those big monster tankers parked across from a smattering of teeny little singles.

PIREP My wife and I flew from Sacramento to Castle. We were awestruck at the enormity of the former Castle AFB in relation to the diminutive Cessna 182 we were flying. Nothing small though about the service from the Gemini folks. They provided a free ride to and from the museum. The restaurant inside the museum entrance is top notch, and the collection of aircraft is outstanding. Well worth the trip.

Auburn, CA (Auburn Muni—AUN)
Wings Grill & Espresso Bar 🍔🍔🍔🍔

Rest. Phone #: (530) 885-0428
Location: On the field
Open:

PIREP Today's Tuesday.
 I had to wait for an outside table, but there was immediate seating inside. Caesar salad was great (fresh and cold), and the burgers got no complaints. Waitresses paid attention, and what scenery!
 The four burger rating stands.

PIREP FBO Comments: 100LL SS and Jet A available.

PIREP Flying into Auburn was a great experience. Once on the ground we were politely directed by the airport manager to the appropriate tie-down spot (free).
 Unfortunately, Wings Grill was closed, but the manager suggested we go just down the street a couple of blocks to Ridge Country Club, which houses the Ridge Bar & Grill. He asked if we wanted a free ride there and called Ridge for us. Within five minutes, a Ridge golf cart arrived at the pilots' lounge, and they happily gave us a lift to the hilltop clubhouse.
 Inside is a luxurious clubhouse restaurant. They also have a full bar for passengers. The restaurant has a great panoramic view of the golf course and beyond. There is also a large outdoor terrace section available overlooking the well-kept grounds. I ordered the Ridge Burger (of course!) which was generous and very good. Prices are below average for a country club. and the service was very good. The ambiance is wonderful. Afterward, we relaxed outside on the deck!
 As soon as we asked, we were promptly given a ride back to the airport, right beside our airplane—what service! All in all, I highly recommend this place. I give Auburn my five-burger rating.
 I will be back!

PIREP We have flown to Auburn a few times. The first time we ate at Wings Grill, where the food was good but the place was crowded as reported, and we had to wait over two hours for our food. The next few times that we visited Auburn we took a cab into the Auburn old town. It is a typical Gold Rush–era town like others along the Sierra Foothills. Lots of charm, small shops, antique stores, and restaurants. There is a large cement statue of Claude Chana, the man who discovered gold near Auburn. We can recommend eating at the Edelweiss restaurant, where the food was excellent and the service was prompt and friendly. It is a full service restaurant, and they specialize in omelets. For the wine enthusiast, located on the nearby corner is Carpe Vino. This is a fairly new wine shop specializing in Sierra Foothill wines, and they also carry a wide range of domestic and international varieties. Be sure to check out their discount cellar behind the main shop. They also offer wine tasting for those not acting as PIC.

PIREP Auburn's excellent airport cafe has recently become even better, I feel. The new owner, Connie Horning, has changed the name to Wings Grill & Espresso Bar, created a new menu, and redecorated the outside covered patio area to make it even nicer than it was before. Soon work will begin to create the espresso bar. Few airports will have a place that can match it, I think. Fly in to Auburn Airport (AUN) to enjoy breakfast or lunch seven days a week.

If you like to golf, bring your clubs because the new 18-hole course, The Ridge, is open and is just across from the airport: 2020 Golf Course Rd., Auburn, CA 95602-9526, (530) 888-7888.

Avalon, CA (Catalina—AVX)

Armstrong's Seafood 🍔🍔🍔🍔

Rest. Phone #: (310) 510-0113
Location: In town
Open: Daily: Lunch and Dinner

PIREP   FBO Comments: No fuel available.

PIREP   Armstrong's Seafood in Avalon has excellent fish and fairly good fries. Good service, and the view from the deck overlooking the harbor can't be beaten.

Note for the airport: it's about seven miles from the town of Avalon, up high on a hill, elevation 1,606 msl. I found that the landing fee is now $20 regardless of aircraft size or passenger count. Runway and taxiway are in decent shape. BYO wheel chocks—I didn't see any tie-downs on the sizeable apron. Many planes were using rocks.

Shuttles run regularly to Avalon, but the cost is $17 per person, round trip. One-way fare is $13. I urge you to make a round-trip reservation—the shuttle back to the airport is frequently full in the afternoon. If you miss the last one (5:30 PM), cab fare to the airport is said to be $85 or more! Yes, the place is expensive, but it's beautiful and the residents are working hard to keep it that way. There are a number of campgrounds, most with running water, johns, lockers, and so on.

Rating: 3

PIREP   This has to be one of the most spectacular island ridge landing sites anywhere! After sliding in, we went into the Runway Cafe, ramp-side, and ordered the specialty buffalo burgers. They are really tasty and generous, too. Wild buffalo roam the island freely and they allow hunting sometimes to thin the population. They have a Mexican-style outdoor patio facing a panoramic view of the island. There is a shuttle available into town (30 minutes) where more good eating opportunities await.

I recommend staying a night or two to enjoy the town. This is truly a great fly-in destination. My only complaint is the $20 landing fee for a single is unreasonable, but their runway is in decent shape. No fuel is available.

That said, I will be back!

PIREP   Stopped here for a bite and to see a little bit of the island during a planned trip to San Diego. The food is good and surprisingly affordable; not much else is on the island. The landing fee is $20 as reported. For a once in a great, great while trip I won't grumble too much, but if I were local I think I'd go elsewhere. On our trip there wasn't a single cloud in the sky except for a low cloud deck over Catalina Island.

It was below minimums when we left but was expected to burn off. I decided to fly over anyway and if it was still below minimums I'd fly back. Fortunately it lifted to 100' above minimum by the time I got there. I'm told this is not unusual due to the steep rising terrain on the island uplifting moisture. The runway seemed to be in better condition than has previously been described. I don't think it's been repaved, but it may have been patched. While I know runway 22 is preferred, they were reporting 020 at 6 when I approached. I announced my intention to use runway 4 due to winds but they said 22 is in use up to 8 knots.

Pulsating visual approach slope indicator on the left side of runway 22. That was a new one on me—I don't remember seeing that one in my pilot training. I looked it up before I left though after seeing this in the airport directory entry. Pulsating white, too high. Pulsating red, too low. Steady white, on glide path. Steady red, slightly low. Watch the altimeter closely, don't get too low on downwind or base, and you should be fine turning final.

PIREP   Just wanted to let you know KAVX has a new landing fee for ALL airplanes regardless of size, # of pax or # of engines: $20.

I'm glad to pay it though, if it helps to pay for the maintenance of this outstanding, unique, and exciting airport! I felt like I was going back in time taxiing up to the 1940s-built terminal. There was also a DC-3 on the side of the runway that took off right after me. Someone there told me it did regular runs between the island and mainland shuttling goods and supplies. It added to the old-time feel of the place!

I didn't have lunch, but I have pilot friends who said the restaurant was good, so I will try it next time.

I'm also a gardener and plant enthusiast. It looks like they recently had a very nice garden/plant display right there by the transient parking in front of the terminal. From what we could gather, its caretaker had perhaps died—too bad. I hope they earmark some of that money for the display garden! It would enhance the fly-in experience.

Big Bear City, CA (Big Bear City—L35)

Barnstormer 🍔🍔🍔🍔

Rest. Phone #: (909) 585-9339
Location: On the field
Open:

PIREP   Barnstormer is definitely one of the nicer airport cafes in SoCal. This is not your typical airport greasy spoon. Excellent breakfast (I'd recommend the Barnstormer Omelet or the bacon frittata), prompt service, and a great location. Watch the high DA.

Rating: 4

PIREP   Just had my first trip to Big Bear, which has a field elevation of 6,748', and parked the Skyhawk in front of the Barnstormer. I had the turkey melt with terrific potatoes. The place was jammed with customers, most of whom had not flown in. The service was friendly and the prices reasonable. Atmosphere was pretty nice, although I've seen better.

We used up about a third of the 5,850' runway on takeoff. The scenery up there is spectacular, especially after the first snowfall. Just be sure to watch that mixture!

PIREP   I fly to California to rent and tour almost biannually. My checkout invariably includes a trip to Big Bear. I wonder why???

Could it be that: a) I enjoy the atmosphere and the food so much, and/or b) my checkout instructors do too?

I can thoroughly recommend Big Bear's Barnstormer Restaurant. I often switch fries for potato salad (healthier, perhaps?), but my tip of the day is try their freshly squeezed lemonade, it's adorable, and just the ticket for the high-altitude flyers who tend to dry out faster than the sea-levelers.

You could say that going to Barnstormer is as good as a trip to your dietician. It's certainly more fun!

Location: 9/10—overlooking the runway.

Service: 8/10—I've never had bad service there.

Food: 8/10—very good!

Ambiance: 8/10—always a good feel to the place.

Price: 8/10—although a foreign visitor, I know good value when I see it.

PIREP   Great food—instead of the french fries with the entree, try the fruit salad instead. Good views of the runway. The owner came out and spoke to each and everyone at their table to see how the food/service was. I have no complaints. This restaurant has won my wife over and made flying to restaurants much more palatable.

The restaurant is closed between 3 PM and 5 PM to prepare for dinner.

When executing the right traffic pattern for runway 08, stay over the "water." The lake level is down, so where there used to be a lake immediately west of the runway threshold is a dry lakebed. No seaplanes on Big Bear Lake.

Bishop, CA (Eastern Sierra Regional—BIH)

Sierra Wings Cafe 🍔🍔🍔

Rest. Phone #: (760) 872-2971
Location: On the field
Open:

PIREP Restaurant is excellent in all respects. Come over to God's Country and try the Sierra Wings Restaurant!!

PIREP Good food, great service, and a beautiful view. Talked with the owner Teresa, who needs aircraft photos (framed only, please) for the walls. The FBO has also moved into the new building, and the old tower building is going to be restored for historical and sentimental reasons.

Boonville, CA (Boonville—D83)

The Buckhorn Saloon 🍔

Rest. Phone #: (707) 895-3369
Location: 1 mile to town
Open:

PIREP The Buckhorn Grill has changed hands a number of times. It is currently closed but is scheduled to reopen in the spring of 2006 per a local I spoke to at the post office.

PIREP FBO Comments: Still no fuel.

PIREP Several restaurants within a mile from the airstrip, but for a burger, the best place is the Buckhorn Saloon. Very good food at a reasonable price, and they offer excellent beers from the local brewery, Anderson Valley Brewing Company. Buckhorn is closed Tuesday and Wednesday. If you don't feel like walking and have planned ahead, the brewery offers a shire horse-drawn carriage ride into town or to their tasting room at the brewery (about two miles from the strip). Two-day notice is required to get a carriage ride; $50 for 1–4 people; phone (707) 895-BEER.

PIREP FBO Comments: No fuel available.

PIREP This little out-of-the-way airstrip is actually one my favorite places to visit. There is nothing going on at the airport itself, just an outhouse shaped like a small control tower. The runway is in good condition; the tie-down area is a bit rough but passable. However, walk about a quarter mile to downtown, and you'll be rewarded with the Buckhorn Saloon. They have locally brewed Anderson Valley Beer for passengers and great big, tasty hamburgers, fries, and steaks. Service is good and prices are average. There is a nice old-fashioned hotel right across the street with a very tastefully designed outside patio/garden area for relaxing. Exploring the rest of the town, there are coffee shops and boutiques to enjoy as well.
  I give it a solid three-burger rating.

PIREP FBO Comments: No fuel or any service at all at Q17!

PIREP Great little airport in Booneville. Very quiet, charming place. Taxi back on runway, park on the south end. Walk to town and enjoy!
  The Buckhorn Saloon, which used to also be the Anderson Valley Brewery shop, is closed Tuesdays and Wednesdays. Your passengers can still sample all of the beer there when they are open, however. The gift shop has moved next door, although the staff says it is moving once again up the road another mile to the actual brewery.
  The deli/cafe next door to the gift shop was excellent! Friendly service, good food, good value.

Boonville General Store 🍔🍔🍔

Rest. Phone #: (707) 895-9477
Location: 17810 Farrer Ln.
Open:

PIREP The Buckhorn is closed, so we asked a local person where to eat. She recommended the Boonville General Store. She used the word "hippy" to describe it. All natural foods only, so you won't find a Diet Coke there. I enjoyed one of the daily specials. There is indoor and outdoor seating right along the main drag.

To get there, follow the highway east from the airport. Take a right at the T intersection. The restaurant will be a block away on the left side of the road.

Airport Note: The parking area has been recently improved with new pavement and tie-downs. No more bumpy taxiing.

Borrego Springs, CA (Borrego Valley—L08)

Crosswinds Steakhouse 🍔🍔🍔🍔

Rest. Phone #: (760) 767-4646
Location: On the field
Open:

PIREP I had heard that the restaurant was opening on Oct 7 for a Friday and Saturday lunch from 11 AM to 1:30 PM and in the evenings of Friday and Saturday from 5:30 to 8 PM. I flew down there for lunch and was shocked. The restaurant is set up very nicely, the decor is great. But the food wasn't too good, and the prices were astonishing. A cheap sandwich began at $8. The regular lunch plate of food was from $17 to $23, plus drinks. Better take a pocketful of money with you.

I rate it a two-burger place at best.

They were doing aerobatic competition, so there were a few people around. Otherwise, I doubt that there is much of a crowd.

PIREP WOW! A REAL restaurant on an airport! I had been here about five years ago, but the new owner has really fixed up the place and the food is GOOD! The owner is a very well-known chef. Lunch and dinner is available Tuesday through Sunday (Sunday is a brunch buffet instead of the lunch menu) and there is a full bar—for your passengers, of course. There's even a rooftop patio to watch the airport action, which can be exciting since this the home of the IAC AkroFest every October. Fly here in February/March to see the spectacular wildflowers! Also, Chef John said he would open for fly-in breakfasts or private parties if any clubs are interested.

PIREP The Crosswinds restaurant is open again, but they only serve lunch on Saturdays and Sundays, starting at 11:30 AM. Open for dinner at 5 PM everyday. This is now a "steak house." Limited lunch menu has one type of hamburger and a few salads. Have not tried the dinner menu. The new owner is said to be a well-known chef, formerly from Ram's Hill. Have only visited once so far, the hamburger was fine. Would have preferred some deli-type sandwiches. They had been open for only a few weeks and may still be in the planning stages.

La Casa Del Zorro 🍔🍔🍔🍔🍔

Rest. Phone #: (760) 767-5323
Location: Call for pickup
Open: Daily: Breakfast, Lunch, and Dinner

PIREP If you're tired of eating at the same type of breakfast cafe and looking in Southern California to enjoy a VERY NICE restaurant with beautiful surroundings, then the La Casa Del Zorro in Borrego Springs is where you should go. The price tends to be on the higher side, but what you get in service and ambience is well worth the trip. I have stayed overnight and can't say enough about the accommodations. This is NOT an airport cafe.

When flying in to Borrego Valley, call unicom (122.80) for advisories and tell them how many in your party for Casa Del Zorro. They'll call the resort for you for pickup; after landing, go into the FBO—the transportation should be there to pick you up (no charge either way—I tip the drivers though). The people that work there look after you and your needs and will open doors for you and so on. Get ready to be pampered. For breakfast, sit inside next to the fireplace or outside looking at the green plants, trees. and beautiful garden. The grounds around the resort are worth walking during your visit; you'll forget you're in the desert. Two swimming pools, one for families with children and the other for adults only, nice and quiet.

First time flying in to Borrego Valley, from the description of this resort, you'll think you made a mistake. The valley from the air is desert and not much to look at, but you arrive at the resort and go in behind the trees and you think you stepped into another land, a beautiful oasis. Everything is green and colorful with flowers and plants. Summertime is hot, but spring, winter, and fall are nice times to visit. Borrego Valley shares the same weather as Palm Springs.

If you can't decide where to fly for breakfast and you have a couple of extra bucks in your pocket and you want to be pampered, take some friends and fly into Borrego Springs to the Casa Del Zorro, and enjoy yourself.

Bridgeport, CA (Bryant Field—O57)

Bridgeport Inn 😋😋😋

Rest. Phone #: (760) 932-7380
Location: 2-minute walk
Open: Daily

PIREP Don't think that winter flying at Bridgeport poses a threat. Although I don't live there anymore I still own the hangar that's on the field and put about 1,500 hours in a Grumman Tiger flying in and out of that field. Winter flying can be great. In the winter of 1979–80 I commuted from Bridgeport to Tonapah every workday (and some weekends). During that whole winter I missed three days due to weather. A word of caution: flight into any precip or clouds in this area should be considered flight into known icing (and there's quite a number of scattered aluminum patches in the backcountry that prove that). Density altitude can be daunting in the summertime, but is never a problem in the winter. Bring a warm coat and hat anytime you go to Bridgeport. Winter is any day of the year. It has snowed there on the Fourth of July, and there have been more Fourths than not when there was frost in the morning.

For my money, the best burger in town can be had at the Jolly Cone, which is on Main Street, just east of the center of town. If you go to Bridgeport, don't miss the bakery that is next door to the Jolly Cone. Their sheep-herder bread is the best in the state.

PIREP There are two restaurants here, a two-minute walk from the airport. The airport is roughly equidistant between South Lake Tahoe and Mono Lake. The scenery and terrain is wonderful, if a little challenging. (The airport is 6,500' and surrounded by mountains.)

The Bridgeport Inn is quite nice and slightly formal, in a western Gold Rush town way. The Victorian & Fries is excellent, and yes, they do have hamburgers.

The Hays Street Cafe is more like a diner, although the menu is varied. This place is actually a little closer to the airport than the Inn.

Really, this place deserves more traffic, although not during the winter, when most of the town is snowed in.

Brownsville, CA (Brownsville—F25)

Pine Tree Eatery 😋😋😋

Rest. Phone #: (530) 675-9150
Location: Right next door
Open: Daily 10:30 AM–7:30 PM

PIREP   If you are looking for an interesting airport in the mountains with a motel and two eating places next door, try Brownsville in the foothills east of Oroville. Great scenery and a fun airport to fly into and out of. At the northeast end of the airport, across the road is a motel and the Mountain Rose Cafe, open every day 7:30 AM to 3 PM. About half a mile to your left is the Pine Tree Eatery, open every day 10:30 AM till 7:30 PM.

Byron, CA (Byron—C83)
The Byron Inn 🥪🥪

Rest. Phone #: (925) 634-9441
Location: 2 miles
Open: Daily

PIREP   The Byron Inn is located about two miles from C83 (Byron airport) The food is great but I don't know how you could get there except walk. They are open from 5:30 AM until 2:30 PM and serve all homemade dishes. Voted best breakfast in the county. Great biscuits and gravy.

Cheers,

Mike Bohlander

Madd Maxx Air Racing Team

P.S. Turn right off the access road and left at the stop. It's just over the tracks along the Byron Bethany Highway.

Calexico, CA (Calexico Intl—CXL)
Rosa's Plane Food 🥪🥪🥪

Rest. Phone #: (760) 357-6660
Location: On the ramp
Open:

PIREP   FBO Comments: Watch out or be prepared for the bumpy runway.

PIREP   Rosa's makes Calexico a great stop on your way somewhere or a destination all its own. Authentic Mexican food in a small border-town cafe. It isn't much to look at, but the food is really good. Their ceviche is excellent. Prices may be a little on the high side. It helps if you speak Spanish, but you can get by without it.

I can't wait to visit again.

Rating: 4

PIREP   While running my Cherokee to California a couple of years ago I stopped into CXL, mainly because it was just about where I needed a stop to fuel the plane and defuel the pilot. The airport is literally a stone's throw from Mexico, and the restaurant is right on the ramp.

I'm from Texas. Mexican food is a staple here. VERY few big restaurants get their Mexican recipes right—too gringo-ized. The best food comes from the places where nobody speaks English or a friend's house where Mexican is the native dish.

Rosa's is almost one of the former, and they get it right. I didn't have time or room for anything but the enchiladas, but my mouth is watering right now as I'm typing this. They are the real thing.

If you like REAL Mexican cooking, you need to go.

If I ever hit the lottery and get a jet, CXL will be will be a regular $2,000 hamburger run. In fact, now that I think about it, there WERE three Citations parked in front when I was there.

PIREP FBO Comments: VERY nice clean small airport with polite, friendly employees. Avgas available. This is a small airport, but it's also an Airport of Entry, with a U.S. Customs office on the field. GA only, no scheduled airlines.

PIREP Rosa's Plane Food is on the airport, just a few steps from the GA terminal. The restaurant is very popular with pilots returning from Mexico, since this is an Airport of Entry for those flying back from Baja California, Mexico. The restaurant is closed Tuesdays (call first anyway). Rosa's Plane Food has mainly Mexican food, but a few burgers and sandwiches on the menu. Priced very reasonably. Lots of locals eat here, too.

California City, CA (California City Muni—L71)
Airport Cafe 🍔🍔🍔

Rest. Phone #: (760) 373-8008
Location: On the field
Open: Daily

PIREP I've flown into there a couple of times recently for breakfast. The remodeled restaurant is way overkill for a small airport restaurant, but it is very nice. The food is good and the price is so low that I was embarrassed. I left a tip that was about the equal of the bill because I'm hoping they can afford to stay in business.

Camarillo, CA (Camarillo—CMA)
Way Point Cafe 🍔🍔🍔

Rest. Phone #: 805-388-2535
Location: On the field
Open:

PIREP FBO Comments: Very fast and good service. They are sitting in the trucks on the ramp looking for anyone needing fuel.

PIREP We went to Camarillo for lunch.
The hot roast beef sandwich was not very good.

PIREP I recently revisited Camarillo's Way Point Cafe. It's still great, deserving of five hamburgers in my opinion. Food and service are consistently good. Tri-tip sandwich alone is worth burning up a few gallons of avgas for. Transient parking is no problem and steps away from the cafe. I do recommend staying away on Saturday morning, however. It seems that the Porsche owners' club has adopted the Way Point as its meeting place and the patio is jammed on that day (8–10 AM) every weekend.

PIREP Stopped at the Way Point Cafe for lunch today, the third time I've been there in the past two months. Tried the tri-tip sandwich and it is great. It was not on the menu, however, and the waitress indicated it was only on weekends. The cafe is absolutely first class in all departments—I have been to Chiriaco Summit on four occasions and, while it is a nice cafe, it does not hold a candle to the Camarillo airport cafe (CMA). The cafe is very popular with the locals and, today being Sunday, both the outside patio and the inside were full of contented diners. Very convenient parking right out front.

Cameron Park, CA (Cameron Airpark—O61)
Cameron Park Liquor & Deli 🍔🍔

Rest. Phone #: (530) 677-4243
Location: Right across the street
Open: 8 AM–8 PM

PIREP Cameron Park Liquor & Deli is a small deli with three tables indoor and three outside. The menu is full of sandwiches with names like Spitfire and B52 Bomber. During summer Paul, the owner, fires up the barbecue and serves tri-tip sandwiches.

It is directly across from the airport.

Chiriaco Summit, CA (Chiriaco Summit—L77)
Chiriaco Summit Cafe 🍔🍔🍔🍔🍔

Rest. Phone #: (760) 227-3227
Location: On the field
Open:

PIREP FBO Comments: No fuel available.

PIREP Stopped at Chiriaco Summit to see the George Patton Museum, pretty impressive. At the Chiriaco Summit Cafe, ordered the General Patton burger, which comes with cheese and all the fixin's, plus a small bowl of good chili. Didn't try the date shake, but the chocolate milkshake was wonderful—very chocolaty, in a way you don't get with bland fast-food shakes.

Chiriaco Summit Cafe was opened in 1933 by Joe Chiriaco, at a time when the place was called Shaver Summit. His family still runs it today. General George Patton set up the Army Desert Training Center there in 1942 to train his tank troops, and you can see a number of WWII and later tanks around the museum there named in his honor. The airstrip was built to support the training center. In 1958, a post office was opened, and the name officially changed to Chiriaco Summit after its first permanent resident.

My rating: 4.

PIREP FBO Comments: No aviation services other than runway and tie-downs.

PIREP Flew into L77 for breakfast. Since it had a five-burger rating, I was looking forward to a good breakfast after a very nice one-hour flight. I was not disappointed. Portions are very generous and the quality was excellent. This place rivals Flo's at CNO for both. All the meals I saw being served looked just as large and delicious as my sausage and eggs with biscuits and gravy and hash brown potatoes. Service was also excellent, with the waitress being very attentive and friendly. Prices quite reasonable. Ambiance is a roadside diner right out of the '50s and seemed very clean. I have no problem recommending this as a breakfast or lunch flight destination.

For the runway, the center repaved portion is really all that's available, as the older section is too overgrown with weeds to be usable. Tie-downs are at the west end of the field, so if your takeoff is westbound, you'll have to taxi east on the runway. CHP aircraft stop here often for meals. I was able to find some chains on the tie-down cables, but it would be a good idea to bring tie-down ropes in case there are more than a couple of planes on the ramp. There's room for perhaps ten planes. I was one of two.

Overall, an easy flight with good food waiting at a very out-of-the-way, back-in-time diner.

Fly safe.

Coalinga, CA (Harris Ranch—3O8)

Harris Ranch Restaurant 🍔🍔🍔🍔

Rest. Phone #: (599) 935-0717
Location: Next to the strip
Open:

PIREP   FBO Comments: 100LL SS available.

PIREP   We went back to Harris Ranch with a good appetite. Their automated fuel pump now takes credit cards directly. As usual, my steak dinner was beautiful—my five-burger rating stands!
We'll be back!!
Ted Jersey

PIREP   FBO Comments: 100LL SS available, pay at the gas station across the street.

PIREP   For steak and beef lovers, this is THE place to land! On approach, the 30' width runway appears challenging, but once on stabilized final, it was no problem at all. There is plenty of good tie-down space and you can either walk a few steps to the restaurant or a free shuttle can take you to the hotel (quite nice). The restaurant and bar has a nice ambiance to it. I ordered the rib-eye and it was unbelievably tasty and juicy. Wow! Service was very good and attentive. Prices are a little high but well worth it. There is a gift shop where souvenirs and Harris Ranch beef can be purchased cold-packed.
Highly recommended. I give Harris Ranch five burgers.
I will be back!

PIREP   FBO Comments: Narrow private, brush up your landing skills. Plenty of ramp parking, self-serve fuel, shuttle on request but it is a short walk to the hotel, restaurant, convenience store.

PIREP   Took my wife for her first $100 hamburger to the Harris Ranch. Very good beef, reasonable prices though not cheap. Short walk to from the ramp to restaurant/steakhouse/butcher shop. Hotel is clean and comfortable. About 100 miles from San Jose on I-5. If you like steak, this is the place.

Columbia, CA (Columbia—O22)

Goldstreet Bakery and Cafe 🍔🍔🍔🍔🍔

Rest. Phone #: (209) 533-2654
Location: Short walk
Open:

PIREP   I think that Columbia has to be one of the best places in California to stop in for great food and some old-fashioned sightseeing. Columbia is an authentic 1849 era Gold Rush town. It makes one feel as if you went back in time. The airport and runway are in good shape. When we landed, we were told there was a nature trail to easily get to town. It takes about ten minutes on foot. The best part for us was the City Hotel Restaurant. I ordered the filet mignon and it was truly superb. Service and ambiance was excellent. The wine list is very good as well. We stayed at the Harlan House B&B. It is very nice and peaceful there too, and what a nice breakfast! This town is a dream come true for those who enjoy truly great eating in an old-time setting.

PIREP   Funky, but in a nice way. That would describe the Goldstreet Bakery and Cafe at 22690 South Gold Street in Columbia, CA. Open Thursday through Monday, a relaxed and inexpensive alternative at a great fly-out destination. About ten minutes from the airport via foot and nature trail, turn left onto Gold Street before reaching town.

The cafe is in a wooded residential area and has outdoor shaded dining area. Very nice sandwiches on home-baked bread for $4.50 as well as salads and specialties, including lots of vegetarian stuff. Can be a bit lighter than the typical burger…good! No beer or Visa.

PIREP   Columbia State Park is a former gold mining town; it has its own website under the name www .columbiacalifornia.com that includes the history of the town. The airport gets very busy with air tankers when the fire season starts. The airport has lots of group fly-ins, and some even park and camp overnight among the trees in the forest. In the winter it becomes a ski area for the SF Bay Area folks by plane and by car. This area is also known as the Sierra Foothills. The town has many events all year long by all sorts of groups and organizations. So much for the propaganda for this neck of the woods.

I will not be flying for a while, it's annual time—Uncle Sam says, if it isn't broke, fix it till it is.

Concord, CA (Buchanan Field—CCR)

Atrium Court 😋😋😋

Rest. Phone #: (925) 825-7700
Location: On the field
Open: Daily

PIREP   Personally, I like going to the Atrium Court Restaurant inside the Sheraton at Concord. Given the usual landing direction, you taxi all the way down to the end of the runway and then hang a right. Nice place for lunch or dinner, with tablecloths, trees, and running brook inside. Also, very reasonable prices. If you want something more Las Vegas-ish with a lot of variety, the Pepper Tree is right across the street.

Corona, CA (Corona Muni—AJO)

Bob's Cafe 😋😋😋

Rest. Phone #: (951) 734-2570
Location: On the field
Open: Daily: Breakfast, Lunch, and Dinner

PIREP   One of the BEST!! Try the tuna melt—yummy! Then have them call Aviation Spruce to come pick you up and take you to the Candy Store.

PIREP   Service is outstanding, food quality and quantity is great, and the parking is plentiful and close.

PIREP   I just tried Bob's Cafe in Corona, CA. My friend swears that it is the absolute best cafe he has ever flown to. I enjoyed it but can't swear to it without a few more visits. The chili burger was great! They make their own potato chips.

Death Valley National Park, CA (Furnace Creek—L06)

Furnace Creek Inn 😋😋😋😋

Rest. Phone #: (760) 786-2345
Location: Call for pickup
Open: Daily: Breakfast, Lunch, and Dinner

PIREP   Airport Comments: Bring your own tie-down ropes. There are no tie-down ropes or chains, only cables along the tarmac to tie to.

PIREP   This is written by a woman pilot who wants to let all of the guys out there know that there is a Sunday brunch to take your nonpilot lady friend or significant other to that she will LOVE! The food is fantastic (+2)! The ambiance is great (+1)! The gift shop has neat clothes and jewelry that aren't too outrageously expensive! The scenery is stark but startling beautiful! For you golfers, the golf course is right next to the airport! And the big pay-off is, you must fly there if you don't want to spend an entire day in a car to get there!!

Imagine this: complimentary transportation from plane-side to hotel (which I consider a big plus, but your rating system dictates –1). Gracious service (+1). A spectacular view of the valley, with a telescope provided for your viewing pleasure. Omelets made to order. Cheese blintzes. Eggs Benedict. Maple-smoked sausage. Prime rib. Lox. Seared tuna with wasabi. Smoked trout. Maple-baked chicken. Asiago-stuffed ravioli. Chilled jumbo Gulf shrimp (well, maybe just very large rather than jumbo). Marinated mushrooms. Tomato, onion, and artichoke salad. Green pea and peanut salad. Sliced melon. Vegetable slaw. Sauteed mixed vegetables. Tiramisu. Tiny fresh fruit torts. Crème brulée. Apple crisp. And more that I can't remember because I didn't have enough room to sample it all.

I'm talking five-star quality food here. AND IT WAS ALL YOU CAN EAT!!

Be advised: if you have a large party, call ahead for reservations, else you may have a long wait because the hotel is quite busy during the cooler months (as in, booked solid every weekend until Mother's Day).

PIREP   There are two airports in Death Valley National Park; the largest is at Furnace Creek, the other is at Stovepipe Wells.

Furnace Creek is where the park service visitor center is, as well as the Borax Museum, where you can see the relics of the 20-Mule Team days. Furnace Creek has the best restaurants, including the Furnace Creek Inn where you can get excellent food. The Inn is about a two-mile walk from the airport, or you can call for a shuttle van. The Furnace Creek Ranch has several less expensive places to eat, including the Steak House, a coffee shop, and the golf course.

Both runways are small and not well maintained, Furnace Creek (L06) is 3065 × 70', but 211 feet BELOW sea level! Stovepipe Wells (L09) is 3265 × 65' and less maintained than Furnace Creek. It's 25 feet above sea level.

Best time to visit is spring and fall; avoid summer as it can be very hot! They don't call it Death Valley for nothing!

Death Valley National Park, CA (Stovepipe Wells—L09)

Stovepipe Wells Motel & Restaurant 🍔🍔🍔

Rest. Phone #: (760) 786-2387
Location: Call for pickup
Open: Daily: Breakfast, Lunch, and Dinner

PIREP   There are two airports in Death Valley National Park, the largest is at Furnace Creek, the other is at Stovepipe Wells.

Stovepipe Wells has only one place, at the motel there. It is only a quarter-mile walk. Food is good, but the views are great!

Both runways are small and not well maintained, Furnace Creek (L06) is 3065 × 70', but 211 feet BELOW sea level! Stovepipe Wells (L09) is 3265 × 65' and less maintained than Furnace Creek. It's 25 feet above sea level.

Best time to visit is spring and fall; avoid summer as it can be very hot! They don't call it Death Valley for nothing.

Delano, CA (Delano Muni—DLO)

El Delfin 🍔🍔🍔

Rest. Phone #: (661) 725-5022
Location: On the field
Open: Daily: Lunch and Dinner

PIREP   FBO Comments: There is a lot of parking, free for the day but $2 overnight. The terminal building is open until 5 PM.

PIREP   Great little place to eat; a full meal including complimentary soup, chips and salsa, and fresh vegetables was a great value. The burrito I had was very good, and the restaurant is probably only 50 feet from the terminal building. Also has a view of the runway, but it's a pretty small airport, so not too much to see.

PIREP   Stopped in Delano for fuel on a trip to SoCal. Got stuck three hours waiting for weather to clear farther south. There is a Mexican restaurant right across the street from the terminal building, so we decided to give it a try. It wasn't busy, but the staff was friendly and the food/service was great. They didn't open until about 11 AM. Lowest fuel prices I've seen in CA.

El Monte, CA (El Monte—EMT)

Annia's 🍔🍔🍔

Rest. Phone #: (626) 401-2422
Location: On the field
Open: Daily

PIREP   There is a new restaurant at KEMT (El Monte, CA), which is my home base, called Annia's.
It is a welcome addition. Good food, large portions, and very reasonable prices. Definitely worth including in the *$100 Hamburger* universe.

Eureka, CA (Murray Field—EKA)

The Rusty Hangar 🍔🍔🍔

Rest. Phone #: (707) 442-2074
Location: On the field
Open: Daily

PIREP   I have found after three recent trips to Eureka that The Rusty Hangar is a great little spot for breakfast or lunch. Friendly staff, good food, and big portions. Cheap. The breakfast burrito is huge and might cause you to recalculate your weight and balance! Located right in the terminal building, or FBO. Great onion rings—real ones, not out of a frozen bag. No credit cards and closed on Mondays. Breakfast and lunch only.

PIREP   The restaurant on the field had some of the best airport food we have eaten in a while. We had Swedish pancakes and a veggie omelet. The restaurant is on the airport with a friendly staff (the owners were cooking and serving) and 100LL available. Because the field is two miles from the coast, we had to fly an instrument approach (Murray has several) and the weather was cool even in late July. The food and service were worth the trip.

Fresno, CA (Fresno Chandler Executive—FCH)

Chandler Runway Cafe 🍔🍔🍔

Rest. Phone #: (559) 459-0630
Location: On the field
Open: Tues.–Sat. 7 AM to 3 PM

PIREP   The Fresno-Chandler Downtown Airport restaurant reopened as Chandler Runway Cafe. The operator is Ida Robinson, who was at Shafter-Minter Field, and some may have tried her good food there. The food is great, the service is fast, the waitresses are friendly and witty, the portions are large, and the restaurant is in the administration building just a few steps from the transient ramp.

Fullerton, CA (Fullerton Muni—FUL)

Tartuffle's 🍔🍔🍔

Rest. Phone #: (714) 870-9235
Location: On the field
Open:

PIREP   There's a fine view of the runway from tables in the "annex" off the main dining area. The place was quiet early on a Thurs. afternoon, but probably busy on weekends. Hours are 7 AM–2 PM. The chicken breast sandwich was good but not exceptional. They follow the SoCal custom of serving meat sandwiches with dressing, either mayo or Thousand Island. The chicken is also heavy on alfalfa sprouts and is topped with avocado slices. Staff is friendly.

PIREP   Very good restaurant in terminal building. Busy at lunch but very good food and service.

Georgetown, CA (Georgetown—E36)

Skyway's Sandwich Shop 🍔🍔🍔

Rest. Phone #: (530) 333-9203
Location: On the field
Open: Daily

PIREP   Georgetown sits in the middle of the California Gold Country. On the field, you'll find a sandwich shop. If that doesn't ring your bell, there are several restaurants in town. You can also camp out here. Call the campground manager at (916) 333-0810 for details.

Fly safe, plan ahead but most of all, enjoy this extraordinary airport. In the travel industry, I think it qualifies to be called a DESTINATION!

PIREP   A great camping getaway!

Georgetown is located about 1-1/2 hrs north of Sacramento. Georgetown offers a great little camping spot right off of the runway. The view getting in is breathtaking! There is one 2980 × 60' runway (in great condition) and the approach for 16 is great! Transitioning from the downwind to base for runway 16, you fly over a beautiful river nestled in a valley right off the end of the runway. Turning to final, the cliff looms up ahead and you see the boysenberry bushes rushing underneath you before you touch down. The area north, south, and east of the runway slopes down into the valley, and there are trees on the west side.

It is probably the closest most of us will come to landing on the U.S.S. Georgetown, but it seems like you are attempting a carrier landing. The camping area is right off the west side of the runway. I pitched a tent about 20 feet from our plane!

There is no landing or tie-down fee there. If you are into camping and are looking for a great little getaway, Georgetown would be a great place to go!!! HAVE FUN!!

Grass Valley, CA (Nevada County Air Park—O17)

The Parsonage 🍔🍔🍔🍔

Rest. Phone #: (530) 265-9478
Location: Call for pickup
Open:

PIREP   Chuck Shea is a commissioner at the airport and also owns The Parsonage. Give him a call! He picked us up at the airport. We had a wonderful night at The Parsonage, and an awesome breakfast! Then he brought us back to the airport. Door-to-door service, and Chuck and Susan are great people.

I highly recommend staying at The Parsonage if you fly in.

PIREP   We flew up here last December. The town puts on a Victorian Christmas that is really fabulous. Fortunately, I had booked early and was able to get a room at The Parsonage. My wife loved the ambiance. I enjoyed the food. Susan Shea creates a special homemade breakfast each morning. It includes fresh-baked muffins and popovers. Eggs are brought in from the Shea's Wandering Star Ranch located in nearby Penn Valley. During the summer months, fresh blueberries and strawberries are on the menu.

For over 80 years, The Parsonage was the home of the Methodist minister for Nevada City. In 1986 the home underwent its first major renovation to begin its new life as a B&B. This home sits in the middle of California history, where the story of thousands who pursued elusive wealth in the gold mines and river beds of the Sierra Nevada foothills came to live.

We had dinner one evening at Citronee Bistro. It rivals any cafe in our hometown, San Francisco. To get to and from the restaurant we used a horse-drawn carriage, a very nice touch.

Ground transportation from the airport is easy. A call to Chuck Shea, the owner of The Parsonage, was all it took. He picked us up and brought us back. By the way, Captain Chuck was an airline pilot before he settled in God's Country. Coffee and conversation with him is a great way to fill the morning.

Groveland, CA (Pine Mountain Lake—E45)

Corsair Coffee Shop 🍔🍔🍔

Rest. Phone #: (206) 962-6793
Location: On the field
Open:

PIREP   I flew into Pine Mountain as part of my first cross-country training flight. We stopped into the Corsair Cafe so I could buy my CFI lunch. We arrived at 2:30 PM. (Note: they close at 3 PM on Saturdays.) They were very busy (there was a 'Swift' fly-in that day), but we didn't have to wait to be seated. The service was good and the food was fantastic.

While you are there, be sure to visit Kent's Hangar on the far end of the strip. You will be BLOWN AWAY by all of the amazing items he has tucked away back there. You'll be able to identify which hangar is Kent's by the beautiful Lockheed Twin and Seabee parked out front.

I'd recommend this $100 burger to anyone!

PIREP   I didn't expect much from the outside look of the place but was pleasantly surprised when I walked in. The smaller front room was full when I was there (around lunchtime on a Thursday), but service was pleasant and prompt. Having started the day late, I opted for the Co-pilot Breakfast, which was simple but well prepared and very tasty. The server was very attentive (and pretty cute), and the whole experience hit the spot.

Half Moon Bay, CA (Half Moon Bay—HAF)

Barbara's Fish Trap 🍔🍔🍔🍔

Rest. Phone #: (650) 728-7049
Location: Short walk
Open: Daily

PIREP   I fly out of Half Moon Bay and have always enjoyed the wonderful restaurants that are nearby. Barbara's Fish Trap has wonderful clam chowder and great fish—but they do not take credit cards (cash only).

The airport cafe (Cafe Three-Zero) has won "Best Breakfast on the Coast" for a good number of years.

Finally, if you continue south a few blocks (beyond Barbara's Fish Trap) and head down the hill into the harbor, you'll come across even more restaurants. You can also buy salmon and crab fresh off the boat.

PIREP   My wife, Anita, and I have been flying to HAF for over 25 years to eat calamari at Barbara's Fish Trap. Nobody makes calamari better, nor does anybody make better clam chowder. Unfortunately, it always has the longest lines waiting to be seated; another indication that many agree it is the best in Princeton.

The atmosphere is informal and very friendly. Oil tablecloths best describes it. But none has a better view of the harbor and none has better and fresher food.

The reopened Brewery and the reopened restaurant nearest the airport are best described as linen tablecloths, and prices are consistent with that. But their food is good too—just not as good. We sample the other restaurants once in a while, especially when we take "snooty" friends, but we always come back to Barbara because of the food and atmosphere.

Half Moon Bay Brewing Co 🍔🍔🍔🍔

Rest. Phone #: (650) 728-2739
Location: 390 Capistrano Rd.
Open:

PIREP   This is a great cafe with indoor and outdoor seating. The outdoor area is dog friendly. You can simply hang out around the fire pit and take in views of the ocean and main drag. There are outdoor heaters and wind screens to help on those chilly days. There were a lot of tempting items on the menu, but I felt inclined to try the burger! It was a delicious Angus burger worthy of my $100. Good beer selection that I was not able to try since I was the pilot that day.

This is just two blocks from the airport's south parking area. Head south along Capistrano, and it will be on your left. I was told by a fellow pilot that the Italian restaurant just outside the airport gates is worth trying.

WARNING: Fog rolls in quickly! I didn't see any until I got to the airport, and it was already building up over the coast. If I stayed a half-hour more, it would have been IMC. Two fatal accident reports at HAF were from noninstrument pilots flying in IMC weather.

PIREP   One of my all-time favorites is Half Moon Bay. After waiting for an uncommon fog-free VFR day, we flew into KHAF to see what was going on. We were not disappointed! The generous runway and ramp is in good shape. We taxied to the far SW end and there were plenty of free tie-downs. Just a short walk ahead there is a small gate that leads to the seaside and right into the Mezza Luna Restaurant (Italian, great place) parking lot. Across the street is a little coffee shop, and next door is the Half Moon Bay Brewing Co. Still another choice is Barbara's Fish Trap (seafood), across from the Brewing Co. There is a pier behind Barbara's with a great view of the harbor, and some private anglers were managing to catch some good crabs and fish there. Fishing, anyone? After much deliberation, we decided on Half Moon Bay Brewing Co. because my passengers wanted to sample their homemade beers. They said the beers were really great. I had the burger, which was very generous and excellent. Prices are average and service was good. In addition to the indoor seating and bar, they have a great outdoor eating/bar area with open fire pits burning propane, which are really nice when it gets chilly (often). I give KHAF and HMB Brewing Co. a solid five-burger rating.

I will be back!

Hawthorne, CA (Jack Northrop Field / Hawthorne Municipal—HHR)

Nat's Airport Cafe 🍔🍔🍔

Rest. Phone #: (310) 973-4152
Location: On the field
Open: Daily: 7 AM–4 PM

PIREP Nat's restaurant, in the admin. building next to the FSS, has been there for a dozen years and is still operating. Nat's has good food—a mix of Mexican and American.

PIREP Hawthorne (HHR) is very close to LAX, and getting in can be intimidating if you aren't used to sharing airspace with the heavy metal. The cafe is overlooking the runway and the flight path to LAX. They recently put in a light rail track right near the airport. This gives you pretty decent access to downtown, the South Bay beaches, or East L.A. to see the famed Watts Towers. This airport is adjacent to Northop, and you'll see some experimental aircraft come out once in a while.

Hayfork, CA (Hayfork—F62)

Irene's 🍔🍔🍔

Rest. Phone #: (530) 628-5385
Location: Short walk
Open: Daily

PIREP Hayfork is the best kept secret in California. The runway is long and in good condition; the pilot's lounge is tidy and comfortable. Irene's is the real treasure though. *Great* home cooking for a reasonable price. A five-minute walk from the airport. The flight is worth it just to walk over the suspension bridge.

Be warned! The airport is closed to night operations.

PIREP The restaurant in Hayfork is nothing special, but the flight and walk to town are enjoyable. From the Central Valley, fly west from Red Bluff through the low spot in the hills. Turn right at the second small valley and descend into Hayfork. Just a touch of mountain flying. The highlight of the trip is the walk into town which traverses the river (small). There is an old suspension bridge used to cross the river. Not very high but keeps you on your toes. Motorcycles and horses are not allowed on the bridge. Walk to Main Street from bridge and the cafe is on the left. Food is always good.

Healdsburg, CA (Healdsburg Muni—O31)

Picnic 🍔

Rest. Phone #:
Location: On the field
Open:

PIREP No food at the field and the walk into town is over a mile. There is no local taxi service.

I would strongly recommend flying into Healdsburg for a picnic and scenic walk. The airport is set on a hill amid beautiful vineyards, and there are picnic tables at the airport. The county road which runs by the airport passes through beautiful vineyard scenery. It's a nice stroll before getting back into the plane and flying somewhere else. Obey the noise abatement procedures in effect for the airport (avoid flying over the town itself).

Hemet, CA (Hemet-Ryan—HMT)
Hangar 1 Cafe 😐😐😐

Rest. Phone #: (951) 218-3698
Location: On the field
Open: Daily

PIREP Nice little one-room cafe with a small ramp-side porch. Great parking. Good food. Pretty laid-back atmosphere. Worth checking out.

Hollister, CA (Hollister Muni—3O7)
Ding-a-Ling Cafe 😐😐😐

Rest. Phone #: (831) 637-1566
Location: On the ramp
Open: Daily

PIREP The Ding-a-Ling Restaurant is still a good place for food. Coffee is now $1.25, and their menu is still good as ever.

Heavy on weekends with glider and jumper traffic.

PIREP I slid into Hollister the other day and wandered into Ding-a-Ling Cafe just off the ramp area for a well-needed breakfast. Service was good and the prices are very reasonable. Nicely prepared food and good portions. Eating out on their patio was nice.

I will be back!

Imperial, CA (Imperial County—IPL)
Hangar 19 😐😐😐

Rest. Phone #: (760) 355-1055
Location: On the field
Open:

PIREP Open seven days a week, breakfast, lunch, and dinner, but call first (760) 355-1055. You know how those things change.

We ate there last week and found a wide selection of American, Mexican, Italian, and Greek fare. This restaurant is inside the airport motel.

Inyokern, CA (Inyokern—IYK)
Bernadino's 😐😐😐

Rest. Phone #: (760) 377-4012
Location: On the field
Open: Daily

PIREP This is the best Mexican food possibly found in California. This is a family-owned restaurant—actually they have two restaurants, one is in the close town of Ridgecrest. On Friday they have a killer buffet with prime rib, also their menu items are just fantastic.

Kernville, CA (Kern Valley—L05)
The Airport Cafe 🍔🍔🍔

Rest. Phone #: (760) 376-2852
Location: On the field
Open: Breakfast, Lunch, and Dinner

PIREP   There is a campground at the Kern Valley Airport managed by the Airport Cafe. It is adjacent to the runway and has paved tie-downs for about ten aircraft. The campground itself is a large grassy area shaded by many large trees.

A picnic table and a fire pit are located at each camping space. Water is available from a spigot and a hose is provided for filling a black barrel above a shower enclosure for taking a solar shower. The sanitary facility is only a portable potty, but it is adequate.

Tie-down/space rental is $15 per day, payable at the cafe. It's a long walk from the campground to the cafe since the campground is at the north end of the runway and the cafe is at the south end, but it's only a short walk over to the Kern River.

The town of Kernville is also within walking distance if you're in good shape, but the airport also has a beater car, a not-so-Grand Fury, available to borrow.

PIREP   The Airport Cafe at Kern Valley (L05) has been added to my personal "recommended" list. The ladies that run the place are great! I've been there twice and both times I found the service, friendliness, and food good enough to return for. Breakfast is nothing special but if you go for lunch I recommend the burgers. They are big and very tasty for the price, which is very reasonable by airport standards. The cafe itself is small and not much to look at, but the friendly, small-town feel is a nice change for a city dweller.

In addition to the cooking and serving, the ladies also work the unicom and are very willing to give basic wind information to help with the runway choice. It's also nice to hear a friendly voice wishing you a safe flight right after you announce departure. I'll be going back, for both the flight and the food.

Lakeport, CA (Lampson Field—1O2)
Skyroom Cafe 🍔🍔🍔

Rest. Phone #: (707) 263-7597
Location: On the field
Open: Daily: Breakfast, Lunch, and Dinner

PIREP   Food and service was great! We stopped in to wait for a storm to move out of the area. Friendly staff. Worth the stop.

PIREP   We ate at the Skyroom Cafe Friday. It was friendly and reasonable. The salads were very good, and the French dip was okay. The new owner was very nice and is a pilot. They are open for lunch and dinner Tuesday through Sunday.

Lancaster, CA (General WM J Fox Airfield—WJF)
Foxy's Landing 🍔🍔🍔

Rest. Phone #: (661) 949-2284
Location: On the field
Open: Daily: Breakfast, Lunch, and Dinner

PIREP After looking at the previous posts, the name has changed to Foxy's Landing Restaurant. My family and I made two stops for breakfast. The food was great. We received ample food for the money. We also were able to sit and watch the traffic taking off and landing, and even more important for me, we had a good view of the windsock, which changed directions 180 degrees during breakfast.

La Verne, CA (Brackett Field—POC)
Norm's Hangar 🍔🍔

Rest. Phone #: (909) 596-6675
Location: On the field
Open: 7 AM–3 PM, 7 days a week

PIREP Ate today at Norm's, and it was good food, as good as any airport restaurant could be. They have "on the porch" seating, right at the runway. I have not seen any others like it. I was happy with the service, and the bill was manageable. Hours are 7 AM to 3 PM, 7 days a week.

PIREP We recently visited Norm's Hangar Restaurant. The restaurant building and atmosphere are very nice. There is ample seating and a nice view of the AOA. I must say, however, that I was quite disappointed at our entire dining experience. First of all, the service was very slow! We were there at the lunch hour; however, there were many wait-staff on duty. Nobody came to our table for about 15 minutes. When they did, they got our order wrong. The food was okay, the usual airport cafe fare.

When I went to pay, I found out that they *do not accept credit cards*! I had to go out to the plane and scrounge up all the loose change from my flight bag. To make matters worse, I went for an after-lunch mint, and I was informed that they charge 15 cents for each one!

Needless to say, it will be a long time before I return to Brackett for lunch……

PIREP I visited Norm's Hangar and I must agree with previous comments. I found the restaurant really not very friendly and the service was terrible. The food was (on a scale of one to ten) about a four. I would not go there for the food. They just don't care if you go there or not!

Lee Vining, CA (Lee Vining—O24)
Nicely's 🍔🍔🍔

Rest. Phone #: (760) 647-6477
Location: Short walk
Open: Daily: Breakfast, Lunch, and Dinner

PIREP FBO Comments: No fuel, no services.

PIREP I've been eating a few times a year at Nicely's for almost 30 years. It is an old Standard stop on 395 in Lee Vining. Nicely's has good home-cooked meals, and you can get a good $100 hamburger if you want. Normally, I'll get a much better meal, like their daily specials. This is a great stop to eat if you don't mind walking about a mile into town (west, down the gravel road, then north for several blocks on 395). Be prepared for the higher elevations as the field is 6802 msl, but I've been flying in-out of O24 for 21 years in my 152, so it can't be too difficult. Great view of Mono Lake just east of the field.

Little River, CA (Little River—O48)
Cafe Beaujolais 🍔🍔🍔

Rest. Phone #: (707) 937-5614
Location: Take the shuttle or rent a car
Open: Daily: Lunch and Dinner

PIREP You'll need to rent a car from the FBO; call Coast Flyers at (707) 937-1224. Drive into Mendocino (about 15 minutes), do some sightseeing and perhaps shopping in this quaint New England–style town, then have breakfast, lunch, or dinner at the Cafe Beaujolais—act like a local or a regular and call it the "Cafe B." I don't have the number here; call directory assistance for Mendocino.

Anyone know why a miniscule little place like this has a 5249 × 150 runway?

PIREP Visited Mendocino on a beautiful weekend. Shuttle is now $7 per person each way. Rental car is $34 for a small car. The shuttle schedule is spotty at best, so plan on waiting to leave the airport and to be picked up in town. If you have two or more people, call ahead to arrange a rental car. You can rent a car when you arrive, but calling ahead is advised, as the proprietors are not always on the field.

Wonderful town to visit, have lunch, maybe stay overnight. Watch the fog morning and night!

Livermore, CA (Livermore Muni—LVK)

Beebe's Grill 🍔🍔🍔

Rest. Phone #: (925) 455-7070
Location: Short walk
Open: Daily: Breakfast, Lunch, and Dinner

PIREP We flew in the other day and tried Beebe's Sports Bar and Grill on the NW side of the airfield at the golf course. Good food and quick service. Worth another trip in the future.

Lodi, CA (Lodi—1O3)

Lodi Airport Cafe 🍔🍔🍔

Rest. Phone #: (209) 369-6144
Location: On the field
Open: Daily

PIREP And after that meal, you can walk over to the Parachute Center and watch skydivers do their thing on the same field. Pretty entertaining.

PIREP My buddy and I stopped by the cafe to have lunch after a round of golf at the Dry Creek course five miles up the road. Great food and the aviation atmosphere is tops—the cafe sits right on the flight line and you have a great view of the aircraft taking off and landing. I would recommend this for a fly-in breakfast or lunch.

Lompoc, CA (Lompoc—LPC)

The Steak House 🍔🍔

Rest. Phone #: (805) 742-0026
Location: Short walk
Open: Daily: Lunch and Dinner

PIREP With a new ramp just added on the south side of the field at Lompoc, it's now convenient to tie down at the southeast corner and walk through the pedestrian gate into town. The gate combo was advertised as "the model number for the world's most popular two-seat trainer." I got it wrong the first time.

There are two restaurants nearby: The Steak House has good food, including a Sunday brunch from 9 to 2 and lunch daily from 11:30. There's a just-opened Denny's a bit further down the block.

I dropped a Citation crew at the Steak House yesterday, and they rated the food excellent.

Lone Pine, CA (Lone Pine—O26)

Bonanza 🍔

Rest. Phone #: (760) 876-4768
Location: Long walk
Open: Daily: Breakfast, Lunch, and Dinner

PIREP Lone Pine is a great place to stop if you are taking up the sights along the Owens. Plus you get to view the great 14s of the Sierras, including Mt. Whitney. Try the Bonanza for breakfast or lunch. They serve up great Mexican food! Breakfast gets a 4.

Los Angeles, CA (Whiteman—WHP)

Rocky V "On the Strip" 🍔🍔🍔

Rest. Phone #: (818) 896-8828
Location: On the field
Open:

PIREP On a recent trip to Whiteman airport I noticed significant changes to the airport restaurant. The place is now called Rocky V "On the Strip" and calls itself the "Home of the Jumbo Burger." As was noted before, folks are friendly, portions are big, food is good. I like the roast beef sandwich, warmed up with all the fixings, and it is obvious that the new management spent several months last year to renovate the place.

I would like to mention the outdoor seating.

PIREP Transient parking is right outside. It's nothing fancy, but the folks running it are really nice and the hamburger was the best I have had in a long time.

I'm not a tremendous burger fan, but this one may actually be *worth* a hundred dollars!

Los Banos, CA (Los Banos Muni—LSN)

Ryan's Place 🍔🍔🍔

Rest. Phone #: (209) 826-2616
Location: Short walk
Open: Daily: Breakfast, Lunch, and Dinner

PIREP Coffee at Ryan's Place is now $1.50.

PIREP The rest is in town and more than a ten-minute walk. Go and fly to Los Baños, almost everybody speaks English. Mentionable is also El Palomar Mexican restaurant on 6th Street, open only Tuesdays through Sundays, various times.

Woolgrower's Exchange 🍔🍔🍔

Rest. Phone #: (209) 826-4593
Location: Walk to town
Open: Daily: Lunch and Dinner

PIREP   Today I was talked into walking into the town of Los Baños to visit a Basque restaurant called the Woolgrower's Exchange. It is in a building that was erected in 1894. It's a 45-minute walk from the airport; by taxi, 5 minutes.

The Woolgrower's restaurant seats its guest on long tables, each with 50 seats. They greet you at the table with a bottle of wine for the nonpilots. The menu consists of soup, salad, sourdough bread, lamb stew, lamb shanks, tri-tips, or chicken. All well served with ice cream dessert, all for $15 per person, including tax. It's worth a trip. Not a fancy place.

Madera, CA (Madera Muni—MAE)

Madera Muni. Golf Course Snack Bar 🍔🍔

Rest. Phone #: (559) 675-3504
Location: Crew car, 2-mile drive
Open: Daily: Lunch

PIREP   The Madera Municipal Golf Course is about two miles from the airport (you'll see it from the pattern just beyond and to the left of runway 30).

To get there, you'll have to borrow a car from one of the FBOs and take Aviation Drive out to Avenue 17 and turn left. Go for about a mile and a half to the corner of Avenue 17 and Road 23, and you're there.

The snack bar is in the back side of the country club with picture windows commanding a view of the driving range and the starting tees. There are about 15 tables or so, and a TV that must be set to the Golfing Channel above the bar. The snack bar is more "bar" than "snack" (after all, this *is* a golf course), but the grill (which closes at 3 PM) serves burgers or sandwiches with french fries and a soft drink or iced tea for about $5. Service was okay, but you may have to wait a bit if the waitress has to take a cell phone order to golfers on the fourth hole. For those who miss the grill hours, there is still a fridge with premade cold-cut sandwiches and bags of chips, but they are not nearly as satisfying to the famished flyer.

The food was good, served hot and quickly. A nice relaxed atmosphere, and the prices were reasonable. The only negative is that it isn't at the airport.

Mariposa, CA (Mariposa-Yosemite—MPI)

Airport Inn 🍔🍔🍔

Rest. Phone #: (209) 377-8444
Location: 1/2 mile
Open: Daily: Lunch and Dinner

PIREP   We have the Airport Inn a half mile from the airport, and they have great hamburgers!

PIREP   The Airport Inn that is a half-mile walk from the airport is a dark dive of a bar that has hamburgers and hot dogs. Unfortunately, the day we were there, they were out of hamburgers. We walked back to the airport, flew ten minutes to Columbia, walked into town and had a great Mexican lunch.

The Mariposa Inn is a hotel several miles away in Mariposa and is reputed to be nice.

Mojave, CA (Mojave—MHV)

Voyager Restaurant 🍔🍔🍔🍔

Rest. Phone #: (661) 824-2048
Location: On the field
Open:

PIREP   FBO Comments: You can get a tour of parked airplanes from the FBO for $5 per person.

PIREP   The Voyager Restaurant offers great food and service. The Mirah Hotel also has good food and a bar. It is within walking distance.

PIREP   Airport comments: Tower open only on weekdays, otherwise standard nontowered field. No weather info, but if you call up the unicom you can get winds/altimeter and traffic info.

PIREP   I was surprised Mojave was in the *$100 hamburger* database. Heard about the parked airliners out here, so decided to visit. Found the Voyager Restaurant right on the flight line, underneath the old control tower. The stored airliners are parked across the field on the west and north ends.

Flanking the terminal building are two BAE hangars with F-4 Phantoms, which I guess are flown just about every day. A few more airworthy F-86 Sabres round out the mix.

Finding the guy to get fuel was confusing at first. But once I tracked him down, fuel was not hard.

Found the restaurant inside the terminal. Serves breakfast and lunch. Friendly and fast service. Bay windows allow you to see the transient parking, weird to see Cessnas and Pipers mixed in with F-4's and an MD-80. all right next to each other. Further on you can see the windmill fields and parked airliners from all different airlines: Virgin, Continental 747-100's, U.S. Airways, Northwest, Delta, United, SwissAir, and so on.

Can't walk on the flight line or out to the parked airliners though, but you can do all the pattern work you want.

Monterey, CA (Monterey Peninsula—MRY)

Tarpy's Road House 🍔🍔🍔🍔🍔

Rest. Phone #: (831) 647-1444
Location: 5-minute cab drive
Open: Daily: Lunch and Dinner

PIREP   For a nicer meal, consider Tarpy's in Monterey. A five-minute cab ride from Del Monte East, the FBO's light-plane tie-down area. Ask them on unicom to call a cab, and it'll come right in and meet you just about when you're done with your plane.

PIREP   I agree that Tarpy's Road House is excellent. However, if it's really a burger you want, take a cab the same distance in the other direction to Fremont Street and The Del Monte Express, "home of the five napkin adult hamburger." Longtime favorite of locals. Railroad decor.

The Golden Tee 🍔🍔🍔

Rest. Phone #: (831) 373-1232
Location: On the field
Open: Daily: Lunch and Dinner

PIREP   The Golden Tee at the Monterey Airport (ILS available) has been there for years but was redecorated. It is located on the second floor of the terminal building itself and has a great view of the Monterey Bay and of takeoffs and landings if they're using runway 10 and of the taxiing airplanes if they're using runway 28.

They serve some of the best sand dabs I've ever tasted—lightly breaded and very delicate—and their Monterey Combo offers both the sand dabs and a calamari steak. Their chef salad is also very good, as is their BLT, which I always order with avocado. Don't miss the house salad dressing and the freshly baked bread, which comes with the full luncheon plate.

They're open for lunch and dinner—don't know about breakfast.

PIREP You have an excellent restaurant on the second floor of the terminal building. It is full service with the greatest calamari in the world. Park at Del Monte east or west—it's an easy walk to the terminal. The view of the runway is also great.

Murrieta/Temecula, CA (French Valley—F70)

French Valley Cafe 😋😋😋

Rest. Phone #: (951) 600-7396
Location: On the field
Open: Daily: Lunch and Dinner

PIREP This is a favorite breakfast stop with my 10-year-old son. He loves their pancakes (the regular stack is $2.95 and is huge, more than most people would care to eat). The breakfast burritos aren't too bad either. Great location on a great uncontrolled airfield.

PIREP Diverted to French Valley (F70) in Temecula, CA because of a low cloud layer at Catalina Island (AVX). Right on the ramp at French Valley is a great little restaurant that not only serves breakfast, but lunch and dinner as well. The prices were right and the food quite tasty. Don't remember the name of the restaurant, however; the menu did contain a wide assortment of burgers with many aviation-related names. The airport was a quick ride up the valley from San Diego and I am sure an easy hop from Orange County.

Napa, CA (Napa County—APC)

Jonesy's Famous Steak House 😋😋😋😋

Rest. Phone #: (707) 255-2003
Location: On the field
Open: Daily: Lunch and Dinner

PIREP Previous raves about their steaks are still valid! And the "special" potatoes are absolutely disgusting; I ate every bite! This was Mother's Day; reservations were suggested for lunch, and I'm glad we got them. Service was friendly and prompt. I feel like they should get a five or six, but then I'm prone to get emotional over food. We enjoyed the whole experience, including quick and easy clearance through SFO's Class B airspace, and look forward to returning.

PIREP Your listing for Napa did not do them justice. I have been flying in to Jonesy's Restaurant since 1960. They had a sirloin steak dinner for two for $11.75, It was a great bargain then, and with inflation covering 37 years, it is now $18.75. They take a huge sirloin and cook it (only one way) and cut it into two portions. Special cheese-covered potato strips and a huge green salad. We either fly in or stop by driving back from Reno. It is the greatest, bar none. And of course, being Napa, a great wine list (*not for the pilot, please*—passengers only).

PIREP The annual "Best of the Napa Valley" voting picks Jonesy's at the Napa County Airport year after year in the steakhouse category. They have coffee shop food also.

New Cuyama, CA (New Cuyama—L88)

The New Cuyama Buckhorn 😋😋😋

Rest. Phone #: (661) 766-2591
Location: On the field
Open: Daily: Lunch and Dinner

PIREP Great little roadhouse coffee shop attached to the front of the Cuyama Buckhorn motel. But if the owner repaved the runway only three years ago they must be using it for aerial target practice. It is usable, but that's about it. Many chunks of asphalt missing, along with the associated loose gravel all over the place. But after inspecting my propeller and finding no new gouges or nicks…yeah, I'd recommend the place! Couldn't do better on Route 66.

PIREP New Cuyama has a great place, the Buckhorn. The landing strip is decent and there is a short walk out to the Buckhorn. The food is good and the folks are friendly. No fuel or services of any sort, just good food.

It is 35 miles north of Santa Paula, CA in the middle of the giant Cuyama Valley. Used to be an oil field airstrip. New Cuyama has a population of about 600 people. The fellow who owns the Buckhorn had the strip repaved about three years ago so people would start flying in again for his food, and it worked!!

Oceano, CA (Oceano County—L52)
Old Juan's Mexican Cantina 🍔🍔🍔

Rest. Phone #: (805) 489-5680
Location: Short walk
Open: Daily: Lunch and Dinner

PIREP Oceano, just south of Pismo Beach, is a fun place to land, just an eighth of a mile from the beach. The tie-down area is paved and free. There are many campsites available. Vehicles are allowed to drive on Oceano Beach.

I slipped into Old Juan's Mexican Cantina, an easy walk from the airport. The place is decorated in the traditional Mexican style. They have indoor and outdoor seating available. The steak fajitas I had were generous and very good. Service was very good and the prices are average. Watch out for fog. I was told that it is fairly common and can roll in unexpectedly.

I will be back!

PIREP We now have four restaurants within two blocks of the airport (Pier Ave. Cafe, Oceano Fish & Chips, Old Juan's Mexican Cantina, and The Rock & Roll Diner)

A trolley operates from the airport to the surrounding communities on weekends during the winter and seven days per week in the summer.

Oceanside, CA (Oceanside Muni—OKB)
Sharon's Cafe 🍔🍔🍔

Rest. Phone #: (760) 757-7500
Location: On the field
Open: Daily: Lunch and Dinner

PIREP Oceanside Airport, south of John Wayne in Orange County, is a good place for a unique fly-in. The burgers are a three-block walk (across a busy highway) at a bar up the street perpendicular to the airport. They're friendly and will set you up with a good, greasy burger at almost any time of the day. Pay close attention to airspace and contact the unicom early for landing advisories, as they're trying desperately to keep the airport open.

Ocotillo Wells, CA (Ocotillo—L90)
Desert Rose Restaurant 🍔🍔

Rest. Phone #: (940) 989-3302
Location: On the field
Open: Sept.–May *only*, Tues. and Wed.

PIREP We landed here last month and the airstrip was fine. The restaurant is open September through May, except closed on Tuesday and Wednesday. The Saturday and Sunday breakfast buffet is great. The folks that work in the restaurant are very nice and friendly, and there's a pilots-only guest book to sign.

PIREP Lots of heavy thunderstorm activity on the low desert has caused the dry lakebed to become much less dry. There is a dry crust about an inch or so thick, covering some pretty soft muddy spots. Probably a good idea to stay away until you can at least see vehicle tracks on the sand.

Oroville, CA (Oroville Muni—OVE)
Table Mountain Golf Club ☺☺☺

Rest. Phone #: (530) 533-3922
Location: On the field
Open: Daily: Breakfast, Lunch, and Dinner

PIREP The restaurant at the golf course adjoining the Oroville Airport is quite good and only about 150 feet from aircraft tie-downs. The decor is typical golf course but they have sit-down service, and the welcome was enthusiastic. Food was good and the portions large. Their hours are 0800 to 1500 Monday to Friday and 0700 to 1500 Saturday and Sunday. The golf course is named Table Mountain, but the airport is in the Sacramento Valley and anything but mountainous. Golf course (and restaurant) is on the opposite side of airport from the FBO. You can taxi right to the golf clubs and there is plenty of parking and tie-downs.

PIREP Dropped in to the golf course restaurant at Oroville (OVE) this morning for breakfast and found the hash browns particularly good. Full breakfast with coffee was less than five bucks. Aircraft parking is west of runway 1-19 and less than 100 yards from the golf course.

Oxnard, CA (Oxnard—OXR)
Buky's ☺☺☺

Rest. Phone #: (805) 984-6311
Location: On the field
Open: Daily: Lunch and Dinner

PIREP Buky's is located on field next to the terminal in Oxnard, CA. It has the best barbecued tri-tip sandwiches and, as a side, you can order "cornados," which are french fries made of corn—they taste great and Buky's is the only restaurant serving them that I know of.

Highly recommended.

Palm Springs, CA (Bermuda Dunes—UDD)
Murphy's ☺☺☺☺

Rest. Phone #: (760) 345-6242
Location: 100 yards
Open: Lunch 11 AM–2:30 PM, Dinner 5–9 PM, 3–9 PM Sunday (call ahead to confirm)

PIREP Just a short walk from the airport. Murphy's does arguably the best fried chicken around. Their Martini burger is another great lunch, but not if you're flying home. They also have daily specials that are really good. Bermuda Dunes is a nice little airport and worth a trip.

PIREP   Finally sampled the famous Murphy's pan fried chicken, and it lives up to its reputation. Open 1100 to 1400 for lunch and again at 1700 for family-style dinner. Arrived at 1230 to a packed house and was seated after a short wait; the service was great and the food was excellent.

PIREP   This airport is about 13 miles east of Palm Springs, CA. It is right next to the 10 freeway, with a 5,002' runway length! From June–September the temperature can soar to well above 100 degrees! Most of the private jets stop coming during this time and use Palm Springs instead. About 100 yards south of the terminal you'll find Murphy's, which is well known for its fried chicken! There is also a wide variety of things on the menu and a nice bar adjacent to the restaurant. Most of the patrons are well into their 70's, but that goes with the territory down there. Many celebrities use this airport because of its convenience to the posh Palm Desert area. I have seen Arnold Palmer's Citation X in there many times, and there are often many private jets on the field. G-3's even land here! During the peak season, November–April, the field is scattered with private jets! Beware, Murphy's closes during the heat of summer!

Palm Springs, CA (Jacqueline Cochran Regional—TRM)
LaQuinta Aviation 🍔

Rest. Phone #: (760) 399-1855
Location: On the field
Open: Daily

PIREP   Had a nice stop over for fuel at Thermal, CA (TRM). Nice small restaurant on field at the FBO (LaQuinta Aviation). It is run by a husband and wife team. Burgers are cooked outside on an open pit. Food is good, assortment of extras (chips and so on). Reasonably priced, friendly and prompt service. Open seven days, exact hours unknown.

Palo Alto, CA (Palo Alto Arpt of Santa Clara Co—PAO)
Abundant Air Cafe 🍔🍔🍔

Rest. Phone #:
Location: On the field
Open: Daily: Breakfast, Lunch, and Dinner

PIREP   Palo Alto is my home airport and I really enjoy hanging my hat there. For those unfamiliar with the area, I suggest reviewing the S.F. terminal chart before heading under the KSFO Class B VFR to Palo Alto. This is a very busy airspace but definitely worth a visit. Landing pattern is usually right traffic 31 (TPA 800'), right over S.F. Bay—great view! The 2400' runway is in good shape and there is plenty of transient tie-down space available.

A couple of minutes walk is Abundant Air Cafe, located on the airfield near the rotating beacon. It is clean with good but somewhat pricey cuisine. They have a nice patio eating area out front facing the parking lot. I like their nice variety of good salads. You can check out the live action at this place at www.abundantair.com/buddha_cam .html (where else but in Silicon Valley?).

For those who want something more upscale, take a short cab ride over to downtown University Ave and check out Evvia (Greek), Zibibbo (Mediterranean), Nola (New Orleans), or Andale (Mexican, inexpensive).

Downtown Palo Alto is a really great eating destination. You won't be disappointed!

PIREP   Finally, a reason to fly to Palo Alto on the weekends. The cafe is now open from 8:30 AM–5 PM on Saturday and Sunday. Have a great breakfast, a latte, fresh grilled sandwiches, homemade soups and salads. Ask for the Pilot's Box lunch. If you're not flying, beer and wine are available!

Dennis McKnew
Owner
Abundant Air Cafe/PAO

PIREP   The Abundant Air is a great stop for coffee/pastry and lunch, located just west of the West Valley Flying Club offices. A couple minutes walk from KPAO's transient tie-downs. Open Monday through Friday, 7 AM–3 PM. Daily soup and chili are very tasty, and the sandwiches are, well… abundant! Fast service, though there may be a wait in line during the lunch rush. Very little indoor seating but plenty of outdoor tables—too bad the views are of the parking lot and not the airplanes. I give Abundant Air four burgers (though they don't serve burgers) for fresh food, quick, friendly service, and good-sized servings.

Bay Cafe 🍔🍔

Rest. Phone #: (650) 856-0999
Location: Short walk
Open: Daily: Breakfast, Lunch, and Dinner

PIREP   Bay Cafe is located adjacent to Palo Alto Municipal Golf Course, just outside the airport boundary. They have a decent view of the golf course and a full bar for passengers. The food is served cafeteria style. I had the hamburger, which I would rate as just average. Prices, service, and ambiance are also average. This is an okay place to eat, but given the choices available I would much rather eat at Abundant Air Cafe at the airport or walk a bit further south on Embarcadero to eat at Ming's Restaurant & Bar.

PIREP   Excellent food, they have a lot of good lunch and breakfast specials. They're only open for lunch and breakfast, and they are a heck of a lot friendlier than the staff at the other food stop. Just a short walk from transient, go to the golf course clubhouse. They also have a great happy hour if you are not flying home.

Ming's Restaurant & Bar 🍔🍔🍔🍔

Rest. Phone #: (650) 856-7700
Location: 1/4-mile walk
Open: Daily: Lunch and Dinner

PIREP   About a quarter mile west on Embarcadero, an easy walk from the airport, is Ming's Restaurant & Bar. It is an upscale Chinese restaurant. Prices are reasonable and they have some of the best dim sum I have ever tasted. There is also a fancy full bar for passengers. The ambiance is nice and service is prompt. Ming's is definitely a four-burger stop. Highly recommended.

   I'll be back!

Paso Robles, CA (Paso Robles Muni—PRB)

Matthew's at the Airport 🍔🍔🍔🍔🍔

Rest. Phone #: (805) 237-2007
Location: On the field
Open: Daily: Lunch and Dinner

PIREP   It cannot get any better than this.

   As it has now been over a year (more likely close to two) since I last ate at Matthew's, my wife and I returned for lunch today. What a delight! We ordered different salads, and both were everything we'd hoped for. Our waitress was friendly and attentive—and amused when most of the patrons were more interested in the departing P-51 than anything else!

   We heartily recommend Matthew's and will return eagerly.

PIREP We spent a weekend in Paso Robles and had a fabulous time.

We started off with a superb lunch at Matthew's, where they now have a $100 hamburger in honor of the $100 Hamburger website! You can get a burger for less than $10 if you want, but if you order it with champagne, you can pay the whole $100!

Matthew is a fantastic chef. I had the burger; my husband shared his lobster quiche with me—unbelievably good. We followed it with a superb raspberry cheesecake with a dark chocolate crisp base and an espresso crème brulée. Both were outstanding. The restaurant has an extensive list of very inexpensive local wine, and as we were staying overnight, we polished off a bottle of Baileyana chardonnay.

Matthew comes out of the kitchen to chat with customers, and he told us that it would be very valuable to him to have a certificate to put up in the restaurant showing him as "Best of the Best" $100 burger restaurants in the West. He tries very hard, so how about it guys? Zagat does! it, so it is actually a compliment that he really wants your official recognition!

We went on a wine tour with the Wine Wrangler ((866) 238-6400)—$45 each for a five-hour tour visiting five wineries of your choice or recommended by your guide if you wish. It was very laid back and enjoyable. Pickup and drop-off is available from your hotel, so you can drink whatever you want.

We stayed at the Paso Robles Inn downtown. At $125, it was a little expensive, but the room was nice (fireplace and refrigerator), and the hotel restaurant was good. They have a free shuttle and will take you to the airport and pick you up from there if you call.

We had a fabulous weekend, and we will certainly go back. Next time we hope to see that official recognition certificate from the *$100 Hamburger* hanging on the wall at Matthew's!

PIREP The #1-rated fly-in restaurant on the West Coast? Of course, we just had to fly there ASAP to see what the excitement was all about.

Paso Robles is a well-kept airport. The only thing to be aware of is the small restricted airspace just to the north. The runways are generous and in very good condition. We taxied to the nearly new terminal and tied down right in front of Matthew's Restaurant. Walking into the terminal, if one goes left there is a nicely appointed bar with large windows that provides a great view of the ramp and runways.

Going to the right took us right into Matthew's Restaurant. Matthew himself was there, greeting customers and keeping a watchful eye on things. The restaurant is cozy and tastefully designed in the Mediterranean style. A large window provides a great view of the ramp and runways. The menu is very comprehensive—they even have a $100 hamburger (the true price is about $6) on the menu. I decided to indulge in the wild mushroom soup (superb), the Caesar salad (average), and the salmon filet over a bed of spinach (wow!). They have a nice variety of quality wines for passengers. The prices are very reasonable considering the caliber of this restaurant. The service was excellent. The ambiance is friendly and classy.

Did Matthew's meet our expectations? The answers is a big yes. This is a true five-burger fly-in, highly recommended.

We'll definitely be back!

Perris, CA (Perris Valley—L65)

Airport Cafe 🍔🍔🍔

Rest. Phone #:
Location: On the field
Open:

PIREP The Airport Cafe is known as the bomb shelter. It is a great place to eat and watch the skydiving. I personally like the cheesesteak! As for the airport, the advisory frequency is 122.775, and make sure that you always stay to the west of the airport (spinning props and skydivers don't mix well). Skydivers land right next to the runway on the east side. Also while you are there you should try a tandem jump or off the parking lot—there is a vertical wind tunnel where you can try indoor skydiving. There is also an ultralight school. Overall, this is a great little airport to spend an afternoon at!

PIREP   The unicom frequency has changed once again to 122.975. No fuel available. And there are about five Twin Otters flying continuously on weekends.

Be careful! Be careful out there.

Use runway 15 most of the time, with all downwind traffic to the west side as mentioned.

PIREP   Perris Valley is another well-kept secret (if that's possible in Southern California). The airport is primarily used for skydiving and ultralights, but "real pilots" are welcome. The skydiving is world class and great to watch. This weekend we watched "sky surfing" (like snowboarding with a parachute) being taped for viewing on ESPN and NBC sports. The kids enjoyed watching the chute packing and ground practice sessions while Dad watched the "jumperette" action.

The restaurant and sports bar were completed just two years ago and are really attractive. Food is American, Mexican, and daily specials. Prices are very reasonable. There is also a large lawn and swimming pool area for your use (free!).

When you fly in, remain west of the field and call unicom on 122.9 at least five miles out. The jump zone is east of the field and the ultralights remain generally south. Runway is in good condition and paved for at least 3,000' and then extends to 5,100' with smooth dirt. This is a great place for a spring or fall half-day excursion.

Petaluma, CA (Petaluma Muni—O69)

29er Diner 🍔🍔🍔

Rest. Phone #: (707) 778-4404
Location: On the field
Open: Closed Monday

PIREP   Had a nice lunch here on a Saturday afternoon. Service and food was very good. There's a great little pilot shop here that gives AOPA discounts. Took a ride on their Segway transporter. What a cool experience. You step on and the gyros take over.

PIREP   Ate at the 29er Diner. Lunch was excellent. Not the fastest in getting the food out, but well worth the wait. Good work.

PIREP   Flew into Petaluma to the 29er Diner twice lately and both times had a good experience. I was giving a ride to the chef from Cafe Prague in SF. He tried the Country Pilot breakfast and thought it was great. Kid friendly, too.

Placerville, CA (Placerville—PVF)

Tomei's Restaurant & Bar 🍔🍔🍔🍔

Rest. Phone #: (530) 626-9766
Location: Call for pickup
Open: Daily: Lunch and Dinner

PIREP   Flying to Placerville in the heart of the Gold Country was a great experience. The runway and ramp area are in good condition, and the generous, paved tie-down space was free. There is a small pilots' lounge with local tourist materials available.

Before we departed from our home airport, Palo Alto, we called Mike Tomei, the owner of Tomei's Restaurant. We let him know we'd be coming and he said he'd gladly give us a ride to his restaurant in the old town. When we landed, we called again and Mike was there to greet us within ten minutes. On the five- to ten-minute drive to downtown Placerville, Mike gave us lots of interesting information about the town. Placerville was originally named Hangtown because of the many hangings of lawbreakers during the Gold Rush days.

Since we arrived a few hours early for dinner, it provided a nice opportunity to do some exploring of the old town. In the beautifully preserved downtown, there are many boutiques, art galleries, antique shops, and cafes to enjoy. We stopped into the Centro Cafe, a nice place for a coffee break. Right across the street from Centro Cafe is Tomei's Restaurant.

The decor is contemporary and classy, and the prices are average for a restaurant of this caliber. We first ordered the oysters on the half shell, which were excellent. That was followed by the minestrone soup and a very tasty, generous rib-eye steak. The food was superb and the service was excellent. They have a full bar and a nice selection of locally made wines. The restaurant has a friendly ambiance and got crowded with mostly locals in the evening.

After thoroughly enjoying ourselves, Mike gladly gave us a lift to the airport. The airport's main gate is locked at night but we quickly discovered a pedestrian access gate to the ramp. The night departure over the Gold Country foothills was spectacular.

All in all, this is a special place. Tomei's Restaurant and Placerville get my five-burger rating. We'll definitely be back!

PIREP   A short distance from the Placerville airport you'll find a very trendy spot, Tomei's Restaurant & Bar.

Tomei's is open for lunch and dinner Tuesday through Saturday with an elegant champagne brunch on Sunday. Full bar and excellent selections of local wines available. If you need to stay, Tomei's offers overnight packages including food and lodging. A wine tour package is currently in the works.

The proprietor, Mike Tomei, currently offers car service to and from the airport. Give him a call in advance to work out the details, (530) 626-9766. It is really quite simple.

Porterville, CA (Porterville Muni—PTV)
Michel's All-American Grill & Spirits 🍔🍔🍔🍔

Rest. Phone #: (559) 784-8208
Location: On the field
Open:

PIREP   We got stranded overnight at Porterville with mag problems late on a Saturday night, and Michel's staff went the extra mile to make sure we got a place to stay, then they dropped us off at the motel. The next morning, Michel himself got on the phone and found us a great mechanic to get us on our way. With that kind of hospitality, we'll definitely be stopping by more often. The food is quite good, nicely served and well attended.

Well worth a stop.

PIREP   Michel's is right on the field at Porterville in the terminal building. The dining room is airy and spacious, with large picture windows filling the entire wall facing the flight line. There were also picnic tables on a grassy lawn and a covered patio, too. The food was good and it appeared to me that over half of the people had driven in rather than flown, which speaks highly of the place given that it is about five miles from town. I met Michel, who was quite personable and thanked us for flying over to patronize his establishment.

In addition to the dining room, there was a small bar attached that opened into the terminal lobby. There was a piano present, which makes me wonder if there might not sometimes be live entertainment. It would be worth a look.

PIREP   I flew to Porterville for breakfast on Sunday morning. Michel's is one of the best on-airport restaurants you will find. The owner came out from the kitchen and introduced himself and made sure everything was okay. Friendly service and great food, this place should not be missed.

It has many amenities normally not seen at airport restaurants including cloth napkins and tablecloths and a varied and interesting menu. I recommend the filet mignon, which is excellent. The restaurant seems to be succeeding in attracting not only the airport crowd but also the locals.

They also have a very respectable bar, useful after the last landing of the day. When the restaurant is closed between lunch and dinner, food is available in the bar.

Quincy, CA (Gansner Field—2O1)
Morning Thunder Cafe 🍔🍔🍔

Rest. Phone #: (530) 283-1310
Location: On the field
Open: Daily: Breakfast and Lunch

PIREP   Just returned from our first breakfast run to Morning Thunder Cafe. We were not disappointed. The town of Quincy and its airport create a picturesque setting from the air, and the airport reminds me of small-town airports from an earlier era. The walk was easy into town and to the cafe. We found it to be charming, with good food, ample portions, and good service. We did a walking tour of town after breakfast, but nothing else is open on Sunday. So if you want to check out the stores, go any day but Sunday.

PIREP   Very nice little airfield to drop into. Sleepy Hollow–type of FBO with friendly guy who gave us a ride to the restaurant. Not really necessary though since, were it not for the airfield fence, it would have been a three-minute walk. It's a small restaurant and we had breakfast on a weekend. Although almost full already, we sat right down. The omelet ingredients were fresh and very creative and they were *huge*. Great side dishes and home fries too. The waitresses were young, cute, and very attentive despite being busy. Reasonable prices, great food, and a ten-minute walk from the ramp. I'm going back to try the lunch menu! They also have espresso, beer, wine, and they cater. They will do takeout too. Darold and Patti DeCoe are the owners.

Red Bluff, CA (Red Bluff Muni—RBL)
Blue Sky 🍔🍔🍔

Rest. Phone #: (530) 529-6420
Location: On the field
Open: Daily: Lunch and Dinner

PIREP   Stopped by the restaurant on the way back from Bend, OR after the Palms to Pines Air Race. I understand that is tradition for the group I was traveling with. My hamburger was a little bit dry, but the onion rings and chocolate shake were terrific! Very reasonable prices and a very friendly staff; quite popular with the locals too!

PIREP   Great views of snow-capped mountains in three directions contrasting with beautiful spring grass. Great service. Restaurant closes at 2 PM.

Redding, CA (Benton Field—O85)
Benton Airpark Cafe 🍔🍔🍔

Rest. Phone #: (530) 241-7934
Location: On the field
Open: Daily: Breakfast and Lunch

PIREP   Visited the restaurant today for lunch. Great friendly service, portions a-plenty, and very reasonable price. I think I paid $6 for a huge mushroom burger, fries, and a soda.

The cafe is located upstairs above the FBO right next to the tie-down ramp, which provides for a great view during a meal.

PIREP   Steve Miller has an excellent cafe, which serves breakfast and lunch. The prices are competitive, coffee is always hot, and the food served fast. There is a shelf above the coffee pots that has cups for the California Highway Patrol Air Force, with the pilots' names on each cup; their fixed wing and helicopters are located on the field so you may be dining with the CHP pilots. This is a favorite hangout for local pilots. The Father's Day Open House Pancake Breakfast is an extremely popular event, held in the Hillside Aviation's Hangar, sponsored by the local EAA.

Redding, CA (Redding Muni—RDD)
Peter Chu's 🍔🍔🍔

Rest. Phone #: (530) 222-1364
Location: On the field
Open: Daily: Lunch and Dinner

PIREP   There were some old pireps concerning Peter Chu's. One person thought he may have the name wrong. It is indeed Peter Chu's. There is one location on RDD and one in town. The one in town I am told is a little cheaper. The one on the field provides some awesome views of the mountains.
   Well worth the trip!

PIREP   I was at Redding recently and discovered one of the first airport restaurants that I've seen offering only Chinese cuisine. It's upstairs from the Pax terminal, has huge windows so you can watch air operations, nice atmosphere. Prices are very reasonable and they give you a lot of good-tasting food for the money. The staff are quick and courteous. Try it out. I want to say it's "Peter Chu's," but I forget the exact name. Well worth the effort to go, especially if you like Chinese food!

PIREP   On our way back to STS from Oregon, we stopped in Redding and checked out Peter Chu's. The food is good! A nice change of pace from burgers and other fried food one always finds at airport diners. Redding can get really hot, and if you stop for lunch and mention you're a pilot, they'll let you hang out in the air conditioning and be casual. Great views from upstairs (there's an elevator). When you contact ground tell them you want restaurant parking (if it's available); it's much closer than regular transient.

Rialto, CA (Rialto Muni/Miro Field—L67)
Elva's Airport Cafe 🍔🍔🍔

Rest. Phone #: (909) 428-6050
Location: On the field
Open: Mon.–Sat. 7 AM–4 PM, Sun. 7 AM–2 PM

PIREP   Elva's, located on the east end of the airport, is a cool 1950s-themed diner. They serve great Mexican and American breakfasts and lunches (the homemade tamales were awesome). There is plenty of parking right next to the restaurant, and they have a refreshing, covered patio.

Riverside, CA (Riverside Muni—RAL)
D&D Cafe 🍔🍔🍔

Rest. Phone #: (951) 688-3337
Location: On the field
Open: Daily: Breakfast, Lunch, and Dinner

PIREP This is a fairly new restaurant in the terminal opened by the folks who ran the Flabob eatery for years. This is the current favorite of the bunch. Regular fare, good service.

PIREP When we asked the tower is the food here any good, he responded with, "You mean the world-famous D&D Cafe." Then someone else on unicom felt the need to back up the claim of great food. Well, no disappointments. The club sandwich was good, and with fries, soup (or salad), and a drink (+$.25 refill) was around $7. The cafe is open until 9 PM local time, making it one of the latest in the basin (that I know of).

Riverside/Rubidoux, CA (Flabob—RIR)
Silver Wings Cafe 🍔🍔🍔

Rest. Phone #: (951) 683-2309
Location: On the field
Open: Daily: Breakfast and Lunch

PIREP Good for lunch, too!

PIREP My son and I arrived early on Saturday and just beat the crowd of locals at the cafe. Plenty of parking. Not much flying going on while we were there but a neat little airport with a great little airport cafe. Really takes you back in time. Friendly service, good food, very reasonable prices.

PIREP Outstanding restaurant and great food. Airport also has a great view of Mount Rubidoux and Riverside. On July Fourth, you can see fireworks from Mt. Rubidoux from the runway, which is located just northwest of Mount Rubidoux.

Rosamond, CA (Rosamond Skypark—L00)
Runway Cafe 🍔🍔🍔

Rest. Phone #: (662) 256-4965
Location: On the field
Open:

PIREP My husband and I flew into Rosamond two months ago. The Mexican restaurant there is excellent. We parked the airplane in front of the restaurant and were seated in the patio area. The food was outstanding (I judge the food based on the salsa), and we could watch the activity on the field. This restaurant deserves at least a four-burger rating.

PIREP You have to give them another burger. Three isn't enough. This is a nice trip out of the L.A. basin, and it is a great restaurant right on the taxiway. We'll go again!

PIREP I can second the phenomenal Sunday brunch. Make sure you fast on Saturday and leave the baggage at home, because you will eat plenty and need all the weight margin you can get on the way out. Food is truly excellent, service is great, there's a large patio right on the runway, and you park immediately in front of the restaurant.

Sacramento, CA (Sacramento Executive—SAC)
Aviator's 🍔🍔🍔

Rest. Phone #: (916) 424-1728
Location: On the field
Open: Daily: Breakfast and Lunch

PIREP   The food is *good*. The filet mignon is one of the best I have tasted. My wife and I have not been disappointed with any of the meals we have had. I give it four hamburgers! Transient parking is in front of the terminal building (at the base of the tower). Walk into the terminal building, turn right and up the stairs to the restaurant. Large windows on two sides of the restaurant allow a good view of the ramp and runways.

Self-fueling island is a few steps north of the terminal building. Truck fueling at your airplane is also available at a higher price.

PIREP   My husband and I walked in while waiting for our FAA exam scores and enjoyed a wonderful lunch. The food was terrific, the service quick and friendly, the view entertaining. You can sit inside with a great view of the airport or you can sit outside on the patio. The prices were very reasonable. We highly recommend it.

Salinas, CA (Salinas Muni—SNS)

Landing Zone 🍔🍔🍔

Rest. Phone #: (831) 758-9663
Location: On the field
Open:

PIREP   FBO Comments: Yikes!!! I don't understand why the fuel prices here are a dollar higher than anywhere else in the area. At least they warned us before pumping.

PIREP   The new owners have made this into a very nice destination and dining experience. The menu is extensive, and they also have a soup/salad bar. Prices were reasonable and service was excellent, just a few steps away from parking on the ramp.

PIREP   Finally! A restaurant at SNS that has good food and nice decor.
Newly renovated, just opened and ready for business. The perfect place to stop for a meal when on a long cross country. Taxi right up to the terminal and step up to a good meal. The pasta is good, as well as the Cobb salad. Try it, you'll like it!

San Carlos, CA (San Carlos—SQL)

Izzy's Steakhouse 🍔🍔🍔🍔

Rest. Phone #: (650) 654-2822
Location: On the field
Open:

PIREP   A new restaurant at San Carlos, Izzy's, is a branch of the famous San Francisco steakhouse. You can park your airplane by the Hiller Museum and it is a very short walk to Izzy's!
Steaks are the best, and for nonmeat eaters, they have very good fish. Not much for vegetarians, but the veggie side dishes are okay, although the famous Izzy's potatoes are quite greasy.
If you are in the mood for a sandwich or burger, I'd recommend the Sky Kitchen, but Izzy's is a terrific destination restaurant for a good meal.

Sky Kitchen Cafe 🍔🍔🍔🍔

Rest. Phone #: (650) 595-0464
Location: 620 Airport Drive (NE side of field, adj.)
Open: 6 AM–3 PM, 7 days/week

PIREP   Lots of my nonpilot friends love eating here. The breakfasts are always fantastic, and the hamburgers are very tasty!

San Diego/El Cajon, CA (Gillespie Field—SEE)
MayDay Cafe 🍔🍔🍔

Rest. Phone #: (619) 448-2707
Location: On the field
Open: Daily: Breakfast and Lunch; closed Sunday

PIREP   A small one-man sandwich shop with a humorous owner. No hamburgers but a great toasted egg, ham, and cheese sandwich called the Mayday Melt. Also ham, cheese, turkey, and pastrami sandwiches and great coffee. Located on the east side of the field. You can park next to High Performance or Commemorative Air Force and enter through the back door.
   Closed on Sundays.

Gillespie Field Cafe 🍔🍔🍔

Rest. Phone #: (619) 448-5909
Location: On the field
Open: Daily: Breakfast and Lunch

PIREP   Nice little ramp-side cafe. Great for a morning out with family or friends.

PIREP   Nice venue with patio and inside seating. Arrived at 1230 and was able to park in front of the cafe—decided to sit in the patio and watch the airport activity; good view of airport operations. The service was excellent and the fish and chips were good. This can be a busy airport with landings on one runway and departures on another.

San Diego, CA (Montgomery Field—MYF)
94th Aero Squadron 🍔🍔🍔🍔

Rest. Phone #: (858) 560-6771
Location: On the field
Open: Daily: Lunch and Dinner

PIREP   MYF has three different places to eat on the field and one other within close walking distance.
   Mexican: Casa Machado, located in the second floor of the terminal building. Ask Ground Control for directions.
   WWI style: 94th Aero Squadron, located on the north side of the airport, near the control tower and FSS. Aircraft parking *is* available next to the fence; again ask Ground Control.
   Deli: B&B Deli, located across the street from Gibbs Flying Service. A wide selection of breakfast and lunch items.
   Four Points Hotel: located on the southwest corner of the field, no direct access from the airport. Best bet is to park at Gibbs (see above) and call the hotel for a ride. Otherwise, the walk is less than a mile. Standard hotel-type restaurant fare.

Casa Machado 🍔🍔🍔

Rest. Phone #: (858) 292-4716
Location: On the field
Open: Daily: Lunch and Dinner

PIREP Casa Machado is easy to find right at transient parking in the terminal. Stopped here on the way back from Catalina. My friends enjoyed the margaritas so I can't comment on how good they were. Restaurant is designed so you get a view of the field, even from the land side. For the tables in back the floor is raised higher. Numerous pictures on the walls of WWII aircraft and Montgomery Field from its founding onward. Signed photos from Air Force 2 and Voyager's round-the-world flight crew.

San Diego, CA (Brown Field Muni—SDM)

The Landing Strip 🍔🍔🍔

Rest. Phone #: (619) 661-6037
Location: On the field
Open: Daily: Lunch and Dinner

PIREP The Landing Strip is not the place to go for a weekend breakfast. The restaurant has two rooms, the cafe and the bar. On the weekend the cafe is closed but you can order breakfast in the bar. They open early (0630), but unless the experience of eating breakfast in a bar is your thing you'll be disappointed. The food was good, the prices are average to high. Not much there to bring you back.

PIREP The Landing Strip located just a stones throw from transient parking offers outstanding fast food fare at good prices with friendly service. I highly recommend the 747, a third-pound cheeseburger—*outstanding*!
   At nearly 8000' runway 26R provides excellent touch and go opportunities.

San Jose, CA (Reid-Hillview of Santa Clara County—RHV)

Eastridge Mall 🍔

Rest. Phone #:
Location: Short walk
Open:

PIREP Although there is *still* no restaurant on the airport, a minor strip mall has recently been built adjacent to the southeast corner of the airport, corner of Tully Rd. and Capitol Expressway. It is about a 10-minute walk from the terminal building, about half the distance of walking to the main Eastridge Mall that you fly over on final. In this strip mall is a Togo's, Starbucks, Jamba Juice, and Panda Express, in addition to Albertson's supermarket. I give it a rating of three burgers because I like Togo's. If you want to walk to Eastridge, the Red Robin restaurant there is good.

San Jose, CA (Norman Y. Mineta San Jose International—SJC)

Penang Village 🍔🍔

Rest. Phone #: (408) 980-0668
Location: Short walk
Open: Daily: Lunch and Dinner

PIREP   This restaurant opened recently in what used to be the Coleman Burger, right next to the now-defunct Coleman Still. This restaurant offers a wonderful selection of "home-style" Malaysian food (think Thai with lots of prawns and peanut sauces) that ranges from spicy to mild. Main course portions are fairly generous, and the appetizers were excellent. I highly recommend the roti, which is like a crepe dipped in a curry sauce. This is a great place to go if you're as adventurous in your dining experiences as you are with your airplane. You may want to bring breath mints for afterward.

The decor is ex-Coleman Burger (complete with country horseshoe motif, uneven wooden floors, and walls) with some flags and Malaysian posters tacked on the walls. The staff is friendly and attentive.

To get there, exit the San Jose Jet Center and turn right onto Coleman Ave. about two blocks.

San Luis Obispo, CA (San Luis County Regional—SBP)
The Spirit of San Luis 🍔🍔🍔

Rest. Phone #: (805) 549-9466
Location: On the field
Open: Daily: Lunch and Dinner

PIREP   Hey, maybe I'm a bit prejudiced, so I'm a Cal Poly alum, so I met my wife at the Spirit of San Luis when it was the SBP terminal in 1975, but really, this is the best fly-in restaurant in the Western Hemisphere. You can park your bird within 100 feet of your table, get treated like a king, and get great food while they sell you 100LL. Try the shrimp and crab sandwich with an O'Doul's and you can still fly out.

PIREP   The Spirit of San Luis Restaurant is a very good place to eat, but my suggestion is to take a cab or the hourly bus into downtown SLO. There you will find numerous five-star eateries to choose from, all within walking distance from the downtown center. There is also the nice creek to sit by if it's too hot. All the people of SLO are nice, and the food is even better. Check out Firestone for excellent burgers and fries, Woodstock's for pizza, Cold Stone for ice cream on the ride back to the airport. There are many other places to choose from. The main terminal at SBP has an info booth if you have any questions.

PIREP   Great place to eat. Very friendly people, very good food. I have been there many times and always enjoy going back for another visit. Fuel prices are in line with the times. I always like the temperature. It can be very hot inland, but San Luis Obispo airport is always cooler.

Santa Ana, CA (John Wayne Airport-Orange County—SNA)
Chanteclair's 🍔🍔🍔🍔

Rest. Phone #: (949) 854-5002
Location: Short walk
Open: Daily: Lunch and Dinner

PIREP   John Wayne Airport is the only commercial airport for Orange County, CA, which makes it one of the ten busiest airports in the country. There is a left and right runway which effectively separates the small fry from the big iron.

For *real* food, you will have to depart John Wayne Airport and go across MacArthur Boulevard (and it may take a taxi to do this) to one of many restaurants or hotels within sight, offering a menu for most any specialty or budget.

If you're throwing caution (and your pocketbook) to the winds, try Chanteclair's on MacArthur Blvd. for excellent French cuisine. Don't skip the wines. Their cellar is fabulous. Stay the night.

Santa Barbara, CA (Santa Barbara Muni—SBA)

Beachside Cafe 🍔🍔🍔

Rest. Phone #: (805) 964-7881
Location: Crew car
Open: Daily: Lunch and Dinner

PIREP FBO Comments: Parked my Cherokee at Mercury and bought eight gallons of fuel to avoid the tie-down fee. The FBO gave us a courtesy car that we used to drive to the Beachside restaurant.

PIREP Fresh seafood at the restaurant, great ambiance (it is at the beachside, after all), and friendly service. The cross-cut fries are great, and the fruit was super fresh. I had an ahi sandwich (sorry, the clam chowder filled me up too much to get a hamburger), and it was just great. I'll be going back next week for some cross-country time and I'll definitely stop by the Beachside again.

PIREP Today we went to the Beachside Cafe. A first for us—we normally end up in the Elephant Bar!

We parked at Mercury, as advised by other $100 consumers. It was not initially obvious where to park, as the Mercury ramp was full of light jets. Eventually it became clear that Mercury wanted us to taxi almost all the way to the end of taxiway bravo, and we parked within a stone's throw of a huge military transport turboprop. The Mercury office people were friendly and helpful. The parking fee is $7 for a light single, waived if you buy 8 gallons of fuel. We made the mistake of buying over 40 gallons at $2.94 per gallon.

The restaurant is a pleasant ten-minute walk along a bike route, through a wetland preserve to the beach area. The location easily gets five hamburgers; it is right on the beach, and you can eat outside on the enclosed terrace next to the pier. No problem on arrival time either, if you habitually run behind (like us!)—the hours are from around 11 AM to 10 PM. I ordered the Beachside burger, which I would give barely a three-burger score, compared with the fives we have had at Chiriaco Summit! Neither the burger nor the fries were hot enough, and the burger was not juicy enough. My husband ordered a bay shrimp quesadilla, which I shared and enjoyed immensely. It was delicately spiced but oven fresh (or stovetop fresh!). The score for the quesadilla was at least a four.

There were also seasonal oysters, oyster shooters, and various other great-looking seafood and salad dishes, which we didn't try but would be worth another trip. Maybe we'll stay overnight next time—they have some fantastic happy hour cocktails!

Hope the update helps others to explore.

The Elephant Bar 🍔🍔🍔

Rest. Phone #: (805) 964-0779
Location: Short walk
Open: Daily: Lunch and Dinner

PIREP I flew my wife and two daughters (ages six and three) to SBA last weekend for lunch. We parked at Signature and ate at the Elephant Bar.

Signature's close proximity to the Elephant Bar makes it very convenient. Their line staff was professional, friendly, and on the spot whenever they were needed. The folks inside the office were also quite nice.

The Elephant Bar's patio with a view of the flight line can't be beat. It's nice to sit there, watch the planes come and go, enjoy a beautiful Santa Barbara day, and chat with other pilots. The service leaves quite a bit to be desired, although the food remains quite good. The kids really like the children's cups which resemble tiger's feet, elephant's feet, and so on. Their children's menu is also quite good. As a renter pilot, I wish they didn't take quite so long to bring us the bill. This happens to us every time we go there.

PIREP   FBO Comments: Signature is a rip-off. They want a $22 "handling fee." Waived if you buy their fuel at a very high price. This bites in a big way. I have never been charged a parking fee for two hours, and I've been eating $100 hamburgers for many years.

PIREP   The Elephant Bar and Restaurant is a delight. Great food, service, and prices. There's got to be another place to park!

Santa Maria, CA (Santa Maria Pub/ Capt G Allan Hancock Field—SMX)
Vintner's Restaurant 🍔🍔🍔

Rest. Phone #: (805) 928-8000
Location: Close by
Open:

PIREP   The Radisson Santa Maria is located just next to the airport. Pilots can park their plane and walk to the hotel. We have a great restaurant and of course serve burgers.

Santa Monica, CA (Santa Monica Muni—SMO)
Spitfire Grill 🍔🍔🍔

Rest. Phone #: (310) 397-3455
Location: On the field
Open: Daily: Lunch and Dinner

PIREP   Four planeloads of my friends and I made a post-Thanksgiving run to Santa Monica for a late lunch on the patio of the Spitfire Grill. It's not immediately obvious how to get from the admin building over to the restaurant, but their sign is visible beyond the chain-link fences around an aircraft parking area on the south side of the field. It's simple enough to figure out how to walk down the driveway to get to the street and cross over. I just wish that there was a sidewalk since it's a bit hazardous to walk in a group along a long, narrow driveway. Also, one couple had a stroller. (They were taking their three-month-old baby for her first flight, to get her started out right!)

The food at the Spitfire Grill is better than the usual airport cafe fare but nothing out of the ordinary for a Santa Monica restaurant. It's the casual atmosphere, moderate prices, and proximity to SMO that make it worth the trip. SMO is very close to the coast so the temperature was very pleasant; we were in short sleeves outside on the patio while we needed jackets up in the Antelope Valley where we started from.

PIREP   Besides the Typhoon and the Hump there is another really great place—you have to walk out of the south side of the airport and across the street to the Spitfire Grill. They have recently redone the entire eating area and added a *great* chef. The food is really good and worth the small walk. Plus they have cool historic aviation paintings and posters adorning the walls. Good place for lunch.

Santa Paula, CA (Santa Paula—SZP)
Longdon's 🍔🍔🍔

Rest. Phone #: (805) 525-1101
Location: On the field
Open: Daily: Lunch and Dinner

PIREP   Longdon's was almost empty, but the service was friendly and the food was excellent. Landing at Santa Paula was a little tricky, due to the tiny runway and strange winds from the surrounding mountains.

Be sure to read the A/FD before landing here. I wouldn't call the airport dangerous, but it's really important to follow correct pattern procedures here, and the runway length might be a problem for larger planes.

Santa Paula hosts an open hangar show on the first Sunday of each month. A great time to drop in and see some neat planes.

I'll definitely fly back to Santa Paula one of these days.

Santa Rosa, CA (Charles M. Schulz–Sonoma County—STS)
Manny's Restaurant 🍔🍔🍔

Rest. Phone #: (707) 573-6900
Location: On the field
Open:

PIREP   I flew here on my first dual cross-country and had a terrific Cobb salad. It was huge and I only ordered a half. The service was great and timely. It is a short walk from tie-downs and a tremendous coastal ambiance. A revisit will be a must.

PIREP   The same Manny that owned the fuel service there for many, many years. He is open seven days a week serving breakfast, lunch and dinner from 8 AM to 10 PM. Food is excellent and you can now park your aircraft just a few steps from the front door of the restaurant.

I had a great fried scallop dinner there last night and my wife enjoyed a big rib-eye steak.

Santa Ynez, CA (Santa Ynez—IZA)
Solvang's Bakeries 🍔🍔

Location: Call the bus. It's a hike!

PIREP   The airport is really fun to fly to—spectacular views coming and going and a very friendly staff. A couple of locals sitting out front tasting some of the local vineyard crop greeted us on arrival. The town of Solvang is absolutely quaint and fun. Excellent food at all the places we tried. Ebelskivers!?! Hey, I didn't name these things. Just try one and you'll be hooked! I recommend the public transportation, as a cab ride is now up to $18 for two (one way), while the bus is still only a couple of bucks.

Very enjoyable 1-hour flight from FUL.

PIREP   Just one ridge northwest of Santa Barbara. The scenery is spectacular, and the quaint little landing strip is just downright fun. I don't remember there being a restaurant on the field itself, but they have taxis, dial-a-ride, and a bus that takes you into Solvang (a Scandinavian township) within a couple of minutes. Spending the day is worth it, if only for the bakeries.

Schellville/Sonoma, CA (Sonoma Valley—0Q3)
Cline Vineyards 🍔

Rest. Phone #: (707) 940-4000
Location: Reasonable walk, tastings only
Open: Daily

PIREP   FBO Comments: 100LL SS available.

PIREP   When we landed on runway 7, the taxi back is via a 180-degree turn and back to the opposite end of the airport for tie-down. The runway here is in good shape, but the transient area is dirt and can get a bit soft and muddy in the rainy season. This field has an active antique airplane rides service. Nice planes, but I suggest keeping an eye out for them performing aerobatics when in the vicinity of the airport. Walking across the street (State Highway 121) from the airport there is a large fruit stand. They have a great selection of fresh produce and they also carry local wines at reasonable prices. Be especially careful when crossing the highway because the folks passing through tend to drive fast. About a quarter-mile south of the fruit stand is the Cline Winery. They have some nice complimentary wine tasting available for passengers and a gift shop. The folks at Cline are very friendly and it's worth the walk, although the minimal shoulder on the side of the highway makes it somewhat dangerous for pedestrians. If you can get a ride, I strongly recommend going another quarter-mile south to the hilltop Viansa Winery. They have a spectacular view, an Italian café, and wine tasting. However, I don't recommend walking to/from Viansa because of the imposing traffic—especially if your passengers have been tasting the great wines.

I give Schellville/Sonoma two burgers.

Viansa 🍔🍔🍔

Rest. Phone #: (707) 935-4700
Location: 1-mile walk
Open: Daily: Lunch

PIREP   If you're a wine snob, or just like to eat well, fly in to Schellville's Sonoma Valley Airport. There are two runways; 07-25 is 2700' and is the main runway in use; 17-35 is 1500' and is marked restricted. Check with the airport office at (707) 938-5382 before you decide to use the shorter runway.

Park on the grassy area in front of the faux ATC tower and walk about one mile south on Route 121 to Viansa Winery. It's on the left just past Cline Vineyards (also open for tasting but no firsthand experience to report).

Viansa has an extensive Italian marketplace with all sorts of gourmet food to sample before you buy. There is also a deli indoors, and of course there is wine tasting—of course, now you are not going to be able to fly home but that is another story for another day.

Outside the marketplace and tasting building, Viansa has a Tuscan grill that has several different specialty sandwiches averaging $6.50. Very tasty grilled chicken, Italian sausage, pork tenderloin, and so on. Enjoy some more wine with your Tuscan grill or deli lunch. You can sit on one of several tables on top of the Viansa hill that overlooks the airport. Then when you are done, you can have some gelato or sorbetto afterward.

The only problem with all this food is you have to walk back to the airport afterward. Be careful of the traffic on Route 121; it is very busy, especially on weekends.

Shafter, CA (Shafter-Minter Field—MIT)

Tailspin 🍔🍔🍔

Rest. Phone #: (661) 393-6095
Location: On the field
Open: Daily: Breakfast and Lunch

PIREP   Airport Comments: Lots of neat stuff to see here.

PIREP   The airport cafe is under new management and all is well once again. The food is excellent and the service is spirited ("sassy" to the locals, and we love it). They are open Monday through Friday 7 to 2, although we are working on them to open up weekends…I'll keep you posted.

Shelter Cove, CA (Shelter Cove—0Q5)

Chart Room Restaurant 🍔🍔🍔

Rest. Phone #: (707) 986-1197
Location: Short walk
Open:

PIREP   Beautiful place right on the water. Airport is surrounded by a nine-hole golf course. There are three restaurants a short walk from the south end of the field. Low end is the RV general store ((707) 986-7474) with great fish and chips. Next up for lunch is Mario's Marina ((707) 986-7595) on the cliff overlooking the cove. High end is dinner at the Chart Room Restaurant ((707) 986-9696). There are also places at the north end of the field.

Better call ahead for schedules, especially in the winter.

Airport is VFR day-only. It can be tricky with wind. There is a part-time weather reporting station (KO87) near the field. You can also call the Chart Room Restaurant for weather at (707) 986-1197.

I can't describe how beautiful this place is.

PIREP   FBO Comments: No FBO, no services, part-time unicom from one of the restaurants.

PIREP   Shelter Cove is on the Lost Coat of California, 40 miles north of Mendocino and 250 miles north of San Francisco. The airstrip is approximately 3400' running 30 and 12. The winds are mostly in line with the runway, but be careful when there is an offshore blow from the northeast.

To determine the actual conditions at Shelter Cove (flight service often does not have up-to-date information and the fog is not as bad as they indicate due to its rural unique location) call the airport manager at (707) 986-7361 or the Chart Room Restaurant at (707) 986-9696.

PIREP   We go to Shelter Cove quite a bit, and our favorite food and atmosphere is at the Chart Room. Friday night we were there and had a steak dinner that was so good, and Saturday was a prime rib special that was excellent. The people are so friendly; it is a lot of fun. The store deli by the boat launch has great fish'n'chips as well as burgers. We have eaten breakfast and dinner at Mario's and the food is good, but you can't be in a hurry. Breakfast for two on Saturday took 1 hour, 20 minutes and it *was not* busy. All in all it is a nice place to visit with plenty of good food around, but our favorite is the Chart Room—steaks, pasta, salads, almost everything homemade!!

Sonoma, CA (Sonoma Skypark—0Q9)

Harmony Club Wine Bar 🍔🍔🍔🍔

Rest. Phone #: (707) 996-9779
Location: Cab
Open: Daily: Lunch and Dinner

PIREP   FBO Comments: 100LL SS available.

PIREP   This nicely kept private airstrip is open to the public. The runway and ramp is in good condition. Watch for other traffic in the area, especially from nearby Sonoma/Schellville Airport. They have an airport manager on duty who directed us to the proper spot to tie down (free). There is no food at the airport, but I suggest having the manager call a cab to downtown Sonoma (five–ten minutes, <$10).

Downtown Sonoma is beautiful and historic and has a wide variety of great eating and drinking places. Wine tasting for passengers is highly recommended. We went into the Ledson Hotel ground floor, which houses their Harmony Club Wine Bar and Restaurant. This is a very classy place and the food is superb. We had soup and salad followed by the fish of the day. Prices are in line with a restaurant of this caliber and the service was excellent. We even purchased a couple of bottles of their Ledson Sangiovese to bring home (excellent wine!). The airport doesn't allow operations at night, so if you have to stay (why not?), you can indulge in some of the great local wines. There are many nice bed and breakfast places in town.

I will be back!

South Lake Tahoe, CA (Lake Tahoe—TVL)
Chase's Oyster Bar and Louisiana Grill 🍔🍔🍔🍔

Rest. Phone #: (530) 544-9080
Location: On the field
Open: Daily: Lunch and Dinner

PIREP   The service, food, and location make this a great lunch or dinner drop-in option. Located in the terminal (two-minute walk from the ramp), the views from the balcony and dining room let you keep up with all the happenings at the airport. The crab cakes starter is recommended, and they usually have a daily chef's special entree that is worth it. Nice bar too.

The website does not have operating hours but a call ahead is not a bad idea since they often host special functions that fill their modest dining area.

PIREP   It's hard to imagine a better destination for a great meal. Lake Tahoe is magical and arguably one of the prettiest spots on the planet. Like many people, we were disappointed when the restaurant in the terminal building closed a few years ago, but it's back open now and perhaps better than ever. Cuisine is uniquely Cajun, reasonably priced, nice variety. The restaurant overlooks the 8500 × 150' runway and has tables outside to enjoy the clean, mountain air. A full bar is available with some terrific local beers and ales. Service was excellent, and the views of the surrounding Sierra Nevada mountains is spectacular. It's also usually cooler at Tahoe than in the central valley of California or Nevada, and it's only an hour away from most areas in northern and central California.

A bonus is the opportunity to brush up on your mountain flying skills. The AWOS computes the density altitude for you which is usually over 8000' or 9000' in the warmer months, sometimes even higher. This is not the place to depart with all the seats and tanks full, especially considering that there will be downdrafts somewhere during the flight.

PIREP   The field is home to the U.S. Forest Service Fox Tanker Base, located at the east end of the field. This base is designed to accommodate large numbers of air-tankers (aircraft that drop aerial fire retardant), as well as lead-planes (Beech Baron 58P's that lead the air-tankers over the fires) and air-attack planes (usually Aero Commanders or 0-2's). The base is manned year-round but is usually busy from May to November. We are in the process of adding new facilities to the base and one day hope to add a viewing area for the general public. I'm told this is still a few years off, due to budget constraints. During large fires, we usually have a few people outside the fence, where they can get good views of DC-4's, DC-6's, DC-7's, C-130's, P-3's, PB4Y's, S-2's, P2V's, SP2H's, and other assorted aircraft. Please be careful around the field during a major fire and yield to our aircraft, as seconds can mean the difference for someone's house. During slow times, give us a call and we will be happy to show you around. Call (805) 948-6082 and ask for Martin or any of the Forest Service pilots.

Stockton, CA (Stockton Metropolitan—SCK)
Top Flight Cafe 🍔🍔🍔

Rest. Phone #: (209) 468-4726
Location: On the field
Open: Daily: Breakfast, Lunch, and Dinner Mon.–Fri.

PIREP   There is a coffee shop/restaurant in the terminal building at Stockton. Not great (maybe two burgers), but it's there if you wind up in that area hungry. Seems like lots of folks head over there for practice instrument approaches, and those can sure make you hungry! You can sit outside on a deck and watch the air traffic. Sometimes the airlines go over there and it's fun to watch DC-10s do touch-and-go's.

PIREP   The restaurant, Top Flight Cafe, must have improved since the last report. It's full service and has specials each day when it's open, which is Monday through Friday. I'd give it an above average rating. With no regularly scheduled airlines, the airport is a pleasure to fly into. The tower seems happy to have someone call. The restaurant is located in the terminal building on the second floor. Transient parking is immediately adjacent to the terminal off runway 11R.

Taft, CA (Taft-Kern County—L17)
Ollie's 🍔🍔

Rest. Phone #: (661) 765-4833
Location: Short walk
Open:

PIREP   FBO Comments: Landings 25 only, watch uphill slope! 122.8 for unicom. Watch and listen for frequent skydiving activities. Tie-downs on south side of runway at far west end (closest to town).

PIREP   With just a short walk to town, Taft is one of our favorite stops. From the tie-down area, just walk straight out of the airport and past the park to Kern Street. A couple of blocks up on the south side of the street, you'll see Ollie's, a '50s-style diner complete with jukebox, carhops, soda fountain, and dining room. The tables are made of sections of real bowling lanes; *very cool!* The place looks brand new, clean, and fresh but the atmosphere is pure 1950s. Service is great (cute waitresses), and food is excellent and reasonably priced. Their hours are 10 AM to 10 PM.

   Note: There are several motels right there in town as well as a breakfast-lunch-dinner place that is good too.

Tehachapi, CA (Mountain Valley—L94)
The Raven's Nest 🍔🍔🍔

Rest. Phone #: (661) 822-1213
Location: On the field
Open: Daily: Breakfast and Lunch; closed Sunday

PIREP   This is an undiscovered *gem*. Right next door to TSP, Mountain Valley, although a glider port is very power plane friendly. With a 5400' runway on the south side for the power planes, of which 1000' is paved on the 27L end and 500' on the 9R, density altitude is not a problem.

   The Raven's Nest makes the best deli sandwiches on their own homemade bread for lunch. Although the menu is limited, breakfast and lunch are excellent and reasonably priced. The Raven's Nest is upstairs and a view of the runways and glider operations is great. The restaurant sports pictures and memorabilia from Edwards as the Flight Test Center uses Mountain Valley for glider training.

   No services for the planes but tie-downs and very friendly folks. Larry and Jane plan on trying to pave 500' a year of their well-kept dirt strip for the power planes. Worth the trip and the food to help them reach their goal. Oh, and by the way, take a glider flight while you're there and see the beauty of this wonderful valley.

Tehachapi, CA (Tehachapi Muni—TSP)
The Apple Shed 🍔🍔🍔

Rest. Phone #: (805) 823-8333
Location: Short walk
Open:

PIREP　Try the Apple Shed, located just a couple blocks south/southwest of Tehachapi Municipal Airport. It's a restaurant, bakery, deli, and country store. They have a full espresso bar and specialty coffee. It gets its name because it is located in an old shed originally constructed in the 1940s for the purposes of packing and shipping local produce and seed. It has since been beautifully restored and has the smell of cedar inside. They are open six days a week from 11 AM to 5 PM (closed Mondays). Breakfast hours are planned so try calling first. They will also pick you up from the airport.

PIREP　Try Las Palmas (authentic Mexican food), about 300 meters west of the Apple Shed. Excellent food and service. There is an all-you-can-eat buffet (brunch type) until 1400 hours on the weekends. It's an easy walk from the airport—less than ten minutes. Some antique and art shops nearby. New camping park on the airport at east end.

Trinity Center, CA (Trinity Center—O86)
Trinity Center Inn 🍔🍔🍔

Rest. Phone #: (530) 266-3223
Location: On the field
Open: Daily: Lunch and Dinner

PIREP　Trinity Center is a great spot to go on a sunny day. The airport is on the edge of Trinity Lake (technically, Claire Engle Lake, but don't call it that around the locals) in the foothills of the Trinity Alps. There is no fuel available, but Redding is only a few minutes away. On the airport is the Trinity Center Inn, a motel/restaurant that serves dinner most of the time and lunch in the spring/summer. It is at the northwest end of the runway. Up the street a couple of blocks is the Sasquatch, a bar and grill that is open for lunch and dinner on weekends. The food at both places is great (for mountain fare), and the location is awesome.

PIREP　The Yellow Jacket is a short walk from the field, but that is probably a good thing since you will need to walk off the effects of the meal. The burgers are pretty hefty—juicy, messy, home-style burgers. Onion rings not too bad either.

　　Trinity Airport is an exceptionally beautiful and challenging approach. Well worth the trip.

Truckee, CA (Truckee-Tahoe—TRK)
Runway Cafe 🍔🍔🍔

Rest. Phone #: (530) 582-9351
Location: On the field
Open: Daily: Breakfast and Lunch

PIREP　Truckee, just north of Lake Tahoe, is a great place to fly into for a fun Sierra getaway or a good bite to eat at the Runway Cafe. The runway and generous tie-down area are in good shape. Just steps from the transient tie-down area is the Runway Cafe. There is a round fireplace seating area for the colder months. They have a great view of the ramp and runway. Hearty sandwiches are the specialty here. I had the Philly cheesesteak, and it was generous and tasty. Prices are average. For skiing or mountain activities, Northstar at Tahoe, my favorite Lake Tahoe ski resort, has a free shuttle that arrives hourly at the airport. The friendly, old-fashioned town of Truckee is a short cab ride from the airport. There are plenty of good restaurants, hotels, and shops. On Friday and Saturday nights the bars in town really get lively.

　　On the way out I suggest reviewing the noise abatement procedures. I recommend flying out over Lake Tahoe. Have your camera ready as the view is truly spectacular.

　　I will be back!

PIREP FBO Comments: Full-serve from truck is five cents more per gallon. Very noise-sensitive areas around airport. Pick up noise abatement brochure at unicom and read friendly departure routes. Take a quick test and get five-cent discount on fuel!

PIREP The Runway Cafe has grown to be a very good place for that $100 hamburger. Located in the main terminal building and close to the self-server fuel island. Transient parking is very close by. Amy and her girls are always bright and cheery and the atmosphere is very laid back. The Philly Flyer cheesesteak sandwich and the Spirit of St. Louis sub are both excellent choices. The grilled cheese is also one to try.

If you want to come up and get some mountain flying experience under your belt, Truckee is an excellent airport to practice. The 7000' runway and big wide valley makes for a wonderful flying experience.

Ukiah, CA (Ukiah Muni—UKI)
Bluebird Cafe 🍔🍔🍔

Rest. Phone #: (707) 462-6640
Location: On the field
Open:

PIREP FBO Comments: Sometimes you can get the fuel truck on 123.0, but the cell number for the fuel truck is (707) 272-5515 and for after-hours fueling call the manager at (707) 467-2817.

PIREP The restaurant is right across the street from the airport. There is one in the town of Hopland also that has been doing a great job for 10 years, and last year they opened their second location in what used to be the Beacon Restaurant. The staff is great, food is excellent. They have a good selection of your basic dinner fare but also have things like alligator, elk, and bear on the menu. Weekend brunches are excellent, but a hundred dollar moose burger, now *that* you can write home about.

Upland, CA (Cable—CCB)
Maniac Mike's 🍔🍔🍔

Rest. Phone #: (909) 982-9886
Location: On the field
Open: Daily: Breakfast and Lunch

PIREP The Cable (CCB) airport in Upland has a restaurant called Maniac Mike's. Mike's coffee is absolutely delicious, followed closely by breakfast and lunch. This place may be an undiscovered gem thus far. Prices are very reasonable, but the breakfasts are to die for!

Great food, great location, clean, neat, and good service. Try one of Mike's BLTs for lunch and you will return many times.

Yeah, he also does hamburgers that rank second to none, but my favorites are breakfast and lunch. This has got to be a five-star pilot restaurant.

PIREP Maniac Mike's is one of my favorites! However, he doesn't serve dinner. If you don't mind a short walk, the Buffalo Chip on Foothill Blvd. is only about a block south of the airport. (You can actually sneak through the airport fence and take the direct route south.) The Buffalo Chip was built in 1929 on what used to be Route 66. There is a large outdoor patio with fire pit and great nostalgic feeling inside the rustic restaurant. Kinda looks like a biker bar, but just a laid-back locals place. Great burgers, buffalo or beef.

Vacaville, CA (Nut Tree—VCB)

Star's Restaurant 😋 😋

Rest. Phone #: (707) 455-7827
Location: Short walk
Open: Lunch

PIREP   The Coffee Tree closed (one year ago). We still have some very nice restaurants just over I-80. About a ten-minute walk: Black Oak, In & Out, Applebee's, Mel's Diner, Fresh Choice, HomeTown Buffet, Tahoe Joe's, and a bunch more. A walk toward the west (10 minutes) will get you to Star's Restaurant. This is a wonderful sports bar atmosphere with four-star food and a great price. The chef's salad will feed two, and iced tea is great.

   Come to the Nut Tree Airport. We are still open.

Watsonville, CA (Watsonville Muni—WVI)

Zuniga's 😋 😋 😋

Rest. Phone #: (831) 724-5788
Location: On the field
Open: Daily: Lunch and Dinner

PIREP   Zuniga's Mexican Restaurant was a nice experience. Upon landing, the runway and taxiways are in good condition. The generous tie-down area is only steps away from Zuniga's. It was a really nice day, so I decided to eat outside and watch the planes take off on runway 20—nice ambiance. The restaurant has a full bar (not for me) and lots of tables with large windows facing the ramp. They first brought out a huge jar of homemade salsa and chips, which were very good. I ordered the chicken tacos, which were very generous and tasty. Service was good and attentive and prices are average.

   I will be back!

PIREP   Located near the Pacific Ocean. Great restaurant featuring Mexican food at Zuniga's in airport terminal; very nice personnel in the restaurant as well as on the flight line, serving as gas passers when you need 'em. The fuel per gallon is 30 cents above pump price.

Willits, CA (Ells Field-Willits Muni—O28)

Ardella's Kitchen 😋

Rest. Phone #: (707) 459-6577
Location: 4 miles
Open:

PIREP   FBO Comments: $3 per night tie-down fee; self-serve 100LL; comfortable pilot's lounge.

PIREP   AutoMart in Willits will rent cars (new Ford Taurus) at $39/day and pick you up at the airport (five miles north of town) by prearrangement. Ardella's Kitchen at 35 E. Commercial Street, half-block off of Main St. (Hwy 101) is fantastic home cooking!! In town and airport is four–five miles away but worth the dial-a-ride.

PIREP   Like trains? Fly into Ells-Willits (O28) and ride the Skunk train to the Mendocino Coast. This is not a trip for those in a hurry, but a round-trip ticket (all-day ride) is cheap. The folks at the airport will give you a ride to the station if you ask them nicely.

Willows, CA (Willows-Glenn County—WLW)

Nancy's Airport Cafe ⊜ ⊜ ⊜

Rest. Phone #: (530) 934-7211
Location: On the field
Open: Daily: Breakfast and Lunch

PIREP Nancy's is awesome. I don't understand why it rates only three burgers. You can taxi right up to the back window; it's open 24 hours a day, 7 days a week; the waitresses call you Honey; and it is frequented by crop dusters and truckers. What more could you want out of a restaurant?

I used to stop in there often on early morning flights on my way to Redding to fly a P-3 air-tanker for the Forest Service. It was the perfect meal before a hard day of firefighting.

Just beware of the one they call "The Goon."

PIREP There is a Wal-Mart across the street from Willows. I'm telling you, there's *nothing* like fly-in shopping for everything from groceries to hardware. I love it. Pick up some stuff for the house, office supplies for work, and a gut-packing meal from Nancy's, all in one stop. One of my favorite destinations. When you leave, keep it at 1000 AGL and cruise the farmlands and wildfowl nesting areas. The snow geese in the winter number in the tens of thousands.

Woodlake, CA (Woodlake—O42)

The Outpost ⊜ ⊜ ⊜

Rest. Phone #: (559) 564-3244
Location: On the field
Open: Daily: Breakfast and Lunch

PIREP Woodlake is a small (3320 × 50') airport nestled at the foot of the Sierras about 15 minutes north of Porterville. The restaurant on field is the Outpost, and it looks like one. It resembles an old wooden cabin. The food is good with a wide selection. Pies are the main dessert and are baked fresh daily at the restaurant. The waitresses are friendly and funny—expect to be kidded a bit if you give them an opening. We've made several trips there and haven't had a bad meal. The pies are delicious.

PIREP The restaurant was recently revamped and cleaned up, The food is excellent with a fairly large selection, and Velma just loves people; she brings a real friendly reception to everyone passing through her doors!

Dessert pies are still the largest part of the menu. Breakfast is the busiest meal of the day, popular with the locals as well as the fly-in crowd! It's open from 6 AM until 2 PM daily; Sunday is the most popular fly-in day!

100LL is available during restaurant operating hours; your waitress will also fill up your plane—go figure! They sure deserve the tips!

Mechanical assistance is available on the field most days. There seems to be a resident mechanic who comes when you stand in the field and wave—I have not been able to figure out where he works from!

For those who wish to spend the night here, Sequoia National Park is only 20 minutes away. There are a few bed and breakfasts within a short distance of the airport.

Woodland, CA (Watts-Woodland—O41)

Flier's Club ⊜ ⊜ ⊜ ⊜

Rest. Phone #: (530) 662-0281
Location: Next door
Open: Daily: Breakfast, Lunch, and Dinner

PIREP   FBO Comments: 100LL SS available, but expensive.

PIREP   Watts-Woodland Airport is located adjacent to a nice country club and golf course. They have a recently paved runway and ramp. The transient area is at the NE end. A short walk past the old white T37 jet on display leads into the clubhouse. There are many autographed photos on the walls of famous aviators who have stopped there—it's worth checking them out. Unfortunately, the dining room was closed for a private party, but we went into the well-appointed bar and had some light fare, which was good but somewhat pricey. The view of the golf course and airport from the bar/dining room is wonderful. This is a nice place to stop, eat, and relax a bit. They said their hours and nonclub member access can vary. I suggest calling ahead to make sure the restaurant is open to fly-ins before going.

PIREP   We visited the Flier's Club for lunch today. Food and service were as described although I feel the prices were in line with the norm at some airport dining establishments.

My main comments concern the club's dress code. They have begun to enforce the code on everyone—members, guests, and drop-in pilots. To quote, "Proper attire for men must include shirts with collars, no T-shirts, tank tops, or short shorts are acceptable. Similarly for women, halter tops, tube tops, tank tops, or short shorts are taboo. Both men and women will not wear cut-offs or ragged jeans. Shoes will be worn."

My wife and I arrived not knowing this, and I had on a T-shirt and Bermuda-type shorts. I was supplied with a clean polo shirt to wear while having lunch. I did not find this a problem and felt that everything was handled very well.

While there, check out the T-37 Tweet on a stick in front of the clubhouse.

PIREP   Watts-Woodland Airport has an excellent little restaurant called the Flier's Club located about 200 yards off of the end of runway 36. A little pricey but excellent sandwiches and light lunches.

Yuba City, CA (Sutter County—O52)

Sundowner Cafe 🍔🍔

Rest. Phone #: (530) 674-7376
Location: 1 mile
Open: Daily: Breakfast, Lunch, and Dinner

PIREP   You can get a good, reasonably priced breakfast (I haven't tried lunch) at the Sundowner Cafe. Follow the airport road to the 2nd Street and Garden Highway intersection. Continue straight on 2nd Street, four blocks to Bridge St. Turn left on Bridge to 208 Bridge St. (just a few doors from the corner). The Sundowner is open 7 days, 6 AM to 2 PM Monday through Friday and 6 AM to 1 PM Saturday, Sunday, and holidays.

From tie-down to the cafe is one mile.

The Hundred Dollar Hamburger

Colorado

Akron, CO (Colorado Plains Regional—AKO)

Crestwood 🍔🍔

Rest. Phone #: (970) 345-6888
Location: 1-mile walk or crew car
Open: Daily: Breakfast, Lunch, and Dinner

PIREP FBO Comments: Fuel is available by self service or by truck. Fuel pricing has been competitive and very competitive on self service.

PIREP A courtesy car is almost always available and the town is only a mile away. The supermarket has a deli and daily lunch specials. The restaurant down the street has a full menu. Both have been good.
The draw is an efficient FBO, a courtesy car ready to go, and quick access to the food service in town.

Aspen, CO (Aspen-Pitkin Co/Sardy Field—ASE)

The Red Onion 🍔🍔🍔🍔

Rest. Phone #: (970) 925-9043
Location: In town
Open: Daily: Lunch and Dinner

PIREP The Red Onion is one of the best places to eat in Aspen. Great not-so-expensive food with TONS of character. It has been in its original location for over 100 years (read back of menu), with some original decor still on display.

PIREP Flying into Aspen is an experience you are not likely to forget. First of all the approach is a thrill a minute, then when you walk around the ramp you will think you have stumbled into a fly-in for jets. I have never seen so many Lears, Falcons, Starships, etc., etc., etc. in one place in my life.
The FBO is like a resort, the people are very friendly, and they will gladly loan you a crew car if one is left (be nice and fill it as well as your airplane). The town is only about five minutes away and is crowded with great places to eat. The flight in and out is over some of the most spectacular country you will ever see...but BE CAREFUL, these are real mountains and you will have to be able to cruise well above 12,000 feet: "check density altitude." Don't forget to take survival gear just in case; a walk out after an engine failure could take weeks.

There are dozens of local "joints" that have character—generally hamburgers, deli sandwiches, chicken breast sandwiches, chili, or Mexican are plentiful. I'm sure there are more elegant places with high prices, but I avoid such places like the plague. Parking can be a problem, but on Sunday the parking meters are free, so just park where you can. No place in town is really too far to walk.

We were in a C-182 out of Jeffco (BJC) across Corona pass, Kremling, Eagle, Lindz int, and up the valley to Aspen.

Buena Vista, CO (Central Colorado Regional—7V1)

Cafe Del Sol 🍔🍔

Rest. Phone #: (719) 395-6472
Location: Crew car
Open: Daily: Breakfast, Lunch, and Dinner

PIREP The Buena Vista, CO airport is a mountain airport that has a courtesy car so that one can go to town (two miles). Many restaurants there, but the Cafe Del Sol has great Mexican food, in the medium price range, and is right on the main drag of town (the N-S highway) and also next door to a wonderful bed and breakfast place. (The B&B is expensive.)

PIREP Buena Vista airport (7V1) is located in the Rocky Mountains of Colorado. Elevation is 7,946', and the airport has a 8,300' paved runway. Fuel prices are the lowest of all the mountain airports. The airport has a large heated hangar and several outside tie-downs.

There are numerous restaurants in town that are within walking distance; however, for those who would rather ride, the airport provides three courtesy cars.

Some of the restaurants in town are Cafe Del Sol (Mexican food, great food, moderately priced) and Buffalo Bar and Grill (dinner and steakhouse, great food, moderately priced).

There are also several bed and breakfasts in the area, one being the Meister House, a historic restored hotel with six unique rooms, all provided with gourmet breakfasts.

The area is loaded with activities in the summer or winter, including skiing, hiking, camping, fishing, rafting (on the Arkansas River). There is also Mount Princeton Hot Springs nearby, with hot spring pools along Chalk Creek at the base of the Chalk Cliffs.

On top of all this, the owners of the FBO are terrific people who will do anything they possibly can to make your visit a memorable one.

Colorado Springs, CO (City of Colorado Springs Muni—COS)

Solo's 🍔🍔🍔

Rest. Phone #: (719) 570-7656
Location: On the field
Open: Daily: Breakfast and Lunch

PIREP Solo's is located about a half-mile west of runway 17R and 35L, definitely in walking distance. The food is better than most airport lounge food, and the atmosphere cannot be beat. The restaurant is located inside an old Air Force Reserve KC97. The cockpit is sometimes open for tours, but you either can sit in the restaurant or you can hike up some stairs and actually eat inside the airplane. There is plenty to look at while you are in there as well.

PIREP

Location—short walk (less than a mile)
Food—very good

Service—good
Ambiance—airplanes and airplane parts hanging from every surface!
Prices—fair

Cortez, CO (Cortez Muni—CEZ)
El Grande Cafe 🍔🍔

Rest. Phone #: (970) 565-9996
Location: Crew car
Open: Daily: Lunch and Dinner

PIREP There's a little Ma and Pop place a few miles north of the airport in Cortez, CO. It's called the El Grande Cafe. Been in the same family for over 40 years. Great burgers, homemade french fries, shakes, a little Mexican food, and a great chicken-fried steak. If you need a meal to go, just call and ask, or even a full deli tray, just let them know.

The guys at the FBO know and they always seem to have a loaner car around. They all know the route.

Denver, CO (Jeffco—BJC)
The Bluesky Bistro 🍔🍔🍔

Rest. Phone #: (303) 460-7228
Location: On the field
Open: Daily: Breakfast and Lunch

PIREP The Bluesky Bistro moved in when the Tailwinds Cafe left. Still cafe style but the food is unbelievable! They basically have normal business hours so they aren't open late like the Runway Grill across the street, but when they're open they have the best food on the field with the service to match.

Denver, CO (Centennial—APA)
The Perfect Landing 🍔🍔🍔

Rest. Phone #: (303) 649-4478
Location: On the field
Open: Daily: Breakfast, Lunch, and Dinner

PIREP This is a great restaurant with an unbelievable view. I wasn't even able to finish my breakfast burrito (which doesn't happen often!). The portions are huge.

It has a great view of the airport's operations. As a new pilot, I had a blast explaining to my nonpilot friend how the airport was operating. The unique thing about this restaurant is that the airport is so busy, there is always something to watch!

PIREP The Perfect Landing restaurant is on the field at Centennial Airport (APA) on the south end of Denver. The restaurant itself is upstairs in the Jet Center FBO building. They serve breakfast and lunch, and the food is very good. The icing on the cake is the view: all the windows face west with an outstanding view of the runways and the Rocky Mountains. Hard to beat.

Denver, CO (Front Range—FTG)
Front Range Cafe 🍔🍔🍔

Rest. Phone #: (303) 261-9100
Location: On the field
Open: Daily

PIREP The Front Range Cafe is a popular place for airport patrons to enjoy a good meal and friendly conversation. The cafe is open daily from 11 AM–2 PM Monday–Saturday and features daily lunch specials and catering.
   Laura Shewmaker is the manager of The Front Range Cafe.

Fort Collins, CO (Fort Collins Downtown—3V5)
Charco Broiler 🍔🍔🍔

Rest. Phone #: (970) 482-1472
Location: Short walk
Open: Daily: Lunch and Dinner

PIREP I highly recommend going to the Charco Broiler from the airport. We have walked it many times now. It is a 10–15-minute walk on side roads. Get good directions the first time. It is not obvious and not right next door, but it's a nice walk on a nice day. I highly recommend the steak sandwich, which is a nice-sized one-inch-plus thick steak with some toasted bread on the side. My wife gets the kabob. It is a little rustic, but a nice steak at a nice price. Kid friendly too.

PIREP FBO Comments: Friendly service. Offered use of the courtesy car (a decent Mercury station wagon) even though we didn't buy fuel and were going to walk.

PIREP The Charco Broiler makes for a good lunch or dinner stop. The burgers aren't fancy but are good. The service and price can't be beat.

Granby, CO (Granby-Grand County—GNB)
Remington's 🍔🍔🍔

Rest. Phone #: (970) 887-3632
Location: Crew car
Open: Daily: Lunch and Dinner

PIREP Bertie's has changed its name to Remington's.

PIREP A telephone call to (970) 725-3347 will get you the combination to a lockbox that holds the keys to the courtesy car. Be sure to tell 'em you're a pilot and ask for directions to Bertie's Restaurant ((970) 887-3632). It is about 1.5 miles away in Granby. There is a woodstove in the dining area and they serve really great STEAKS!

Greeley, CO (Greeley-Weld County—GXY)
Barnstormer Restaurant 🍔🍔🍔

Rest. Phone #: (970) 336-3020
Location: On the field
Open: Daily: Breakfast, Lunch, and Dinner

PIREP The greasy spoon at the airport had marvelous chicken-fried steaks (good food and cheap fuel). Buffalo burgers are good too if the thought of gravy plugging your arteries scares you.

PIREP   A great place to eat and watch aircraft while doing so. The Barnstormer Restaurant has expanded its menu. The service is always great, and on weekends the place is full of pilots. An excellent fuel and food stop. They have hamburgers, Mexican food, and recently added buffalo to the menu. Definitely worth four burgers.

Gunnison, CO (Gunnison-Crested Butte Regional—GUC)

Mario's Italian Restaurant 🍔🍔

Rest. Phone #: (970) 641-1374
Location: Crew car
Open: Daily: Lunch and Dinner

PIREP   As photo surveyors, we see many different locales within and out of the state, and Mario's Italian Restaurant in GUC is the best we have been to (try the calzone/soup lunch special—cheap and good). It is too far to walk from the airport, but with a fuel purchase, you can borrow a blue Jeep Cherokee from the FBO and go to town. Mario's is on Main Street next door to a really nice sporting goods store (hunters—wink wink).

Kremmling, CO (McElroy Airfield—20V)

Mrs. Z's Burger Barn 🍔🍔

Rest. Phone #: (970) 724-9300
Location: Crew car
Open: Daily: Lunch and Dinner

PIREP   My wife Julie and I flew into Kremmling from Erie Airpark for the first time.

We were very pleased with a hamburger place called Mrs. Z's Burger Barn. It's right on the way into town on the north side. The food quality was excellent with a variety of different types of burgers from regular to buffalo to a jalapeno-blended beef patty! For those into old video games, they have Pac Man (my wife enjoyed that!).

Doug at the FBO was very friendly, and a courtesy car is available.

PIREP   McElroy Airfield sits in the middle of three really nice restaurants. Each will come and get you for the price of a phone call. I don't have a favorite. They're all pretty good.

The Wild Rose Restaurant, prime rib, steaks, free shuttle: (719) 724-9407.

The Wagon Restaurant, prime rib, steaks, free shuttle: (719) 724-9219.

Bob's Western Motel, free airport shuttle: (719) 724-3266.

Courtesy car available; ask for directions to a local restaurant/motel.

La Junta, CO (La Junta Muni—LHX)

Hogsbreath Saloon and Restaurant 🍔🍔

Rest. Phone #: (719) 384-7879
Location: Crew car
Open: Daily: Lunch and Dinner

PIREP   FBO Comments: I called ahead for a crew car and had a breeze getting one to get me into town. There was no charge, but donations were accepted.

PIREP   We didn't have much time for lunch, so we got a recommendation for one of the closest restaurants and a local favorite, the Hogsbreath Saloon and Restaurant. The ambiance was quite unique and included money (dollar bills up to 20 dollar bills) covering the walls. Apparently it was a local tradition to put the money on the wall in case you had to come back in for a drink to help cure a hangover the next morning. We ordered barbeque roast beef sandwiches and were in for a real surprise. They were great and very reasonably priced.

Limon, CO (Limon Muni—LIC)

Flying J 🍔🍔

Rest. Phone #: (719) 775-2725
Location: Short walk
Open: 24/7

PIREP Flying J truck stop cafe, good ol' American food at low prices. There's also a Pizza Hut, Wendy's, and Dairy Queen. All are an easy walk from the airport, less than half a mile.

Pagosa Springs, CO (Stevens Field—2V1)

Green Mountain Inn 🍔🍔

Rest. Phone #: (970) 724-3812
Location: Short walk
Open: Daily

PIREP FBO Comments: Nice FBO. AVJET Corporation (970) 731-2127. Friendly, pilot-oriented people.

PIREP Green Mountain Inn is right off the airport within walking distance. Great food and decent prices.

Pueblo, CO (Pueblo Memorial—PUB)

The Parachute Grill 🍔🍔🍔

Rest. Phone #: (719) 948-4185
Location: On the field
Open: 6 AM–8 PM, 7 days a week

PIREP FBO Comments: Friendly linemen at Flower Aviation. Parked us and gave us directions to the Parachute Grill. There's a small aviation museum right across from the FBO.

PIREP Stopped in for breakfast and was pleased with the service and the food.

Salida, CO (Harriet Alexander Field—0V2)

The Country Bounty Restaurant 🍔🍔

Rest. Phone #: (719) 539-3546
Location: Several miles crew car
Open: Daily

PIREP The Country Bounty Restaurant has great home style meals and rhubarb strawberry cobbler to die for. It is several miles into town, but they usually have a car to borrow, or call ahead to put your name on it.

Telluride, CO (Telluride Regional—TEX)

Rustico Ristorante 🍔🍔🍔🍔

Rest. Phone #: (970) 728-4046
Location: In town
Open: Daily: Lunch and Dinner

PIREP   FBO Comments: Excellent FBO. There was a landing fee and tie-down fee. They sell really cool T-shirts that say, "Some fly at 9,078'—we land."

PIREP   We dined at Rustico for lunch. There is one window table that sits privately in a corner on Main Street by the entrance. Very romantic and quaint. Request this one, or the one by the fireplace. Otherwise, there are many tables through various rooms of this antique building. The restaurant has super food. Prices are moderate. Very large wine selection. It is located downtown on the main strip.

The Hundred Dollar Hamburger

Connecticut

Danbury, CT (Danbury Muni—DXR)

McNally's Steak House 🍔🍔🍔

Rest. Phone #: (203) 730-1466
Location: On the field
Open: Daily: Lunch and Dinner

PIREP Landed on clean runway with 7-foot snow banks. Tower was very friendly and guided us directly to Reliant Air to park. Parked all the way to the left, as the transient sign states, and was promptly greeted by a very friendly Reliant employee who chocked our 172SP. He complimented us on reading the sign and parking in the correct area and told us how to get to McNally's Steak House. It is directly above Reliant. Nice atmosphere, tower radio on in the background with a great view of the runway. Angus burgers were great, along with the crab cakes and clam chowder. Nice place to stop in. No parking fee.

PIREP FBO Comments: Parked at Reliant Aviation and needed a little help inspecting an oil filler problem. The line mechanic could not have been nicer—came right out to my plane and helped put our minds at ease.

PIREP The restaurant is very nice—the server was great and the food was surprisingly good and well designed. They have a great location where you can sit and watch the runway and look over your aircraft at the same time. Food was excellent, service was perfect, and ambiance was better than average. Good value for the money.

East Haddam, CT (Goodspeed—42B)

Gelston House 🍔🍔🍔

Rest. Phone #: (860) 873-1411
Location: Short walk
Open: Daily

PIREP FBO Comments: We did not buy fuel. The landing and parking fee is $5. Runway paved, parking is right on the Connecticut River.

I took my wife to Goodspeed Airport today for lunch, and we had a very pleasant surprise when we had lunch at the Gelston House, which is a very short walk from the airport. The best part of all is that my wife came away having enjoyed a very pleasant day in our Citabria, and it doesn't get any better than that.

PIREP   In addition to the coffee house there is a great (but expensive) restaurant called the Gelston House. We have dined there many times (breakfast, lunch, and dinner) and always found it wonderful (no hamburgers, though). And you could also see a show next door at the Goodspeed Opera House, which is a historical site that puts on several musicals a season. Both are in easy walking distance of the airport.

Groton (New London), CT (Groton-New London—GON)
Abbott's Lobsters 🍔🍔🍔🍔

Rest. Phone #: (860) 536-7719
Location: Cab ride 10 minutes
Open: Daily: Lunch and Dinner

PIREP   If you want good seafood, a 10–15 minute cab ride to Abbott's Lobsters in the rough in Noank, CT is a great deal. They have a lobster dinner that includes steamed clams or mussels, chowder, shrimp cocktail, and chips for $20. Downtown Mystic CT is also just a short ride from Noank with a lot of interesting stores and restaurants. The Nautilus submarine museum is only a short ride from the airport also. I rate this airport five flying burgers.

PIREP   This controlled airport is a $16 cab ride from Mystic Seaport and Foxwood's Casino. Located in Connecticut on the very edge of Long Island Sound, there are numerous restaurants and places to stay, swim, and sun. A great lobster pound (Abbott's) is located between the airport and Mystic.

Hartford, CT (Hartford-Brainard—HFD)
Wings 🍔🍔🍔

Rest. Phone #: (860) 241-1114
Location: On the field
Open: Daily: Lunch and Dinner

PIREP   FBO Comments: Our fly-in transient parked (seven planes) at Atlantic Aviation ((860) 548-9334) located at the north end of the airport by the approach end of runway 20. Very nicely appointed FBO attached to the Wings Restaurant. FBO personnel could not have been more helpful and welcoming. Plenty of additional transient parking was available.

PIREP   Wings had very reasonable prices for a varied menu. It is adjacent to the ramp so you can watch takeoffs and landings. Motif is a sports bar atmosphere with multiple monitors and a pool table.
Staff very helpful and pleasant rearranging tables for our group. Felt very welcome. Definitely will return.

New Haven, CT (Tweed-New Haven—HVN)
Sandpiper 🍔🍔🍔🍔🍔

Rest. Phone #: (203) 469-7544
Location: New Haven
Open: Daily: Lunch and Dinner

PIREP   Sandpiper Restaurant is in New Haven. I have been there many times and it's always great. Fly into Robinson. Cheap fuel and a shuttle van, typically 10 minutes to Sandpiper. It's on the beach (with a playground, so great for kids). Great seafood. Affordable, clean. Great beach. GREAT FBO.

Willimantic, CT (Windham—IJD)

New England Pizza 🍔🍔

Rest. Phone #: (860) 456-8770
Location: Short walk
Open: Daily: Lunch and Dinner

PIREP   New England Pizza is right up the hill behind the airport (south of the main strip and buildings). Maybe a 100-yard walk. I ate there this summer and the food was excellent, and the prices are very inexpensive. If you want fancier food, walk across Route 6 and up to Ruby Tuesdays. Maybe another 150 yards. Since it is a chain, you all know what you can get there. There is even a Wendy's right in the plaza. And a Wal-Mart.

One-stop shopping and dining at IJD.

Windsor Locks, CT (Bradley Intl—BDL)

New England Air Museum 🍔🍔🍔

Rest. Phone #: (860) 623-3305
Location: On the field
Open: Daily

PIREP   New England Air Museum has some truly rare aircraft on display. My all-time favorite plane is the Republic RC-3 Seabee. I never got to fly one, and if I could find one on the market I'd quickly offer my Cherokee in trade. They have one here, as well as a Bleriot X1 Monoplane—this is not a replica! The museum offers an audio tour. This is really novel for aviation museums. I hope others are encouraged to follow New England's lead, as this really makes the exhibits come alive. The tour gear consists of a digital handset and a map of the two indoor hangars. You use a supplied map to locate the numbered pylons around the museum. Next, you type the pylon number into the handset and listen to the in-depth narration. This fascinating tour takes visitors through the history of flight chronologically from man's earliest recorded attempts at engineering a flying machine through the present. The tour has 30 stops and takes approximately one hour to complete. The $3 rental fee is well worth the price. I completed the audio tour, turned in my handset, and went back through the museum to see all of the displays that weren't covered.

There are three best times to come to this museum each year called Open Cockpit Days. You will be invited to climb into the pilot's seat of 12! WWII fighters; jet fighter(s), an airliner, helicopters, and a civilian plane are normally included. Bring your camera; this is a great chance to get some photos of these old birds' interiors. You may never have a chance like this again—don't let it pass you by.

They have food, but it isn't the reason to come.

The Hundred Dollar Hamburger

Delaware

Georgetown, DE (Sussex County—GED)

Jimmy's Fly-In Restaurant 🍔🍔🍔🍔🍔

Rest. Phone #: (302) 854-9010
Location: On the airport
Open: Mon.–Thurs. 7 AM–8 PM, Fri.–Sun. 7 AM–10 PM

PIREP Cloth napkins (yes, you can take your wife) with a casual atmosphere and reasonable prices. Open Friday–Saturday 7:30 AM–9 PM, Sunday 8 AM–7 PM, Monday–Thursday 8 AM–8 PM. Veggies were fresh, and portions were good. Excellent stop! This restaurant should be upgraded to 4 or 5 hamburgers.

PIREP Stopped here on the way back to PNE from a drop-off at SBY last evening. Happy to find Jimmy's open. Food was plentiful and very good. Seafood and down-home cooking. I had the crab cakes, which were great. No fillers, just crab. The corn pudding was awesome as well. I'll be back.

PIREP My wife and I had lunch there on a Saturday afternoon just after Christmas. The tuna sandwich and hamburger were excellent, and the service was excellent also. The restaurant is only about 50 feet from the parking ramp. The restaurant, FBO, and flight planning rooms were all clean and nicely appointed. We'll be going back again!! The restaurant is open Monday–Thursday 8 AM until 8 PM, Friday 8 AM until 9 PM, Saturday 7 AM until 9 PM, and Sunday 7 AM until 8 PM.

PIREP FBO Comments: Excellent facility. Restaurant is just of the west side of runway 4/22.

PIREP Good food, but we found it a bit on the pricey side. Pasta was fantastic, as were the sandwiches. Good field, easy approaches. Have to give it four burgers.

PIREP Really good food at modest prices with good service. Local folks eat here, so it will not disappear.

PIREP Stopped there and had an EXCELLENT chicken dinner, and my wife had the fish special. We all had crab cakes and fresh-baked bread (still warm!), and everything we had was excellent. One of the best airport restaurants I've been to, and very clean and friendly.

PIREP   Ditto on Jimmy's. Stopped to fuel on the way to the Outer Banks and found a new terminal and a restaurant: Jimmy's. Had the crab soup—very good. The hamburger was also good. Didn't try the pies, but wish we had.

PIREP   I had a chance in August to visit the NEW Jimmy's Restaurant at Sussex County Airport. Yes, it is BRAND NEW! The smell of fresh wood and paint is still there. The eating room faces the ramp and runway 4/22. You can watch your bird as you eat. I had breakfast and only paid $2.50 for two scrambled eggs with toast and hash browns. I don't know how they can do it.

   Jimmy also has a restaurant in Bridgeville, DE and has impressed Sussex County to allow him to open at the airport. Don't forget to try the pies. There must be ten different choices!

PIREP   My wife and I stopped in Georgetown Airport. They have a new terminal with a nice restaurant. A sharp looking place. They have just recently opened, and we were on a weather delay and needed a place to hang out. They were very nice and we had a sandwich. Lunch was very reasonable, served until 4 PM. Dinner was in the $12–16 range, and a Sunday brunch was also available. $4 landing fee if no fuel is purchased. With cloth napkins—give them four burgers.

Wilmington, DE (New Castle—ILG)

The Air Transport Command 🍔🍔🍔🍔

Rest. Phone #: (302) 328-3527
Location: On the airport
Open:

PIREP   Nice and friendly airport. Ask to park at the GA/Terminal ramp. No parking fee there if you're not staying overnight! You walk out through a gate and then it's a short walk to the Air Transport Command. I've been to a number of the chain and this one is about the same. You have to try the BEER CHEESE SOUP. That alone makes the trip worthwhile. I took a few nonflying friends and they got a kick out of watching planes land and take off. The restaurant is at the threshold of 1 and 32. Friendly service, good food with relatively large portions.

PIREP   AvCenter FBO at Wilmington-New Castle County provided their courtesy car to us to use to drive to the Air Transport Command Restaurant. We did top off the tanks on our C172. Bad news was a $14 charge to de-ice a slight coat of frost.

PIREP   I made an unscheduled stop at Wilmington (ILG) the other day and was pleased to find a cool restaurant right on the field.

   The Air Transport Command Restaurant overlooks runway 1 at ILG, and is decorated in a WWII motif. Lots of neat old black and white pictures, WWII memorabilia, and several fireplaces, as well as '40s-era big band music. This place is owned by Specialty Restaurants, the same company that owns the 94th Aero Squadron, and other similar theme restaurants. While I generally dislike chain restaurants, this one is better than most.

   The food is typical corporate/chain restaurant fare; decent, but nothing to write home about. In all fairness however, this seems more like a dinner place than a lunch place, and the dinner menu at least sounded better. I would eat there again. Overall, I would re-affirm the four flying burger rating.

PIREP   The Air Transport Command is located on the edge of New Castle County Airport, and its motif is WWII/1940s. The entrance to the restaurant is lined with WWII hardware—jeeps and the like—and the sound of planes taking off and landing nearby adds to the ambiance! Most tables offer a view of the runways.

   You can either leave the plane at the terminal and walk to the restaurant (about 1/2 mile along a busy road), or park the plane right by the restaurant. To park alongside the restaurant, however, requires special approval from airport security—you must either call ahead or stop at the terminal first for approval (approval cannot be given over the radio). They can be reached at (302) 328-4632.

The restaurant serves a full menu, with entrees starting at about $12. Their specialty is beef entrees, but they also have a nice selection of veal and seafood. Their warm, home-baked cracked wheat bread was delicious! Dinner for two will probably run about $35–$40.

There are a variety of other restaurants bordering the airport, including a TGI Friday's and a Wendy's.

PIREP Another plug for the Air Transport Command: excellent food, good prices, and Beer Cheese Soup—yummy!

PIREP I stopped at Air Transport Command recently for Sunday brunch. Found no tables would be available for brunch this day. No problem for me as I hangar at ILG. However, if I had already invested half that $100 hamburger price, I may have wished I had called for reservations.

If you find there are no reservations available and your heart is set on Sunday brunch, try (N14) Flying W in NJ. You won't give anything away for brunch except the 30 minutes it takes from ILG.

PIREP If you are anywhere near ILG, you have to eat at Air Transport Command. This is my favorite restaurant...period! Same goes for my wife. The Beer Cheese Soup, fresh cracked wheat bread, juicy steaks, cinnamon ice cream, and great atmosphere is only matched by the view!

You will not be disappointed, I promise!!!!

Arner's Restaurant 🍔🍔🍔

Rest. Phone #: (302) 322-3279
Location: On the airport
Open: Sun. 6 AM–10 PM, Mon.–Thurs. 6 AM–11 PM, Fri. and Sat. 6 AM–12 AM

PIREP I thought I would add my comments about Arner's Restaurant. We parked by the terminal (no fee) and walked along the highway north for about 0.4 miles. The walk does you good! Most any entree is good with a reasonable price. Veal parmesan with spaghetti is my favorite, though the crab cakes are a very close second.

Arner's is a very small chain of restaurants with only five locations, four of which are in Reading, PA where the original restaurant was located, so it still has that family-operated style.

Try the strawberry pie.

If you go away hungry it's your own fault.

Florida

Apalachicola, FL (Apalachicola Muni—AAF)

The Owl Cafe ⬤ ⬤ ⬤

Rest. Phone #: (850) 653-9888
Location: Cab ride
Open: Daily: Lunch and Dinner

PIREP In beautiful downtown Apalachicola, FL is a really nice restaurant called The Owl Cafe. The food is wonderfully prepared and when served, it is a piece of art, almost too beautiful to eat. The wait-staff are friendly, don't hover and bother you, but stand near enough to not let your water glass run dry. I recommend a coffee after, out in the garden and calling ahead for reservations on the weekend. To get there, fly in to Apalachicola muni (AAF) N29.43.45 W85.1.44. The people at the FBO are friendly and if not busy can drive you into town (four miles) or call the only cab ($6), but he is a little slow. The Gibson Inn is at the base of the bridge and is a real treat to spend the night. We go there for the weekend occasionally and you don't need a car to walk around town.

Arcadia, FL (Arcadia Muni—X06)

Wheeler's Cafe ⬤ ⬤ ⬤

Rest. Phone #: (863) 993-1555
Location: Cab ride to town
Open: Daily: Breakfast, Lunch, and Dinner

PIREP Wheeler's Cafe opened for business in 1929 and is still darn good. It is a small Mom and Pop cafe serving Southern home cooking.
Try their homemade desserts. Worth the trip.

PIREP Arcadia is about 20 miles west of Sebring in the middle of the state. It is a $3–4 cab ride into town. The older part of town includes numerous antique stores. One of the more popular spots is the old opera house. This upstairs facility is packed with antiques including many of the old projectors (silent and early sound) used in the theater. The deli up the street has excellent (large) sandwiches and delicious fresh pies ($1.50 a slice). All the people were very friendly.

PIREP   There was virtually no one at the airport to be friendly with. We had to hunt for a telephone book to look up the number of a local cab company for the ride into town. After a half-hour wait, the driver finally arrived but had no advice as to where we should head for breakfast. We had her drop us off in town amid the numerous antique shops, only two of which were open. We found no restaurants open and walked back toward the airport.

Cute town. Friendly folks, possibly. We just didn't see it. Perhaps Sunday was not the best day to visit.

Bartow, FL (Bartow Muni—BOW)

Kathy's Too 😋 😋

Rest. Phone #: (863) 519-5484
Location: 1/2 mile off field
Open: Daily: Breakfast and Lunch

PIREP   Restaurant off field about 1/2 mile. Not open on Sunday. I left hungry and went to Lakeland to eat.

Bunnell, FL (Flagler County—X47)

High Jackers 😋 😋 😋

Rest. Phone #: (386) 586-6078
Location: On the field
Open: Daily: Lunch and Dinner

PIREP   My family and I have been there several times in the last year, most recently March 4. The food and service has been excellent each time. On weekends the airfield is lively.

PIREP   A correction to an earlier pirep. High Jackers is not open for breakfast but only lunch and dinner. We got there Sunday morning at 10 AM and were told they did not open for lunch till 11. However, the excellent staff rustled something up for us a bit early that was delicious! Lots of traffic from nearby Embry Riddle University, so keep your ears open when approaching this small field. Park 50 feet away, great atmosphere and food…a great choice!!

PIREP   What a heaping amount of food! I almost had to redo my weight and balance.

This restaurant is only a few feet from the ramp with both outdoor and indoor seating. With a bunch of TVs, you can watch the game *and* watch the planes. What more can a guy ask for? I had the proverbial $100 hamburger which I could not possibly finish. It was simply awesome! Cute waitress, too. I will definitely be returning to HighJackers.

PIREP   I was looking for hangar space in the St. Augustine area and went to check out Flagler County (any excuse to fly works). We left St. Augustine headed south direct to X49. About 15 miles out tuned the unicom to 123.0; I heard no less than four aircraft in the pattern. For an uncontrolled airport, and not too clear a day, this was one busy place. At one time there was everything from a Beach Jet, a King Air, a C182, a couple Embry Riddle C172's and two Army OH-58's in the pattern. It was really worth getting in, though.

HighJackers was excellent. Great Cuban sandwich, great price, and more food than you should eat. (Earlier comment about weight and balance after eating is worth noting!) Unfortunately no hangar space and over 60 on a wait list with no plans for further expansion. High Jackers will definitely be on our list of places to frequent. Yes, the waitresses were cute and very friendly.

PIREP   Stopped by Flager County Airport (Bunnell, FL) today. Good food, good service, good-looking friendly cute waitresses—what more could one want!!!!

Well worth a stop when in the Daytona Beach area. Try it, you will like it!

PIREP Went to High Jackers at Flagler (Bunnell) Airport last week, was very impressed, huge improvement over the previous restaurant. They closed in the porch outside and the whole place was very busy for lunch. Great varied menu from hamburger to fish to chicken. Great staff, good service, and good food at reasonable prices.

Carrabelle, FL (Carrabelle-Thompson—X13)

Summers at the Bridge 🥪🥪🥪

Rest. Phone #: (850) 697-3663
Location: Close by; they will pick you up
Open:

PIREP Summers at the Bridge will pick you up if you call ahead ((850) 697-3663) and take you to their restaurant. The same cook who used to cook at Julie May's is now at Summers. As we all know, Julie May's is now closed.

I have not eaten there personally but was told by a friend that the food is good and reasonably priced.

Cedar Key, FL (George T Lewis—CDK)

Sea Breeze 🥪🥪🥪

Rest. Phone #: (352) 543-5738
Location: 2 miles to town, cab meets you as you land
Open:

PIREP FBO Comments: No fuel, so plan accordingly. 2300' runway 75' wide I think. No problems, no obstructions, and well lit at night. Road parallel looks like a taxiway but isn't. Surprised there aren't any noise abatement procedures for this airport.

PIREP Parking area and ramp are rough, as is the old Checker Cab that will pick you up. Held together with a few bumper stickers and a couple rolls of aluminum tape, but still runs good! The lady listens to the CTAF—she tried to respond with her handheld when she heard us but we never heard her. She operates the taxi during daylight hours for the most part (the headlights are a little finicky), $5 per head per way into town. Quite an interesting ride if you're not used to old cars and small-town FL waterfront homes. The food at Anne's was delicious, mostly seafood of your choosing, but there is a burger on the menu. Highly recommended are the two-way or three-way platters where you can combo shrimp or grouper or whatever you like. Meals come with a good-sized salad or side-sized serving of coleslaw. They have seating inside or out with views of the Gulf. There was a pack/herd/pod of manatee that swam near the dock as we were eating—pretty neat. About $16–20 a person for food at Anne's.

Beautiful scenery in the afternoon and evening. Kind of a touristy area so the downtown shops look like it. Two or three blocks to walk the whole downtown strip once at Anne's or Frogs (the other restaurant).

No reports on lodging as we were only there for the afternoon. Looks expensive though.

Leaving Cedar Key on a moonless night, be good on your basic instrument skills as taking off in either direction there are few lights for horizon or ground reference.

Had a great afternoon and evening at Cedar Key. I'm really glad the days are getting longer!

PIREP We flew VFR to CDK for the first time. There's a windsock at the end of each runway (05/23). Judy the local cabby monitors the CTAF and will pick you up in her white Checker Cab. She's pleasant and a wealth of local info. Cab fare is $5 p/p each way. We wanted seafood. Judy recommended Anne's Restaurant. We waited about five minutes to be seated though the entire restaurant, including the outdoor deck, was no more than three-quarters full. All the servers appeared a bit overwhelmed, perhaps short-staffed? Our young, pleasant, preoccupied server finally got our drink order straight after repeating it three times. We both ordered fries with our meal; we got the worse tasting coleslaw either of us ever had. When our oyster sandwiches (delicious and hot) arrived,

I received barely warm fries. My wife received none—her fries arrived after she had finished her sandwich, and she's a slow eater, but at least they were hot. The tab was $30 plus tip. I also had oyster stew, which was delicious at $6.95 a bowl. The fish chowder is a pricy $9.95 per bowl though the water views are priceless, and remember: Cedar Key is a tourist trap.

Enough said.

PIREP   Don't worry about how you'll get into town. Judy the cab driver has her radio on and when she hears you call, she'll radio back with the runway in use and also ask if you need a ride into town. The ride is $10 per person, round trip and is well worth it. It's way too far to walk (maybe three miles) in the heat. Besides, Judy is an interesting person and she needs to make a living. Her service is one that is rarely seen anymore, so let's keep her working. Judy is a great source of local trivia and current restaurant recommendations. Almost any place along the waterfront will be good. For a special treat, have Judy drop you off on Main Street and visit the Island Hotel for lunch. This place is on the National Register of Historical Places (or whatever it is called) and the new owners (within a year) are taking it back to what it used to be like. Great food and service. A bit more expensive than run of the mill places but not bad at lunch.

PIREP   Flew into Cedar Key on a Saturday. Even with the strong crosswind it was an easy landing. Plenty of tie-downs. When we called on the radio, the cab lady gave us a wind report. Very nice service. She also picked us up and took us to town. $5 each one way. She recommended Sea Breeze or Anne's. We chose Anne's and the food was great. Excellent fried oyster sandwich! Nice little town. Lots of tourist trap shops. Great place to check out!

PIREP   Everything that the pilots have written above is correct. Don't let the smaller runway keep you from enjoying this wonderful place. The most important thing to remember is to save this airport for clear days. MVFR is not recommended because of the intense likeliness to lose the horizon. However, when it is VFR it is a great place to fly. Try the Island Hotel & Restaurant. It is right next to all the shops and action and it serves the best food on the island (or key). It is also a great place to stay the night. While not the cheapest place in town, it is the best for dining and staying overnight. The owner will come out and pick you up for free at the airport if you let him know you are coming.

PIREP   My wife and I have been flying to CDK for several years. Besides the Captain's Table, there are many other restaurants on the island that have great food too. There is a little Mom and Pop place on Second Street that is open for breakfast and lunch that is inexpensive and where locals eat. If you want to really impress your sweetie, the Island Room is the only place for dinner. It is located at the Cedar Cove Resort and Yacht Club. You will be impressed! Pilots, don't let the short runway scare you away. I have flown in and out with a Cherokee Six with all seats full without a problem.

A couple of weeks ago there was a Pilatus in and out with eight aboard. Just remember that nighttime operations, unless instrument rated, should not be attempted. You can buzz the town with your gear down to alert the taxi or give a call on unicom five minutes out to get the cab there. The cab used to have an aircraft radio in it, I don't know if it still does or not. The brochures and chamber tell you to go to the airport to see a great sunset. Unfortunately, there is a county maintenance worker there that thinks the airport belongs to him and will try to chase you away. Don't park where you can't just to appease the yahoo. There is a public road on the northwest side of the airport that should provide excellent photo opportunities. If you are staying overnight, the L&M bar is a good place to wet your whistle.

PIREP   Cedar Key is a small town on the west coast of Florida about 80 miles north of Tampa. The scenery flying into Cedar Key is great, with miles of empty shore and palm trees. The area is dotted with small islands with pure white sand. The runway is 2400' and both ends of the runway end in the ocean. Landing on runway 5 brings you in over the gulf.

Note: you have to back taxi on the runway if you land on 5—this is a change from when you could taxi off the end of 23 onto the road. Now they have a stop sign at the end of the turnoff from 23 and it is difficult to taxi past.

The town is an old fishing village with many quaint shops and restaurants. It would be worth spending the night here at one of the bed and breakfast–type places. You can take a boat out to the islands about one mile offshore and explore for an afternoon. Locals in town were very friendly and accommodating—a couple of nice bars with pool tables, but this town does not get rowdy.

I give this destination 7 clams out of 10 clams and most of that is for the cool runway. You can also camp out on the side of the runway if you want!

PIREP   The wife and I did an overnight trip to Cedar Key and enjoyed it. "The" cab driver (Lester Ridgeway) monitors CTAF in his cab and at home and will suggest a runway. His fee for the five-minute trip to town in the Checker Cab is "whatever you think is right." If he is busy, his wife may come get you. The short field over some low trees onto a displaced threshold with a Gulf of Mexico overrun is fun. Three cars came out from town to watch me land and drove off while I taxied in. There is no fuel or any services available.

There are four or five seafood restaurants in town with similar pricing, and some serve breakfast, lunch, and dinner. After a tedious process to get a seat, we ate at the Captain's Table for about $50 including a drink each, tax and tip. The seafood was excellent. I would suggest calling ahead for a reservation and attempting to get a window seat for sunset (the restaurant is located on the second floor above a lounge). There is another restaurant (Brown Pelican), which offers a second floor screened porch view of the Gulf.

There are also several hotels on the island. The hotels, restaurants, and gift shops are all within a very short walk. Most are on the dock. If you want to, you can rent a golf cart to ride around, rent a boat to go out to the outer islands, or go on an airboat tour of the area. The big events are an art festival and a seafood festival that occur in April and October, respectively. There is no real beach on this island, although it looked like there were some nice ones on the outer islands, which you may be able to get to by boat.

The taxi man has an attitude with pedestrian tourists who walk in the middle of the road, but he's very entertaining!! Ate at the Sea Breeze. The food and service were excellent. I had an outstanding broiled seafood platter. If you ever are in the CDK area, don't miss the chance to stop in.

Clearwater, FL (Clearwater Air Park—CLW)

The Expo Restaurant 🍔🍔🍔

Rest. Phone #: (727) 461-4140
Location: 5-minute walk
Open:

PIREP   The Expo is less than a five-minute walk from the FBO at CLW. Go out to the street, and walk a block north.

The Expo serves home-style food at good prices. There are daily lunch specials. The breakfasts can't be beat. My personal favorite is the gyro and feta cheese omelet. Portions are large and everything is delicious. It is open seven days a week, although Sundays they serve only breakfast and close at 2 PM.

Clearwater Airpark is a good airport. Runway 34 has right traffic. The city has made it illegal to fly after 9 PM.

Crystal River, FL (Crystal River—CGC)

Plantation Inn 🍔🍔🍔

Rest. Phone #: (352) 795-4211
Location: You could walk, or they will pick you up
Open: Daily

PIREP   Call the Plantation Inn from inside the terminal for a pickup; the phone number is next to the wall phone. Lunch menu includes both sandwiches and excellent entrees at very reasonable prices. Walking would not be fun as it is almost a mile to the north along a very busy highway.

Franchise Heaven 😋😋

Rest. Phone #:
Location: Walking distance
Open:

PIREP   Just had lunch at the Olive Tree Restaurant, only a short walk past the Dairy Queen about 200 yards north on the east side of US 19. Good, clean, solid Greek food (I had the gyro; it was great), friendly service, and we stopped by the DQ for an ice cream cone on the way back. Lunch for two with tip was $20.

PIREP   Get a Blizzard at the Dairy Queen (visual distance from ramp) or hop a ride over to the plantation with two restaurants, boat rentals, and golfing. Don't rent a canoe if it's windy—get a flat bottom and go look for manatees on the river.

Nicole's 😋😋😋😋

Rest. Phone #: (352) 564-2300
Location: Short walk
Open:

PIREP   Nicole's is across the street from the airport. It's a short hike from the ramp but very well worth it. It's an upscale restaurant with a nice atmosphere and excellent food. Expect $50 to $60 plus gratuity for two people. If you like grouper, try the Grouper Bostonian—delicious!

Defuniak Springs, FL (Defuniak Springs—54J)

Vince's Ciao Italia Cafe 😋😋

Rest. Phone #: (850) 892-2114
Location: 3 miles
Open:

PIREP   Vince's Ciao Italia Cafe's new location and phone number: 171 Country Club Lane, (850) 892-2114.

PIREP   The restaurant in Defuniak Springs, FL that has all kinds of great homemade foods is a place called Vince's Ciao Italia Cafe at the Hotel De Funiak. Great homemade desserts, too. In August it will be open for breakfast, lunch, dinner, and also for coffee and dessert in between the meals. Everyone, check it out !! The hotel is small and plush and is listed as a bed and breakfast. We enjoy this stop very much!!

PIREP   Almost antebellum town located halfway between Ft. Walton Beach (and Eglin AFB) and Panama City Beach—a little north of beach view but right off I-10 (so IFR pilots have no trouble). The city airport still had a mechanic last time I was there. You can hitch a ride into town for eats at any number of delightfully quaint or rugged places (as well as fast fare, as everywhere).

Deland, FL (Deland Muni-Sidney H Taylor Field—DED)

The Perfect Spot 😋😋😋

Rest. Phone #: (386) 734-0088
Location: On the field
Open: 7:30 AM–Dinner

PIREP   I met friends for lunch at the Perfect Spot on January 16, 2006. Thanks to a previous pirep, I had no trouble in locating it and parking reasonably close by. The restaurant did provide a nice view of the skydivers; however, the food and service was at best average. I probably won't stop there again; there are too many better places.

PIREP   We flew over to Deland and went to the Perfect Spot and had lunch. They have a great selection of hamburgers, sandwiches, and salads; they even have some veggie dishes. There was a lot of energy in the place with all the skydivers around. After a great lunch we went out to the deck and watched the skydivers practice on the trapeze. While inside it was very entertaining to watch the videos of all the skydiving. Personally I cannot see the point of jumping out of a perfectly good airplane. It is a bit confusing as to where to park to go to the restaurant; we parked in front of the row of T-hangars. You have to look for the skydiving flags to spot the restaurant at the east end of the field. You must try one of the smoothies made with ice cream—what a nice finish to a good meal. Since the four of us were in a 172SP we had to jog for about 30 minutes to burn off some food to stay within our weight limits!!! (Only kidding, the runway is plenty long!)

PIREP   The old snack bar at Skydive DeLand is no more. It has been replaced with a restaurant and bar called The Perfect Spot. It opens seven days a week at 7:30 AM and is open through dinner. It seats over 100 inside and has a 5000-square-foot outside deck.

Come and check it out—the food is great and the view is fantastic.

Airport Restaurant and Gin Mill 🍔🍔🍔

Rest. Phone #: (386) 734-9755
Location: On the airport
Open: Daily: Lunch

PIREP   The Airport Restaurant and Gin Mill has simply the best burgers in this part of the state. They are fresh, tasty, and juicy—two napkins won't be enough. It is adjacent to runway 5, south of the jump area. You can sit indoors or outside on the deck, where you'll have a good view of the crazy parachutists.

Destin, FL (Destin-Fort Walton Beach—DTS)

Back Porch 🍔🍔🍔🍔

Rest. Phone #: (850) 837-9775
Location: 2 miles from the airport
Open:

PIREP   I had lunch at the Back Porch during my long solo cross-country flight and I loved every minute of it! I had no problems getting a ride from the FBO desk staff and it's only about a three-minute car ride. The restaurant's right on the beach and my table was right on the water. With the windows open and the sea breeze lazily blowing on my face, I enjoyed my delicious grilled mahi-mahi sandwich and fries, which came out very quickly. I was highly impressed with the service and quality of food and would definitely go back. If you are flying anywhere around the Panhandle, please do yourself a favor and stop by the Back Porch!

PIREP   The Back Porch is great. Another outstanding restaurant is McGuire's. You will never go wrong with ordering a burger. They have seven to choose from.

PIREP   When flying into Destin, be aware that there can be congestion from the air bases as well as the airport itself depending upon the time of year, so plan that into your arrival times. As for the $100 hamburger, having lived and served there myself, I can say that the Back Porch *is* great, as well as several other locations. The great thing about the area is that the local sheriff's office is very aviation minded and responsive to your needs. Ask them where the best deals to eat in town are that day.

PIREP   One of the best places in town for a fish sandwich is the Back Porch, located right on the Gulf of Mexico and about two miles from the airport. There is a courtesy car available, or call a taxi—cost is around $4 round trip. Truly a fine place to kick back and enjoy the sights as well as the food.

PIREP   Back Porch? Definitely! The service was extraordinary but I kept getting distracted by those 22-year-old ladies in their bikinis walking by about six feet from me. I think the food was great.

Everglades, FL (Everglades Airpark—X01)
The Oyster House 🍔🍔🍔🍔

Rest. Phone #: (941) 695-2073
Location: 3/4 mile, they'll pick you up
Open:

PIREP   Just got back from Everglades Airpark (X01), which is located on the edge of the Everglades on the Florida west coast. My wife Maria and I have gone there several times, and I take my students there for XC training and short field landing practice. The runway is 2400 × 50' and surrounded by water on three sides. Full flaps and touchdown on the numbers and on airspeed is a must. This is a great place to visit. The FBO is county managed. Very helpful and accommodating folks.

Several good restaurants within biking distance, or they'll send transportation to get you. Ate lunch at the Oyster House once. It was pretty good. I highly recommend this location.

PIREP   The nice thing about this place is its proximity to the airport. One could walk from the approach end of runway 33, but in the summertime the mosquitoes would get you before you got your grub.

It is better to call the restaurant, (941) 695-2073, and ask them to pick you up. Across the street from the restaurant is a national park offering sightseeing boat rides and other fun things.

The food here is about average for places you can find on or close to an airport, which is to say, barely adequate. When we were there, the place was not at all busy, but the wait service was slow, although the cook got the food out quickly. Sample prices: hamburger $5.95, grouper sandwich $8.95, fried oysters $16.95, and fried or steamed shrimp $17.95. Nearby is a national park, all kinds of charter fishing outfits, and some places to stay. Come in the winter and don't expect posh.

PIREP   Three planes in our Phantom Squadron group went to Everglades Airpark Airport (X01) this past Easter Sunday. The attendant at the airport was most willing to help us with complimentary transportation to the restaurant of our choice, but we opted to walk the very short distance through the woods to The Oyster House.

Sample food and prices: fresh grouper sandwich $7.95, seafood basket (shrimp, oysters, clams, fish) $11.95, bay scallops basket $9.95, charbroiled boneless chicken breast sandwich $6.95. All included fries, coleslaw, and a hush puppy. Endless pitchers of pink lemonade and ice tea for $1. We were all happy with the food.

PIREP   Everglades Seafood Fest was great with lots of food and crafts. Everglades Airpark (X01) is the last airport on the west side of Florida traveling south. The airport has greatly improved—there's now an FBO (of sorts) in a brand new building, clean bathrooms, 24-hour gas, phone, Coke machine, fenced-in airport, adjacent police department. David Blalock is the airport supervisor. They really like airplanes to visit!!!!

Go through the bushes—there is a recently constructed gate there marked to be open from 8 AM to 5 PM on the approach end of runway 33—and cross the street to the Oyster House Restaurant. You'll get the freshest seafood around, or a burger if you like. I recommend the grouper sandwich. It is delicious!!

PIREP   I just returned from Everglades City and must say that I had a wonderful time. I ate at the Oyster House. Transportation provided by Mr. Miller, the owner. Just call once you arrive at the airport ((941) 695-2073). If no telephone is available (no cellular service or office is closed) it is a three-quarter-mile walk. To my surprise, the restaurant also owns quaint cabins that are adjacent to the restaurant. They offer boat rentals. Slightly pricey but well worth the ambiance! If you are looking to relax in a quiet island-like atmosphere, this is the place!

City Seafood 😋😋😋

Rest. Phone #: (239) 695-4700
Location: They'll pick you up
Open: Daily

PIREP   You can fly to Everglades City, FL (X01). 2500' × 50' asphalt, taxi to the pay phone, and call City Seafood (the number is permanently affixed to the booth). They'll pick ya' up in five minutes and take you to the docks, where you can buy fresh Florida Stone Crab claws, *right off the boat*! They sell refreshments and dipping sauces too!

Bring a cooler and take some home with you. Just watch out for the gross weight! The airport is located on the edge of Everglades National Park, and there are many local attractions in Naples and Miami, not to mention the Keys…but that's another burger.

Fernandina Beach, FL (Fernandina Beach Muni—55J)

Brett's Waterfront Cafe 😋😋😋😋

Rest. Phone #: (904) 261-2660
Location: 3-mile cab ride
Open: Daily: Lunch and Dinner

PIREP   It was a great burger run. We landed at Fernandina Airport (55J) and were greeted by the FBO: McGill Aviation (904) 261-7890. They were extremely nice and helpful. They called a cab, which ran eight bucks, to the historical district. We ate at Brett's Waterfront Cafe, which is reasonably priced—between 6 and 12 clams on an average. The historical district is extremely nice and dotted with numerous gift shops and other restaurants.

The Intercoastal Waterway goes right by the historical district and adds an old-world fishing charm to the place. I would give Brett's Waterfront Cafe four burgers on the rating scale—really nice.

Fort Lauderdale, FL (Fort Lauderdale/Hollywood Intl—FLL)

Aviators Tavern & Grill 😋😋😋😋

Rest. Phone #: (954) 359-0044
Location: On the airport
Open: 7 days a week, opens 7 AM

PIREP   I am forever stopping at FLL to clear customs on my way back from the islands and have found a very friendly stop at Jet Center. It is a very easy airport to negotiate (always get 9R-27L), and I must say the restaurant there is top notch. Parking at Jet Center Terminal includes Customs and the restaurant. The facility is professional and caters to C-150's to 747's. The greeting is the best and I always feel welcome. The restaurant always has daily specials and the prices are very reasonable. The tastes are great. While dining, you are looking out on the ramp areas.

PIREP   Fort Lauderdale International Aviators in Jet Center opens 11 AM 7 days. I vote four burgers, very nice.

Fort Myers, FL (Page Field—FMY)

Mel's Diner 😋😋😋

Rest. Phone #: (239) 275-7850
Location: 1 mile off the field
Open: Breakfast, Lunch, and Dinner

PIREP   Mel's Diner remains the best local/convenient burger joint by the field. Despite growth and remodeling, it still maintains great value, fast service, and long loitering time allowed. Head north if it's hot, about half a mile, and you will find Rita's Ices (indescribable). Once you're full, you can go north one mile to Edison Mall, half-mile north to Page Field commons (Hops, Best Buy, Books A Million) or approximately five miles south to several barbecue/sports bars and Barnes & Noble.

If you're really full, go back to the aviation center pilot lounge for a nap in big overstuffed recliners, watch satellite TV, or access high-speed Internet. On Fridays, there are free hot dogs, sodas, and Hilton Hotel cookies. Be warned that the local airport bums come out that day also. You might learn something about aviation if they do, as some are war veterans, fliers from the '30s (a couple are over 80), as well as hangar builders, bums, and millionaires. The trick is to correctly guess who is who. If you forget to get a gift for the family, the well-stocked aeromart at the center has a nice selection, including jewelry.

There is also EAA Chapter 66 on the field. Meets second Tuesday at 7 PM. Pancake breakfast first Sunday at 8 AM. Warbirds Sq 24 members are also with Ch 66. Civil Air Patrol FL 040 Squadron is based on the field too.

PIREP   FMY is growing by leaps and bounds, but come on down. Your best bet is Mel's Diner down the road from Page on US 41. Also, try Rita's Ices. Edison Mall is down about a mile. EAA Ch 66 has a $5 pancake breakfast/Young Eagle Fly Day, first Sunday of each month. You can get seconds and thirds. The Outback is busy and great, and there are over 15 Chinese restaurants as well as Hispanic eateries all within a taxi drive. You can contact EAA 66 for assistance. We also have a CAP unit in town.

Call the FBO for wheels or take a taxi or walk about five minutes to Hwy 41, then take a bus or five-minute walk to restaurants on either side of town.

Most of the time, people at the field will be willing to give you a ride. If you come in on Young Eagle Day, call or come over to EAA Chapter 66 and we will drive you.

PIREP   Not only is Mike's Landing closed, but the building has been knocked down. We now have no restaurant on the field. However, there are several great diners just blocks from the field: Mel's Diner, Hoops Restaurant, and so on.

Fort Pierce, FL (St Lucie County Intl—FPR)

Frankie's Flight Deck Cafe 🍔🍔🍔

Rest. Phone #: (772) 489-4418
Location: On the field near departure end of 14
Open: Lunch Mon.–Sat. 11–3, Dinner Fri., Sat. nights

PIREP   Closed. No longer in operation according to the Airport Tiki.

PIREP   What a great little place for lunch! Stopped by to check it out—heard it was good, thought I'd find out for myself. Not only is the po' boy sandwich done right but great service to boot. They keep the ice tea flowing also.

PIREP   FBO Comments: Frankie's Flight Deck Cafe is right next to Chevron self-serve, just north of the departure end of runway 14.

PIREP   I have eaten at Frankie's Flight Deck Cafe several times since he opened. I must say the menu is superb. Originally from East Texas, I have had my share of Tex-Mex and South Louisiana Cajun cooking. Frankie has definitely mastered both. I recommend the Muffalotta sandwich or the sausage and pepper po' boy. But no matter what you get, you have to have the Bananas Foster for dessert. (NOTAM: You may have to be carried back to your plane after dessert.) I haven't eaten there in the evening, but I hear the live jazz bands are good and the beer is cold.

As a nice touch, Frankie has about a hundred different pepper (chili) sauces to choose from. Nice view of the airport.

PIREP   Frankie is the new chef *and* owner of what was once the Flight Deck Cafe. Frankie's logo is "First class view, first class food, with a South Louisiana touch." From hamburgers to crawfish, everything we've ever had is always fresh and plentiful, not to mention absolutely mouthwatering! His family are all originally Louisianans but have been here 20 years. They are open for lunch Monday through Saturday 11–3 and on Friday and Saturday nights. We love to listen to the jazz jam on Friday nights and enjoy the sweeping view of the airport lights. My family strongly recommends Frankie's for delicious food and really good service. Everyone seems to either be related or a close member of the family. On Friday nights the locals line up to say hi to the Lott family and have a great meal. The place is all newly glassed in with a cool bar that serves beer and also wine.

The Airport Tiki 😋😋😋😋

Rest. Phone #: (772) 489-2285
Location: On the field, next to Customs
Open: Breakfast and Lunch

PIREP   Very nice place. Good view of the runway. Lunch was fast, tasty, and inexpensive. Service was excellent. I highly recommend the blackened mahi sandwich followed by the peanut butter pie. Mmm mmmm good. After landing, ask to "taxi to the Tiki."

PIREP   I fly into FPR frequently. Good dinner-style food and inexpensive. Reminds me of a Perkin's or Denny's style of restaurant. Service has always been good and friendly. At lunchtime the place is busy because they have two large flight schools on the field plus customs next door, but they somehow handle the crowds very well. The food is awesome!

PIREP   FBO Comments: Full service, next door to Customs, Port of Entry, rent cars, charts, plates, fresh Florida oranges by the sack. Located south of the departure end of runway 14, taxiway C. Just ask ground for the Tiki; it's behind customs by the BP sign.

PIREP   The Tiki restaurant is always a good choice. The menu is diverse. Breakfast is excellent, good coffee (bottomless cup). For lunch, I recommend the hamburgers (your choice of regular or spicy fries), crab cakes, mahi-mahi (dolphin, dorado), or the Caesar salad. Daily specials.
   NOTAM: Breakfast is always on time, but expect some delays during the lunch rush—it's a busy place, popular with the locals and with students and employees at two large flight schools based at KFPR.
   There is a pizza and sub shop inside also. You can call or radio your order in if you're in a hurry. (Sorry, I don't know the number or frequency.) Service can be slow at times but always friendly. Nice view of airport.

PIREP   Just got back from having lunch at Tiki, Ft. Pierce, FL (FPR). Well worth the trip. Excellent home cooking. Plastered on the wall is a quote from *The Wall Street Journal*, "Amazing hamburger." One of our party had the burger and confirmed the appraisal. We had Maryland crab cakes. They were delicious. Look for the big BP sign.

PIREP   There is another eating place at FPR: Tiki. Has become well known as the kickoff point for trips to the Bahamas. They will rent you a raft and life jackets, and provide advice on the paperwork for entry into the Bahamas. The food is tasty and reasonable—always seems more reasonable on the way back from the Bahamas than before you leave. Located next to the Customs Office.

PIREP   The Airport Tiki is next to U.S. Customs at FPR and has the best burgers around. Open breakfast and lunch; just walk next door after clearing Customs. Clean and bright, with wheelchair access.

Gainesville, FL (Gainesville Regional—GNV)
The Original Sonny's BBQ 😋😋😋

Rest. Phone #: (866) 697-2872
Location: Just outside the airport
Open: Daily: Lunch and Dinner

PIREP FBO Comments: Clean FBO, courtesy car available with fuel purchase.

PIREP Probably the closest restaurant to the field is Sonny's Barbeque. This restaurant has become a chain in the south, but the one on Waldo Road that borders the airport is *the original*!

Sonny's has some of the best barbecue and prices around. When I was in school at UF we would go to Sonny's pretty regularly. One idyllic Saturday we even got to see the main man Sonny in person, cooking in the kitchen with his cowboy boots and fist-sized belt buckle.

The Copper Monkey 🍔🍔

Rest. Phone #: (352) 374-4984
Location: 5 miles
Open: Daily

PIREP FBO Comments: University Air Center is the only FBO located on the field. They are very professional and have a beautiful building. They have a van to drive you anywhere you need to go.

PIREP In my opinion, the Copper Monkey has the best burgers in the state. I have been eating them since 1990. I recommend the cheeseburger with fries. The portions are huge. The burgers are made with 100 percent Angus beef and weigh half a pound. You will be very filled after eating. The Monkey is located about six miles from the field on University Avenue across from the University of Florida. The FBO will give you a ride with no problem.

Indiantown, FL (Indiantown—X58)

Seminole Country Inn 🍔🍔🍔🍔

Rest. Phone #: (772) 597-3777
Location: Off the airport; they'll send a van.
Open: Sunday Brunch is *best*

PIREP The Inn is a very good place to eat Sunday Brunch. The staff is friendly and attentive, the food is great, and the place itself is neat and clean. It is an actual inn (hotel) that has the Duke and Duchess of Edinburgh in its history.

A call to the Inn from the FBO will get a van dispatched to pick you up in short order. The hours are "normal business hours," but the brunch ends around 1500.

Inverness, FL (Inverness—X40)

Kimberly's Ice Cream Parlor & Grill 🍔🍔🍔

Rest. Phone #: (352) 344-4222
Location: Off field, 15-minute walk
Open:

PIREP Stopped in last night. I would call it more like a 15-minute hike, but close enough. Wings were good with a home-cooked special each day. Open seven days a week and breakfast till 11 AM.

PIREP   Next to the airfield there is a speedway practice ground where they test their homemade racing cars. Walk five minutes into town, opposite the funeral parlor, and check out Kimberly's. Super chicken wings. Sundaes and other delights at very reasonable prices. Friendly service, nicely decorated place reminding us of the good old '60s.

Key West, FL (Key West Intl—EYW)

Crabby Dick's 😋😋😋😋

Rest. Phone #: (305) 294-7229
Location: Off airport, take the bus!
Open:

PIREP   When the price of fresh oysters in Naples gets stratospheric, you can fly 45 minutes to Key West and eat them all day long at Crabby Dick's for $3.95/dozen. They serve ice-cold O'Douls (light only, no amber), and the service is great. Sit on the second floor porch over the sidewalk and watch the natives and tourists amble by. The Duval Street restaurants are an interesting 30-minute walk or an overpriced $15 taxi ride.

PIREP   Key West (EWY) certainly boasts more restaurants and bars than most anyplace in such a small area! Most are located in Old Town which is a $6 cab ride (3 miles) from the airport. Have been going there for years; we like some of the less expensive off-Duval or locals places like PT's Late Night on Carolyn Street, 7 Fish just off Simonton, and Alfresco's on Appleruth Lane; all officially in Old Town.

There is little doubt KW is one of the best tropical $100 hamburger places to go in the U.S., bar none! Countless world-class haute cuisine eateries exist, as well as street-side vendors like B.O.'s Fish Wagon, which servers memorable fish sandwiches. Few flying folks I know only go there once!

PIREP   Key West has it all, including polite Americans plus the best freak show every evening at Mallory Square. Park your plane, get a room for the night, go to Duval Street, stop by Fat Tuesday, and get a cooler of pina colada and head down to the old Navy Pier (Mallory Square) and watch the show and the sunset. If you have time, take a snorkeling trip or cruise. During the evening, have a beer at Sloppy Joe's and a cheeseburger at Jimmy Buffet's Margaritaville. Don't buy too many T-shirts or you will not get off the ground.

PIREP   FBO Comments: Island City Flying Service charges a $10 ramp fee, waived with topoff. *But:* fuel usually 50 cents more expensive than on the mainland.

PIREP   If you fly into EYW for $100 burger, don't spend your valuable burger money on a taxi—take the bus. The grossly under-advertised city bus service runs the Red, Green, and Blue buses through the airport to downtown for $1.25! The bus stop is at the far east end of the airline terminal (the end where the airport restaurant is) and is very poorly marked—the sign faces away from the terminal.

Schedules can usually be found in the airport restaurant or at www.KeyWestCity.com.

Conch Flyer 😋😋😋😋

Rest. Phone #: (305) 296-6333
Location: Cab ride
Open: 24 hours a day

PIREP   After a trip to Key West from Fort Lauderdale, weather service advised returning ASAP due to incoming storms from the southwest. A quick meal from the airport restaurant, Conch Flyer, was a delicious and modestly priced surprise. Soup, fried conchs, atmosphere, and great service made the first time visit a pleasure.

PIREP   Yes, there is a restaurant at the Key West airport, the Conch Flyer. Open 24 hours a day (like the Boca Chica Lounge on Stock Island).

La Belle, FL (La Belle Muni—X14)

Flora and Ella's 🍔🍔🍔

Rest. Phone #: (863) 675-2891
Location: A couple of miles off the field
Open: 10 AM on weekdays and 8 AM on weekends

PIREP   FBO Comments: Friendly. Courtesy car was available.

PIREP   You will need transportation to the restaurant unless you would like to walk the two and one-half to three miles. Food is very good. Pies are delicious. Saturdays they may have music. Worthwhile trip.

PIREP   You won't need the loaner car to get here. It's right across the street. This very clean and pleasant Mexican restaurant is open at 10 AM on weekdays and 8 AM weekends for the best breakfast buy in Southwest Florida! $4 to $5 will get you a four-course breakfast, and it's more than you can ever finish! They'll even let you share one breakfast for two! And that gets the price down to ridiculously minimal. I hope you'll help that deal out with a generous tip, now that I've let this one out of the bag!

PIREP   Had a fantastic time at Flora and Ella's Restaurant in La Belle. Don't miss the meringue pies. They'll pack one to go if you wish. The FBO has good fuel prices and the owners were nice enough to let us use their van to drive to the restaurant. The restaurant is now open Sundays from 7 AM to 3 PM.

PIREP   Near the airport (the restaurant will pick up or the FBO can arrange a ride) is Flora and Ella's Restaurant, known for the gigantic pies piled with meringue. Located in a new Florida Cracker–style building on State Road 80 in La Belle, it is a landmark in the area for good eating and where the local politicos congregate daily for breakfast and lunch. There is a nice gift shop in the lobby. Open daily except Sunday.

Lake City, FL (Lake City Muni—LCQ)

Ken's Bar-B-Que 🍔🍔🍔

Rest. Phone #: (386) 752-5919
Location: 1 mile off the field
Open: Daily: Lunch and Dinner

PIREP   Ken's is a chain of barbecue restaurants in North Central Florida. I am not alone in preferring another barbecue chain in the area called Sonny's. Neither Ken's, Sonny's, nor any other restaurant is conveniently located to the FBO at LCQ.

PIREP   Lake City is definitely worth a stop to take in Ken's Bar-B-Que. The airport has a nonfederal control tower that operates part time Monday through Friday. Even though they have a tower, it is not Class D airspace. They have some huge runways, for example 10/28, 8002 × 150', since, like many Florida fields, it was a WWII base. You may even see a Value Jet plane or two parked there since Aero Corp is one of the remaining approved maintenance facilities.

Ken's Bar-B-Que is located approximately one mile from the airport and just off the campus of Lake City Community College. This is the best barbecue I have had outside of Memphis. The prices are phenomenal, and Wednesday is all-you-can-eat chicken day. The service is good and everyone is very friendly. You will certainly get your fill. This definitely is worth a stop.

Lakeland, FL (Lakeland Linder Regional—LAL)

Tony's Airside 🍔🍔🍔🍔

Rest. Phone #: (863) 644-8684
Location: On the field
Open: Breakfast, Lunch, and Dinner, 7 days a week

PIREP   We make dinner runs several times each week, weather permitting, and we try to fill the plane with any and all willing participants. We usually make the stop at Tony's in Lakeland as it is good, a quick flight from X14 and FMY, and they are open until 9 PM. Especially during season here, it is just as quick and easy to call our friends after work and hop in the planes to meet at Tony's as it is to try to get into the local Outback!

PIREP   FBO Comments: Great team, friendly, cooperative, tower crew great.

PIREP   Give it four to five hamburgers. Just great value, fabulous service, food better then 98 percent of airport, on-field restaurants I have eaten at. The tuna steak salad was superb—would cost $18 anyplace else; at Tony's, $7.50.

PIREP   FBO Comments: Piedmont Aviation is one of two FBOs on the field. Recommend Piedmont due to the easy access from the taxiways.

PIREP   Tony's has great food. I recommend the Top Gun Burger with cheese and fries. This will run around $7. The staff is polite and very quick in making the food. The restaurant is located upstairs in the Piedmont building. You overlook the airfield as you are eating.

PIREP   After a long weekend of mostly flying (insert harrowing tales of waiting for T-storms to dissipate, then landing to refuel only to find the FBO abandoned for the night—if you ever visit Savannah, GA after 0030, be reassured there is a hotel within walking distance), the next day we finally got to Key West (yay!) and had a lovely vacation of less than 24 hours. We were headed homeward when an airplane-eating storm chased us to ground at Lakeland. (Checking the radar later, the entire state of Florida was one big t-storm!) So we made the best of it and decided to catch dinner while we waited (for what turned out to be something in the way of 5 hours, but that's another story). The taxiing seemed to take forever (the restaurant is on the other side of the tower) but, as in most cases, it was worth it. The new terminal is very nice—lots and lots of space. The restaurant is upstairs and so very aviation themed! There are murals on every table, as well as a huge one on a wall. There is even a propeller shape in the glass over the door as you walk in.

   Our server's name was Airic, which we thought was either going with the aviation theme, or his way of being funny, but that's actually how he spells his name. He was a good server, though, and had a way cool belt (but again, that's another story).

   The three of us (two handsome pilots and myself) all ordered the same thing: french fries and the Top Gun burger. And believe me, it was tops. As in huge and tasty! I got through most of mine but had to give up at some point. One of the pilots I was with got dessert too, something along the lines of Death by Chocolate in very moist cake form that was so rich. Pricing wasn't bad: my bill was $7.88.

   Unfortunately, we finished dinner long before the storms decided to call it a night, so we made good use of the "quiet room," complete with reclining chairs and footstools. All in all a good place to make a pit stop. Just maybe not for five hours.

PIREP   May be the best fly-in restaurant in Florida. The prices are very reasonable, the service and food are good, and I have never had to wait. I have been there many times, as a result of the need to fly my plane weekly in order to keep it healthy.

   The new terminal is really nice. The restaurant is on the second floor with large windows that provide visibility of the entire airport and your plane too. Park your plane at the door.

   They usually have reduced fuel prices on the weekend.

PIREP   Tony's has been ruined! Sure, they've moved to a new terminal building. But the entire old staff is gone, and the atmosphere is creepy. Tony's used to have the best sausage gravy anywhere. Now it's simply awful. The eggs were murdered. That breakfast must have been prepared by a Yankee!

PIREP   Two of us flew in today. The restaurant was packed. We both thought that the food and service was great. I think they have the bugs out now being in a new location. I'd recommend it to my pilot friends.

Life is simple: eat, sleep, fly.

PIREP   For a fly-in restaurant, it's pretty good. Service is usually excellent. The Sunday noon crowd is there, but don't shy away from the place because that is when you expect to arrive. We have been there many times at noon on Sunday and have always gotten a seat without waiting.

Prices are 6-ounce burger $3.25; smothered chicken breast dinner $5.95; tortellini $4.95; grouper sandwich $5.25; and taco salad $5.25 (it is more like chili than taco seasoned meat and includes Doritos instead of the shell bowl you might expect).

Okay, it isn't Bern's Steak House, but it is cheap and is on a good airport.

Cafe Panino 🍔🍔🍔🍔🍔

Rest. Phone #: (863) 619-2100
Location: On field; ask for progressive taxi
Open: Mon.–Fri. 11 AM–2 PM and 5–10 PM, Sat. 5–10 PM, closed Sunday

PIREP   We went to Cafe Panino at Lakeland last night. They really have a nice restaurant and view of the airport grounds. The prices are quite low, such as a 12-ounce steak for 10 clams. Not bad! The meat quality was okay but had a great taste. Other dishes they serve are really above the standard you might expect, and the blonde waitress was really nice and quite pretty. She could have served me Alpo and I would have been fine with it at any price…ha.

It's actually easy to find with the tower's help. I landed on 9 and asked for a long landing, which put me at the end of the runway and right at the beginning of taxiway L, which goes to the restaurant. I left there after sunset and it was quite dark but had no problem getting back to the end of 9.

PIREP   I just wanted to write in to you about the Cafe Panino, the "other" restaurant at Lakeland.

What a treat! We went last night as an alternative to the regular run to Tony's. Cafe Panino is a really good Italian restaurant with a good variety on the menu. There are the usual Italian but salads and other dishes as well. We all started with good *hot* rolls. My husband had a pork tenderloin with sweet potato casserole and veggies. He did not leave even a crumb on his plate. I quickly snatched a piece of pork, which was heavenly. I had cannelloni which was very good as well; and we all had desserts ranging from chocolate espresso cheesecake to a banana tort to a seriously chocolate fudge cake. We all rolled out the door quite full and happy. The nice part, as well, is that it was very reasonably priced. For our party of four (no alcohol) with salad, dinner, and dessert was right at $80. We enjoyed the atmosphere and they do accept reservations so we had no wait for our table, although it was not a packed house anyway.

As for getting there by plane: we landed short on 27, took the first left and looked for the round Darth Vader–looking building. Park basically anywhere (there were no other planes on that part of the ramp); through gate 36 and up the stairs. The only word of caution is to make sure you have a taxi light to get back to the runway at night. Right now, because of the construction, there is a portion of the taxiway where there is no lighting. We had no problem at all negotiating the taxiways. I encourage others to make the effort for this alternative—they will not be disappointed.

PIREP   FBO Comments: Piedmont is top notch all the way.

PIREP   This may be heresy but there is another restaurant at the Lakeland airport. I have been flying and driving to Tony's for years. I truly enjoy it, and it is my wife's favorite restaurant. We live 19 miles from Tony's and many times we are driving around near our house thinking of a place to eat and my wife will say, "What about Tony's?" So we jump on I-4, and while we would still be rocking at Cracker Barrel we are being seated at Tony's. But enough about Tony's.

I found another very good restaurant on the south side of LAL. It is near the old Piper, Flight Safety buildings. It too is upstairs and actually seats more than Tony's. It is called Cafe Panino and has a very large and varied menu. The day I was there I had the seared tuna on Caesar and a lobster bisque. It was fantastic and the prices very reasonable. It has only been there since 2003 and is open for lunch Monday through Friday from 11 to 2, then for dinner 5 to 10, Saturday just 5 to 10, and it is closed Sunday. They have hamburgers, steak, pizza, all kinds of salads and wraps, oven-baked subs, deli classics, and a full bar for those riding not flying or driving.

It is quite a bit more difficult to get there by plane than by car. If you tell the tower you want to go to the restaurant, they will direct you to Tony's. You would have to ask to go to the ACA ramp and probably get progressive taxi instructions. There is a large ramp area just west of the building, and Cafe Panino is located upstairs in the round part of the building on the far west side of a long hangar. After you park your plane (you may want to bring your own chocks—no lineman here), just walk over to the small gate 35, go through the door on the west side of building, and take the elevator to the second floor. If you made it this far you can find it from here. I believe you will find it worth the extra effort to get there.

I would rate Cafe Panino five hamburgers!

Lake Wales, FL (Chalet Suzanne Air Strip—X25)

Chalet Suzanne 🍔🍔🍔🍔🍔

Rest. Phone #: (863) 676-6011
Location: This is their airstrip
Open: Breakfast, Lunch, and Dinner

PIREP   Just thought I'd drop you a pirep on the Chalet. It was absolutely great. One of the previous pireps noted the cost and they are absolutely correct. A sack of gold is about right—but really great. My wife had a ham steak and I had a King Crab casserole—no alcohol, of course. Our bill was $85 clams. My foremost concern was the grass strip. We have a Cherokee 140 (150 hp) that needed about 1800 rpm to taxi, but we were off of the ground (20 degrees of flaps and short-soft procedures according to the manual) at about halfway down the yard. Our gross weight was about 1950 lbs., and the weather was dry with a crosswind of about 9 mph. The temp was about 65. The turf seemed to be in good shape but a bit squishy from all of the recent rain. We really loved it and will do it again. Maybe we'll get the soup and sandwich for $19 clams instead of the Alaskan King Crab deal.

PIREP   I flew to Chalet Suzanne to drop off my son to see his grandparents. The turf was in excellent shape but requires a lot of power to just taxi (very thick). I would be careful in a low-powered airplane. I've seen comments on other websites where folks use the entire runway to get airborne. I have a RV4 with 180 hp, so no problem with the length but it did take longer to get airborne due to the thick turf. Ate lunch, and the cheapest thing on the menu is $19. The average price is $31. I also landed there for the cheap gas but was unable to get any due to the pump being locked up. I called ahead and they said gas was available, but when I got there, no one knew where to find the key to unlock the pump. I was not pleased with this since I had planned on getting gas and called ahead to make sure. So be careful if you are low on gas when you land, because you might have to go somewhere else to refuel.

PIREP   Chalet Suzanne is very expensive, with prices running around $100 for dinner for two. The food is outstanding. They have a Lobster Newburg type dish that should not be missed.

PIREP   Ask about pilots special, which generally includes dinner for two, room for the night and breakfast.

Food: An excellent restaurant (known for their Chalet Suzanne soups sold in many food store gourmet sections): (863) 676-6011.

Lodging: Chalet Suzanne has its own very unique rooms, each furnished with different antiques.

Transportation: None required—private 2450' turf runway adjacent to restaurant.

Alternate: Lake Wales Airport, 4000' paved, four miles south of Chalet Suzanne. Courtesy car (ask Chalet Suzanne) or taxi.

Lake Wales, FL (Lake Wales Muni—X07)

Chalet Suzanne 😋😋😋

Rest. Phone #: (863) 676-6011
Location: 3 miles
Open: Breakfast, Lunch, and Dinner, 7 days a week

PIREP We landed at X07. Call ahead, the Chalet staff will meet you at the airport. The food is a gourmet delight. Lunch was a two-and-a-half-hour experience. Take a sack of gold with you, it is very expensive. Lunch was followed by a guided tour of the adjacent Chalets.

LekaricA Country Inn 😋😋😋😋

Rest. Phone #: (863) 676-8281
Location: 3 miles away
Open: Normal

PIREP LekaricA Country Inn, Restaurant, and Golf Course is located on Lake Easy in Lake Wales, FL. We sponsor several splash-ins a year for float and amphibian aircraft. We are close to Lake Wales Muni (X07) (under 3 miles) and are happy to pick up golfing guests.

LekaricA also features the area's finest Sunday brunch menu, offering tableside service from a menu. Unique items, coffee beverages, and chef-designed items make Sunday brunch one of our favorite days.

Lunch is terrific, and you can build your own burger for only $6.95 in a setting you'll be proud to be a part of.

For dinner, there is a good wine list. I even found my favorite chardonnay: Cakebread Cellars 1997 and 1998.

For airport pickup, call (863) 676-8281, toll free: (888) 676-8281, fax: (863) 676-8492.

Leesburg, FL (Leesburg Regional—LEE)

Franchise Row 😋😋

Rest. Phone #:
Location:
Open:

PIREP The Cracker Barrel is now open, just adjacent to the airport. Much better than Captain Bell's. There is also a TGI Fridays and Outback within two blocks of the airport.

PIREP Captain Bell's is good. If you park at the FBO, it's about a half-mile hike up and on the opposite side of a four-lane highway. The restaurant is not classy but serves decent food at reasonable prices (although I highly recommend passing on the water). The service was very good. I give them three of five hamburgers.

PIREP Captain Bell's seafood restaurant about an eighth mile from airport gate serves very good, reasonably priced seafood seven days a week. It's crowded from 12 to 1 PM with local traffic. There is a place to park near the airport gate that is about the midpoint of 13-31 and next to some really old low hangars. Go out the gate and walk left up the hill.

PIREP Park at far northeast end of field. Walk toward the giant American flag, where you will find Capt. Bell's seafood. Can't beat the prices and quantity of food.

Marathon, FL (The Florida Keys Marathon—MTH)

Buccaneer Lodge 😋😋

Rest. Phone #: (305) 743-9071
Location: Cab ride away
Open: Daily

PIREP   Landed MTH today and used Marathon Jet Center FBO. FBO folks were very nice and greeted the airplane with bottles of cold water. 100LL = $2.08 self service.

Took a $3 cab ride to Buccaneer Lodge. Spectacular view of Gulf with boardwalks and charter boats next door. Food and service were below par but view made up the difference. Two out of five burgers.

Cabana Breezes 🍔🍔🍔

Rest. Phone #: (305) 743-4443
Location: Cab ride
Open:

PIREP   Visited on weekend of March 19, 2005. People at Marathon Jet are friendly as usual. Ramp was busy, but plenty of AC parking. Cab ride about six minutes, $5 to Cabana Breezes Rest. Call for exact hours, normally open at 11:30 AM seven days a week. Great view of beach and ocean. Seating inside or outside. Good menu, seafood, and other. Had a Cuban sandwich, great choice. Service was excellent. They now have hotel suites available, but only 12 so they book up fast. Check website for more info. Waited five minutes for cab back to airport, still $5. Best cabs are On-Time Taxi: (305) 289-5656.

PIREP   FBO Comments: Marathon Jet Center is excellent, greets you with cold water, great customer service.

PIREP $5 cab ride to a very nice waterfront restaurant with excellent food and great service. Highly recommend for breakfast or lunch. Marathon has a pretty approach and one of the few NDB approaches left in the state.

Keys Fisheries Market & Marina 🍔🍔🍔

Rest. Phone #: (866) 743-4353
Location: Cab ride
Open:

PIREP   FBO Comments: Paradise offers great service and attitude. They lost about 15 planes and 40 cars in Wilma but are cheery and happy to have business and give super support.

PIREP   Took the $5 (each way) cab to Keys and had a great lunch. Won't repeat all the other good stuff, but wanted to say they are open and the quality is very good. Fresh fish, varied menu for even non fish eaters and decent prices.

Rating: 3

PIREP   FBO Comments: Marathon Jet Center. They meet you with a bottle of cold water when you land. Very friendly operation. They will waive the $10 ramp fee if you buy gas, which is well cheaper than most home fields.

PIREP   Take a $5 taxi ride to the Keys Fisheries Market. The food is great and the seafood cannot be fresher (they own the boats and you can even go fishing or lobstering with them). Extensive menu and a great spot.

PIREP   Very good food, cab was cheap and pleasant. It is outdoor eating on picnic benches, so if it rains you get wet, when it's hot out you're hot. It is also on the bay—we had 50 seagulls trying to get a bite to eat, too. Very good food, but we would have preferred a bit classier of a place after a long trip.

Island Tiki Bar 🍔🍔🍔

Rest. Phone #: (305) 743-4191
Location: Cab ride
Open: Daily: Lunch and Dinner

PIREP Tiki Restaurant is just as advertised by previous pilots. Still $4 taxi ride. Seafood and salads are great. The airport has a Top Gun operation using a simulator and L39 aircraft. Watch for them in the pattern.

PIREP Parked at Paradise Aviation. Very courteous but never seem to have a courtesy car anymore. The cab ride to Island Tiki is only $4 each way.

Island Tiki has the best fresh fish I have had anywhere in Florida. Scenic outdoor dining on the water under thatched roofs. Highly recommend.

PIREP The Island Tiki Bar on Marathon Key (MTH), well worth the trip. A $4 cab ride from the aerodrome. Picturesque view of the Gulf. Excellent seafood, good service with a smile. Innovative menu.

PIREP We parked at Grantair, a very friendly FBO; they called a cab for us, and we had lunch at the Island Tiki Bar. It's an inexpensive five-minute ride from the airport. The sign on the restaurant proclaims, "What you came to the Keys for."

The Island Tiki is directly on the water; the architecture is "eclectic island," with thatched roof over the bar, tin roofing elsewhere, and a very open, laid-back feeling.

The fresh fish is outstanding. You will not be disappointed if you are looking for a local eatery with good food and atmosphere.

Stout's 🍔🍔🍔

Rest. Phone #: (305) 743-6437
Location: Across the street, short walk
Open: Daily: Lunch and Dinner

PIREP You will now be paying $110 for your $100 hamburger! Grant Air now charges $10 for a tie-down even if you are walking across the street to Stout's. You might want to try Paradise Air at the other end of the field.

PIREP Visited Marathon Airport in the Florida Keys today. Parked at Grant Air. Walked across US-1 and ate at a luncheonette called Stout's.

Clean home cooking. It was above average at a reasonable price but nothing exceptional.

Marco Island, FL (Marco Island—MKY)

The Snook Inn 🍔🍔🍔🍔

Rest. Phone #: (239) 394-3313
Location: Short drive by airport car
Open:

PIREP FBO Comments: Great service and friendly welcome. Parking is interesting, but they are helpful. Courtesy van available and best popcorn in Florida. Pattern can be busy and proximity to Naples means you take off almost directly into Class B space, so be alert if you are VFR.

PIREP Snook Inn was fun, and seating on main harbor channel is great place to relax and enjoy. Other restaurants on the road to Snook Inn looked good for evening dining.

PIREP   FBO Comments: Great and friendly folks. They are very polite and helpful. The airport is owned by the county. The county has vehicles that you can use free of charge at the airport.

PIREP   The Snook Inn has an indoor and outdoor dining area. Recommend sitting outside on deck. The deck overlooks a bay area with emerald green water. The burgers are great and reasonably priced. I took a date and Katie absolutely loved it.

Melbourne, FL (Melbourne Intl—MLB)

Phil's Casbah 😋 😋 😋

Rest. Phone #: (321) 956-3434
Location: Less than a half mile off airport
Open:

PIREP   Great home-style Southern food with a good selection of seafood, sandwiches, and plate lunches and dinners. Open early for breakfast as well as lunch and dinner most days. Best to call for current dinner hours as some seasonal changes occur. Short taxi ride from the airport, less than a mile away.

Naples, FL (Naples Muni—APF)

Chrissy's 😋 😋 😋

Rest. Phone #: (239) 643-6631
Location: Short walk
Open: Daily: Lunch and Dinner

PIREP   Landed at Naples Airport the other day due to convective activity along the route of flight to the Everglades. Since I was on the ground I wanted to get some lunch. The FBO at Naples recommended Chrissy's. They even offered a ride in their courtesy van. The restaurant was excellent, with fresh generous portions of food. I had the coconut shrimp salad. It had fresh pineapple, large shrimp, fresh mixed greens, and was a very full bowl of food. My pax had the mushroom burger; it was good and freshly made. I saw other food, and it all look liked good home cooking. Price for our lunch was just under $15.

PIREP   My husband and I took my step-daughter and her two sons (3 years old and 18 months old) to Chrissy's for breakfast on a Sunday morning. The restaurant is located just off the airport (walking distance, even with two toddlers) in a small strip shopping center. We waited about ten minutes for a table but enjoyed complimentary coffee and read the menu while doing so. Service was pleasant and efficient, the menu offered some variety, the food was excellent, and prices were moderate. This place gets my recommendation.

The Landings Cafe 😋 😋 😋 😋

Rest. Phone #: (239) 435-0510
Location: In the terminal
Open:

PIREP   Nice place, not particularly exciting. It is in the commercial terminal. A drive from the GA terminal, but the airport staff ferried us to and back. Wife had shrimp cocktail and Caesar salad, I had a hamburger. The food was great. The ambiance is not too great. Not a bad place but we have been to more exciting places we would rather go back to. I will not fly back for it, but if you are already there it was very good food.

PIREP   FBO Comments: The FBO was nice. Courtesy vehicles are available but the FBO (NPA—Naples Port Authority) is pretty busy so you may not get one. Lots of turbine traffic operate out of this airport. FBO staff was friendly and helpful.

PIREP   The Landings Cafe, located in the passenger terminal, is a nice cafe. We had their roast beef and Philly steak and cheese sandwiches, and they were excellent. GA cannot park on the commercial ramp (that is, the passenger terminal area). You must park at the FBO (NPA—Naples Port Authority). Transient parking may only park in the yellow parking areas; blue marked areas are reserved parking. The FBO will be glad to shuttle you to and from the passenger terminal.

FYI: You cannot walk to the passenger terminal.

PIREP   The Landings Cafe has the most *amazing* sandwiches. I had the club sandwich and my sister had a cheese-burger. *Wow!* The food was fantastic, the best lunch we ever had. The service perfect! We went back the next morning and had breakfast—the *best* breakfast we ever had. I highly recommend the Landings Cafe to everyone!

PIREP   I recently had a breakfast at The Landings Cafe, which is in the terminal, at Naples Airport (KAPF). I was pleasantly surprised at the quality and quantity. Even the coffee was excellent. I was there for a commuter flight to Key West, but I would fly there just for the breakfast again. I understand the hamburgers are equally good. I'm trying it on my next trip.

Michelbob's 🍔🍔🍔

Rest. Phone #: (239) 643-7427
Location: Short walk
Open: Daily: Lunch and Dinner

PIREP   Michelbob's is about a ten-minute walk on the road running behind the airport. The FBO will gladly give you a ride there. Michelbob's claims to serve the best babyback ribs in Florida and they do so tongue-in-cheek: "Since baby pigs tend to rest on their right side, that side is tender before we cook it. We leave the left side to others: ever heard the term 'leftovers'?" I think they are wrong with their claim: their ribs are the best in the world!

PIREP   The ribs just keep getting better and better. There is a greasy spoon named Chrissy's along the road. Do your stomach a favor and head for Michelbob's.

New Smyrna Beach, FL (New Smyrna Beach Muni—EVB)

Gumbo's 🍔🍔🍔

Rest. Phone #: (386) 426-5777
Location: Just off the airport, 15-minute walk
Open: Lunch and Dinner

PIREP   Gumbo's has great food and is a fun place to fly to. You can park in front of Epic Aviation and walk across the parking lot to the restaurant. From there, it's only a three-minute walk. You can see it from the ramp. The restaurant sits in a hammock of oak trees and Spanish moss. Eating outside on a cool day is a real treat as you can see airplanes doing their thing while you are seated on a wooden deck surrounded by trees and overlooking the airport. They can be a couple of bucks over the usual lunch menu but there's no cab fare to add to the lunch experience so it's not a bad deal.

PIREP   Gumbo's, which replaced Abaco's, is now open and has pricey New Orleans food. Depending on where you park it can be a long walk.

Ocala, FL (Greystone—17FL)
Jumbolair Banquet Hall 😑😑😑😑😑

Rest. Phone #: (352) 401-1990
Location: On the airport
Open: First Sunday of the month

PIREP   The monthly Sunday brunch at Jumbolair should not be missed. The owners are very cordial and go all out for their guests' enjoyment.

One note of caution. Greystone is a strange airport. The 7500' length is really much less because of the displaced threshold. This is due to a tower directly on the final approach path. There are also no taxiways, so aircraft are back taxiing on a cordoned off section of the runway.

There are also many aircraft in the area, because they could have over 50 planes flying in for the brunch.

If you are careful, the effort is well worth it. This is much more than the usual luncheonette fare at most airports.

PIREP   Fabulous! Bring your family and be proud you're a pilot. A *$100 Hamburger* adventure simply can't get any better than the First Sunday Brunch at Jumbolair! If you truly want and feel you deserve the very finest, this is the place for you.

Brunch is served in the Jumbolair Banquet Hall. The reflection pool and twin fountains just inside the front door greet you and trumpet the treat to follow. Your hosts Jeremy and Terri Thayer have instructed their staff to pull out all the stops. No wonder it can only be done once a month.

You'll want to take a rest from the food and enjoy the grounds—550 acres of splendor set in the heart of Florida's horse country. Take a walk or enjoy a carriage ride under the umbrella of 200-year-old oaks. This month 65 aircraft showed up.

Don't miss the next Jumbolair First Sunday Brunch—I know I won't!

PIREP   Very unique place; touch down and turn off quickly anywhere to taxiway. People on ground directing you to park. Parking is against concrete blast fence or on sides of runway. No fuel. Very busy!

We had friends come down from Michigan this last weekend and thought a great outing would be Jumbolair. It had great ratings and was just a short half-hour flight, so it seemed perfect for Sunday brunch. I e-mailed a reservation for 10 AM. Weather was nice and we were off. This is probably the busiest noncontrolled airport I have been into with my husband. Planes land consecutively and everyone seems to know the drill. Land and get off the runway quickly, which is no problem since there is a side taxiway with excess no matter where you turn.

Now the food. We were seated immediately, which was kind of cruise style where you are seated in open seats at larger tables. The couple at our table had just driven from Lakeland on their Harleys. A little chilly so I was glad we were in the plane. The food was very good and lots to choose from. Roast beef or ham, shrimp, homemade omelets, grits, salads, waffles (but not made to order), stuffed mushrooms (my favorite!), fresh fruits, breads, and last but not least, fabulous desserts. The waitress and waiters were very attentive. This place was very busy and we got there early.

The setting is beautiful, with picket fences and horses grazing. We walked around after eating and did some horse petting and just enjoying the grounds. It is amazing watching all the planes land one after another and park by the big concrete wall situated just before the planes land. This place is truly unique and a fun place to fly on a Sunday morning. The food was very good but I have to say, West Palm Beach's 391st Bomb Squad has them beat in my book—but you can't go wrong with either.

PIREP   This is a private airport that is only available to the public on Brunch Day! There are no facilities for your aircraft except parking. OCF is a five-minute flight away.

This was a superb Sunday getaway. The staff is very courteous, the food is excellent, and the atmosphere was the best. No wonder John Travolta is a resident here.

Ocala, FL (Ocala Intl–Jim Taylor Field—OCF)
Tailwind Cafe 🍔🍔🍔🍔

Rest. Phone #: (352) 291-0283
Location: In the terminal
Open: 7 till 7, 7 days a week!

PIREP FBO Comments: I am not sure what the gas prices were. The FBO folks are great.

PIREP Tailwind Cafe has some great food with awesome service. I recommend a cheeseburger with fries and an ice tea.
 Rating: +2, +2, +1, 0, +1

PIREP The location is nestled next to the FBO; the owner is there to take your drink order right away. Price for a burger is cheap for the amount of food you get. I had the mushroom burger, and the mushrooms were freshly sautéed and not overdone like most places I have gone to! The owner had some time to talk with us—she was very friendly. We felt like we were the only people there, even though the place was 85 percent full when we got there. The Tailwind Cafe gets a six in my book.

PIREP FBO Comments: Very friendly and nice.

PIREP I have been to the restaurant many times. The parking is very convenient and the line boys are helpful. Tailwind's has good food at inexpensive prices. Sam the waitress is friendly and a real delight. This makes for a very pleasant meal. I recommend it highly.

PIREP One of the best hamburgers I've ever eaten. Food +2, service −2, location +2.

PIREP I don't get to go to all of the restaurants on this website. Wish I did! Every now and then I drop in on the best ones. This is truly the number 1 everyday stop for Florida pilots.
 Recently, I had the rare privilege of making this burger run with Florida's number one aviator, John Painter from Air Orlando. The Local 6 News Crew from Orlando showed up to film our meal. John Painter's presence demands that kind of attention.
 The food is terrific, the crowd congenial, and Sam? Well, she makes the trip worth it at twice the price.
 Editor
 The $100 Hamburger

PIREP This is one of my favorite stops. JAX Approach will let you fly the full ILS, no vectors—either the DME arc to final or to the IAF compass locator, track outbound, procedure turn, and so on. This is fun and very good practice. On the ground you'll see some interesting airplanes. Restaurant owners Curt and Sam are good people.

PIREP This is a great restaurant with new owners now. The restaurant's new name is Tailwind Cafe. Using the fresh ground beef instead of frozen patties is a great improvement in the food. They serve food till 7 PM. The cook Curt makes the best food, not to mention the fish fry on Fridays. They have low prices compared to other airports, and the main waitress, Sam, is the nicest, prettiest waitress I have ever seen. The cook for you ladies is not an eyesore either. You want breakfast for $2.25, visit Tailwind's Monday through Friday, 7:30 AM to 10 AM. You won't be disappointed—trust me on this one. You will have a fun time with some fun and sarcasm from the hired help. Not a boring restaurant. All staff good.

Okeechobee, FL (Okeechobee County—OBE)
The Landing Strip 🍔🍔🍔

Rest. Phone #: (863) 467-6828
Location: On the field
Open: 8 AM–5 PM every day

PIREP   FBO Comments: 24-hour self-serve Chevron.

PIREP   Restaurant now open from 0800 until 1700 every day with daily specials a bargain and tasty. Service still somewhat slow but worth the wait.

PIREP   Most aircraft in the south Florida area fill up here as they have the cheapest fuel anywhere.

PIREP   I flew from KTMB to KOBE for lunch at the Landing Strip. It's a nice restaurant with a reasonable menu. Prices are good and service good but a little slow. Probably due to the traffic of eaters.

I parked my plane right on the ramp outside the restaurant. I couldn't raise St. Pete's FSS on the ground (122.25), but there is a planning room in the FBO where you can close out flight plans by telephone.

All in all a pleasant flight/meal.

PIREP   I have flown to Okeechobee twice to eat now. Once was for breakfast and once for lunch. The food was great, the service was great, and I just had a good time at the airport. You can hang around inside or outside and talk to other pilots and just enjoy the airport atmosphere. Great place to eat, hang out, and take a break.

PIREP   The Landing Strip Cafe is excellent. Last Saturday there were so many planes you would have thought it was a fly-in.

The cafe looks out on the field, and you can watch the TOs and landings as you have your hamburger. Afterwards you can fill up at the cheapest gas in Florida.

If you really want to pig out, ask for the Pilot's Special.

PIREP   We had breakfast last Saturday; good food, good service, good prices. Parking was tight with lots of aircraft on the ground. Worth the trip from Melbourne!

PIREP   The people at the Okeechobee restaurant were very nice, and there is plenty of parking. I would recommend the fried fish sandwich—it looked tremendous. We had the open-face roast beef and turkey sandwiches. Oh man, unless you like really salty, processed meat, I would recommend ordering anything else. One word for both of these: yuck.

PIREP   Just opened this past week in Okeechobee. Located in the terminal building, it gives us another excuse to go to Okeechobee.

We had six planes from our West Palm Beach breakfast group on Sunday morning. Another six or so showed up from other locations. Service was a bit slow, but I think that they were swamped with all the opening weekend traffic.

While the restaurant was probably busier than normal, the food was excellent and as everything is brand new the facilities were excellent. Prices were very reasonable.

They are open Tuesday through Friday and Sunday until 5 PM and Friday and Saturday until 8 PM. Closed Monday.

Menu prices look great for breakfast lunch and dinner, and if the food is as good as our breakfasts were then we have a great fly-in food destination.

Orlando, FL (Executive—ORL)

Franchise Row 🥪🥪

Rest. Phone #:
Location: Looooong walk
Open:

PIREP   Flew into Orlando Executive, parked at Showalter. Received excellent service. No parking fee with fuel purchase. Very nice facility, high class. They provided a crew car to run out to East Colonial Dr. for dinner.

Several good restaurants within half mile of Showalter on East Colonial Dr., north of airport: Smokey Bones, Olive Garden, and Red Lobster, along with other fast food.

Rating: 3

PIREP   FBO Comments: Sheltair is a great FBO. They have a very nice facility to relax and flight plan. They will waive the parking fee with a topoff or a purchase in the pilot shop next door. The staff is very helpful and friendly. He has quite a few good stories from working at FBOs across the state.

PIREP   Houlihan's is about a 12-minute walk from Sheltair. They have some awesome burgers. The burgers are made from 100 percent Angus beef. Additionally, the french fries are to die for. The service is quick and prices are on target.

PIREP   There are several great restaurants around and near ORL. Key to a good flying experience is finding the right FBO or avoiding *bad* ones. Showalter is definitely on our avoid list.

Management sets the tone at any business, and management at Showalter is rude and arrogant. "They strike me as not really being compatible with the direction we are heading."

We won't do business with them and neither should you! Go to Sheltair instead.

Editor

The $100 Hamburger

PIREP   Two half-blocks due west of Sheltair on Primrose at Robinson St. is Sandwich King. It's a good deli that's open five days a week until 3 PM. They usually have a soup of the day, and a sandwich and soda usually cost me $6–$7.

ORL is located in the city with tons of places to eat nearby. Also on the west/northwest side of the field are Einstein Bros Bagels, Baja Burrito Kitchen, Little Anthony's (pizza/Italian), and Toojays (expensive deli)—though those are more like a mile away from Executive Air Center. There are a lot of places along Colonial Drive, along the north side closer to Showalter, but most of those are chain restaurants or fast food.

Approaching ORL is a lot of fun for passengers since the runway in use 80 percent of the time takes you right over downtown. The VOR and ILS approaches will bring Disney in sight as well (for passengers!).

Ormond Beach, FL (Ormond Beach Muni—OMN)
River Bend Golf Club 🍔🍔🍔

Rest. Phone #: (386) 673-6000
Location: Just outside the gate
Open:

PIREP   There is a golf course right outside the airport at Ormond Beach. Haven't played it yet but will soon! It is less than 100 yards from the tie-downs.

The telephone number for River Bend Golf Club is (386) 673-6000. There is a snack bar with a pretty good burger.

Pensacola, FL (Coastal—83J)
Beulah Land BBQ 🍔🍔🍔

Rest. Phone #: (850) 941-2933
Location: Across the street
Open:

PIREP FBO Comments: Field is 2526' × 230' grass field that is in excellent condition. They have glider and ultralight operations at the field that are really cool to watch!

PIREP I stopped in at Coastal Airport and saw Beulah Land BBQ across the street. I decided to walk over and check it out and was pleasantly greeted by the owner Doug Allen (super nice guy, by the way). I didn't have time to eat and only got some ice tea (unsweet and with fresh lemon—hey, I'm a Yankee), but whatever was cooking smelled awesome.

They have a great menu, and I plan on making the trip back real soon to get some food. They seem very eager to make customers happy. I will update my pirep as soon as I try some of their stuff.

Pensacola, FL (Ferguson—82J)
Naval Aviation Museum 🍔🍔🍔🍔

Rest. Phone #: (850) 453-2389
Location: Free FBO shuttle
Open: Daily: 9 AM–5 PM, closed Christmas, Thanksgiving, and New Year's

PIREP FBO Comments: (850) 453-4301, (850) 453-4181. Provides a free shuttle to the Naval Aviation Museum.

PIREP This is the friendliest airport in Pensacola!!! Slips 98 is nearby. They have seafood baskets and sandwiches. If you're going to the Naval Aviation Museum, and you should, eat on board the base. The food's pretty good.

Pierson, FL (Pierson Muni—2J8)
Carters' Country Kitchen 🍔🍔🍔

Rest. Phone #: (386) 749-4983
Location: Right across the street
Open: Mon.–Sat. 6 AM–2 PM, closed Sun.

PIREP We took a ride to Pierson this weekend. The turf was nice and solid, but the parking area was a little squishy. It would be better to park nearest to the highway. The restaurant was really nice. It was clean, friendly, and the food was absolutely great. I had the catfish and it was excellent. This was a fun place to go and I would recommend it to anyone. The owners of the restaurant have been in business for 15 years and now I know why. I had the best apple pie I have chomped on in a long time. As you leave the airstrip, you walk south for about 350 feet to the end of the shopping center—and there they are!

PIREP A new place just discovered is the grass airstrip at Pierson Airport (2J8), 14 nm northwest of Deland. Carter's Country Kitchen is a block from the airfield. Open 6 AM–2 PM and closed Sundays. Good home cooking with the Wednesday special, fried chicken $3.65—renowned by locals and pilots who fly there especially for the chicken. Outstanding food and reasonable prices!

PIREP Fly in to Pierson Muni (2J8) 14nm nw of Deland (ded). Carter's Country Kitchen located approximately 150 yards from apt. Hamburgers, cheese burgers, grilled cheese, patty melt, hot dog, turkey, ham and cheese, roast beef, or club sandwich, all priced from $1.95 to $5.35. Hours of operation, Monday through Saturday 6 AM to 2 PM and closed on Sunday. Large a/c turnouts on Wednesday and Saturday.

Punta Gorda, FL (Charlotte County—PGD)
Skyview Cafe 🍔🍔🍔

Rest. Phone #: (941) 637-6004
Location: On the field
Open:

PIREP   Excellent breakfast with specialty omelets and one pan breakfast. Lunch menu includes sandwiches, wraps, homemade soups, daily specials, and salads, including an excellent chicken walnut salad.
New owners, Ed and Caroline Gallagher.

PIREP   FBO Comments: Fantastic airport, rebounding from Hurricane Charley. Great flight schools, maintenance shop and avionics shop. They are coming back strong.

PIREP   The restaurant just reopened with a new, clean, fresh look and new owners. The food and presentation is great. Prices are very reasonable and the service is terrific.

PIREP   FBO Comments: Great airport!!!! Self-service fuel.

PIREP   Just opened!!!! The Skyview Cafe, completely remodeled restaurant. Fantastic food, very clean and friendly, new owners. Great menu and reasonable prices. Friday night dinners until 8 PM. Great fly-in restaurant.

River Ranch, FL (River Ranch Resort—2RR)

The Fisherman's Cove Restaurant 🍔🍔🍔🍔

Rest. Phone #: (863) 692-1321
Location: On the field
Open: Daily: Lunch and Dinner

PIREP   Skip the breakfast. The service and resort make it a great meal outing, but skip the breakfast. I imagine the lunch and dinner are better. They gave us a free golf cart to get around, have a petting area with various farm animals, a rodeo on Saturdays, and hotel rooms if you spend too much time in the saloon. We also went trap shooting during our two-hour visit; they supply everything. Next time I'll go for dinner.

PIREP   River Ranch is always a great place to go. The food is good and the people are really nice—and the golf cart at the airport still costs $5.36 for all day. Only the marina is open for food, and don't try looking for food during the week because they're only open on weekends or special occasions—call first during the week. The marina restaurant has a great screened-in porch to eat on, or you can sit in the air conditioned interior. There is an old western town array of shops to visit located just around the corner from the marina, and they also have an ice cream shop.

PIREP   Flew into River Ranch for a Mother's Day breakfast at the Fisherman's Cove Restaurant. It was great! A full breakfast buffet for $7.95 plus good service and great coffee. Rent a golf cart for $5 and you can drive all around the resort and check out all of the sites.
Pilots beware! Many pilots that fly in here don't seem to know how to use their radios. Extra precautions on takeoff and climbout are in order.

PIREP   We stayed for the weekend at Westgate River Ranch. The courtyard area rooms had two double beds, refrigerator, microwave, oven, and stove. We had a screen porch facing the tennis courts. We had breakfast Saturday and Sunday at the Fisherman's Cove (marina restaurant). The highlight was the Saturday night hayride and barbecue in the East Corral. The $14 included a one-hour hayride followed by all-you-can-eat hamburgers, hot dogs, barbecued chicken, bean, corn on the cob, and sodas. We then watched the rodeo, which drew quite a crowd. Don't miss the airboat tour of Lake Kissimmee.

PIREP   Neat place since it has been purchased by the timeshare king. We spent some time on the skeet range and shot a few clay pigeons—what a hoot!

PIREP   Da' Ranch is open for business again.

I flew a 172 with 3 pax and trailed a buddy of mine in a Piper Cherokee with 2 pax out of PCM to 2RR on January 9, '03.

The only gut check was staying out of the MOA's which were "hot" that day and who got to 34 first. Plan on staying in the pattern while the coast clears, or orbiting over the river five miles out.

You can get a golf cart for $5 and ride over to the restaurant about a half mile or so away. Food is reasonable: $5.50 sandwich with the fries, pickle, and slaw sides. Yakking it up is free. Nice outdoor deck to eat and look over the river.

PIREP   This is really a great place for a quick lunch and great atmosphere. They have purchased a large fortune's worth of new golf carts that are available to the aviation community for 5 clams. Absolutely wonderful and entertaining. They have a nice hotel as well. The only drawback is the 122.8 unicom frequency, which is so crowded that you are lucky to get a word in between squeals. Also, not everyone is aware of the right-hand traffic on runway 34, so keep looking for competition on a left base while you're on a right base to 34. The winner will land and the runner-up gets about five more minutes of flying to go around for another look.

St Augustine, FL (St Augustine—SGJ)
Fly-by Cafe 🍔🍔🍔🍔

Rest. Phone #: (904) 824-3495
Location: Located above FBO
Open: 6:30 AM–3:30 PM Mon.–Fri., 6:30 AM–4:30 PM Sat., Sun.

PIREP   Fly-by Cafe now has their act together. They produce a great burger—one of the best—and have a full breakfast menu as well as lunch menu. They also have a fresh salad and soup bar. Outstanding food at moderate prices. Highly recommended.

PIREP   KSGJ is my home base so I frequent the cafe at least once a week or so. Fly-by Cafe just reopened after being closed a couple of months. They still seem to be getting their act together, but it definitely meets acceptable standards at this point. I expect it to improve as the management and staff gain more experience with its clientele.

St Petersburg, FL (Albert Whitted—SPG)
Renaissance Vinoy Resort and Golf Club 🍔🍔🍔🍔🍔

Rest. Phone #: (727) 894-1000
Location: Close by, call for limo
Open: Breakfast, Lunch, and Dinner

PIREP   The Renaissance Vinoy Resort is a AAA 4 Diamond Florida getaway located very near Albert Whitted. The original hotel was built in 1925. It reopened in 1992 after a $92 million restoration.

It has five restaurants that range from casual to elegant:

Alfresco's, located adjacent to the pool deck, offering casual dining featuring specialty sandwiches, salads, and grilled items, Floribbean style.

The Terrace Room, located in our historic main dining room, featuring a variety of cuisine with an emphasis on fresh Florida seafood, pasta, and USDA prime beef.

Marchand's Bar & Grill, the Vinoy's signature restaurant, featuring Mediterranean cuisine with an emphasis on grilling, roasting, and sauteing using fresh herbs and olive oil. Note: a spectacular Sunday market brunch is offered in the Terrace Room and Marchand's.

Fred's, an elegant private dining club experience reserved for resort guests and Vinoy Club members, featuring USDA certified prime beef, fresh seafood, and lamb.

The Clubhouse Restaurant, located at the golf club on nearby Snell Isle, offers casual dining to Vinoy Club members and resort guests overlooking our exclusive 18-hole championship golf course.

Be aware that it is more expensive than your average airport restaurant, $6–$10 per person for lunch at Alfresco's or $35 for the Sunday brunch in the Terrace Room and Marchand's. If you call the hotel from SPG, they will send the hotel limousine to pick you up and will deliver you back to SPG after eating.

If you want to RON, the resort offers a private 74-slip marina, 18-hole golf course totally renovated in 2003, 12-court Har-Tru tennis complex, fitness center, salon, and day spa.

There are some wonderful museums nearby, including the Florida International Museum, the Museum of Fine Arts, and the Salvador Dali Museum, which houses the world's most comprehensive collection of the Spanish artist's works.

The Pier 🍔🍔🍔

Rest. Phone #:
Location: Short walk
Open:

PIREP Just wanted to pass along a spot that I think a lot of pilots might enjoy. Albert Whitted (SPG) in St. Petersburg, FL is a great place to fly in. The airport is directly on the water in downtown St. Pete. The only FBO, I forgot their name, will provide free transportation, or it's a beautiful walk to the pier. There are several restaurants ranging from very fancy (read: expensive) to hamburger places.

At the end of the pier there is a good-sized mall. It's a great place to take a few pictures, do a little shopping, and enjoy the beautiful scenery while you dine.

As I am from HPN Westchester Co. Airport, in White Plains, NY, I find the change of pace another reason for enjoying my visits. Hope everyone will give it a try.

PIREP From the restaurant on top of the pier called Coconuts, there is a fantastic view of the airport itself. I love that place!!!!

St Petersburg-Clearwater, FL (St Petersburg-Clearwater Intl—PIE)

Hooters - The Original One 🍔🍔🍔

Rest. Phone #: (727) 797-4008
Location: 5 miles from the airport
Open: Lunch

PIREP FBO Comments: You can park at Signature Aviation. They have a very nice facility.

PIREP Hooters is located at 2800 Gulf to Bay Blvd in Clearwater. The restaurant is approximately five miles from the airport. Signature can drive you over for free. I recommend this particular Hooters because it is the original one. They have lots of memorabilia from the years on the wall. The food is good and as always they have great burgers. This Hooters has the best food out of all of them, because the corporate headquarters is located behind the restaurant.

Sebastian, FL (Sebastian Muni—X26)
Zoobar 🍔

Rest. Phone #: (561) 581-0111
Location: On the field
Open: Daily

PIREP   The owner of the restaurant was really nice, but the cuisine is strictly microwave goodies and probably should have a rating of a couple less hamburgers on the rating schedule.

PIREP   Full-time skydiving center located at the site so don't fly through the overhead.

Full-time restaurant and bar located on the west side of the airfield called the Zoobar. Family run with great inexpensive food. Sit outside on the deck and watch all the skydiving action or relax inside.

Jet and avgas available, paved runways.

Sebring, FL (Sebring Regional—SEF)
Runway Cafe 🍔🍔🍔🍔

Rest. Phone #: (863) 655-6444
Location: On the field
Open: Tues.–Sun., 7 AM–2:30 PM

PIREP   Flew in to meet friends for lunch. The name has changed to Runway Cafe. Excellent burger, service was very pleasant, and prices were very reasonable and best of all, 100 avgas was full service and reasonably priced.

Asked the FBO desk just across the lobby from the entrance to the cafe to have my plane gassed up on my way into lunch—gave him my tail number, had lunch, and paid for the gas in less time than it usually takes when I fill up my car.

Runways in great shape and great parking.

PIREP   FBO Comments: Runway 36/18 has a new asphalt paving and marking and is very nice.

PIREP   Flew to Sebring (SEF) for lunch. We both had the chicken Caesar salad, which was quite good ($6.25 each). The restaurant had booklets with $1 off coupons at the check-out counter, so we took advantage of that.

The restaurant is open Tuesday through Sunday, closed Monday. Hours are Tuesday, Wednesday, Thursday 10 AM to 2:30 PM. Friday, Saturday, Sunday 7 AM to 2:30 PM. On Saturday and Sunday, they have a breakfast buffet from 7 AM to 11 AM.

It was a cool morning, so the fireplace in the restaurant was operating. Felt good and nice atmosphere. On warmer days, sitting on the patio for lunch gives you an excellent view of runway 36/18, which is typically the active. Generally some ultralights in the pattern for variety.

There happened to be a B-25 and B-17 on display on the ramp and offering rides when we were there as an extra bonus.

PIREP   The restaurant has changed management. They have an AWOS now, can't remember the frequency.

Flew in a few days ago and had a barbecued chicken sandwich and it was delicious, but the french fries were bland. A wonderful porch with slight view of the speedway was a plus.

Stuart, FL (Witham Field—SUA)
Chantel's Restaurant 🍔🍔🍔

Rest. Phone #: (772) 286-1988
Location: On the field
Open:

PIREP   FBO Comments: Galaxy Aviation will provide transportation to Chantel's Restaurant located in the old Grumman Complex. It's within walking distance a few hundred feet from Galaxy Aviation: (772) 781-4720.

PIREP   Chantel's has great food and is priced well, too. It's a breakfast and lunch spot, but the owners Richard and Patti will host evening scheduled events too.
You gotta' try this place—the Angus burger is big and worth the wait.

Tallahassee, FL (Tallahassee Regional—TLH)
Gordo's 🍔🍔

Rest. Phone #: (850) 576-5767
Location: 5-minute drive
Open:

PIREP   FBO Comments: FBO is nice. Staff friendly. Didn't have to wait long for a courtesy car to arrive (brand new Ford). Easy to get from runway 18/36 to FBO. Look for a large cottage-style building and an Avitat sign. All taxiways were closed, so had to back taxi.

PIREP   Gordo's serves excellent authentic Cuban food, and the staff is extremely friendly. Prices are great and there is seating on an outdoor patio. It's only about a 5-minute drive from the FBO. Take Capitol Circle to Blountstown Hwy (right) and it's down just past Ocala on the right next to a gas station. It's small, so if you're not looking for it you'll miss it.

PIREP   I flew in two days ago, only had to wait 15 minutes for a crew car. Brand-new Ford Escape, leather. Fuel prices about average/a little high. Yeah, it's about four or five miles into town, but there's everything you could ever want. Probably 30 places to choose from on the strip, some Mom and Pop shops and some fast food. Easy airport to fly into and good service.

Tampa, FL (Tampa Intl—TPA)
International Mall 🍔🍔🍔

Rest. Phone #:
Location: 5-minute walk from FBO
Open:

PIREP   One of the best spots to fly for lunch or dinner? TPA! I live in Sarasota and fly to Tampa International for dinner all the time. There is one FBO—Raytheon and the folks there are great. I've never purchased fuel there (don't need it flying from SRQ), but they are always very friendly and helpful. From there it's about a five-minute walk to the International Mall (or Raytheon is always happy to drive you there). At the mall you will find several great restaurants like the Cheesecake Factory, Profusion (sushi restaurant), Bamboo Grill (great Asian food), and many more! Bear in mind though, there is a huge mall attached to these restaurants and some spouses may find it too hard to resist!

Tampa, FL (Peter O Knight—TPF)
Tate Brothers 🍔🍔🍔

Rest. Phone #: (813) 251-2767
Location: Close by, FBO will drive you
Open: Daily: Lunch and Dinner

PIREP   FBO Comments: Atlas Aviation was awarded the FBO contract this summer. Tampa Flying Service is no longer in business. Atlas is in the same building as Tampa Flying. Atlas is taking a customer service approach. They aim to expand to other airports.

PIREP   Tate Brothers is an awesome place to eat. They have a great cheeseburger and the best pizza around. The owners are from New York City and make all their food New York style. The restaurant is located a mile from the airport. The folks at Atlas will give you a ride and pick you up.

PIREP   FBO Comments: Tampa Flying Service is the only FBO at Peter O. Their number is (813) 251-1752. They have a large staff with experience on all sorts of airplanes.

PIREP   The Channel side district is located approximately three miles from Peter O. The FBO can call a cab for you. The district has a Hooters, IMAX theater, and several other restaurants, including Stumps. I highly recommend Stumps. They have a large menu with very good food.

Titusville, FL (Arthur Dunn Air Park—X21)

Dixie Crossroads 😋😋😋😋

Rest. Phone #: (321) 268-5000
Location: 1 mile, short cab ride
Open: Daily: Lunch and Dinner

PIREP   Dixie Crossroads is as good ever but no more free taxi rides.

PIREP   The best fly-in eatery in the state of Florida? My vote is for Dixie Crossroads. They offer a full menu of terrific seafood. All you can eat rock shrimp for $16.95. The restaurant will even pay for your cab ride to and from the airpark!

Go early, as the wait can often be over two hours.

PIREP   Dixie Crossroads in Titusville, FL is just a short taxi ride from Author Dunn (X21). The restaurant even picks up the taxi tab. This is an excellent place for a great seafood or steak at a reasonable price. Walkwitz FBO has a clean facility and friendly personnel. The skydivers put on a good show also.

PIREP   The Dixie Crossroads pays for the cab, approximately one mile from airport—get the broiled rock shrimp!!!

Titusville, FL (Space Coast Regional—TIX)

Outer Marker Cafe 😋😋

Rest. Phone #: (321) 264-9644
Location: On the field
Open: Irregular

PIREP   Stopped at the outer marker for the first time on the way back to Melbourne from a great weekend in Cedar Key. I had the best grouper salad I have ever had. The grouper filet was huge and the whole thing only cost $6.75!! Nice view of the airport too.

PIREP   Restaurant now open after hurricane repairs. Very good, large burger. I give them a four-burger rating. Open seven days, breakfast and lunch. Located on west side of airport next to Discovery FBO.

PIREP   I don't get to go to all of the restaurants listed on the $100 Hamburger website and our companion book. At least not every year. We have more than 1600 of them. That's a lot of burgers.

Every Wednesday a group of aviators from all over the central Florida area descend on TIX and the Outer Marker. I happened to be there this past Wednesday, the day before Thanksgiving.

This is a great towered airfield with long wide runways. NASA's Spaceport is just across a narrow bay—or is it a wide intracoastal canal? No matter. You'll get a wonderful view of the launch facility and tourist center as you land and depart.

The restaurant is all that you expect from a ramp-side diner and more. The prices are unbelievably low. The food and service were a treat.

Do yourself a favor and head to TIX some Wednesday morning—you'll be with a *large* congenial crowd.

Venice, FL (Venice Muni—VNC)

The Cockpit Cafe 🍔🍔🍔🍔

Rest. Phone #: (941) 484-5428
Location: On the field
Open: Tues.–Sun. 7 AM–2 PM; if they see you coming and it's 2 PM they'll keep the door open for you!

PIREP   Great little surprise. We did not know that Aero Squadron was closed. Local FBO recommended Cockpit Cafe. Very friendly staff. Excellent atmosphere and homemade burgers. Open 0700-1400. Breakfast all day on Sunday.

PIREP   FBO Comments: Very busy airport, especially since Hurricane Charlie wreaked destruction on nearby Charlotte Co (Punta Gorda). Runway 13 is right-hand traffic (noise abatement). Keep head on swivel: planes were taking off and landing in both directions on two of three runways. Very courteous pilots.

PIREP   Tuesday through Saturday 7 AM–2 PM. Located NE corner of field. Transient parking is on N ramp. My kind of place—a classic on-airport diner with good food and friendly service. Although this was my first trip there, I felt like I was part of the local crowd. I agree with the "must visit" recommendation.

PIREP   This was my first $100 hamburger! The Cockpit Cafe is easy to get to from either of the runways, and there was plenty of parking available when I arrived. Despite the available parking, the restaurant was very full at the time; however, the chef definitely knows what he is doing. We sat at the counter and watched him work his magic. Amazing what one chef can do. The food looked and tasted great. This may have been my first $100 hamburger, but I don't think too many places are going to beat this one. I highly recommend it to anyone and everyone.

PIREP   I keep flying back for more. The place is run by a wonderfully innovative chef/owner; the waitress is cute, aspiring model; the wife of owner, new mom, helps out; the food is *greeeaaatt*. Try the specials of the day— the potato salad is superb, meatloaf will make you overgross at takeoff. All in all, the best value and food around.

PIREP   Went to the Cockpit Cafe today for the first time. The food is *excellent*, and the staff and owners are very friendly. I look forward to going back many times. They are just about to add outside tables, so if you like to fly with Fido, call ahead to see if the outside tables are in yet. Dogs will be welcome (outside only) and Kirsten and Jarda (the owners) are going to put some doggie specialties on the menu.

The Cockpit Cafe, a *must* place to go.

PIREP   The Cockpit Cafe is open again. The chef has taken over management, and the food is great. The service is also friendly and quick. Open 8 AM–2 PM Tuesday through Saturday and 8 AM–1 PM on Sunday. On Fridays they serve a superb clam chowder and they have a soup of the day every other day. They are all homemade, of course. The location is at the old Huffman building, and you can park your plane next to the restaurant.

P.S. The FBO is now at the Triple Diamond Jet Center.

PIREP　They only take cash.

Sharky's on the Pier Restaurant 🍔🍔🍔

Rest. Phone #: (941) 488-1456
Location: Walkable—on the beach!
Open: Daily: Lunch and Dinner

PIREP　Sharky's is great. They have two menus. One menu is for the deck area outside and the other is for inside dining. I highly recommend the Bimini Burger with fries.

I walked from Triple Diamond Jet Center. The walk was maybe half a mile. As you leave the airport, you can cut through a field at the west end of the airport. This maneuver cuts down on the trip by half a mile.

PIREP　A follow-up to the "crossing the field to save time" idea: we did it and spent ten minutes picking sandspurs thorns out of our socks, shoes, and pants. More of those prickly things than I have ever seen in one place before. Walk *around* the field, not through it. The distance savings is *not* worth the time or the pain!! The food was great, by the way. Very fast service, reasonable prices, and you cannot beat the view from the patio.

PIREP　Sharky's is very nice. Menu is limited to surf and turf, and a vegetarian friend was hard pressed to find something suitable, but the staff was very nice. The food quality is fair to good, and the price reasonable. The restaurant is located on the beach and has an attached pier where local fisherman gather. There is also an outdoor patio with bar. A steel drum band was playing once. In all, it's a very good experience, as you can eat, watch the sunset, walk the beach and pier.

It's a half-mile walk to Sharky's from the airport, but every time I came in, one of the large FBOs, Suncoast Flying Services, drove my party to and from the restaurant in a rental car. Once the restaurant was full, but somehow our driver got us a table. The FBO even picked up some Dramamine for a pax who was a little ill, and refused tip.

If you use this service, at least buy some fuel at the FBO, as that's how they make their money. I cannot say enough nice things about the service and people at Suncoast Flying.

PIREP　My wife, son, and I recently made the flight from Tampa and had a super time. We could not have asked the people at the restaurant to have been any nicer. They are used to fly-ins and really cater to you. We finished supper about 9 PM, and the shuttle was no longer running so one of the waitresses piled us into her car for transport back to VNC (approximately one mile). Really a great family place to go, walk the beach, and see a beautiful sunset.

Vero Beach, FL (Vero Beach Muni—VRB)

CJ Cannon's 🍔🍔🍔🍔

Rest. Phone #: (772) 567-7727
Location: On the field
Open:

PIREP　Excellent dining experience. Taxi up to the door, friendly service, good food at a reasonable price. Dining outside or inside you will find a combination of locals and retirees. A real enjoyable experience.

PIREP　Since retiring to Florida over nine years ago, I have frequented C.J. Cannon's twice a month. Always have a great meal, good service, and reasonable prices. Well worth my 200-mile round trip.

PIREP   Vero Beach, FL's airport restaurant, C.J. Cannon's is superb!!! Great lunch menu within a very short walk from aircraft parking. The restaurant fills up during lunch as it is a popular place for all, not just the airport crowd.

PIREP   Vero Beach VRB. Park in front of the restaurant. Sit in a window seat and watch the flight school trainers hitting nose wheel first. Good food and reasonable prices.

West Palm Beach, FL (Palm Beach County Park—LNA)

Atlantis Grill 😋😋

Rest. Phone #: (561) 641-3330
Location: Within walking distance
Open: Daily: Lunch and Dinner

PIREP   Palm Beach County Park, commonly referred to as Lantana (LNA) on the airways.

Convenient restaurant within walking distance (100 yards) from the FBO called the Ark.

Huge, all-you-can-eat lunch buffet, fast service, clean, and friendly. The FBO also has a small snack bar at this busy uncontrolled field and fascinating old-time decor, complete with a Stearman parked outside.

PIREP   The outdoor cafe at Lantana has been closed down forever! Thanks to the county code guys. There are a couple of good restaurants less than a mile from the field: Rosalita's, excellent Tex-Mex fare, and Atlantis Grill, also quite good.

West Palm Beach, FL (Palm Beach Intl—PBI)

391st Bomb Group 😋😋😋😋

Rest. Phone #: (561) 683-3919
Location: On the field
Open:

PIREP   FBO Comments: When we landed we told them we wanted to go over to The 391st Restaurant. They immediately had a courtesy van come to the plane and take us over. We gave our fuel order to the line crew and were off. What they did not tell us is we needed to check in at the front desk. Arriving two hours later, our plane had not been filled and after waiting a half-hour for a fuel truck, we decided to just leave because we had more than enough fuel to get home. As a courtesy, we just like to buy fuel when we use the facilities, but our Cessna 310 obviously did not need the amount of fuel as all the jets on the ramp. Go inside and fill out a fuel slip or you could wait a while. Our fault for not asking.

PIREP   We picked this restaurant because of the current review and it was about an hour flight for a Sunday morning brunch. When we landed Galaxy was kind to have a courtesy van take us over to the restaurant, but after eating so much we took the short walk back. Signature Flight Support is even closer for those that need to roll back to the plane. We were fortunate: we did not have a reservation and they squeezed us in upon arrival. I would suggest calling ahead at (561) 683-3919, for it is a busy place on Sunday morning. Of course, the atmosphere is perfect for a restaurant on the field. We took friends of ours and they loved it.

Now the food—you name it, they have it. For those not piloting, they have champagne or mimosas. So much food so I only have room to mention the highlights: perfectly done roast beef, ham, cooked and smoked salmon, blackened fish and chicken, made to order omelets, eggs Benedict, Belgian waffles, fresh fruit, pasta bar, Italian dishes, fish (!) dishes, salads. Please leave room for desserts: pies, cakes, ice cream, and my favorite Bananas Foster! The servers were friendly, the atmosphere perfect, the location couldn't be closer to an FBO's, and the food was well worth the trip!!

Enjoy.

PIREP   FBO Comments: Excellent facilities catering to both the high-end charter jet types and the Cessna 172/Piper 140 crowd.

PIREP   The 391st Bomb Group is a WWII-themed restaurant that adjoins Palm Beach International Airport. Outside the restaurant there are WWII-vintage Jeeps, transport vehicles, and a couple of WWII planes. You enter through a sand bag–walled bunker. Many of the tables overlook the runway. It offers a reasonably priced and excellent Sunday brunch. Best to go early before the church crowd fills the place. Park your plane at Galaxy Aviation FBO, and they'll give you a to-and-from limo ride.

Zephyrhills, FL (Zephyrhills Muni—ZPH)
Caddyshack Grill 🍔🍔🍔

Rest. Phone #: (813) 598-6598
Location: At the golf course adjacent to the field
Open:

PIREP   Lupton's Barbecue Buffet ((813) 715-1511) just opened and is great. All you can eat for one price ($7.99 lunch, dinner is $7.99 before 1800, $8.99 from 1800 to closing). Exactly one mile from the airport office building, it is worth the walk. For lunch and dinner it's barbecue ribs, pork, and chicken among others, country veggies, a salad bar, spaghetti bar, fruit bar, bread bar, dessert bar plus hand-dipped ice cream, iced tea or coffee included free. Complete breakfast bar from 0700–1000 too at $5.99 on weekdays, $6.99 weekends. Friendly staff, good service. It is already popular, so at peak times is a bit crowded and there can be a wait for a table, but it's worth the wait. I eat there once or twice a week. You do have to watch your takeoff weight after eating, though.

Directions: from the airport go west to stoplight at Rte. 301, turn right, and the restaurant will be about 200 feet ahead on the right, before the Village Inn.

PIREP   FBO Comments: Best fuel prices in area.

PIREP   Caddyshack Grill located at golf course due south of the terminal office at edge of field is an easy walk. It will close for summer in May, open back up in mid-October, and remain open all year after that. It is open 8 AM to 5 PM daily. It is not fancy, just a little sandwich shop at the golf course office. It has daily specials and the prices are really reasonable.

PIREP   Several pilots raved about the jambalaya they had for lunch. Others in the past have talked about the hot breakfast and good homemade sandwiches for lunch. Now that the Tampa Bay Soaring Association has located next to the skydiving facility, pilot gets to watch the skydivers plus the glides while enjoying a good meal.

The Hundred Dollar Hamburger

Georgia

Alma, GA (Bacon County—AMG)

Pizza Hut 🍔🍔

Rest. Phone #: Unknown
Location: Short drive. FREE crew car
Hours: Daily: Lunch and Dinner

PIREP Alma, GA is everyone's favorite fuel stop with one of the lowest 100LL prices in the country.

Well, I can see why we have not had much action on the Burger reports from Alma. We took the courtesy car into town—about a 5-minute drive. Turned right and started hunting.

Basically, there is Dairy Queen, Pizza Hut, McD's, and the usual array of fast foods. There is also LB's Family Steakhouse which, unfortunately, is only open from 5–9 PM, Thursday–Saturday. As we were determined to eat, we opted for Pizza Hut which, as Pizza Hut goes, was consistent with about all the other Pizza Hut's at which I have ever eaten.

So, keep stopping for fuel, and if you need a quick bite, by all means, stop. The people are friendly and the keys are in the car 24/7.

Atlanta, GA (Fulton County Airport-Brown Field—FTY)

Flight Deck Cafe 🍔🍔🍔

Rest. Phone #: (404) 699-7730
Location: On the field
Hours: Mon.–Fri. 11 AM–3 AM

PIREP Fare is good, and it is located right on the airport. The only problem I have is the hours of operation, Monday–Friday 11 AM–3 AM. It seems like I have been getting there before they open or after they close.

PIREP At Fulton County (FTY) in Atlanta there is the Flight Deck Cafe. They serve standard airport fare, breakfast through early dinner! I base there and I like it!

Atlanta, GA (Dekalb-Peachtree—PDK)

57th Fighter Group 😋😋😋😋

Rest. Phone #: (770) 457-7227
Location: On the field
Hours: Daily: Lunch and Dinner

FBO Comments: Epps. busy but gave us a crew car. Expensive gas but cheapest on the field.

PIREP   The 57th Fighter Group has an excellent Sunday brunch. It is served as a buffet. It includes an omelet station and a carving station, a variety of breakfast foods, seafood selections, many hot entrees and sides, and a fabulous selection of desserts. Bubbly is included (after 12 noon) for those who aren't PIC. We thought the food and selection was excellent. The price was $25 per adult, which is about right for Atlanta. This brunch was very busy when we were there. Most of the other diners were locals and did not fly in.

Although the place is on the field, they don't have aircraft parking available. They recommended that I park at Mercury or Epps FBO (Signature and a couple of others are also on this field) and that either one would give us a ride over. I parked at Epps and they handed me keys to a crew car! Pretty nice considering that they are really busy and I showed up unannounced in a spam-can. I noticed, however, that if I had parked at Mercury the restaurant would have been within reasonable walking distance.

PIREP   My wife and I ate at the 57th Fighter Group. It apparently re-opened some time ago under new owner-ship. It has been several years since I ate there. The decor is still basically the same, with the outside displays being somewhat changed. The food was very good, although the service was a little slow and the prices may be a little high. However, it was very enjoyable to be able to eat a nice meal in a relaxed atmosphere while watching the planes through the large picture windows facing the runways. It is worth a trip there every now and then.

The Downwind 😋😋😋😋

Rest. Phone #: (770) 452-0973
Location: On the field
Hours: Daily: Breakfast, Lunch, and Dinner

FBO Comments: Landed at Mercury, no junk fees if there less than two hours and no obligation to buy fuel like at the other FBOs at PDK.

PIREP   I have wanted to see what all the fuss was about at the Downwind. We arrived on a Saturday around 1 PM. Although they were busy, there was plenty of wait-staff to take care of us. We had our food inside of 15 minutes. I highly recommend the Greek salad with blackened (Cajun) chicken on top (no brown lettuce). Their sweet tea isn't that great, but the scenery made up for it. Reasonable prices considering where in Atlanta I was. I guess they've been reading this website and heeded the advice.

To save money, park at Mercury Air Center when flying into PDK, as there is no charge to park if there under two hours.

PIREP   The Downwind at PDK could be ten times better than what it is. Food is okay but overpriced. The same stands for the ambiance/decoration. The only justification for the price is the view over the ramp and runways. One of the worst things about the Downwind is that it is difficult to find a table outside. The owner decided a year ago to cover the beautiful terrace, making about 80 percent of it a big dark glass-enclosed room. Now there are only about six tables on the very small terrace. Guess what? Everybody wants to sit outside—very difficult to get a table when the weather is good. Sitting inside kills the purpose of going to the Downwind, which is to sit on a terrace over the ramps. I hope the owner will understand that covering his terrace was a waste of money and will uncover it again, providing sun umbrellas for the hot summer days. In winter, he already has a big room inside! By the way, the hamburgers are not that great.

Augusta, GA (Daniel Field—DNL)

Village Deli 😋😋😋

Rest. Phone #: (706) 736-3691
Location: Across the street
Hours: Daily: Lunch and Dinner

PIREP The Village Deli is just across the street (a good walk) from Daniel Field. Inside, the west wall is covered in aviation-themed signs, pictures, models, and even airplane parts. The east side has a golf theme (what else in Augusta, GA?). Pricing is moderate, the service very friendly/casual, and the food quality is very good (be sure to try a basket of their seasoned fries). Weekday lunchtime is usually very busy, so plan to take your time—but worth the wait.

There are several places to eat (and choices) in the Daniel Village Shopping Center (the first shopping center in the U. S.). Our favorite is the Deli.

PIREP Wing Stop is a new restaurant in Daniel Village across the highway from Daniel Field. Obviously they serve wings, some of the best you'll find. Just fly into Daniel Field and park by Popeye's, walk across the highway into the corner of Daniel Village, and you'll find it. There are several very good places to eat in Daniel Village.

Barbarido's is a new restaurant across the road from Daniel Field in Daniel Village. Just taxi over next to the fence by Popeye's (which you can also eat at) and walk across the highway. Excellent food with generous portions.

Augusta, GA (Augusta Regional at Bush Field—AGS)

Tailwinds 😋😋😋

Rest. Phone #: (706) 842-0954
Location: On the field
Hours: Daily: Breakfast, Lunch, and Dinner

PIREP Bar-and-grill style restaurant located in passenger lounge of commercial portion of airport. Nothing fancy, but our barbecue sandwiches were tasty and reasonably priced, and service was very pleasant.

Brunswick, GA (Malcolm McKinnon—SSI)

Dresser's Village Cafe 😋😋😋

Rest. Phone #: (912) 634-1217
Location: Short drive, crew car available
Hours: Daily: Breakfast, Lunch, and Dinner

FBO Comments: Nice FBO. Like the island, it has a charming, rundown-but-clean-neat-and-cozy feel to it. Courtesy car (three available, call ahead).

PIREP Take the courtesy car to the village. There is a large selection of restaurants and cafes. We tried Dresser's Village Cafe, which is obviously a local favorite and serves breakfast all day.

Barbara Jean's 😋😋😋

Rest. Phone #: (912) 634-6500
Location: Short drive, crew car available
Hours: Daily: Lunch and Dinner

FBO Comments: The FBO will loan you a crew car to get to town. I usually purchase some fuel as a courtesy.

PIREP   The BEST crab cakes in the world! No filler, just crab. Don't forget to try the crab soup as well. If you love seafood, you'll love Barbara Jean's. It's only five minutes from the airport.

Cornelia, GA (Habersham County—AJR)
Runway Fish House 🍔🍔🍔🍔

Rest. Phone #: (706) 776-1238
Location: 1/2 mile
Hours: Fri., Sat. 5 PM–9:30 PM; Sun. 11:30 AM–2:30 PM

PIREP   The Runway Fish House is still dishing out delicious meals. It is open 5 to 9 on Friday and Saturday and 11 to 2 on Sunday. On Friday and Saturday they have a seafood buffet and on Sunday it is a buffet with plenty of Southern-type dishes.

It is located just off the south end of the runway and is an easy half-mile walk. You especially need the walk back after the delicious all-you-can-eat buffet.

PIREP   It's an easy flight from the Atlanta area. After parking your aircraft, walk out to the road and turn right. The walk is about a half-mile to the Runway Fish House. After eating, you'll be thankful for a little walk to pack down the excellent food!

It's a buffet! They are only open on Fridays and weekends from what I understand. We went on a Sunday morning thinking that this was a quaint little out-of-the-way restaurant. After seeing the parking lot in front of the restaurant, I was expecting tour buses to pull up.

The food sure draws 'em in. It's a "come on in" restaurant, so you can relax and enjoy your meal. There was a long buffet with plenty of vegetables, catfish, fried chicken, and fixin's to satisfy any craving. They also had boiled shrimp when we were there, which is a favorite of mine. After two notches on the beltline, we topped lunch off with ice cream.

Thank goodness for the walk back, we needed it. This is a great place to eat for a home-cooked type meal.

Dahlonega, GA (Lumpkin County-Wimpys—9A0)
The Smith House 🍔

Rest. Phone #: (706) 867-7000
Location: 6 miles, NO pickup, no taxi, no crew car
Hours: Daily: Lunch and Dinner

PIREP   The Smith House no longer sends a van to the airport. If you decide to go, make sure to call a shuttle service before you arrive.

PIREP   The Smith House is known far and wide as a good family restaurant to visit when enjoying the fall foliage in the North Georgia Mountains. It is several miles from the airport, but they will send a van for you and are VERY used to doing this. Auto traffic in the mountains in the fall is awful, so lots of their customers actually do fly in. The ride is great because since you're not driving, you get to see the foliage.

The restaurant is one of those big family dining experiences. The food is so-so but reasonable in price (less than $8 for a BIG lunch). It's basic meat-and-potatoes. The three times I've been there, the service has been outstanding. Very friendly and fast.

The town is worth visiting; it's all an easy walk. Georgia had a mini-gold rush in the late 1800s, and Dahlonega was its center. The gold dome on the courthouse (great landmark for aerial navigation) is real gold lamé. You can pan for gold right on Main Street and do other goofy tourist stuff. Be sure to tell the restaurant van driver when you'll be back (or ask if they are on a schedule). I've never had a problem coming and going when I liked. Yes! Gold Leaf is correct.

Overall, I give the restaurant two burgers for the food, but I give the airport-town-restaurant experience a solid four out of five. Do it in the fall on a clear day.

BTW, there are several eateries that offer shuttle service, but I've not tried them so I can't comment. The last time I flew into Wimpey's, the shack was open with a phone, and the Yellow Page listings told who had shuttle service.

Dawson, GA (Dawson Muni—16J)
Paul's 🍔🍔🍔

Rest. Phone #: (229) 995-2816
Location: Crew car
Hours: Daily: Lunch and Dinner

PIREP   Take a trip back in time with a visit to Paul's Restaurant in Dawson. It's a great country buffet with fresh veggies and even hoecake corn bread.

The husband and wife team at the airport are friendly and have a courtesy car available.

This has become one of our favorites.

Douglas, GA (Douglas Muni—DQH)
Holiday Inn 🍔🍔🍔

Rest. Phone #: (912) 384-9100
Location: Across the street
Hours: Daily: Breakfast, Lunch, and Dinner

PIREP   There is a Holiday Inn across the street with a great buffet.

You can park on the ramp by runway 22 and you're only 2000' from Wendy's. There are two other restaurants, hotels, and a Wal-Mart within half a mile. Great place to spend the night and wait out the weather.

Dublin, GA (W H 'Bud' Barron—DBN)
Casa Maria 🍔🍔🍔

Rest. Phone #: (478) 275-7722
Location: Crew car, short drive
Hours: Daily: Lunch and Dinner

PIREP   Excellent location to stop and rest or get a quick meal. The (DBN) airport is one of the better in middle Georgia, with 6000+ feet of uncontrolled runway with an ILS.

Excellent FBO and friendly staff. Courtesy car for getting to a WIDE variety of restaurants within three minutes of the field. And, if you need to spend the night, there is a host of hotels/motels within five minutes of the field.

One of our favorite restaurants is Casa Maria. They have great Mexican food with excellent service.

Griffin, GA (Griffin-Spalding County—6A2)
Frank's Country Cooking 🍔🍔

Rest. Phone #: (478) 994-2160
Location: Short walk
Hours: Daily: Breakfast, Lunch, and Dinner

PIREP Frank's Country Cooking is now within walking distance of the terminal. I think Monday through Saturday. All-you-can-eat buffet or order from menu.

Good food.

Jekyll Island, GA (Jekyll Island—09J)
Jekyll Island Club 🍔🍔🍔🍔🍔

Rest. Phone #: (912) 635-2600
Location: Short walk
Hours: Daily: Breakfast, Lunch, and Dinner

PIREP If you are interested in impressing someone with some great food, the Sunday morning brunch at the old hotel (I think Radisson owns it now) is one of the classiest eating experiences you can hope to have. The hotel will send a van to the airport and return you when you're done. The hotel was once a vacation retreat for the Rockefellers, Vanderbilts, etc., and the food is spectacular. We recommend the sampler plate, but with the omelet station, salad buffet, and dessert bar, you can expect to leave the airport with a slightly different center of gravity. There is no FBO or fuel available at Jekyll, but Malcolm-McKinnon (SSI) offers a full-service FBO and is five minutes away. Malcolm-McKinnon, by the way, has one of the friendliest FBOs around and also provides a courtesy car with a lot of nice restaurants around.

PIREP We had lunch at the Jekyll Island Club 09J and it was really great. The prices are reasonable, and the food was excellent. It is only about 1,500 feet from the airstrip and is a great short walk, or you can call the club and have a car pick you up at no charge.

PIREP It was great flying in to a desolate airport only a mile from the resort. There was only one plane there, a Piper Archer or something like that…could have been a Cherokee 140! It was raining lightly when we arrived after a 50-minute flight (190 kts groundspeed due to 20+ tailwind). The Club Hotel was very nice, and the staff was excellent. We stayed in Crane Cottage and had a large room with fireplace, private balcony, wood floors. This is adjacent to the main (huge) club building. We had a snack at Café Solterra midafternoon. This is more like a well-stocked snack bar open from 8 AM–10 PM. Dinner was in the Grand Dining Room. Jumbo blackened scallops were excellent. Then lobster bisque (I have had better). Then a salad. I had a salmon entrée and Carrie had pheasant. These were okay. Presentation and service were excellent. Atmosphere was excellent. Afterward, we had a chocolate mousse for dessert.

The next day we toured the historic district on the red tram. The tour guide was very knowledgeable and friendly and took us into two of the "cottages." One was 8,500 square feet and the other 10,000 square feet. These were the vacation homes of the rich around the turn of the century. Our stay was very relaxing. We were surprised how quiet it was there.

It was fun leaving, as I had to wait for a clearance and void time to get out of there. Unexpected fog rolled in and two planes were on their way and couldn't get in. Once they diverted, we were cleared to depart and punched through the overcast at about 3,000'.

We made it back to Orlando in one hour—the only way to travel!

Lagrange, GA (Lagrange-Callaway—LGC)
Hog Heaven 🍔🍔

Rest. Phone #: (706) 882-7227
Location: In town; call for courtesy van
Hours: Daily: Lunch and Dinner

PIREP   A courtesy van plus many local restaurants within five to ten minutes. If you need a cab, $3 will get you anywhere in the city. Hog Heaven is a great barbecue place, a one-mile drive. The best hot dogs in the world are at Charlie Joseph's ((706) 884-0379), next door to Hog Heaven. For a sit-down lunch or dinner, try Cleve's Place ((706) 884-2222).

Listed below are other choices:
Cisco's—Mexican (706) 883-6100—1-1/4 miles, 5-minute drive.
Los Nopales—Mexican (706) 883-8547—near the mall, 15-minute drive.
Ruby Tuesday—American 5 miles downtown, 10-minute drive.
Hog Heaven—BBQ (706) 882-7227—2 miles, 3-minute drive.
Bamboo Garden—Chinese (706) 884-7701—not that good.
Banzai—Japanese (706) 882-0750—expensive but entertaining.
Foxes Pizza—(706) 885-0005—good pizza, 10 minutes.
Zaxby's Chicken Fingers and Wings (706) 882-0440.
Jimmy's—Southern (706) 845-0406—downtown.
Milano's—Italian (706) 884-6100—other side of town, 15 minutes.
Spring House Inn—formal dining (706) 812-1546—15 minutes.
Katie's Deli Cafe—(706) 884-0267—only open for lunch, great buffet, good food, Southern cooking.
Basil Leaf—downtown, fine dining, lunch and dinner.
Pizza Villa—unique Italian restaurant in a romantic atmosphere.
Waffle House—near I-85.
Higher Grounds—coffee shop near the mall (like a local Starbuck's).

Lawrenceville, GA (Gwinnett County–Briscoe Field—LZU)

The Flying Machine 😋😋😋

Rest. Phone #: (770) 962-2262
Location: On the field
Hours: Daily: Breakfast and Lunch

FBO Comments: Piedmont Hawthorn is very fast and has friendly staff, but they are on the other side of the field. You can, however, park in front of the restaurant, and they will bring the fuel to you. The airport is in a great location to avoid the heavy air traffic in Atlanta.

PIREP   The Flying Machine, located on the airport, is a great stop to make on a long flight. The food is great, you get your money's worth, and the staff is very friendly. I have been there many times and have tried just about everything. The salads are huge and their hamburgers—well, let's just say you better ask for extra napkins!

FBO Comments: Filled up with fuel while we were there. Fuel is kind of high compared to our area, but FBO folks were very friendly.

PIREP   Flew our 1953 Cessna 180 down for lunch again Sunday. This is the third or fourth time we have been to the Flying Machine, and it is always excellent. Wife had fried shrimp and I had the hamburger steak, which was huge and on a pile of rice. Very good. Service is very good and friendly.

Olde Towne Grille 😋😋😋😋

Rest. Phone #: (770) 499-7878
Location: 1 mile, use crew car
Hours: Daily: Lunch and Dinner

FBO Comments: Buy your fuel at LZU, and then scoot down to Olde Towne!

PIREP   Now this is eating! You have to use a crew car to get there. About one mile down 316 from LZU is one of the greatest restaurants around. The food is fantastic, lots of screens for sports and news, and a classy atmosphere.

Lawrenceville is nonsmoking, so you can also eat without wheezing!!!

The wait-staff is very prompt, and you will enjoy fantastic service from these people!

McRae, GA (Telfair-Wheeler—MQW)
19th Hole Deli 🍔🍔🍔🍔🍔

Rest. Phone #: (229) 868-7474
Location: Less than 1/2 mile
Hours: Daily: Lunch

PIREP   The McRae Airport (MQW) is located on site at the Little Ocmulgee State Park in McRae, Georgia. The park features a 60-room lodge with restaurant, the 18-hole Wallace Adams Golf Course, tennis, camping, and fishing.

The 19th Hole Deli at the golf pro shop, operated by Eunice (a spunky 70-something lady), offers great sandwiches at low prices. You can walk to the park very easily. I can personally testify that it is 4/10 of a mile (statute) and takes about 10–15 minutes. If you prefer, you may pick up the phone in the FBO; it automatically dials the front desk at the park. They will run over and pick you and your gear up.

This is more than a burger run.

Macon, GA (Macon Downtown—MAC)
El Sombrero 🍔🍔

Rest. Phone #: (478) 750-8159
Location: In town; crew car available
Hours: Daily: Lunch and Dinner

PIREP   The folks at the FBO are very, very nice and will offer (or you may ask) for their courtesy car, NO CHARGE!!

Just a short distance (they will give you a map) is an excellent Mexican restaurant, along with a variety of fast food places.

Milledgeville, GA (Baldwin County—MLJ)
Cornbread Cafe 🍔🍔

Rest. Phone #: (478) 452-4812
Location: Crew car available
Hours: Daily: Lunch and Dinner

PIREP   Another place to eat is at Choby's. It is located on the lake and a short drive. Every time we have called them, the owner has picked us up and taken us back. Very good seafood.

FBO Comments: Very friendly, helpful service. A car was provided for our lunch run. On our return from lunch, our aircraft had been refueled and the FBO had checked with others to help us locate our photo objectives.

PIREP   We had flown over to photograph Rock Eagle and had not found it from the air. Food at Cornbread Cafe was good and plentiful and cost very little.

Pine Mountain, GA (Callaway Gardens-Harris County—PIM)
Callaway Gardens 🍔🍔🍔🍔🍔

Rest. Phone #: (706) 663-2281
Location: On the field
Hours: Daily: Breakfast, Lunch, and Dinner

PIREP   Pine Mountain is a great place to stop by if you are looking for a fly-to place specifically for breakfast, lunch, or dinner. PIM is 22 miles north of Columbus, GA. When we flew in, the FBO was most courteous, asking us if we would like a courtesy car to come pick us up while we were five miles out. The attendant came out, guided us to parking, and chocked our wheels for us as we pulled up. The van came directly to the plane and drove us to the Country Kitchen for lunch. Along with being a quaint country store where you can pick up knick-knacks while you are being seated or checking out, the dining area overlooks the valley, which gives a terrific view while you eat. Service is good and the prices reasonable. After ordering, they bring you a basket of hot biscuits to munch on while you wait for your meal. The hamburgers for lunch were thick and delicious. I've also been there for breakfast. The scrambled eggs, bacon, grits, and biscuits are a great way to start a day of flying. They also have an inn/hotel at Callaway that serves a buffet for dinner that I have tried with the wife; it was quite nice also. As you check out, you can have the cashier call the van for you, which will pick you up at the door and drive you directly to your plane. If you want to get a buy-in from the wife or girlfriend, this would be a good place to take 'em.

PIREP   Flew in this weekend with my wife for lunch. Airport is manned by a retired Army helio inspector named Jes. Great guy. He tied our plane down and called for ground transportation to Callaway. There are a number of restaurants there, but only one has free entry—the Country Kitchen. The others require you to buy a park admission since they are in the park. We decided to go into the park and had a great time. Nice place to visit and we will go back.

  FBO Comments: Great FBO. Friendly and quick service.

PIREP   Flew in on a Sunday for lunch and had a great experience. Standard Southern food prepared very well. Be careful arriving on Sunday just after noon, due to high volume and potential 30 minutes wait time.

Rome, GA (Richard B Russell—RMG)
The Prop Stop 🍔🍔🍔

Rest. Phone #: (706) 235-3452
Location: On the field
Hours: 10:30 AM–3:30 PM, 7 days a week on the posted sign

FBO Comments: Some of the nicest FBO employees in the state.

PIREP   The Prop Stop Deli has a new owner, Stacy Tumblin, who brings her enthusiasm and more great home-cooked food to the Rome Airport. Try the chicken salad sandwich bursting with flavor from her secret ingredients (it might be nuts and raisins). Burgers and cheeseburgers can't be beat, and the fries are always fresh too. I'd up the star rating a burger notch for Stacy and her crew. The restaurant is still open 7 days a week, too.

  FBO Comments: Easy airport to navigate to, with VOR and NDB within a few miles of the airport.

PIREP   Flew to RMG with my family today (it's my birthday, so we flew there for lunch). Airport is easy to navigate to (VOR and NDB are both very close). Service was extremely friendly. The new management is in the process of renovating the place, so it's not the prettiest $100 hamburger joint I've seen. But the food was very good and the prices were about average. My family enjoyed the trip—and the meal! We'll be back.

Four burgers, because my wife said it was the best burger she'd had in a while!

St Marys, GA (St Marys—4J6)

St. Mary's Steak and Seafood House 🍔🍔🍔

Rest. Phone #: (912) 882-6875
Location: Short walk
Hours: Daily: Lunch and Dinner

PIREP   St. Mary's Steak and Seafood House is within walking distance, about a half-mile southeast of the airport from St. Mary's Airport (4J6). They serve good seafood, inexpensive and plentiful, but there is a bit of a wait. It is not terribly fancy.

Savannah, GA (Savannah/Hilton Head International—SAV)

Muther's BBQ 🍔🍔🍔

Rest. Phone #: (912) 964-2845
Location: Short drive, crew car available
Hours: Daily: Lunch and Dinner

FBO Comments: Great service at Signature, but the gas is expensive. When you consider the service, the warm cookies, fresh coffee, and courtesy car though, even poor GA pilots come out ahead.

PIREP

Location—0, crew car required
Food—2, dang near perfect BBQ sandwich
Service—1, the owner's family runs the place
Ambiance—0, nice for what it is and where it is
Price—$5 and some change for a Coke, huge sandwich, and side of coleslaw
   To get to Muther's, hang a left at the main road and follow to next major intersection (look for the Applebee's). Turn right at the light and Muther's is about a mile or two down on the left (next to a McDonald's). Great ambiance for a "on the main road in a not great part of town" kind of place. They have a screened-in outdoor area to eat and outstanding BBQ. Not the kind of place you'd fly to just for the meal, but a great option for good eats on a X-country—we ate while the plane was fueled on the way back from Sun-n-Fun.

Vidalia, GA (Vidalia Regional—VDI)

Benton Lee's Steakhouse 🍔🍔🍔🍔

Rest. Phone #: (912) 594-6751
Location: It's a drive—beg for the crew car!
Hours: Daily: Dinner

PIREP   Vidalia, GA (VDI) in the heart of onion country. Ask Ken Nobles for the courtesy car at the airport and take the 1.5-mile or so trip to the Vidalia Onion Factory on US 280. The onion burger is tops.

   There is a Japanese restaurant also on US 280 past the Onion Factory in Lyons, GA. Nearby is Brewton-Parker College.

   If the FBO will let you have the courtesy car long enough and a map to Benton Lee's Steakhouse at Gray's Landing on the Altamaha River, you are in for a treat! Beg! Best be planning to spend the night in Vidalia.

There are motels about a mile away—Shoney's, Holiday Express are nearest. Benton Lee's does not advertise, but I did a piece on them for Fox Television and here's the scoop.

They do not serve on plates but TV trays. A small steak misses the sides of the TV tray by a couple of inches all around. The medium steak touches the sides of the tray, the large hangs OVER the sides of the tray. If you know of ANYBODY living around Vidalia, ask them to take you to Benton Lee's. As a result of the TV coverage they get people from all over the world, but you won't see the place advertised. The steaks are well cut from selected Angus beef, and while you might have better, more artistically cooked steaks, you will forget all about that!

Waycross, GA (Waycross-Ware County—AYS)
B & E Bar-B-Que 🍔🍔🍔

Rest. Phone #: (912) 287-0077
Location: 1-1/2 miles, crew car available
Hours: Daily: Lunch and Dinner

PIREP   Excellent little airport run by the county there. They have a crew car available (no charge). Free popcorn in the FBO and, believe it or not, Coke in the bottles in a vintage Coke machine.

All restaurants are a short drive but the best barbecue is a place called B&E, located 1.5 miles from the airport, and Ocean Galley Seafood, located four miles from the airport—prices are excellent.

There's also a public golf course that is really nice located just outside of town. The personnel at the airport are very nice and will help you with whatever your needs might be.

Williamson, GA (Peach State—GA2)
Barnstormer's Grill 🍔🍔🍔

Rest. Phone #: (770) 227-8282
Location: On the field
Hours: Lunch and Dinner

PIREP   Food was great; good bit of activity on the Saturday we arrived. I recommend it for group trips.

PIREP   Most airport restaurants exist only in lieu of something better. These folks are the best and their food is superb and a real bargain. The friendly atmosphere, view of the airport, and museum-like motif combine to make this the best airport restaurant experience I've had. Many show up on weekends, but I plan to make this a regular weekday jaunt.

Well worth the trip wherever you are.

PIREP   The owners, Johann and Jennifer, are working hard to have the best airport restaurant anywhere. Their burgers are the best. The menu is being tweaked with salads and sandwiches and even stuff for vegetarians. We went to a dinner party in their banquet room; the food was great. Plenty of ramp and grass parking for all types of aircraft. On low wind days expect hot air balloon operations at the northwest end of the field.

Oh yes, Jennifer is always making homemade pies; save room and recalculate the weight and balance!

Winder, GA (Winder-Barrow—WDR)
Spitfire Deli 🍔🍔🍔🍔

Rest. Phone #: (770) 867-0086
Location: On the field
Hours: Daily: Lunch and Dinner

PIREP   The best airport restaurant I stopped at in years is the Spitfire Deli at KWDR, in Winder, GA!! The owners, Celia and Ewan Lockwood, are delightful folks. Celia is a professionally trained chef. The menu is very ample for a small airport deli. The Rueben is fantastic. If you find a bigger burger, you shouldn't eat it and fly. Put this one on your "gotta try it out" list.

PIREP   My wife and I visited the Spitfire Deli at the Winder-Barrow airport (WDR) for the first time. A wider variety of menu choices were available than at most on-airport restaurants, including many homemade items. Very friendly service, good food, and a great view of the airport combined for a delightful lunch experience. The restaurant is easily accessible at the FBO ramp at the approach end of runway 5.

PIREP   Spitfire Deli, right in the terminal, just steps from the ramp. Great food, even greater people!

The Hundred Dollar Hamburger

Hawaii

Honolulu, HI (Honolulu Intl—HNL)

L&L BBQ 😋😋😋

Rest. Phone #: (808) 924-7888
Location: 5-minute drive
Open: Daily: Breakfast, Lunch, and Dinner

PIREP I know that this will make for a long flight for most of you, but if you are ever in the islands, make sure you stop in at L&L BBQ. You can get the most ONO (delicious) burgers and food in the world. Order the teri chicken or hamburger steak. It comes with one scoop rice and one scoop mac salad!! Cheap and good! L&L is on OAHU about five minutes out of Honolulu Int (HNL). There are others all over the island as well as on other islands. I flew out of HNL and Dillingham, which is a little sailplane and GA airport on the North Shore. I grew up in Hawaii and am now attending The University of North Dakota majoring in Commercial Aviation and a minor in Meteorology. It is much colder here.

ALOHA

Kaunakakai, HI (Molokai—MKK)

Airport Cafe 😋😋😋

Rest. Phone #: (808) 872-3808
Location: On the field
Open: Daily: Lunch

PIREP I lived in Honolulu for a year and was able to rent planes from the Hickham-Wheeler Aero Club. It is no longer in existence, but I did make a few trips to Molokai and Lanai for lunch! First, I landed at Kalaupapa (leper colony) (LUP) on Molokai, and there wasn't anyone around so we took off and flew up to the main airport at Kaunakakai (MKK). We had a great hamburger there! And they sell the famous Molokai bread, which you can take with you! Then, we flew to Lanai (LNY), where we caught a van that took us up to the city, which has a neat little town up in the pines at the 3,000-foot level. The air is clean and fresh, people are friendly, and there are a couple of small cafes where you can get lunch. The strip at Hana, Maui has no eating places, but Lindberg use to fly in there, and it is buried ten miles away. It's difficult to find an FBO to rent planes, but on Lagoon Road near Honolulu Int'l (HNL), they have some.

Lihue, HI (Lihue—LIH)

Marriott Hotel 🍔🍔🍔🍔

Rest. Phone #: (808) 245-5050
Location: Call for pickup
Open: Daily: Breakfast, Lunch, and Dinner

PIREP   A good day trip I'd highly recommend from Honolulu is over to Lihue (LIH) on the island of Kauai. You'll have a little bit of overwater flying, about 70 nm worth—as long as that doesn't bother you and you have flotation gear as a good backup. On your trip over there, take the time to fly around the north side of the island to see beautiful Hanalai Bay and the spectacular cliffs. Shoot over the western side of the island down Waimea Canyon dubbed The Grand Canyon of the Pacific. Keep your eyeballs out of the cockpit for the numerous tour helicopters that fly around here.

Anyway, $100 hamburger…right. Once you land over at Lihue, park the plane, go out the commuter terminal, and give the Marriott Hotel a call ((808) 245-5050) for the complimentary shuttle. They'll come pick you up in about ten minutes and take you over to the hotel, which is about five to ten minutes away. Once there, hit Duke's for lunch, which is at the hotel right on the beach. Fantastic food, great views, reasonable prices, beautiful and Yes friendly waitresses. You can't go wrong. After lunch/dinner, the shuttle will take you back to the airport (tip the driver a couple bucks—live aloha). Word of warning, if you're planning on coming back after dark, I highly recommend having an instrument rating. As LIH is right on the ocean, right after you rotate you're almost immediately in a "black hole" over the water for the long flight back, with no land in sight. I've convinced more than one student that there's no such thing as VFR at night on a moonless night over the water.

Aloha.

The Hundred Dollar Hamburger

Idaho

Big Creek, ID (Big Creek—U60)
Big Creek Lodge 🍔🍔🍔🍔

Rest. Phone #: (208) 375-4921
Location: On the field
Open: Daily: Lunch and Dinner

PIREP Big Creek offers two rustic lodges with food, trail rides, and lodging. If you like tall pine trees, spectacular mountain scenery, and a comfortable remote setting, this is it. I have only been to the Big Creek Lodge, but I have heard that the other one, Gillihan's Lodge, is just as nice. Breakfast at Big Creek is fantastic, and I have been told that lunch and dinner are just as good. Saturday mornings you can usually find many backcountry flyers show up for breakfast. It is recommended that you get reservations for meals and lodging, as both are small lodges. This airstrip is in a mountainous canyon, and mountain flying experience is highly recommended. Be aware of high density altitude, turbulent air, and ever-changing weather conditions.

Caldwell, ID (Caldwell Industrial—EUL)
Cockpit Restaurant 🍔🍔🍔

Rest. Phone #: (208) 453-2121
Location: On the field
Open: Daily: Breakfast and Lunch

PIREP The small on-field restaurant has hamburgers and a few sandwiches. Expect to do some hangar flying. BTW, the airstrip is adequate for any aircraft other than the "heavies."

Cavanaugh Bay, ID (Tanglefoot—D28)
Cavanaugh Bay Resort 🍔🍔🍔🍔

Rest. Phone #: (208) 443-2095
Location: On the field
Open: Daily: Breakfast, Lunch, and Dinner

PIREP   This grass airstrip is 150' from Priest Lake (23 miles long). The bay is also a sea base. There is camping on the airstrip with facilities. Cavanaugh Bay Resort is 100' from the airstrip and has lodging, boat rentals, and great food: hamburgers, barbecued ribs, pizza—they also serve breakfast. You just can't beat the view, food, and accommodations in this remote little place.

Challis, ID (Challis—LLJ)
Y Inn Cafe 🍔🍔

Rest. Phone #: (208) 879-4426
Location: Crew car
Open: Daily: Breakfast and Lunch

PIREP   The trip to Challis is breathtaking, to say the least! It is surrounded by spectacular, tall mountains on just about every side. On the other side of town sits the Y Inn Cafe. Although the town of Challis isn't that big, it's a little far to walk.

The food is great, and the portions are generous and the service is fast. What more can you ask from a back-country restaurant?

Getting there requires some planning. One should probably call the FBO ahead of time and check on courtesy car availability, or see if someone will give you a lift. Don't count on your cell phone; mine didn't work. There is a credit card telephone on the field and fuel in a cool red antique fuel truck in case you need some.

Driggs, ID (Driggs-Reed Memoreal—U59)
Warbirds Cafe 🍔🍔🍔🍔

Rest. Phone #: (208) 354-2550
Location: On the field
Open: Daily: Lunch and Dinner

PIREP   Our first visit to Warbirds was for the Sunday brunch. The quality and selection of food was outstanding and reasonably priced. I can't say enough about the friendly and professional staff, who actually thanked us for choosing to join them the minute we walked in the door and sincerely invited us back to visit again once we had eaten more than we will openly admit.

We really enjoyed the aircraft/hangar decor, which fostered all the adventures of aviation, with a pleasingly simplistic and upscale feel. My husband especially enjoyed the aircraft photos and historical information under the glass tabletops at the booth seating and, of course, the fact that we were practically seated on the taxiway with a beautiful mountain view just beyond. The clever menu, printed on sectionals and including aviation term specials, such as Lower Your Flaps Jack pancakes, the Biplane Bagel featuring Norwegian smoked salmon, 3G Granola from the locals, and the B-1 Buffalo Burger, was a favorite of mine. We ran a quick reconnaissance on the dinner menu and plan on returning soon for the lobster brie quesadilla appetizers, pepper-seared duck breast, prime rib, or a Rocky Mountain elk burger. By the way, there are plenty of tempting choices on the menus for those who prefer a more vegetarian-type meal.

If you find yourself anywhere near the western base of the Teton Mountain range in Idaho, navigate to the airport north of Driggs, follow the signs to the Warbirds Cafe, and do exactly what the emblem inscribed in the stained concrete floor as you enter indicates: check your attitude and get ready for a dining experience you won't soon forget.

Elk City, ID (Elk City—S90)
Tiffy's 🍔🍔🍔

Rest. Phone #: (208) 842-2459
Location: Short walk
Open: Daily: Breakfast and Lunch

PIREP I spent a lot of this past summer "airplane camping" in the Idaho backcountry, and we enjoyed flying into Elk City for breakfast. There is a little restaurant in the main part of town that we always go to for a good meal. I think the name is Molly's, but I cannot be sure of that. If you go, it is the first building on the left side of the road as you walk up the hill into town. They serve a good breakfast and also have great pies. The General Store, across the street, is a good place to load up on supplies or snacks for the flight home.

Fairfield, ID (Camas County—U86)

Iron Mountain Inn 🍔🍔🍔

Rest. Phone #: (208) 764-2577
Location: Short walk
Open: Daily: Breakfast, Lunch, and Dinner

PIREP Camas County is not very careful about runway maintenance. In spring the runway may be mud; in summer it may be soft sand. Either way it makes for a very short landing rollout and a difficult takeoff at best. Look at the runway surface before you land: if you see ruts, add power and keep going! Watch out for the power lines at the east end of the runway! The food is great.

PIREP Fairfield, ID is a small farming and ranching town whose airport has a hard gravel strip, suitable for any SEL or small MEL plane.

Across the street is the Country Inn, which has the standard roadside restaurant fare, including some home-cooked entries and homemade soup. I find it to be a convenient stop on the way out of Sun Valley (SUN) or on the way home from a weekend.

Galena, ID (Smiley Creek—U87)

Smiley Creek Lodge 🍔🍔🍔🍔

Rest. Phone #: (208) 774-3547
Location: Across the road
Open: Daily: Breakfast, Lunch, and Dinner

PIREP This is a grass strip north of SUN. Across the road is a nice but plain restaurant. The surrounding country is some of the most astonishing (awesome, if you like the jargon) in the US of A. This is mostly a summer strip and mostly small SEL's, but I have seen a Beech Starship put down there!

As Idaho backcountry strips go, this is an easy one, with plenty of opportunity for go-arounds.

PIREP The campsites have covered picnic tables and are right by a stream. There was an assortment of things to do and see there. I recommend taking a trail ride with Pioneer Mountain Outfitters. Great price, and they picked us up for a very enjoyable day ride. I was at 300 pounds below gross on takeoff and had no problems with the 4,900' grass strip and 7,000'-plus elevation.

Glenns Ferry, ID (Glenns Ferry Muni—U89)

Carmella Vineyards 🍔🍔🍔

Rest. Phone #: (208) 366-2539
Location: 1-mile walk
Open: Daily: Lunch

PIREP   Glenns Ferry Municipal Airport is a small, unattended rural airport located adjacent to the Snake River. There is no FBO there, just a couple of hangars, an airplane or two, and sagebrush. Don't plan on fueling there—in fact, I don't remember even seeing a telephone there! It's okay, cell phones work well in Glenns Ferry. Carmella Vineyards is located about a mile east of the airport. You can call ahead and arrange for someone to pick you up at the airport, or you can get your mile-long walk for the day. Just start down the road located north of the airport and about 20 minutes later you'll either be there or can see the entrance to the winery in the distance, depending on how fast you can walk.

Breakfast is excellent! I looked over the lunch and dinner menus and they looked good as well. They also have banquet facilities, great atmosphere, a great view, and a golf course behind to help work off your meal. Carmella Vineyards is definitely worth the trip.

Hailey, ID (Friedman Memorial—SUN)
Sun Valley Brewery 🍔🍔🍔

Rest. Phone #: (208) 788-5777
Location: 1-mile walk
Open: Daily: Lunch and Dinner

PIREP   SUN is really the Friedman Memorial Airport in Hailey, where it is not unusual to find more jets on the transient ramp than recips. One time I counted 21(!!) jets belonging to I think the Robins investment firm. But the people are very cordial no matter what you fly.

Tie down and walk a mile to Hailey. The choices include: 1) The Cafe at the Brewery (Sun Valley Brewery); 2) a pizza place; and 3) a small restaurant above a bookshop across the side street from the Cafe at the Brewery. All are on the east side of the road. All are worth the walk if you can't get a ride.

If you have time, drink a home brew at the Brewery and arrange to spend the night in Sun Valley, 11 miles north. The food there is world class.

PIREP   We visited the Full Moon Restaurant. The food was fabulous and the folks friendly.

Lewiston, ID (Lewiston-Nez Perce County—LWS)
Airport Restaurant and Lounge 🍔🍔🍔

Rest. Phone #: (208) 746-7962
Location: On the field
Open: Daily: Lunch

PIREP   A good ol' American restaurant—steak and mashed potatoes.
LWS is a favorite ramp check, so don't talk to strangers.

PIREP   Went to Lewiston LWS yesterday and had some really good homemade lasagna at the restaurant above the terminal. Reasonably priced, and the controller on the way in was the friendliest ever.

McCall, ID (McCall Municipal—MYL)
Common Ground Cafe 🍔🍔🍔

Rest. Phone #: (208) 634-2846
Location: Walking distance—it's in McCall
Open: Daily: Lunch

PIREP FBO Comments: Beautiful place! McCall Aviation is the place to go for outstanding service and hospitality. They can give you a ride into town, or they have rental cars available as well. Self-serve 100LL is the cheapest.

The McCall airport is the gateway to the Idaho backcountry with its many airstrips that offer unlimited fly-in fun: www.mccall.id.us/services/airport.html.

PIREP GREAT coffee, music, company, and service! A nice change of pace from the Starbucks (tenbucks) of the planet. They offer a variety of coffees, teas, mochas, espressos, and such from around the globe. Bagels, breads, and cookies also make the menu. DSL Internet access is available, and they have an eclectic selection of music to choose from. Frequent live music with beer and wine for your passengers to enjoy. Drop in and say hello to Brian, the genius behind it all.

Si Bueno 🍔🍔🍔

Rest. Phone #: (208) 634-2128
Location: Short walk
Open: Daily: Lunch and Dinner

PIREP If you are at McCall at lunchtime or want to walk only a short distance, Si Bueno Southside Grill & Cantina is just across the street behind the FBO. Look for the building that looks like a plane crashed into it: 335 Deinhard Lane.

The airplane on the roof is just a quirky thing the owner put up over 10 years ago. The owner's husband is a professional backcountry pilot.

McCall is walking distance from the field. The pancake house and Maria's Restaurant, both on the northbound side of the street, are favorites. The town is pretty devoid of good eating establishments when you consider that it is a resort, but the selection is better than average for airport eateries.

Mountain Home, ID (Mountain Home Muni—U76)
The German Haus 🍔🍔🍔

Rest. Phone #: (208) 580-1850
Location: Short walk
Open: Closed Sundays

PIREP The German Haus is less than a quarter mile from the airport. Great German food.
Closed Sundays.

Murphy, ID (Murphy—1U3)
Wagon Wheel Cafe 🍔🍔🍔

Rest. Phone #: (208) 875-1067
Location: Short walk
Open: Daily: Lunch

PIREP   Murphy, ID has a population of about ten, a grocery store, two bars/restaurants, the obligatory house of ill repute, and the county offices. They emphasize their desert atmosphere with a single parking meter on the gravel parking lot in front of the county building. Oh yes, they have three airstrips.

Local pilots fly in for their Murphyburgers. The original restaurant that had Murphyburgers changed hands, so you can take your choice of two establishments to get the burger.

Land from the west, wind permitting. The strip is slightly uphill. A wadi precedes the approach, so land with power and be prepared for downdrafts over the wadi. On takeoff, turn right to fly the wadi until you have altitude. The asphalt strip is used to train students in the early fundamentals of backcountry techniques.

PIREP   I work as a reserve sheriff's deputy and pilot for the Owyhee County Sheriff's Department, so I am in and out of Murphy quite a bit. The airstrip and the local wind conditions are to be respected, and density altitude is a factor during the summer. To my knowledge, there hasn't been a house of ill repute in the town since the 1920s.

Nampa, ID (Nampa Muni—S67)
Runway Cafe 🍔🍔🍔

Rest. Phone #: (208) 468-3033
Location: On the field
Open: Daily: Lunch

PIREP   FBO Comments: Three crew cars and overnight shade hangars available. Two maintenance facilities; oxygen recharging; avionics sales, installation, and repair; hangars, shade hangars, tie-downs for lease; ground leases to build your own hangar available.

PIREP   The folks who used to operate the Emmett, ID airport restaurant, Chris and Hellen Goff, moved to the Nampa Airport. The food is excellent and the cafe has a great view of the runways so diners can watch aircraft land and take off while they eat. The daily specials are great, and they have excellent soup, homemade pies, and pastries. Their version of the $100 hamburger is really tasty!

Pocatello, ID (Pocatello Regional—PIH)
Snack Bar 🍔🍔🍔

Rest. Phone #: (208) 232-1101
Location: On the field
Open: Daily: Breakfast, Lunch, and Dinner

PIREP   Sadly, the Blue Ribbon has been closed. I was told that it is now in town about seven miles away. Nothing much on field but a snack bar at the commercial terminal.

Sandpoint, ID (Sandpoint—SZT)
Pend Oreille Brewing Company 🍔🍔🍔

Rest. Phone #: (208) 263-7837
Location: Short walk
Open: Daily: Lunch

PIREP   In Sandpoint, ID (SZT), there is a brewpub with the best burger ever! The Pend Oreille Brewing Company offers a half-pound ground sirloin on a fresh roll with sautéed mushrooms and blue cheese! And great Cajun fries!!

About a mile from the airport, the pub is downtown, two blocks from the lakeside public park. Free courtesy cars available at the airport, so it's easy to get around. This is a lakeside tourist town with many good restaurants, shops, and community events occurring year round.

And don't forget the best $100 hamburger on the planet. It's at Pend Oreille Brewing Company!!

PIREP   Courtesy cars have minimal charges—I believe it's $5, plus 30 cents/mile. Still cheap, and four of us Varieze/Long-EZ drivers partook of the Pend Oreille Brewing Co., and wholeheartedly agree with the listing.

Food and service were great, but alas, we could not bring ourselves to sample the Idaho Pale Ale and Hefeweitzen that are made on the premises—something to do with that eight-hour-bottle-to-throttle rule or something like that. The worker bees at the Brewing Co. were very pleased to learn of their listing.

Stanley, ID (Stanley—2U7)

The Knotty Pine Cafe 🍔🍔

Rest. Phone #: (208) 774-2208
Location: Short walk
Open: Daily: Lunch and Dinner

PIREP   Another easy backcountry fly-in. In town are at least two good restaurants, offering hamburgers and a few homemade goodies. This is almost always open summers and may also be open other times. (Check Idaho Department of Aeronautics in Boise for off-season conditions.)

The airport slopes gently to the north. We are talking high density altitude, so check with local pilots before takeoff, if unfamiliar. If it is excessively hot, you need only wait until evening. But expect reasonably cool temperatures.

The Knotty is next to the Knotty Ore House in Lower Stanley. You'll have to mooch a ride or borrow a car. It has really great garlic burgers for lunch and a limited but exquisitely done dinner menu.

Chef Tom looks a bit like a hippy, not out of place for Stanley, and comes from a little further down river. Get Dia to play the piano for you at Stanley Air Taxi to stir up your appetite.

Warren, ID (Warren/USFS—3U1)

Winter Inn 🍔🍔🍔🍔

Rest. Phone #: (208) 382-4336
Location: On the field
Open: Daily: Lunch

PIREP   Warren, ID is a rustic, historical mining community that hasn't changed much in the last 100 years. The runway parallels main street and ends in "downtown." Winter Inn, about 100 yards away, is a hotel/restaurant/bar that offers breakfast, lunch, and dinner. I have only been there for lunch and had a real good hamburger and a BLT. Hanging in front of the Inn is a plastic spotted owl. Warren takes prides in its annual "Spotted Owl Shoot" held over the Fourth of July weekend. This is in jest, of course, but remember this is a mining/logging area.

Illinois

Alton/St Louis, IL (St Louis Regional—ALN)

Moon Light Inn 🍔🍔

Rest. Phone #: (618) 462-4620
Location: Crew car
Open: Daily: Lunch and Dinner

PIREP FBO Comments: Great accommodations, no tie-down with fuel. Courtesy car upon request.

PIREP The best fried chicken I ever had. My wife and I fly there just for dinner at least once every five or six weeks. You just can't eat it all. When you order a full chicken dinner, check your weight and balance—you will be bringing extra luggage.

Bloomington/Normal, IL (Central Il Regl Arpt at Bloomington-Norm—BMI)

Bolingbrook, IL (Clow Intl—1C5)

Charlie's 🍔🍔🍔

Rest. Phone #: (630) 771-0501
Location: On the field
Open: Daily: Breakfast and Lunch

PIREP My wife and I had breakfast at Charlie's Restaurant yesterday, Sunday. The place was busy and filled with families! The service and the food was great! Everything was freshly made. A very good place for breakfast or lunch.

PIREP Charlie's, located at Clow International (1C5), a class E asphalt strip southwest of Chicago just outside of O'Hare's class B, is a wonderful cafe located in the FBO building, with picture windows overlooking the ramp area. Parked outside—in addition to the usual assortment of Cessnas and Beechcraft, there were beautiful examples of Cubs, a like-new Ercoupe, and a cherry Aeronca. The FBO looks to have been built in the last few years, and the airport appears to have seen significant upgrades in the last decade. I had a great breakfast (really one of the better in recent memory) of eggs, bacon, and pancakes for under $6. The menu was extensive and there was a variety of customers from mothers with kids to businessmen to a group of ten senior aviators sharing stories.

Just down the access road about a quarter mile there is a wide assortment of every chain restaurant you could name: Friday's, Chili's, Bennigan's, Steak and Shake, etc. But I'd stick with Charlie's.

Cahokia/St Louis, IL (St Louis Downtown—CPS)
Oliver's Restaurant 😀😀😀

Rest. Phone #: (618) 337-8222
Location: On the field
Open: Daily: Breakfast, Lunch, and Dinner

PIREP   East St. Louis, IL has a very nice restaurant located on the field.
   Ramp parking next to the entrance.
   Good food.

Casey, IL (Casey Muni—1H8)
Richard's Farmhouse 😀😀😀

Rest. Phone #: (217) 932-5300
Location: Call for pickup
Open: Daily: Lunch and Dinner

PIREP   FBO Comments: FBO very friendly, waited just for us to return to get fuel. FBO called Richard's for a ride (they also have a crew car if Richard's is too busy to pick you up).

PIREP   Very, very good. They had excellent lunch sandwiches and excellent prices to match. This will definitely be a favorite stop.

PIREP   Very clean, friendly FBO! Personnel were very prompt and helpful! By the time the prop stopped spinning, they had asked what they could do for us. At our inquiry, they had the people at Richard's Farm restaurant on the way to pick us up for the one-mile ride to eat.

   Probably one of the best places we have ever flown to eat! We WILL be back! Richard's Farm restaurant is decorated like a Cracker Barrel (lots of antiques, etc). Very nice. Oh, and the people at the FBO and the restaurant are VERY friendly and accommodating!

   Highly recommended.

Champaign/Urbana, IL (University of Illinois-Willard—CMI)
Zorba's 😀😀

Rest. Phone #: (217) 344-0710
Location: Crew car
Open: Daily: Lunch

PIREP   FBO Comments: Clean, accommodating facility. Sizable snack shop also has sectional charts available. Nice briefing/planning room. Flightstar requires $10 for the crew car or a 10-gallon purchase for avgas. If you've seen their fuel prices lately, it's obvious that the former is the better deal. The car has a CD player.

PIREP   My friend Kathy and I flew down to her alma mater for the day for Zorba's beef and chicken gyros. We borrowed Flightstar's (FBO) sporty PT Cruiser (2-1/2-hour time limit) and hit the little college town, where it was surprisingly quiet despite the gorgeous, Saturday afternoon weather. Zorba's is easy to find at 627 E. Green.

There are plenty of places to walk around and plenty of interesting shops to browse. The old, well-kept campus is magnificent.

This was a fantastic way to spend the day. Highly recommended!

Chicago, IL (Chicago Midway Intl—MDW)
White Castle 🍔🍔🍔

Rest. Phone #: (773) 227-5127
Location: Very short walk
Open: Daily: Lunch and Dinner

PIREP   This restaurant is a classic, known for its greasy onion burgers called sliders. The slider is so small that three of them is considered a snack. Here's the best part: this White Castle is right on the southeast corner of Midway Airport, and it is so close to Midway Airport runway 31C that aircraft are only 200' AGL as they pass overhead. Also, it is walking distance from Monarch South, which is where transient GA airplanes may park.

Chicago/Prospect Heights/Wheeling, IL (Palwaukee Muni—PWK)
The Compass Rose 🍔🍔🍔

Rest. Phone #: (847) 537-6710
Location: On the field
Open: Daily: Lunch and Dinner

PIREP   Located on the field at Palwaukee airport (PWK), this is part of a national chain of aviation-themed restaurants. The food is always excellent and the service good. The dress ranges but casual is acceptable. The restaurant also features a Sunday morning breakfast buffet with a twist: the earlier you arrive, the lower the cost!

This restaurant is popular, even among nonaviators. If you go on Friday or Saturday nights (or for Sunday brunch), expect a wait.

The restaurant can be reached by van from either FBO or on foot (if you're willing to walk a bit). Unfortunately, there is no transient aircraft parking close to the restaurant.

Palwaukee airport is located approximately eight miles north of O'Hare airport. It is towered with two instrument approaches (ILS and VOR).

PIREP   The Compass Rose Restaurant in the Palwaukee Inn is on the airport. Breakfast, lunch, and dinner.

On Saturday mornings you'll usually see several PAPA (Palwaukee Airport Pilots Association) members having breakfast and deciding where to fly for lunch.

Chicago/Lake in the Hills, IL (Lake in the Hills—3CK)
Nick's Pizza 🍔🍔

Rest. Phone #: (815) 356-5550
Location: Short walk
Open: Daily: Lunch and Dinner

PIREP My husband and I have eaten at Nick's many times over the last three years. The pizza is excellent, but the atmosphere of the pub is even better! Get your basket of peanuts and not only can you enjoy eating them, but also you can then throw the shells on the floor. Fun place!

PIREP Excellent pizza, the thin crust deluxe is the way to go. The Chicago area is well known for their stuffed pizza, but the thin crisp crust is also very good. The decor at Nick's is interesting—it is taxidermy chic!

It is a little bit of a walk north of the airport, but there is a bicycle path that goes right past the airport and the restaurant. It takes about ten minutes to walk.

Chicago/West Chicago, IL (Dupage—DPA)
Kitty Hawk Cafe 🍔🍔🍔

Rest. Phone #: (630) 208-6189
Location: On the field
Open: Mon.–Fri., Breakfast and Lunch

PIREP This is a really great find in a restaurant. It is inside the terminal and has excellent breakfasts and good sandwiches for lunch. There is lots of seating that is near or next to 20-foot-high picture windows looking out over their main tarmac and two primary runways. There is always an array of corporate jets, business twins, and private aircraft parked out front.

We have made breakfasts at the Kitty Hawk a Saturday morning tradition. You always feel welcome, it is spacious and quiet inside, and the action outside the window is hard to beat!

PIREP I fly out of DuPage Airport (DPA) and, in the rather new terminal building, they have a restaurant called Kitty Hawk Cafe. It has a counter area in front of the kitchen and they also have tables by the windows. The staff will bring your food out once you order at the register. They have soups, hot and cold sandwiches, desserts, coffee and teas, and soft drinks.

After going there for lunch a few times, I found the food and the service to be very good. The pricing is about average for what you get. The food is not fancy, but so far they do a good job and I would not hesitate to recommend them. The tables by the windows let you watch the flight operations on several of the runways and the main ramp, but the terminal building is not that close to the runways.

DPA is the third busiest airport in Illinois, with a lot of corporate jets to watch.

Chicago/Schaumburg, IL (Schaumburg Regional—06C)
Pilot Pete's 🍔🍔🍔🍔🍔

Rest. Phone #: (847) 891-5100
Location: On the field
Open: Daily: Breakfast, Lunch, and Dinner

PIREP FBO Comments: Nice FBO! Brand new—really clean inside, first rate place. The staff is really friendly and ready to help. They said they hang around until 7 PM during the winter months, so plan your arrival. The restaurant is located above the FBO, gate on the side.

PIREP Pilot Pete's deserves six stars, if you ask me! I flew myself and a date there last night and for once dinner was as exciting as the flight! The atmosphere is a Jimmy Buffet-meets-the-classic international-Juan Trippe/PanAm–era, with a mix of every aviation generation to follow! A funky mix of mood lighting creatively illuminates all of the wall murals and both large and small model airplanes hanging from the ceiling! The food was excellent—Pete's has a traditional American menu with its own unique twist. The prices are very reasonable, and the proportions are not disappointing. I had no problem getting into the airport—I flew in from Michigan VFR, followed the lake shore from Gary north, right past the Chicago skyline—passengers are plastered to the window! I went direct DEERE-OBK-06C, north of PWK, no problems, Chicago ATC was friendly getting in and out. I highly recommend Pilot Pete's for a date with the lady or a trip out with the guys—it's a great time!

PIREP   While home on leave from Iraq and visiting in the Chicago area, I decided to fly to 06C for lunch. Pilot Pete's has to be the best lunch at an airport I have ever had. The food is outstanding; the service was excellent; the employees great. The atmosphere is aviation oriented, as it should be at an airport. If you are in the area, I strongly recommend you stop by and partake. You will not be disappointed.

PIREP   I had the pleasure of having dinner at Pilot Pete's last night and it was fabulous! The atmosphere was incredible, with miniature hanging airplanes suspended from the ceiling, wonderfully done wall murals, and uniquely interesting artifacts depicting scenes related to aviation all over the world. The food was superb and reasonably priced. I had the grilled vegetable and asiago cheese pizza, which I thoroughly enjoyed. My friends got the barbecue chicken and margherita pizzas and said that both were very good. I would highly recommend this restaurant to anyone who likes to eat as well as to anyone who is even remotely interested in aviation. You won't be disappointed!

Chicago, IL (Lansing Muni—IGQ)
Shannon's Landing 🍔🍔🍔

Rest. Phone #: (708) 895-6919
Location: On the field
Open: Daily: Lunch and Dinner

PIREP   Shannon's Landing is a traditional Irish pub located on the airport above the FBO's at Lansing Municipal Airport (IGQ), just south of Chicago and next to Indiana. The restaurant has a bank of windows that overlooks the ramp and runway, so it's great to watch the takeoffs and landings. The inside is decorated reminiscent of many pubs in Ireland, with lots of signs and pictures on the walls. The menu consists of a variety of excellent sandwiches none of which exceed $8. You have to try Shannon's Irish chips!! They just started serving a traditional Irish breakfast all day for $7.95 that I dare you to try to finish (Irish bacon, Irish sausage, black pudding, white pudding, baked beans, lamb cutlet, fried tomato, fried potatoes, and an egg). You'll definitely have to recalculate your weight and balance when you leave. At night they have Celtic and country bands and serve the coldest beer in town, so it's worth a stay overnight.

PIREP   This is a small diner catering to the aviation community right in the airport at Lansing, IL. I usually stop there for a bite to eat and have a smoke when I am tired of doing touch-and-go's. Food is good—going by your scheme I would probably give it two hamburgers. Service is extremely friendly, and they let you park your aircraft on a grass patch right in front of the diner.

De Kalb, IL (De Kalb Muni—DKB)
The Egg Haven 🍔🍔🍔

Rest. Phone #: (815) 748-1200
Location: 3 miles via crew car
Open: Daily 7 AM–4 PM

PIREP   I reference the "$100 Burger" before every flight, I fly for a construction co. out of the ATL area, and my passengers love it when I can find airport diners and restaurants. I want to refer you to 'The Egg Haven' in Dekalb, IL…

It's a must eat. The FBO at KDKB is a friendly courteous operation, eager to help every time I've flown in. They have a couple of crew cars and the restaurant is only about 3 miles from the FBO.

The food is diverse and incredible. You can have breakfast all day; go early though, as they are only open from 7 AM–4 PM. Huge portions are standard, and be sure to get the fresh squeezed juice.

Top score from this Pilot.

Geneseo, IL (Gen-Airpark—3G8)
The Cellar 🍔🍔

Rest. Phone #: (309) 944-2117
Location: Call for pickup
Open: Daily: Dinner

PIREP The airport is a public-use, privately owned airport that is owned by 40+ local pilots. It boasts a 2,600' lighted and well-maintained grass strip.

The Cellar is in the basement of a old hotel and has a real charcoal grill. The barbecue shrimp with fried–in-butter hash browns is my favorite. They also have fried mushrooms that are like no other.

Great atmosphere! It was ranked five stars by Mobil. With a little notice to Bob Schaffer, the owner, I am sure he can arrange transportation into the restaurant. There is a phone at the airport.

Another local restaurant is the Victorian Manor. It is also downtown and is open for breakfast, lunch, and dinner. It is located in a old Victorian-style home. They have great home-cooked meals with different specials every day. Walt Musser is the owner—call him ahead of time to arrange a ride into the restaurant: (309) 944-5683.

Hinckley, IL (Hinckley—0C2)
The Cafe 9/27 🍔

Rest. Phone #: (815) 286-9200
Location: Two miles
Open: Daily: Breakfast and Lunch

PIREP Airport Comments: Fun place to fly. Constant sailplane and skydiver operations.

PIREP Nice place with average food. Pilots from 0C2 go there all the time and will gladly give you a free ride down. It's about two miles.

Jacksonville, IL (Jacksonville Muni—IJX)
Lonzerotti's Italia Restaurant 🍔🍔

Rest. Phone #: (217) 243-7151
Location: Crew car
Open: Daily: Lunch and Dinner

PIREP Nice facility and helpful staff, offered us their courtesy car. Drove to Lonzerotti's Italia Restaurant in town. Restaurant located at 600 East State Street, east of downtown, located in old Chicago and Alton Railroad Station. Well-appointed interior and good food, nightly special under $10 Monday–Thursday. Reservations recommended. Airport courtesy car due back by sunset.

Neat experience.

PIREP I flew into to Jacksonville about a week ago. Very nice FBO with a great staff. They recommended I take a crew car and drive into town to a local Italian restaurant. The food was amazing!!! I recommend anyone try it out.

The FBO is open from sunrise to sunset, but if you call them they might stay open a little later: (217) 243-5824. Their fuel prices are also the cheapest I have found in Illinois. Jacksonville is about 35 nm west of Springfield.

Joliet, IL (Joliet Regional—JOT)
McDonald's 🍔🍔

Rest. Phone #:
Location: 1/4-mile walk
Open: Daily: Breakfast, Lunch, and Dinner

PIREP FBO Comments: Tie-down under cover was free.

PIREP Traveled to Joliet, IL. The airport has a crew car available. The service was great. You can drive or walk to McDonald's just a quarter mile away.

Lacon, IL (Marshall County—C75)
Kenyon's Place 🍔🍔

Rest. Phone #: (309) 246-3663
Location: 1/2 mile
Open: Daily: Lunch and Dinner

PIREP In Lacon about a half mile from the airport is a restaurant called Kenyon's Place. The food is excellent, and a free courtesy car is available at the airport for transportation. The runways at the airport have recently been resurfaced, and the FBO is very friendly and helpful. A fun place to fly into right next to the Illinois River.

Litchfield, IL (Litchfield Muni—3LF)
The Ariston Cafe 🍔🍔🍔

Rest. Phone #: (217) 324 2023
Location: Crew car
Open: Daily: Breakfast, Lunch, and Dinner

PIREP FBO Comments: Crew car available.

PIREP Excellent food and service. We will return.

PIREP The Ariston Cafe is one mile north of the Litchfield, IL airport (3LF). If the courtesy car is not available, the restaurant will often pick you up, or you can take a casual stroll down the "Mother Road," the original old Route 66.

In business since 1924, this is the kind of place you would take your mom on Sunday. A large menu, a very nice soup/salad bar, very impressive pies and cakes for dessert. Oh, and did I say a hamburger that any restaurant would be proud to serve?

Open Sun.–Fri. at 11 AM, Sat. at 4 PM.

No smoking.

Immediately adjoining the Ariston is Jubelt's Bakery and Restaurant. More of a full-service bakery that serves breakfast/lunch/dinner in a sort of fast-food way, Jubelt's will make the folks back home wish they'd come along when you take home a sack of cookies.

Open for breakfast at 5:30 AM(!!!), closes at 8 PM, later on weekends.
Smoking and nonsmoking sections.

Macomb, IL (Macomb Muni—MQB)
Macomb Dining Company 🍔🍔

Rest. Phone #: (309) 833-3000
Location: Crew car
Open: Daily: Lunch and Dinner

PIREP FBO Comments: They have a concrete strip 9/27 and a turf strip 18/36. They do have fuel on field but I can't remember the price. They don't have a truck, so to get fuel you must pull up to the fuel farm. They have one courtesy car that you can use even if they will close before you will return. They are not too picky.

PIREP This pirep is about Macomb Airport in Illinois; it's a small airport about two miles north of a college town.

As far as restaurants, there are a few chains: McDonald's, Hardee's, Pizza Hut, Subway, etc.

I went to school at WIU in Macomb and I recommend three restaurants. The first, Jackson Street Pub (Jackson Street on the west side of town), is a total hole in the wall, but the food is great! The service is decent; it is a bar, and it is bar-type service, so don't expect the best. Pricing is great, except for drinks—as it is a bar, you must pay for soft drink refills—90 cents, I think. I recommend the Horseshoe or the Pony, they are great. Bread, meat, fries, onions, and cheese piled high. Also try Jack Stix or Nacho Nuggets—heart attacks you won't want to miss.

The next restaurant I recommend is Rocky's Bar (just past the Jackson Street Pub, on West Jackson). This is a restaurant with a bar in it. Built new just two years ago, Rocky's has a nice, fun atmosphere. The food here is also very good—their signature Wisconsin Cheese Soup is to die for. They also have mozzarella sticks that are wrapped like egg rolls—very nice flavor. The pricing here is a little more expensive, but refills are free.

For those of you that are looking for higher class dining, there is the Macomb Dining Company. Located in the town square, this restaurant has great food and a laid-back, somewhat upscale atmosphere. Any attire is acceptable, but I find khakis and a polo work better. The food is well priced and delicious; no recommendations here as it is all so good. The view overlooks the quaint town square, taking you back 50 years to a quieter time.

Other restaurants to look out for are Yen Ching for tasty but expensive Chinese and Diamond Dave's for great Mexican and awesome margaritas, for a price.

Marion, IL (Williamson County Regional—MWA)
La Fiesta 🍔🍔

Rest. Phone #: (618) 993-0028
Location: Crew car
Open: Daily: Lunch and Dinner

PIREP Located on field, although a (very) short drive from the FBO. They will kindly give you the crew car to get there. All kinds of grilled items on the menu. The grilled pork sandwich is excellent.

PIREP Now open on the field: La Fiesta, as the name implies, a Mexican restaurant. One of two owned by the same family. Authentic ethnic food at great prices, portion size will send your plane over gross! I highly recommend you stop in for a fill-up. You will be treated courteously by all the staff.

Mattoon/Charleston, IL (Coles County Memorial—MTO)
The Airport Steak House 🍔🍔🍔

Rest. Phone #: (217) 234-9433
Location: On the field
Open: Daily: Lunch and Dinner

PIREP   Very good, simple menu. Food is very good and reasonably priced. On airport, you can park in front of the door.

PIREP   Airport Comments: Runway 6/24 closed, but 18/36 turf is great, so you still have two options.

PIREP   This restaurant continues to be a favorite of area pilots. Good food, good prices, and lots of airplanes. Four of us went over last Saturday and landed in the grass on 18/36, a great option. And don't forget the pies.

PIREP   Best GIANT tenderloins anywhere. HUGE—hand breaded. Fried mushrooms taste like you've done them at home. Also lots of pilots in on Sunday morning for breakfast.
    Not sure of your rating system, but I would rate this high—lots of food, relatively low prices.

Moline, IL (Quad City Intl—MLI)
Bud's Skyline Inn 🍔🍔🍔

Rest. Phone #: (309) 764-9128
Location: On the field
Open: Daily: Lunch and Dinner

PIREP   Bud's Skyline Inn is a great restaurant. The place has some of the best food around—definitely a top rating. It is on the north boundary of the airport right next to the threshold of runway 23. It has a great view of the airport. It is decorated in an aviation theme. You can usually find local pilots there and always on Wednesdays at lunch. Bud bought this place as an old tavern and cleaned it up, expanding every couple of years as the business has grown. It is very busy most nights, but he can always squeeze you in somewhere. One neat thing he has is etched glass windows that depict local corporate aircraft.

Paris, IL (Edgar County—PRG)
Tuscany 🍔🍔🍔🍔

Rest. Phone #: (217) 466-1610
Location: Short drive, crew car available
Open: Dinner 7 days, Brunch Sunday 11 AM–2 PM

PIREP   Tuscany is a very good restaurant located in Paris, IL. The menu is expansive and the food is OUTSTANDING! The service is prompt and friendly.
    There is a crew car available. It is always a good idea to call ahead to confirm and reserve it.

Pittsfield, IL (Pittsfield Penstone Muni—PPQ)
Red Dome Inn 🍔🍔

Rest. Phone #: (877) 797-3979
Location: Crew car
Open: Daily: Lunch and Dinner

PIREP   A FREE crew car is available. Several restaurants are within a three-minute drive. My pick is the Red Dome Inn.

Rantoul, IL (Rantoul Natl Avn Cntr-Frank Elliott Field—TIP)

The Caddyshack 😋😋😋

Rest. Phone #: (217) 893-3165
Location: On the field
Open: Daily: Lunch and Dinner

PIREP I recently flew into the airport at Rantoul, IL and was delighted by what I found. Flightstar is the FBO there, and they have a great facility for pilots and multiple crew cars that they offer out to all flyers free of charge. The airport is located on the now-closed Chanute Air Force Base and has two excellent restaurants, The Caddyshack, located on the base golf course, and The Fanmarker, an aviation-themed restaurant housed in the old officers' club. After your meal, be sure to stop by the Air Force Museum adjacent to the Flightstar ramp. They have over 20 aircraft on display.

Quincy, IL (Quincy Regional-Baldwin Field—UIN)

The Tail Winds Restaurant 😋😋😋

Rest. Phone #: (217) 885-3500
Location: On the field
Open: Daily: Lunch

PIREP The Tail Winds Restaurant is located in the terminal building that is immediately adjacent to the FBO with a very short walk from the parking area. The restaurant has a very relaxing atmosphere with a very nice view of the ramp and runways. The waitress was very nice and took my order promptly. I had a half-pound cheeseburger basket that included seasoned fries and coleslaw for under $6. The food was very well prepared and quite tasty. I highly recommend the Tail Winds for those wanting a good lunch destination. I know I won't hesitate to go there again.

PIREP In our ever-widening search for a Saturday morning airport breakfast, two Long-EZs from Cedar Rapids took the short 45-minute flight from Cedar Rapids, IA (CID) to Quincy, IL (UIN). The Skyway restaurant is located in the terminal, and you may park your plane on the ramp in front of the FBO next to the terminal. The food and service were first rate.

Highly recommended.

Rockford, IL (Greater Rockford—RFD)

The Airport Coffee Shop 😋😋

Rest. Phone #: (815) 986-1086
Location: Inside the main passenger terminal
Open: 2 hours before each commercial flight

PIREP This is not the most wonderful airport restaurant you've ever been to and its hours are odd. It is totally synced to the airlines departure schedule. Call them before you arrive to see what their schedule will be that day.

They have a breakfast and lunch menu, read sandwiches even breakfast sandwiches. That's it!

Not great but it is here and fuel prices at Rockford's Emery can't be beat.

Sparta, IL (Sparta Community-Hunter Field—SAR)

Brandy's Bar and Grille 😋😋

Rest. Phone #: (618) 443-3899
Location: Short walk
Open: Daily: Lunch and Dinner

PIREP Brandy's Bar and Grille: Friday night fish fry, Saturday night live music. BP automatic pay at the pump 100LL and Jet A. Two buffet restaurants and even a Super Wal-Mart and Best Western, all in walking distance on IL Rte 4. Good steaks, too. Highly recommended if you want to be near a big city but away from the big city (40 SSE of St Louis). Four burgers for the food and ease of access.

PIREP A nice spot just across the highway that runs by the airport in Sparta, IL is Brandy's Bar and Grille. You can leave the courtesy car in the parking lot. You won't need it.

The Hundred Dollar Hamburger

Indiana

Bloomington, IN (Monroe County—BMG)

Casa Brava 🍔🍔

Rest. Phone #: (812) 339-1453
Location: Crew car
Open: Daily: Lunch and Dinner

PIREP The folks at Cook Aviation were very nice to us on our visit. I called ahead to find a restaurant to visit in Bloomington for my wife's birthday. They suggested Casa Brava (which was very nice and affordable as well). They had a car waiting at the ramp with the air conditioner on! Treated like royalty and we only bought six gallons of gas!

Columbus, IN (Columbus Muni—BAK)

Hangar V Restaurant 🍔🍔🍔

Rest. Phone #: (614) 378-4047
Location: On the field
Open: Daily: Breakfast and Lunch

PIREP FBO Comments: Free transient parking outside the terminal building where the restaurant is located.

PIREP Flew in around 11 AM on a Saturday morning. Both the breakfast and lunch menus were available. The sausage gravy and biscuits were very good. There is a great view of the transient ramp and runways. Service was good and fast.

PIREP The Hangar V Restaurant serves breakfast and lunch (until 2 PM) and is located in the terminal at Columbus Municipal (BAK). The lunch menu includes a wide range of items including light meals and the standard hamburger. The food has always been good and service is very friendly. The large windows overlook the ramp, so you can keep an eye on your wings.

BAK is tower controlled but not very busy and has lots of tie-down space.

Elwood, IN (Elwood—3I1)

Landings Airport Restaurant 🍔🍔🍔

Rest. Phone #: (765) 552-6400
Location: On the field
Open: Daily: Breakfast, Lunch, and Dinner

PIREP   I have eaten at the Landings Airport Restaurant for breakfast, lunch, and dinner. I really like this place. The food is great, the prices are great, and service is friendly. The grass strip has several ultralights, tail-draggers, biplanes, home-builts, and modern airplanes there every flying day. The restaurant is located on State Road 37 at the airport. It claims to be "The World's First Fly-in Drive-in."

PIREP   Best fish dinner around on Friday night! I have had breakfast, lunch, and dinner here and they are great!

This is my favorite place to fly in for great food! And wonderful people serve you too!

I would give them the MAX rating; the food is always great!

Evansville, IN (Evansville Regional—EVV)

Wolf's Barbecue 🍔🍔

Rest. Phone #: (812) 477-5604
Location: Crew car
Open: Daily: Lunch and Dinner

PIREP   Grab the courtesy car from Tri State Aero and head to Wolf's Barbecue. It's ten minutes away and easy to find. The food was awesome, and you can even take home a bottle of one of their many different types of barbecue sauce with you.

Tri State Aero was awesome as well—they even met us with a red carpet on the ramp.

PIREP   The DEFINITIVE barbecued ribs are at Wolf's Barbecue. Borrow a courtesy car from either FBO and get directions to the best barbecued ribs on this planet.

Nice airport with all of the amenities except the big price tag. I've never been charged for a tie-down, nor was I charged when the folks at Million Air pulled my airplane into a hangar (unsolicited!) when a storm was approaching. There are two FBOs, Tri-State Aero and Million Air. Tri-State Aero gave me a free heated hangar one night when the temperature went down in the 20s, and I needed it because I have a warm-weather Texas airplane. It is hard to choose between the two FBOs.

Shyler's BBQ 🍔🍔

Rest. Phone #: (812) 476-4599
Location: Crew car
Open: Daily: Lunch and Dinner

PIREP   A great place to have barbecue. It is approximately 15 minutes from the airport. Tri-State Aero, the local FBO, can provide you with a car and you're off to Barbecue Heaven.

Unfortunately, Shyler's does not take reservations. The wait tends to be long but well worth it.

To get there, head south on 41 to Lynch Road. Take a left on Lynch Road to Green River Road. Take a right there and head south until you get to Shyler's on the right side.

I guarantee you will enjoy it!

Fort Wayne, IN (Smith Field—SMD)
Cork 'N Cleaver 🍔🍔🍔

Rest. Phone #: (260) 484-7772
Location: Crew car
Open: Daily: Lunch and Dinner

PIREP FBO Comments: Excellent! Friendly FBO and well-maintained airport.

PIREP A very short ride from the airport is a local favorite, the Cork 'N Cleaver. This restaurant is located next to the Fort Wayne Marriot Hotel on Washington Center Boulevard. Enjoy an amazing salad bar and the best dark fresh-baked bread I've ever eaten. I must also give high marks for the wait-staff that is perhaps the finest in the Midwest.

I believe that Smith Field is the field where the first Cruise Missile test flew. This is a great gem of a field.

If the family is on-board, be sure to check out the Fort Wayne Zoo and/or Botanical Conservatory. The Zoo is a very short drive from the airport. The Botanical Conservatory is a bit farther, located downtown. Both are well worth the flight.

Zesto's 🍔

Rest. Phone #: (574) 269-6682
Location: Crew car
Open: Afternoons

PIREP FBO Comments: Great airport. Very friendly staff there to help you. This is an historic airport that was used for the old airmail routes of the 1920s and '30s. The hangar on the field is one of the oldest in the U.S. The FBO is well equipped with online weather and is located near several chain restaurants.

PIREP Zesto's Ice Cream is several miles away, but it has great frozen custard and ice cream. Borrow the crew car to get a cone during a hot summer stopover.

French Lick, IN (French Lick Muni—FRH)
French Lick Resort Inn 🍔🍔🍔🍔

Rest. Phone #: (812) 936-9300
Location: They'll send a van for you
Open: Daily: Breakfast, Lunch, and Dinner

PIREP FBO Comments: Cash gets a five-cent discount; no service after hours.

PIREP The resort is very well priced at $95/night for me and my ten-year-old. The menu is good, and the buffet was excellent and priced well. We even did horseback riding—can't do that at most greasy spoons!

PIREP For those flying in/around southern Indiana, a pleasant stop would be French Lick Springs. At one time home to a casino, this resort is now run as a getaway location. French Lick Airport (FRH) is not exactly big or close to civilization, but the gentlemen in the barn next to the strip will talk your arm off! The resort is always glad to send a van over for those visiting. This may tend to attract the middle-aged to older crowd—not exactly the greasy spoon associated with some airports.

Gary, IN (Gary/Chicago International—GYY)
The Casinos 🍔🍔

Rest. Phone #: (219) 949 9722
Location: Crew car
Open: 24/7

PIREP   At Gary/Chicago airport, if you go to the Gary Jet Center, you can get a courtesy car for up to two hours. You may NOT use the car to go to the boats, but the desk crew will tell you to go to McDonald's and have fun at the casinos. I wanted to list the casino buffet, very pricey, but great food! The buffet has prime rib, chicken, ham, and other mains, with a vast selection of sides.

Greencastle, IN (Putnam County—4I7)
Final Approach Bar and Grill 🍔🍔🍔

Rest. Phone #: (765) 655-1600
Location: On the field
Open: Closed Sun. and Mon., Lunch Tues.–Fri., Dinner Fri. and Sat.

PIREP   FBO Comments: Open 24/7.

PIREP   New restaurant. Open Thursday, Friday, and Saturday nights for dinner, Tuesday through Friday for lunch, and Saturday morning for their very popular breakfast buffet! Great food and fair prices. Prime rib special Saturday nights.

Greensburg, IN (Greensburg-Decatur County—I34)
Market Area 🍔🍔🍔

Rest. Phone #: (812) 663 5733
Location: On the field
Open: Daily: Lunch and Dinner

PIREP   Stopped by Greensburg for a snack. Mini-golf, bowling alley, pool, park with basketball court adjacent to ramp. Ice cream stand with burgers, hot dogs, and so on. Outside dining only—would be no fun in the rain. Fuel was available, said to inquire at bowling alley. Unique small-town atmosphere.

PIREP   This is a nice little airport located 45 nm SE of Indianapolis, usually unattended, no fuel, but a nice ramp and the runway is in good shape—watch the crop dusters at low altitude.

You will find, right at the airport, a "soft serve" ice cream place that also serves hot dogs, Coney dogs, and hamburgers. There are bowling lanes, a miniature golf course, and the city swimming pool. Certainly not Disneyland, but a good time for kids—ours love it. Right next door there is the county fairgrounds where, if you are lucky like we were, you might find an antique tractor show or the like.

Don't plan on going into town; there is no way to get there that we could discover. (If you can find a way into town, Greensburg is the home of the county courthouse that has a tree growing out of the steeple. When you leave, if you fly over the downtown area—not too low—and look carefully, you will be able to see the tree.)

The bowling lanes are open during the winter, but the rest of it is only open during the summer months.

Griffith, IN (Griffith-Merrillville—05C)
Mi Tierra 🍔🍔🍔

Rest. Phone #: (219) 922-3633
Location: On the airport
Open: Daily: Lunch and Dinner

PIREP   Food was great. Outstanding first date. She even paid for my meal! What a great experience. Of course, the romance was in the flying, and it was a tad risky eating Mexican what with the beans and all, but I've been back several times and love the girl and the airport.

PIREP   FBO Comments: Didn't get gas, don't know the prices of aviation fuel. But the FBO was very nice and open until 8 PM. Staff was very friendly and helpful. The airport was very well maintained.

PIREP   Thought I would update the pirep for Mi Tierra's. Was very satisfied with the restaurant. The food was excellent and the staff was very friendly. The bulk of the menu was Mexican, but noticed that they also had halibut and red snapper, which I absolutely love. They have four different Mexican soups! She explained them to me— I wasn't brave enough the try the one with tripe, but the soup I picked out was excellent. It is NOT a Taco Bell, Don Pablo's, or Chi-Chi's style restaurant, it is real authentic Mexican food. They are about ten miles from Lake Michigan as the crow flies. The restaurant is connected to the FBO and you have to go inside the FBO because the flight line is blocked off by fences and you can't get out. Once inside the FBO, go down to the end of the hall and you will see a door with a sign that says Mi Tierra. The restaurant is also available to people from outside the restaurant. We noticed the highway patrol in there eating. They eat there all the time. It is not a really big place—probably holds about 40 to 50 people. We absolutely love the salsa there, even better than Don Pablo's. Meal portion size, more than you can eat! You will not be dissatisfied, in my opinion.

We will come again.

PIREP   Excellent and courteous airport staff. The restaurant on the field only complements the camaraderie of the airport. The owners have been heavily involved in the history of the area's aviation startup since the '60s to '80s. The food is excellent with daily chef specials. The airport also has a racing theme, as the owner of the FBO has an Indy 500 background.

Hagerstown, IN (Hagerstown—I61)
Guy Welliver's Famous Smorgasbord 🍔 🍔 🍔 🍔

Rest. Phone #: (765) 489-4131
Location: Call for pickup
Open: Daily: Lunch and Dinner

PIREP   FBO Comments: 4,000' grass strip w/lights.

PIREP   This is a wonderful 60+-year-old restaurant, family-owned and operated since they opened. Great atmosphere—copious quantities of every kind of good ol' Midwestern food. It is hard to adequately describe the amounts of food, other than to say that there is as much or more selections to choose from here than you will ever find. Carved roast beef, fried chicken, shrimp (fried and boiled), crab legs, numerous casseroles, every kind of salad and vegetable known to mankind, at least 15 dessert selections. Leave your cholesterol, fat, and carb counter at home!

Land and use the phone at the ramp to call them, and a family member will come and get you. (Chances are that your cell phone will not work here in the boonies.)

PIREP   One of our favorites places to fly for a meal was Hagerstown, IN. The airport is I61 located approx. 10 miles NW of Richmond, IN. It is a grass strip. When you land, there is a phone booth with a listing for Guy Welliver's Famous Smorgasbord. Call the restaurant and they will send a car to pick you up and, after you're through, deliver you back to your plane (about a five-minute ride).

Excellent food and atmosphere, plus all you can eat for very reasonable price. Ask for the Mill Room!

Indianapolis, IN (Eagle Creek Airpark—EYE)
Kazablanka - Grill and Bar 🍔 🍔

Rest. Phone #: (317) 244-7638
Location: Close by
Open: Daily: Dinner

PIREP FBO Comments: The people here charge a $5 parking fee, but it's free with fuel, and they have rental cars on property.

PIREP Kazablanka has a great mix of Middle Eastern, Greek, Italian, and American food, which should please everyone in your party. Oh, and a full bar. For an appetizer we had dolmas and saganaki, which were both very well prepared and flamed at our table. They have a great assortment of soups and salads, and the Greek salad that came with my dinner was as large as the entrees and very good. I had the souvlaki (shish kebab) for dinner, and my daughter tried the Kazablanka Magic Chicken, which I tasted and found to be more than pleasing to the taste. They also have a large assortment of veggie dishes for those who do not eat meat or just want to try something different. Tell them you are a PILOT for your 10 percent discount. You will take home as much as you eat ! Great place just between EYE and IND right on Interstate 465 at Rockville Road or Hwy. 36

Indy Rick's Boatyard Cafe 🍔🍔🍔🍔

Rest. Phone #: (317) 290-9300
Location: Across the street
Open: Daily: Lunch and Dinner

PIREP FBO Comments: Nice people, friendly line staff, gorgeous FBO. Ramp fee has increased to ten bucks unless one buys avgas. They are competitive, and fuel was cheaper than at my home base. So what's the big deal? Buy some gas.

PIREP Rick's warrants a solid five-burger rating (would be six but prices are not exactly cheap). What a wonderful fly-in experience. Ask for a table near the fireplace, which also has a commanding view of the lake. Three friends and I flew in for lunch. We immediately ordered two dozen blue point oysters on the half shell. They are as fresh as they can possibly be and are not shucked until the order is placed. Every entree is superb, and there is nothing on the menu that is not worth going back for. Service is excellent and very attentive. This was my fourth visit and it reconfirmed that Rick's at Eagle Creek is the best $100 hamburger I've found.

PIREP FBO Comments: Ramp fee and slightly high fuel prices. Fee waived and good discount provided for Angel Flight and Lifeline Pilots, making this a stop of choice when linked with the need for a good meal.

PIREP Rick's Boatyard Cafe is an excellent restaurant, attracting customers from nearby areas. It is about 200' from the FBO. When weather permits you can take a table on the veranda and watch the sailboats on the lake. Food is excellent. The hamburgers are so good that I've not tried anything else, but understand from those with me that their meals were great, too. Price is a little high but fair considering the surroundings and food quality and quantity. Service is quick when requested. This is one of my favorite airports for linking flights with Angel Flight.

PIREP I have been a pilot for about nine years and have frequented Rick's Boatyard with different friends, acquaintances, and business associates regularly for about eight years. My European associates love the outdoor atmosphere spring through fall and request we go to the Boatyard over other restaurants.

One of my favorite dishes is French-cut chops. Food is abundant and excellent and the atmosphere is good. Music every evening. Save room for the mud pie, which is also good to take home in the cold winter months, as you will not be able to eat it all there (or take a cooler for those back home)!

Even with the $5 ramp fee, it is still worth the trip!

Jeffersonville, IN (Clark Regional—JVY)
Neil's & Patty's Place 🍔🍔🍔

Rest. Phone #: (812) 246-5457
Location: Crew car
Open: Mon. – Thurs. 11 AM–10 PM, Fri. and Sat. 11 AM–11 PM, Sun. 11 AM–9 PM

PIREP   The menu is quite varied and full and the food and service are excellent. The chicken livers are very good and the desserts include a list of wonderful cakes. I would rate it four burgers.

PIREP   FBO Comment: Top off your tanks at the FBO with the Phillips 66 sign, Aircraft Specialists. They have the best price on avgas in the region. The FBO is open 24 hours. They also have a courtesy car for you to drive the two miles to Sellersburg.

PIREP   Neil's & Patty's Place is a great family restaurant with really good food. They have a large selection of soups, salads, sandwiches, chicken, seafood, steaks, chops, and pasta. The key here is fresh and homemade. They make their own hamburger patties and hand-bread their chicken, seafood, and tenderloins. This is not the kind of place that uses that frozen, prebreaded stuff. Even the steamed vegetables were nice and fresh—you can tell they didn't come out of the freezer, either. They also have desserts, but I didn't make it that far. If the desserts are anything like the rest of the menu, it's gotta be good too.

Kentland, IN (Kentland Muni—50I)

Joe's Filling Station Restaurant 🍔🍔🍔

Rest. Phone #: (219) 474-9939
Location: Short walk
Open: Daily: Breakfast, Lunch, and Dinner

PIREP   Joe's Filling Station is a truck stop diner, good food, nothing fancy. Good milkshakes. Short walk from the airport ramp, no need for the courtesy car. Hard-surface runway in good shape.

PIREP   FBO Comments: You'll never meet a nicer person than Harold! Harold runs the FBO.

PIREP   Joe's Filling Station Restaurant is one-block north of Kentland Airport, an easy, easy walk!

Open for breakfast, lunch, and dinner. Breakfast served all day.

Kentland, IN is located approximately halfway between Chicago, IL and Indianapolis, IN. The airport has a real nice asphalt runway 09-27 and is approximately 3,500 feet long with two nonprecision instrument approaches.

Joe's Filling Station is a nice stop and is easy to get to from Kentland Airport (50I). Val and I enjoyed the relaxing one-block walk to the restaurant. The first thing we noticed walking out of the airport road toward a gas station is an old antique Phillips 66 advertisement. Entering Joe's right under the "EAT" sign, you can't help but notice the truck and automobile memorabilia on the walls and the antique Sinclair fuel pump on the main floor. This is a small truck stop with approximately 15 tables.

Norman, our waiter, introduced us to Dean, Joe's son (Joe, the owner, was not in). Norman started working at Joe's Filling Station eight years ago. He came in one day just to help out and has been there ever since.

The staff is very friendly and helpful, service is excellent, and the food freshly made to order (they claim it might take a little longer to prepare, but it certainly is fresh and worth the wait). Pricing is very reasonable; Val and I had two breakfast items and ice tea for lunch and paid under $15 including tip. Each day they offer home-cooked specials, and the day we stopped in they had a ham and sweet potato plate for only $6.95.

Give Joe's Filling Station a try, I'm sure you'll enjoy it. And, by the way, don't forget to say hi to Norman and Dean for us.

Lafayette, IN (Purdue University—LAF)

Seattle Beanerie 🍔

Rest. Phone #: (317) 746-3918
Location: Short walk
Open: Daily: Lunch

PIREP In Lafayette, IN, the Seattle Beanerie is about three-quarters of a mile north of the Purdue University airport (LAF). It has specialty coffees and deli-type sandwiches.

Muncie, IN (Delaware County–Johnson Field—MIE)

Vince's at the Airport 🍔🍔🍔🍔🍔

Rest. Phone #: (765) 284-6364
Location: On the airport
Open: Mon.–Thurs. 11 AM–9 PM, Fri. and Sat. 11 AM–10 PM, Sun. 11 AM–8 PM

PIREP Fabulous! Great food, service, atmosphere! Wireless Internet in the restaurant.
  We celebrated our 24th anniversary there, and it was just perfect.

PIREP FBO Comments: Free daytime transient parking right outside the terminal/restaurant.

PIREP One of the best meals I have ever had at an airport. This is not your average greasy spoon, far from it! Highly recommended.

PIREP Wonderful experience, excellent food, very attentive wait-staff. Restaurant is convenient as can be, with huge windows that look right out on the runways. Seafood selections appear to be their specialty, and the proof was in the scallops. They were as big, juicy, and sweet as any I've had. But don't pass up the temptation to try Vince's World Famous Turtle Pie for dessert. It's a huge portion and wonderful. This is a $100 hamburger destination worthy of regular returns.

PIREP Vince's provides an outstanding view of runway 32 while you dine. Their decor is an aviation theme. The service was remarkably friendly and the food was excellent!

PIREP Vince's is the place to take anyone to impress them. The food deserves five stars, or five hamburgers, but hamburgers are going to be the last item you will order there. Their menu is top notch and the food is great. I flew there tonight, in fact, with wife and mother-in-law. I did not make a reservation, but it would be a good idea, especially at night or on weekends. The place was full, and we waited half an hour for a table. It is worth the wait.

Nappanee, IN (Nappanee Muni—C03)

Amish Acres 🍔🍔🍔🍔

Rest. Phone #: (800) 800-4942
Location: Call for pickup
Open: Daily: Breakfast, Lunch, and Dinner

PIREP Land here on the paved 3,000' strip, 27-09. The friendly FBO will let you use the phone to call Amish Acres. They will be glad to send a car for you. On the national historic register, Amish Acres is an "authentic" look at early Amish life and history. Spread out over 100 or so acres, it is a perfect place to fly with the family. Two restaurants, one a lunch spot and the other serving a Thresherman's Supper, are located in a converted barn. There is even a stage show in an old round barn. Ice cream parlor, blacksmith, butcher, and many gift shops are on the site. Nice spot, quality food, great for mom and the children.

PIREP Last weekend my wife and I flew to Nappanee, IN (C03) and enjoyed the delicious "Thresher's Dinner" at Amish Acres. We called Amish Acres to advise them of our arrival and they were happy to pick us up and drop us off at the airport, a five-minute drive.

The dinner is served family style in the historic restaurant, which is really an original Amish barn constructed in 1870. The place is huge and easily could seat 300 people, although the day we were there it wasn't crowded at all.

Cost for the meal is $14.95/person and is all you can eat. You first get an iron kettle of thick ham and bean soup, bread and fresh yellow butter and apple butter, relish, and coleslaw. When you're through with that they bring out the main course. You get to choose two meat entrees (turkey, chicken, ham, or roast beef). We chose the chicken and roast beef. These are accompanied by green beans, beef and noodles, mashed potatoes, dressing and gravy.

Dessert is also included. You're presented with a huge tray of pies (maybe 20 varieties), including their own Shoo-fly pie (I didn't try it, but it looked delicious). Coffee, tea, milk, or lemonade are also included during your meal.

In addition to the restaurant, Amish Acres also has shopping (gift barn, cow shed, soda shop and fudgery, bakery, and meat and cheese shop), a buggy ride, a historic tour of the farm, a theater where plays are presented, lodging, and a short movie about the history of the Amish people.

Another nice feature of the trip was that our fuel cost (taxes included) was low, and the airport manager said he was considering lowering the cost still further.

Best meal I've had in months!

Plymouth, IN (Plymouth Muni—C65)
El Camino Real 🍔🍔🍔

Rest. Phone #: (574) 936-5152
Location: On the field
Open: Daily: Lunch and Dinner

PIREP The Damon's I reported on 5 years or so ago is gone, as are a couple of successors... and the motel isn't a Ramada any more either... but it's still a good place to fly into. The current restaurant occupant is El Camino Real, a pretty good Mexican restaurant with a lot of food for your dollar. I especially like the "Especial Camino Real", which includes steak, chicken, Chorizo, shrimp, and more. A paved path from the ramp leads straight over to the motel and restaurant.

Portland, IN (Portland Muni—PLD)
Richard's Restaurant 🍔🍔

Rest. Phone #: (260) 726-7433
Location: Short walk
Open: Daily: Breakfast, Lunch, and Dinner

PIREP If it is dry, park at the east end of the taxiway (old runway) and walk a couple hundred yards east, across the field to Richard's Restaurant. Good home-style food at very reasonable prices. If the field is wet or planted, plan on about a three-quarter-mile walk.

Rensselaer, IN (Jasper County—RZL)
Fair Oaks Dairy 🍔🍔

Rest. Phone #: (219) 345-6074
Location: Call for pickup
Open: Daily: Lunch

PIREP   FBO Comments: Very nice facility, with all conveniences of a small-town airport. Airport manager was very personable.

PIREP   Fair Oaks Dairy Adventure is a dairy information center and cheese manufacturing plant, right between Indianapolis and Chicago off HWY 65 in Indiana. I used the crew car, but you might call the center—there was talk at the time of setting up a pickup car.

Starbucks coffee shop and a nice deli in the cheese plant. You can watch cheese or ice cream being created as you eat lunch.

Rochester, IN (Fulton County—RCR)

Karen's 🍔🍔🍔🍔

Rest. Phone #: (574) 223 5384
Location: On the field
Open: Daily: Breakfast and Lunch

PIREP   Have never gotten past the lunch buffet, except one breakfast that could have fed four people. Very good fried chicken. Prices very reasonable. And they have declared at least half the building smoke-free. We go there a lot.

PIREP   We flew up for breakfast and had a great time. The portions at the restaurant are huge and the food is really good. You will not go home hungry. People at the restaurant were very nice. Small-town atmosphere!!!

PIREP   First fly-out of the year for the Schaumburg Pilots Association and all of us parked RIGHT IN FRONT of the restaurant in the grass (approximately 30 steps to the front door)!! If you were there, you know the fun we had with the all-you-can eat buffet for ONLY $7.95 per person. Boy, do they put out a spread or what? I know they lost money on at least couple of us (probably all of us)!! Great salad bar (did everyone try the cheese spread?), fried chicken, meatloaf, mashed potatoes, and so on (just like homemade)! I'm really surprised that the airplanes could leave the ground with all the extra weight after brunch. For those of you who are familiar with Karen's, the restaurant has been remodeled with a newer and larger dining area. No wonder this location has a three-hamburger rating on the *$100 Hamburger*, but I'm absolutely sure it will move up to a four-hamburger rating soon! If you have the chance, don't miss this one.

PIREP   I was kind of asleep at the switch and didn't realize that you could just taxi across the front of the airport property toward/along State Road 14 until we were walking over to Karen's. So be advised, taxi south in front of the corporate hangars and look for the matted down grass/snow off to the left about two-thirds of the way down, then just trundle across the frozen tundra and pull up 50 feet away from the front door.

Nice little mom-and-pop sort of small-town diner (I love those places). Food was really good, half of us brought leftovers home because there was so much.

All-you-can-eat perch for $9.50—can't beat that with a stick. My rib-eye was pretty good, as well.

There might be some of us from the office taking long lunches to come up here in the future!

Terre Haute, IN (Sky King—3I3)

Fox's Market 🍔🍔🍔

Rest. Phone #: (812) 466-5825
Location: On the field
Open: Daily: Breakfast and Lunch

PIREP   This airport is not for novices, with one of the shortest paved runways in the state of Indiana. However, once you have landed at this airport, back taxi to the office and park on the ramp.

The market is just a short walk away, across a two-lane road that is literally off the end of RW 26. Walk into Fox's for a sandwich and side, or order a pizza from the Pit Stop just a quarter-mile south along the same road.

Before you arrive, be forewarned that this is a flight school—during the summer, things might be slower than normal. During the school year it might be busier than many controlled airports.

Terre Haute, IN (Terre Haute International-Hulman Field—HUF)

The Hangar Restaurant 😋😋😋

Rest. Phone #: (812) 877-6777
Location: On the field
Open: Daily: Breakfast and Lunch

PIREP   The Hangar Restaurant is indeed open and better than ever. Breakfast and lunch, seven days a week. Good food and good prices. Great chili, Italian beef, and chicken salad, along with lots of other items. And great pies, too. Easy ramp access and a very short walk from the FBO. The restaurant is in the east end of the main terminal building.

PIREP   Just thought I would let all you guys out there know that George's at the airport has a new owner and new name. It's called The Hangar. Wanda is the owner, heard of her? She worked for George last year. The food is even better and the restaurant is clean. Now with "new oil" in the fryer. Good home cooking with daily specials. Open seven days a week. Mon.–Fri. 7 AM–3 PM and Sat. and Sun. 7 AM–2 PM. The workers are all friendly; you'll definitely have a GREAT experience when you walk in the door. Hope to see you soon.

PIREP   Located in the terminal building just below the tower. Parked at the GA ramp and walked through the FBO and across the parking lot to the terminal building. You can't enter directly due to "security concerns." We had the lunch special for $7.50. I had the grilled tenderloin sandwich with everything, heaping helping of fries, and Coke. Hit the spot in the middle of a cross-country. Not a gourmet restaurant, just good food and great service with a smile. Easy access and a great view of the airport operations from the restaurant. Hulman approach and Tower personnel are easy to work with, and I encourage the use of their under-utilized class D airport with three enormous runways.

Valparaiso, IN (Porter County Muni—VPZ)

The Strongbow Inn 😋😋😋😋

Rest. Phone #: (219) 464-8643
Location: Short walk or call for pickup
Open: Daily: Lunch and Dinner + Brunch on Sun.

PIREP   The Strongbow Inn specializes in turkey and will pick you up or the airport will give you their courtesy car. In fact, it used to be an actual turkey farm, and we still call it that.

PIREP   A short half-mile from Porter County Airport (VPZ) in Valparaiso. Great food. Pickup service available.

PIREP   None of the pilot reports so far have mentioned the best part of Strongbow: the Blue Yonder Lounge. The former owner of Strongbow was a big aviation buff. This room is decorated with many aircraft models hanging from the ceiling and numerous local flying-related photos. The entire room is actually designed to look like the inside of an African bush plane, complete with instrument panel. I just stumbled across the room after taking part in the Sunday brunch. Ask them to seat you in the Blue Yonder Lounge!

PIREP   My wife and I flew to Valparaiso, IN (Porter Co.)(VPZ) to visit friends of ours. It is only a one-hour flight (with a Cessna 172), from Indianapolis, IN. Valparaiso (VPZ) is a good dual-airstripped (north-south and east-west) airport, just a few minutes from Chicago Airspace.

We landed about 10 AM on a Sunday morning. Our friends immediately told us they were taking us to the Strongbow Inn for their Sunday brunch. It is only about one mile away from the airport.

Even though we had four adults and two children (with no reservation), we got seated within five minutes. If you don't make reservations come early or late to brunch, because it fills up fast.

They offered either menus or the buffet. Believe me, go for the buffet. It has every possible hot breakfast available: eggs, waffles, potatoes, sausage, gravy, biscuits, barbecued spare ribs, corn, and more, including (carved in front of you), a meat or fowl (they had turkey the day we were there).

The next table contained all the cold breakfast you could imagine (rolls, muffins, fruits, pasta salads, green salads, and more). Then the dessert table!! This included small cakes, brownies (the best I have ever had), pastries, candies, and more.

I have eaten a lot of Sunday brunch buffet meals, but there has been NO BETTER (usually not even close) to the Strongbow Inn's Sunday brunch.

My friends refused to let me see the bill (you know how that goes), but they said the prices are reasonable. They also said the regular menu meals are quite good, too. They stated that the only flaw with the Strongbow Inn is that the service is usually slow for regular menu meals.

But believe me, if you want really good food at reasonable prices, try the Stongbow Inn in Valparaiso, IN. And if you can, come to the Sunday brunch buffet—you will be glad you did.

I wish I could give it ten hamburgers, but I will give it a 4+ rating.

Washington, IN (Daviess County—DCY)

Black Buggy Restaurant 🍔🍔

Rest. Phone #: (812) 254 8966
Location: Crew car
Open: Daily: Breakfast, Lunch, and Dinner

PIREP   There are several restaurants in town and with the airport courtesy car, you can choose any one of them. The airport manager recommended the Black Buggy, and I agree. It's an Amish restaurant with breakfast, lunch, and dinner buffets. The food was good, and there was a large selection. Lunch cost $9.75.

Breakfast, 6 AM to 10:30 AM, lunch 10:30 to 4 PM, dinner 4 PM to 9 PM; Monday–Saturday 6 AM to 9 PM.

Winchester, IN (Randolph County—I22)

Mrs. Wick's Restaurant 🍔🍔

Rest. Phone #: (765) 584-7437
Location: Crew car
Open: Daily—Dessert—This is a pie shop!

PIREP   There is nowhere to eat at the airport, but they loaned us a car. We put a few dollars of gas in it to go to Mrs. Wick's Restaurant about three miles away in town. It turns out that Mrs. Wick's is a manufacturer of pies distributed all over that part of the country.

The restaurant has large sandwiches at low prices. Of at least equal interest, they usually have about 20 to 30 different kinds of pies for $2.25 a slice. They also sell pies to go. Some of the pies for sale are "seconds." I was not able to determine why. The seconds cost as little as $3. I will tell you from personal experience that they tasted just fine.

I highly recommend a stop at Winchester if you are passing by the area.

The Hundred Dollar Hamburger

Iowa

Amana, IA (Amana—C11)

The Ox Yoke Inn 🍔🍔🍔🍔

Rest. Phone #: (319) 622-3441
Location: Short walk
Open: Daily: Breakfast, Lunch, and Dinner

PIREP Probably the finest fly-in eateries in Iowa are located in the Amana Colonies, south of Cedar Rapids in Eastern Iowa. CID approach will even give you a vector.

Amana is a small grass strip. Plenty long for most singles. Many restaurants and shops are within easy walking distance of the field. The Amanas are an old German colony and the restaurants reflect the homeland.

Amana Barn Restaurant

The Brick Haus

The Colony Haus

The Ox Yoke Inn

Ronneburg Restaurant

This is a small list. Try the Ox Yoke for starters. There are MANY shops as well.

Ames, IA (Ames Muni—AMW)

Hickory Park 🍔🍔🍔

Rest. Phone #: (515) 232-8940
Location: 1 mile
Open: Daily: Lunch and Dinner

PIREP Hickory Park is a short drive (about one mile east) from AMW. Midwest Aviation has a free courtesy car. Food and atmosphere and service are all excellent. Lunch or dinner, sandwich to full meal.

Ask Harrison Ford; he ate there when he stopped at AMW.

Anita, IA (Anita Muni-Kevin Burke Memorial Field—Y43)
The Redwood Steakhouse 🍔🍔🍔🍔

Rest. Phone #: (712) 762-4105
Location: Short walk
Open: Daily: Breakfast, Lunch, and Dinner

PIREP Nice grass strip, good food. Park on north end for the Redwood Steakhouse. For breakfast, park on the south end, then take a short walk into town, approximately one block. The route to the Redwood Steakhouse involves a "rustic" trek up a slight embankment, across the railroad tracks, then a highway. Altogether it is about one block also.

Belle Plaine, IA (Belle Plaine Muni—TZT)
Lincoln Cafe 🍔🍔

Rest. Phone #: (319) 444-2013
Location: Crew car
Open: Daily: Breakfast, Lunch, and Dinner

PIREP In Belle Plaine, visit the Lincoln Cafe—so named because it is located on the old Lincoln Highway—for a great breakfast. Stepping into this restaurant is like traveling back in time to the days when the Lincoln Highway was America's main highway. The decor is unchanged from the '50s, prices are reasonable, food and coffee are good, and the service is fast and friendly. We overfed our family of four for less than $15, including tip!

Call ahead to reserve the FBO's courtesy car. You will need it, as town is a bit of a trip.

Burlington, IA (Southeast Iowa Regional—BRL)
The Tender Trap 🍔🍔🍔

Rest. Phone #: (319) 754-4640
Location: A walk or crew car
Open: Daily: Breakfast, Lunch, and Dinner

PIREP The Tender Trap, just off the Burlington Airport, has good homemade food at great prices. Buy some fuel from the nice folks at Remmer's Aviation and they may let you use the courtesy car.

Cedar Rapids, IA (The Eastern Iowa—CID)
Airport Cafeteria 🍔🍔🍔

Rest. Phone #: (515) 256-5342
Location: On the field
Open: Daily: Breakfast, Lunch, and Dinner

PIREP A miniature O'Hare-style semi-cafeteria food service with separately located coffee and juice bar now graces CID's terminal. The sandwich menu is dominated by croissant sandwiches. Prices are reasonable, portions are large, food is a notch above a food court at the mall. Paper plates and plastic "silverware" have replaced china and metal eating utensils. The area is well-lit and cheerful. Walk from Signature (where you might pay a landing fee) or catch a ride from PS Air (at the far end of the field).

Clinton, IA (Clinton Muni—CWI)

Airport Snack Machines 🍔

Rest. Phone #: (563) 242-3292
Location: On the field
Open: 24/7

PIREP Just last week, I flew into Clinton, IA (CWI) for a quick snack. I didn't ask for a courtesy car, so I don't know if they had one. The small terminal building had a microwave, a sandwich machine, pop machine, and candy machine. I felt like I took a step back to the '60s (even though I wasn't born then) when I went inside the terminal. The people there are very friendly, and I will probably fly there again soon.

Decorah, IA (Decorah Muni—DEH)

Mabe's Pizza 🍔🍔

Rest. Phone #: (563) 382-4297
Location: Crew car
Open: Daily: Lunch and Dinner

PIREP I took a ride in "the beast" of a courtesy car (imagine a big red rusty conversion van, so fitting, pictures were taken in front of it for the memory) and, after making it into town (couple of minutes drive), I ate at Mabe's with my sister, who attended Luther College. It is a great place for great-tasting pizza. There are a couple of other restaurants in town and some nice parks and waterfalls on the east side. A great town to go visit, even if you don't know anyone there like I did.

Dubuque, IA (Dubuque Regional—DBQ)

Bev's 🍔🍔🍔

Rest. Phone #: (563) 557-8888
Location: On the field
Open: Daily: Breakfast, Lunch, and Dinner

PIREP Bev's is located in the terminal building at the Dubuque Regional Airport. It looks like a typical cafeteria-style cafe—but it's not! Bev makes all of her burgers from FRESH ground buffalo meat! The buns are fabulously fresh, the fries are excellent, and the buffalo burgers are HUGE.

We were very pleasantly surprised to discover Bev's place—she is a very nice lady, an excellent cook, and she runs the place with her entire family.

They serve breakfast and other sandwiches, too—my wife and I highly recommend the place!

Ida Grove, IA (Ida Grove Muni—IDG)

Boz Wellz Pub & Eatery 🍔🍔🍔

Rest. Phone #: (712) 364-3606
Location: 100 yards from the ramp
Open: Daily: Lunch and Dinner

PIREP Boz Wellz in Ida Grove, IA "KIDG" close to the airport—fantastic menu of steak, seafood, chicken, Mexican, pizza, and so on; 100-yard walk from the ramp and a motel next door—ya gotta try it.

Lighted runway and fuel through the city of Ida Grove.

Reservations most weekends but not required—large dining area.

Iowa City, IA (Iowa City Muni—IOW)

El Ranchero's 🍔🍔🍔🍔

Rest. Phone #: (319) 338-4324
Location: On the field
Open: Daily: Lunch and Dinner

PIREP FBO Comments: Great people. Meg went out of her way to welcome us and give us a ride to and from El Ranchero's. Thank you, Meg!!

PIREP We chose El Ranchero's from the $100 site because one of our passengers was from the local Mexican Consulate. Everyone was impressed with the food. It was excellent. We asked the waiter for his recommendations on the most authentic dishes, and they came through in spades.

Try the carnitas!

PIREP Fly into the finest GA airport in the Midwest (IOW—Iowa City Municipal—three long, wide, intersecting runways, and three instrument approaches) to enjoy the best Mexican restaurant in Iowa, El Ranchero's.

Located just four blocks from the airport, it's an easy walk on a nice day—or borrow the car. The best item on their menu (in my opinion) is the burrito ranchero—a gigantic chicken burrito smothered in enchilada sauce, served with sour cream and guacamole, beans, rice, and unlimited chips for around six bucks! Run by a Mexican family, you won't get any more authentic. (In fact, many of the employees speak no English at all!)

There are actually over 50 fine restaurants within one mile of the Iowa City airport, by the way.

Alexis Park Inn & Suites 🍔🍔🍔🍔

Rest. Phone #: (319) 337-8665
Location: On the field
Open: Not a restaurant

PIREP While you're in Iowa City, stay at the new Alexis Park Inn & Suites, located adjacent to RWY 25. (You'll walk past it on your way to El Ranchero's.) Owned and run by a couple of private pilots, the entire hotel has been remodeled to reflect an aviation theme—with Jacuzzi whirlpool suites and a delivered-to-your-suite continental breakfast!

Mondo's Tomato Pie 🍔🍔🍔

Rest. Phone #: (319) 337-3000
Location: It's a drive, borrow the crew car
Open: Daily: Lunch and Dinner

PIREP It is just a little over 3 miles from IOW, but the FBO is very good about loaning the courtesy car. It is my wife's favorite restaurant in the area and makes for a very nice stopover.

Iowa Falls, IA (Iowa Falls Muni—IFA)

Princess Cafe 🍔🍔🍔🍔

Rest. Phone #: (641) 648-9602
Location: Crew car
Open: Daily: Lunch and Dinner

PIREP FBO Comments: When I called to check on a courtesy car, we found that we would arrive after FBO normal hours, so the FBO came back out and met us to let us use the car! Excellent service. Terminal is very, very nice for a small-town muni.

PIREP The Princess Cafe is a 90-year-old establishment with original Italian slab marble soda fountain and African mahogany high-back booths. The food is average to above average. A must-have is a malt—the cafe history is as an ice cream and sweets shop. Plan for a day and go to see a movie at the restored Metropolitan theater. You will not be disappointed by the ambiance of the day.

Keokuk, IA (Keokuk Muni—EOK)
Hawkeye 🍔🍔

Rest. Phone #: (319) 524-7549
Location: Crew car
Open: Daily: Lunch and Dinner

PIREP Stopped overnight because of weather. The folks at Linder Aviation were very gracious and loaned me their courtesy van to get lunch and also overnight when they closed.

I ate at the Hawkeye restaurant in town. Had an excellent sandwich for dinner consisting of real prime rib (not the stuff like Arby's) for about $10.

I stayed at the Chief Motel, which has the most reasonable rates in town; very nice and clean; rooms included fridge and microwave.

Most of their business is repeat customers.

I visited the boat locks at the power station there and chatted with the operators about powered parachutes.

Also checked out Grand Avenue for the historic housing and very relaxing park there.

A very pleasant experience overall.

Marshalltown, IA (Marshalltown Muni—MIW)
Maid Rite 🍔🍔🍔🍔

Rest. Phone #: (641) 753-9684
Location: Call a cab
Open: Daily: Lunch and Dinner

PIREP It just wouldn't be a trip through Iowa if you don't stop for a Maid Rite sandwich at the Maid Rite restaurant in the center of town. The restaurant has been a local landmark for decades and is known throughout the Midwest. Celebrities from all over stop through Marshalltown to pick up a bag of Maid Rites. The sandwiches were made famous in the sitcom *Roseanne*, but they were called "loose meat sandwiches" in the show. They are tasty, absolutely unique, and go great with a homemade chocolate shake. The restaurant is like something out of the '50s (and, no, it's not a "retro" look, it just never changed!). Take a cab from the airport. Every cabby knows the way. You will love 'em. Definitely worth a quick stop.

Mason City, IA (Mason City Muni—MCW)
Airport Cafe 🍔🍔🍔

Rest. Phone #: (641) 421-9553
Location: On the field
Open: Daily: Breakfast and Lunch

PIREP FBO Comments: North Iowa Air Service.

PIREP The Airport Cafe is located next to the FBO. Every Saturday morning, local pilots gather there for breakfast around 8 AM, a real friendly bunch. The breakfast special is eggs, hash browns, bacon or sausage, and toast. They also offer noon specials Monday–Friday. We had the hot beef sandwich. The FBO also has a crew car available to take in to Clear Lake or Mason City.

PIREP We flew into Mason City, IA (MCW) on Saturday and had breakfast at the Airport Cafe at about 0900. A breakfast with two eggs, bacon (or sausage), toast and jelly was less than $5 and was great.

Milford, IA (Fuller—4D8)
The American Legion 😋😋😋

Rest. Phone #: (712) 336-5942
Location: Across the street
Open: Daily: Lunch and Dinner; closed Mondays

PIREP Nice airport on the south edge of town. The American Legion is across the street with a steak, burger, and fish menu, and they've got nightly specials throughout the week. Friday night's the all-you-can-eat fish fry. Good food, friendly people. Closed on Mondays. Come early in the summer, and there's a water park about a quarter-mile from the airport to go splash around in.

Mount Ayr, IA (Judge Lewis Fld Mt Ayr Muni—1Y3)
Handi-Mart 😋

Rest. Phone #: Unknown
Location: Short walk
Open: Daily: Lunch and Dinner

PIREP There is a gas station/Handi-Mart just 50 yards west of the north end of the runway with a sub sandwich shop attached.

Muscatine, IA (Muscatine Muni—MUT)
The Good Earth 😋😋😋

Rest. Phone #: (563) 288-9626
Location: Short walk
Open: Daily: Breakfast, Lunch, and Dinner

PIREP There is a relatively new little restaurant close to the Muscatine airport. It is located about a quarter-mile to the west. It is called the Good Earth. Great breakfasts, lunches, and dinners. Attached to it is a vegetable market. It has fresh tomatoes, Muscatine melons, sweet corn, squashes, and so on (all in season, of course).

Pella, IA (Pella Muni—PEA)
La Cabana 😋😋

Rest. Phone #: (641) 628-2307
Location: Crew car
Open: Daily: Lunch and Dinner

PIREP   Excellent Mexican food, two blocks off the square. Yep, the keys to the courtesy van were hanging there ready!

Red Oak, IA (Red Oak Muni—RDK)
Firehouse Brewery 😋😋😋

Rest. Phone #: (712) 623-4799
Location: Short walk
Open: Daily: Lunch and Dinner

PIREP   FBO Comments: Best fuel price around and self-service pump to boot!

PIREP   Very nice airport with the cleanest terminal in the Midwest. Frequently has actual homemade cookies and tarts made by grandma!

Use the courtesy car to drive 8 blocks to the Firehouse Brewery. Used to be a real firehouse, and the walls are lined with antiques. Excellent food at Midwest prices.

Sheldon, IA (Sheldon Muni—SHL)
Kinbrae Supper Club 😋😋

Rest. Phone #: (712) 324-4411
Location: Crew car
Open: Daily: Lunch and Dinner

PIREP   Very good "surf and turf" dining, approximately 1-1/2 miles from airport. Courtesy car at airport. Also, city campground on the way.

Sully, IA (Sully Muni—8C2)
Coffee Cup Cafe 😋😋😋

Rest. Phone #: (641) 594-3765
Location: 2 blocks
Open: Daily: Breakfast, Lunch, and Dinner

PIREP   Sully, IA has a grass strip just two blocks south of the town square. There are a couple of "down home" cafes there that have decent food and a comfortable atmosphere. Nice change of pace to put the tires on grass.

The Hundred Dollar Hamburger

Kansas

Abilene, KS (Abilene Muni—K78)
Brookville Hotel 🍔🍔

Rest. Phone #: (785) 263-2244
Location: Crew car
Open: Daily: Breakfast, Lunch, and Dinner

PIREP Two courtesy cars available here—drove one to the motel and then dinner at the Brookville Hotel. Great place—pan-fried chicken served family style, all you can eat. Went the next day to the Farm House restaurant—another great place where Ike used to go when in his hometown. Full menu with great salads—drop in here for a couple of days. I stopped here on the way to Fort Smith and it was a really nice experience.

Cottonwood Falls, KS (Chase County—9K0)
Emma Chase Cafe 🍔🍔🍔

Rest. Phone #: (620) 273-6020
Location: Short walk
Open: Daily: Lunch and Dinner

PIREP FBO Comments: Turf field (very nice turf field).

PIREP Called the restaurant for conditions and gave an estimated ETA. The owner's husband was waiting to take us to eat. We were treated like royalty. We both had the best chicken-fried steak with all the fixin's. We ate one homemade piece of blueberry pie and took home a piece of peach pie. The food and company of this small-town restaurant was great! We walked Main Street and found interesting stores. Only one storefront did not have an active business in it. When we were done looking around, we went back and bought some homemade fudge and got a ride back to the airstrip. Said our good-byes and left with a full belly and good feelings until our return. I would recommend to all.

PIREP Grass strip, three-quarter-mile walk (until chamber will comes up with phone) into town to refurbished Grande Hotel—nice atmosphere.

Council Grove, KS (Council Grove Muni—K63)

Hays House 😋😋

Rest. Phone #: (620) 767-5911
Location: Call for pickup
Open: Daily: Lunch and Dinner

PIREP We called, and Dan the manager drove 7.5 miles to come pick us up! The food was excellent, the Hays House tour he gave us was awesome. Council Grove has 18 historical monuments, and the Hays House is one of them. Leave time to see the rest, or talk with Dan about borrowing a car and sticking around.

PIREP Council Grove, KS has a City Lake three miles from town. The airport is next to the lake. There are two short grass strips, well maintained. You will find a windsock and painted barrels marking the strips. The strip is on a well-traveled county road with friendly people. Try the Hays House in town for excellent meals: 112 W. Main. Closed Sunday night and Mondays.

Garden City, KS (Garden City Regional—GCK)

The Flight Deck Restaurant 😋😋😋

Rest. Phone #: (620) 271-1490
Location: On the field
Open: Daily: Breakfast, Lunch, and Dinner; closed Sat. and Sun.

PIREP The Flight Deck Restaurant is really nice. It brings a lot of town people out to eat. That says something, considering the airport is several miles out of town.

PIREP The Flight Deck was closed on Saturday when we visited and the guy in the FBO reckoned it was only open Monday through Friday. Doesn't make much sense for a fly-in restaurant, but that's what he said. So we got the courtesy car for the long, eight-mile drive into town.

We ate a good barbecue lunch at Adam's Rib BBQ, 1135 College Drive ((316) 275-7427). Good pork ribs with a quality sauce. Eat-in/take-out type of place. Plenty of food for a good price.

As with almost all Kansas towns, there's a strong Mexican presence on the restaurant scene. We took supper home in the plane's baggage compartment from El Zarape West, 2212 Jones Ave. (Continue on US 50 into town and through town; when US 50 Business turns right, go straight on. El Zarape West is on the left (south) a little way farther on.) It's in a rundown neighborhood beside the railroad tracks, but the food was really good. Staff non-English speaking although the señora/owner speaks English fine. Standouts were smothered enchiladas with green chile sauce and guacamole along with a fine kick-in-the-tail medium-hot salsa for dressing the food. You can tell it's an authentic place because chips don't automatically come to your table unless ordered. Not only that, but the traditional tripe-stew menudo is served Saturdays only. It's worth trying if you've never had it.

Other places spotted en route from the airport: a promising little Guatemalan restaurant on the north side of Highway 50 before you reach the town proper and several other Mexican eateries that look worth a checkout. A fancy steakhouse supper club called the Grain Bin is easy to find on the main drag, but it's not open for lunch.

Courtesy car is a Pontiac in only "fair" shape (no rear-view mirror and the driver's seat back at a permanently 60-degree reclined angle), but it does the job. They may have multiple courtesy cars, not sure. Friendly service. NFCT tower Class D.

Goodland, KS (Renner Field /Goodland Muni—GLD)

The Butterfly Cafe 😋😋😋

Rest. Phone #: (785) 890-2085
Location: On the field
Open: Daily: Breakfast, Lunch, and Dinner

PIREP   The Butterfly Cafe is located on the ramp in the old FSS Building. Great food at very reasonable prices—and lots of locals come out from town for breakfast and lunch. Homemade breads and pies are terrific. Ask about the homemade strawberry-rhubarb jam (in season).

PIREP   The Butterfly Cafe at Goodland, KS is a nice stop. The people are very friendly and the food is good enough to attract the locals from Goodland for lunch. The FBO there even gives out their "goodie bags" to student pilots when you top your tanks with a few gallons of gas on X-C's.

Greensburg, KS (Paul Windle Muni—8K7)
The Kansan Restaurant 🍔🍔🍔🍔

Rest. Phone #: (620) 723-3077
Location: Very short walk
Open: Daily: Breakfast, Lunch, and Dinner

PIREP   Greensburg, KS has one of the nicest grass strips to be found anywhere! The Kansan Restaurant is an easy walk (maybe a half-mile) straight down the road that goes by the airport or across the golf course adjacent to the airport. Food is good, portions generous, and prices reasonable. Greensburg is also home of the "World's Largest Hand-Dug Well," which is well worth the extra walk to downtown.

Hays, KS (Hays Regional—HYS)
Al's Chickenette 🍔🍔

Rest. Phone #: (785) 625-7414
Location: Crew car
Open: Daily: Lunch and Dinner

PIREP   If you are ever in Hays, KS, you have to stop by the legendary Al's Chickenette. This is a 1950s-style chicken house that has never changed. You will find huge pieces of chicken and baskets of long fries for a cheap price. The aroma is free and will stay in your clothes long after you take off. To get to Al's, head west from the airport on Highway 40 to US 183. It is one block north of the intersection on the right side.

PIREP   Anyone within flying distance should definitely plan a visit to the Sternberg Museum at Fort Hays State University. There is an excellent steak (Angus beef) restaurant in the museum.

The museum/restaurant is a short drive into town via courtesy car from one of the local FBOs at the Hays Airport.

Bring the kids, it's educational and fun. There is one particularly neat fossil of a fish skeleton with another fish skeleton inside its rib cage. As Western, KS was once on the bottom of a prehistoric ocean, many fossils were found in that area. Plan to spend at least a couple of hours.

Hutchinson, KS (Hutchinson Muni—HUT)
The Airport Steak House 🍔🍔🍔🍔🍔

Rest. Phone #: (620) 662-4281
Location: On the field
Open: Mon.–Thurs. 9 AM–9 PM, Fri. 9 AM–10 PM, Sat. 5 PM–10 PM, Sun. 11 AM–9 PM

PIREP   Don't miss the Sunday buffet. It begins at 11 AM! It is really very nicely done!

PIREP   This has to be the best airport restaurant in Kansas. We flew in for lunch; they have a great buffet from 11–2 on Sunday. I need to get their regular hours, as we will be coming back soon.

PIREP   The Airport Steak House is still serving great food, and the proportions are just too much! The wait-staff is so friendly and the ambiance is outstanding! They have lunch and dinner and even breakfast on some days.

We will definitely be back. It has now become one of our favorite fly-in restaurants. As others have mentioned before, you park right in front of the restaurant and it's a few steps to great food!

PIREP   Hutchinson has a great restaurant right in the terminal building about 50 steps from the ramp area. It has just been beautifully remodeled and has a warm and friendly atmosphere. The decor has the kind of ambiance that both pilots and others will enjoy. At the entrance above the doorway leading into the eating area is the tail section of a half-scale biplane positioned as though it has just flown through the wall. You can eat smorgasbord style and choose from several types of meats and vegetables as well as salads and wonderful desserts, or you can order from a complete menu. If you choose a steak from the menu, you can watch it being cooked on an indoor charcoal pit that is arranged to allow viewing from the dining area. The quality and taste of everything I have had to date has been superb and at reasonable prices. The large glass windows allow a view of the three runways. Military trainers from Vance AFB stop in frequently for food and fuel. The FBO has top-notch service, the kind where they actually smile when they tank you, with reasonable prices. This restaurant stop has got to rate at least a five in anyone's book!

Lawrence, KS (Lawrence Muni—LWC)

Quinton's 🍔🍔

Rest. Phone #: (785) 842-6560
Location: Crew car
Open: Daily: Lunch and Dinner

PIREP   FBO Comments: Gentleman that greeted me at the plane was extremely friendly and pumped the fuel quickly—even noticed a couple of screws missing on the tail cone.

PIREP   I've been to Lawrence many times to see friends and I always have to make a stop at Quinton's. (They are getting tired of me suggesting the place.) It's a soup/sandwich place that has awesome sandwiches and potato-bacon soup in a bread bowl. I also recommend the chips and queso as an appetizer. If you're going to be an over-nighter they also have daily drink specials.

Hetrick FBO has friendly staff and a crew car you can borrow.

Everyone knows Quinton's, so they can give you directions. It's about five miles. 615 Mass Street.

The airport is in great shape, and the crosswind runway is now open.

Liberal, KS (Liberal Muni—LBL)

Cattleman's 🍔🍔🍔

Rest. Phone #: Unknown
Location: Crew car
Open: 10:30 AM–10:30 PM

PIREP   Confirming previous posts, this is a great place for a quick meal (the service is extremely efficient). The food is wholesome and nicely presented. Someone in charge knows what they are doing. We were there for Sunday lunch, and the place was bustling like a New York brasserie. I had a rib-eye steak, and my wife had an excellent burger. I asked for the steak cooked rare and that's exactly how it came. The FBO at LBL provided a trusty Grand Marquis (about 1990 vintage) with functional air conditioning (it was 75F in mid-December at Liberal!) for no charge with fuel purchase. It's easy to find but more difficult to find your way back to the airport. Take notice as you leave! The FBO is to the north. Comfortable lounge area and good vending machines. Three airport dogs (one large, two small) are inseparable canine buddies and patrol the airport like sentries, but they are very friendly pooches.

PIREP   Awesome Kansas City strip steak! Probably the best-tasting steak we ever had; it was seasoned and cooked just right. There is a reasonable choice of sides, the fries were good, the veggies were a small portion, okay. It is on the south east side of town, on the main east/west road, across from the Motel 8.

Lucas, KS (Lucas—38K)
Airport Diner 🍔🍔🍔

Rest. Phone #: (785) 525-6425
Location: On the field
Open: Daily: Breakfast and Lunch

PIREP   38K is a small airstrip with an adjacent cafe that provides some of the best "home-cookin'" style breakfast. If you have a fat or cholesterol problem and avoid great homemade pie on Saturday morning, stay away!

Oakley, KS (Oakley Muni—OEL)
Don's Drive-in 🍔🍔

Rest. Phone #: (785) 672-3965
Location: Crew car
Open: Daily: Lunch and Dinner

PIREP   The best hamburger you can find is waiting for you at Don's Drive-in at the Junction of US 40 and US 83 in Oakley, KS (OEL). There is an old car at the airport with the key under the mat. The drive-in is just about a mile west of the airport. Good curly fries, too. If the car is in use, it is just walking distance to Mitten's Truck Stop. They have a restaurant there so you don't get skunked. I have not tried the food.

Paola, KS (Miami County—K81)
We-B-Smokin 🍔🍔

Rest. Phone #: (913) 256-6802
Location: On the airport
Open: Daily: Breakfast and Lunch

PIREP   This restaurant deserves a higher hamburger rating than a two, maybe three or four. The only thing is they need to have longer hours on the weekends.

PIREP   FBO Comments: Fuel is always the lowest price in the area.

PIREP   Have eaten here many times. Breakfast and lunch is always great and reasonably priced. Staff is friendly and service has always been good.

PIREP   Had breakfast this weekend. Great food. Great aviation setting with all types of experimentals, as well as certified craft coming and going. Great food at very reasonable prices! We'll be heading back for the ribs.

St Francis, KS (Cheyenne County Muni—SYF)

The Dusty Farmer 🍔🍔

Rest. Phone #: (785) 332-2200
Location: Short walk
Open: Daily: Breakfast and Lunch

PIREP   FBO Comments: Fuel until noon Saturday. Best turf runways I have ever used. Friendly people. Home of the famous Stearman Fly-in (second weekend in June).

PIREP   The Dusty Farmer is a five-minute walk north from your tie-down. The food is pretty Western Kansas. It was better when Darrell ran the place, but it's just fine for the halfway point of a nice flight over Eastern Colorado or Western Kansas.

Portions are large and the salad is always fresh.

If you get weathered in, there is a good bar and motel on the property.

Topeka, KS (Philip Billard Muni—TOP)

Philip Billard's Airport Cafe 🍔🍔🍔

Rest. Phone #: (785) 232-3669
Location: On the field
Open: Daily: Breakfast, Lunch, and Dinner

PIREP   FBO Comments: Have made several stops here, en route and overnight —always courteous and accommodating. Crew car available (at no charge) for both our overnight stays; arranged our hotel room downtown at special courtesy rate.

PIREP   Restaurant (walk-in from ramp or FBO) isn't fancy, but food is good and reasonably priced and service is friendly. Open for breakfast and lunch. Buffet (lunch) looks good but haven't tried (yet).

PIREP   I am a student at the University of Kansas and decided to take an excursion to a restaurant in Kansas with my grandfather on some Friday night. We originally intended to go to Paola (K81), but no instrument approaches there and IFR conditions killed that plan in a hurry. We decided on Topeka instead. Still close to Johnson County Executive (KOJC), where I've been flying out of.

The place is right next to the ramp, and as you walk inside it looks a bit rundown, but don't be discouraged! The food was surprisingly good, especially if you order off the menu. I had the smoked pork chops, which were wonderful, and my grandfather could not stop raving about the deep-fried catfish. The service left something to be desired, as did the atmosphere, but the food makes up for it. A solid three stars.

Wichita, KS (Colonel James Jabara—AAO)

Jimmie's Diner 🍔🍔🍔

Rest. Phone #: (316) 636-1818
Location: Long walk
Open: Daily: Breakfast, Lunch, and Dinner

PIREP   Jimmie's Diner is located about one mile west of the airport. Midwest Corporate Aviation at Jabara has two crew vans available. Naturally, they expect you to purchase fuel. Once at Jimmie's Diner, the old-fashioned style dinner with reasonable prices and the best food in the Midwest will win you over. They are open late and can usually get you in with little or no wait. The waitresses even wear poodle skirts and provide awesome service.

Johnny Carino's Country Italian 🍔🍔🍔

Rest. Phone #: (316) 636-4411
Location: Short walk
Open: Daily: Lunch and Dinner

PIREP   FBO Comments: Midwest Corporate Aviation is a nice corporate full-service FBO. Fuel prices are a bit high, and there are several smaller airports near with fuel for less. Crew car is available. The airport has a newly installed ILS/DME approach to runway 18. Minimums are 200' 1/2 mile for Cat A aircraft.

PIREP   Johnny Carino's is a very nice Italian restaurant across the street from the south end of Colonel James Jabara Airport (KAAO). The food and service are quite good and they provide a broad fare of country Italian food. I have tried the eggplant parmigiana and the chicken scaloppini, which were both fantastic. The Tuscan rib-eye is also really great. The restaurant is open for both lunch and dinner.

Ted's Montana Grill 🍔🍔🍔🍔

Rest. Phone #: (316) 634-8337
Location: Short walk
Open: Daily: Lunch and Dinner

PIREP   FBO Comments: Midwest Corporate Aviation is a full-service FBO. Crew car is available—they would expect a fuel or other purchase if it is used.

PIREP   Ted's Montana Grill is the restaurant founded by media entrepreneur Ted Turner. The restaurant features bison and beef from Mr. Turner's ranches in Montana, Wyoming, and Kansas.

The restaurant features comfort food for the twenty-first century that brings the spirit of the American West to the restaurant.

Steak offerings on the menu are available in both beef and bison, and include a tenderloin filet, Kansas City strip, the Delmonico, and prime rib.

Expect to pay well for a bison burger, and the KC strip runs about what you'd expect. All in all, a very nice place for a really nice lunch or dinner.

The restaurant is across the street from the south end of Colonel James Jabara airport (AAO) and is within easy walking distance of the FBO.

Winfield/Arkansas City, KS (Strother Field—WLD)

Landing Strip 🍔🍔🍔

Rest. Phone #: (620) 442-2800
Location: On the field
Open: Daily: Breakfast and Lunch

PIREP   Excellent food only open for breakfast and lunch on WEEKDAYS. Bummer!

PIREP   FBO Comments: Self-serve 100LL and Jet A fuel.

PIREP   Restaurant in front of main terminal. Good food and friendly people.

The Hundred Dollar Hamburger

Kentucky

Ashland, KY (Ashland-Boyd County—DWU)

The Dairy Bar 🍔🍔

Rest. Phone #: (740) 532-6420
Location: Short walk
Open: Daily: Lunch

PIREP FBO Comments: Airport is located on the southern edge of the Ohio River and is a very beautiful approach in the fall or spring. FBO has a really friendly staff. Courtesy car is available.

PIREP The Dairy Bar is only a quarter-mile down the road. It has some great ice creams, hamburgers, malts, and so on, and all at great prices. I took my daughter and three of her friends down there for her birthday. They had a great time. A great place to stop for gas and to get a refreshing treat.

Bowling Green, KY (Bowling Green-Warren County Regional—BWG)

Rafferty's 🍔🍔

Rest. Phone #: (270) 842-0123
Location: 1/2 mile
Open: Daily: Lunch and Dinner

PIREP We flew into Bowling Green (BWG) on a late Saturday afternoon. The friendly FBO employee drove us just up the hill (half mile) to Rafferty's Restaurant.

The restaurant was very nice with a comfortable atmosphere. We arrived at 5:15 PM and had to wait only 15 minutes. (If you come between 6 or 7, you might have a good wait; it is a popular place around prime time.) All the food looked good on the menu, and the food we had was no exemption.

After the meal, we simply called the FBO and they promptly picked us up.

I strongly recommend flying into BWG for a good lunch or supper meal at Rafferty's (at least a four-burgers rating). Also, the FBO employees are some of the nicest I have met.

PIREP   Super place to eat! A short walk, or CO-MAR FBO personnel will drop you off and return to drive you back to the airport. Small but varied menu.

The fried shrimp dinner is excellent! Did not have to wait to be seated, but this was at 3 PM—I understand it's fairly crowded at dinner.

PIREP   Think we caught them on a bad day. The FBO did not have a crew car available. So we walked across the cemetery and the highway to get to the restaurant. Food wasn't bad, but pretty similar to a Chili's or TGI Fridays.

Cadiz, KY (Lake Barkley State Park—1M9)

Lake Barkley State Park Lodge 🍔🍔🍔🍔

Rest. Phone #: (800) 325-1708
Location: Call for the van
Open: Daily: Breakfast, Lunch, and Dinner

PIREP   FBO Comments: Fuel will be available soon through self-serve credit card. It is available now through maintenance.

PIREP   Lake Barkley Lodge is an absolutely great place to eat. The lodge will pick you up from the state park airport (1M9). If you want to stay three days or more, an option separate from the lodge is to stay at Alvy's Hideaway on Lake Barkley in the woods, 10 minutes from the airport or lodge.

PIREP   Great news. Lake Barley is a great place to visit. The Lodge is three miles from the airport, but a shuttle whisks you away for a great buffet meal. Walk around the well-kept grounds and enjoy the lake.

We camped at the campground. Had a wonderful view of the lake with clean rest facilities.

A place to visit again. Kentucky has three state parks that have airports. We have visited two with enjoyable times and are looking forward to trying the third.

Falls-of-Rough, KY (Rough River State Park—2I3)

Rough River Lodge 🍔🍔🍔🍔

Rest. Phone #: (800) 325-1713
Location: On the field
Open: Daily: Breakfast, Lunch, and Dinner

PIREP   FBO Comments: Nice pilot lounge; airport is unattended; you are directly across from a putt-putt golf course.

PIREP   The restaurant serves wonderful food at reasonable prices, nice view of the lake. Be aware that the airport is unattended. When we were taxiing to leave, a grandmother was watching her two grandchildren climb up the struts of a Cessna 172!

PIREP   Had a great time visiting this airport and lodge/restaurant today. Runway is in good condition, well marked, and ample paved parking. Have chocks but no ropes. Easy short walk to the restaurant. Lunch buffet was very good and reasonable prices—great service. Breakfast menu looked good too. No instrument approaches, so save for a nice day. A good site for group visits, as lodging is walking distance and camping is to the right on the airfield with baths/showers.

PIREP   All the rooms have an excellent view of the lake. The restaurant is open for all three meals and you can't beat the food—it's great!

They have a beautiful golf course. And best of all, they have a 3800' runway in short walking distance of the lodge/restaurant!

PLEASE share this, it's a wonderful place.

Frankfort, KY (Capital City—FFT)

Casa Fiesta 🍔🍔

Rest. Phone #: (502) 226-5010
Location: Crew car
Open: Daily: Breakfast, Lunch, and Dinner

PIREP My wife and I flew into this airport today for a change of pace. We were very pleasantly surprised. The airport is very clean and well kept. The crew at the FBO was very attentive to our needs. They gave us a ride into town to the Casa Fiesta Mexican Restaurant, just a short trip down the street, and returned to pick us up when we were done. The food was great and I would recommend it to anyone wanting a $100 chalupa.

PIREP FBO Comments: Good, friendly service.

PIREP Good, authentic Mexican restaurant just a few miles down the road. Ask one of the friendly people at the FBO, and they will give you a ride to the restaurant and pick you up when you are ready.

Gilbertsville, KY (Kentucky Dam State Park—M34)

Patti's 🍔🍔🍔

Rest. Phone #: (270) 362-7446
Location: Call for pickup
Open: Daily: Lunch and Dinner

PIREP Patti's is excellent. Their lunches are tasty and varied. The coconut cake is wonderful.

PIREP My favorite place in Kentucky is located at Grand Rivers, KY. Patti's Restaurant is located only five miles from Kentucky Dam Airport (M34). This is a 1880-vintage decor restaurant, with a two-inch thick pork chop as their specialty. I have eaten there many times and have never had a bad meal. They also feature homemade bread, served piping hot in a flower pot. Their pies are absolutely outstanding.

Reservations are needed most times for dinner, even during the week. They have a large local clientele trade, plus they get a large crowd from the boat docks on Lake Barkley in the summertime. They have a zoo for the kids and grown-ups also. If you want to gain weight but have fun doing it, this is the place to go. They will come to the airport to pick you up, then deliver you when you are done eating.

Great folks!

Jamestown, KY (Russell County—K24)

Jamestown Marina Restaurant 🍔🍔

Rest. Phone #: (270) 343-5253
Location: Crew car
Open: Daily: Lunch and Dinner

PIREP I recently flew into Russell County Airport in Jamestown, KY. The quaint airport is located about two miles from beautiful Lake Cumberland and the Jamestown Marina. The warm and friendly people at the busy FBO allowed us use of their crew car. We had a fabulous outdoor dinner at the marina restaurant overlooking the lake. While there, we found family-style houseboats for rent by the hour, day, weekend, or longer.

Lake Cumberland is one of the largest manmade bodies of water east of the Mississippi River. A vacationer's paradise, Lake Cumberland covers over 63,000 acres. Its waters touch over 1,255 miles of shoreline, with an average depth of 90 feet. With an abundance of commercial marinas, launching ramps, motels, restaurants, boat rentals, Lake Cumberland State Resort Park, and a variety of campsites.

Lexington, KY (Blue Grass—LEX)
Beef O'Brady's 🍔🍔

Rest. Phone #: (859) 223-0017
Location: Crew car
Open: Daily: Breakfast, Lunch, and Dinner

PIREP FBO Comments: TAC Air is very nice and friendly FBO. I've been there three times and have had good experiences each time. Fuel is reasonable for a big FBO, and the facilities are immaculate. As far as Bluegrass Airport goes, it is pretty calm for a class C. Ground and tower control have been friendly and helpful each time I've been there in the Katana. Besides, it's always neat to take off just before a commercial jet! Overall, I found Bluegrass Field to be a nice place to visit.

PIREP Beef O'Brady's is a fun little "family" sports bar in Palomar Center, about five minutes from the airport. The food is pretty good, prices are very reasonable, and the restaurant has a lot of TVs to watch sports! The first time I went there was on my long solo cross-country, so it will always be special to me. To get there, you have to drive because the restaurant is a few miles away. TAC Air's website says they have complimentary crew cars, but I'm not sure about them because each time I've been I had a friend pick me up at the FBO. There are actually a lot of decent places around Lexington but you'd have to drive to all of them. If you have questions, I'd say calling the FBO might be helpful to see about their ground transportation services.

Louisville, KY (Bowman Field—LOU)
le Relais 🍔🍔🍔🍔🍔

Rest. Phone #: 502-451-9020
Location: On the field
Open: Dinner Tues.–Sun. from 5:30 PM

PIREP For those of you who prefer French food, I highly recommend le Relais at Bowman Field just southeast of downtown Louisville. They also have a great wine selection.

PIREP le Relais at Bowman Field in Louisville, KY is within 30 feet of my Champ. Long been heralded as a favorite in Louisville, and I dare say that no other airport in the nation has a higher quality airport restaurant.

PIREP Great restaurant!!!!!!!
A little pricey, but well worth it. The old terminal building is worth looking at too. It's like going back in time to the 1930s.

Middlesboro, KY (Middlesboro-Bell County—1A6)
McDonald's 🍔🍔

Location: Short walk
Open: Daily: Breakfast, Lunch, and Dinner

PIREP   There is a McDonald's about a block away from the Middlesboro airport. You can check out the P-38 project, then walk to Mickey Dee's for lunch.

Murray, KY (Kyle-Oakley Field—CEY)
Rudy's 🍔🍔

Rest. Phone #: (270) 753-1632
Location: City shuttle
Open: Daily: Lunch

PIREP   Rudy's in downtown Murray is world-famous for their hamburgers—at least worth *Southern Living* magazine taking a look! Their Friday all-you-can-eat catfish is excellent!

The Dutch Essenhaus is home-cooking cafeteria-style. Bring a healthy appetite, you will not go away hungry!

Fly into Kyle-Oakley Field (CEY); someone will see to your transportation needs! Our airport doesn't have a courtesy car any longer—it was never replaced after the wreck! There are a few transportation choices. Usually, either Covenant Aero, the maintenance operation on-field, or the airport manager will be glad to drop off and pick up. Otherwise, there is a city shuttle and a taxi service for reasonable rates.

The trip to Essenhaus is about five minutes. Ten minutes will get you to Rudy's.

Owensboro, KY (Owensboro-Daviess County—OWB)
Moonlite Bar-b-que 🍔🍔🍔🍔🍔

Rest. Phone #: (800) 322-8989
Location: Call for pickup
Open: Daily: Lunch and Dinner

PIREP   Fly into OWB (Owensboro-Daviess Co.) and ask unicom to call. Moonlite will send over a van (or at least they did last fall when we flew down from Louisville). The food is great, and they offer a lunch buffet for around $9. They offer several kinds of barbecue (pork, beef, and so on) to match any taste. The buffet includes a small dessert bar with ice cream and hot apple pie.

Moonlite would qualify as a hole in the wall. Not much to look at, but lots of atmosphere and good food. A downside for some might be the lack of a nonsmoking section.

Well known throughout the region! When we opened our flight plan, FSS commented, "Enjoy the barbecue!" Even the ATC can recognize planes that frequent the Moonlite.

PIREP   The Moonlite Barbecue still rates five burgers.

We called ahead on unicom, and Martin Aviation (FBO) answered that the courtesy van was ready and waiting. They gave us the keys and a map and said, "Enjoy!"

It was only about five minutes away. The food was great! Not only was the barbecue some of the best we've had (we used to live down South, so we've tasted some good ones), but the vegetables, bread, and desserts were from old-time country recipes we haven't tasted in years. They have lunch and evening buffets Monday through Saturday, but Sunday is only from 10 AM to 3:30 PM CST, so don't be late! They also do complete take-out and advertised a mail-order service, but we did not get a chance to ask about that.

Good prices, friendly service, and as testimony to the quality, packed with local clientele.

Don't miss it!

Prestonsburg, KY (Big Sandy Regional—K22)
Cloud 9 Cafe 🍔

Rest. Phone #: (606) 298-2799
Location: On the field
Open: Daily: Breakfast, Lunch, and Dinner

PIREP Excellent, friendly service. Good food, unique decor, and very clean. Even the local patrons were very friendly. The FBO service was excellent and extremely helpful and nice. Very clean facilities there as well. We will go back. We give them a rating of five.

PIREP We have been there three times and have enjoyed great food at reasonable prices; the hospitality is excellent. Fuel is reasonable and service outstanding. We wish we lived closer. One more note: in the evening you might see a nice herd of elk! Keep your eyes open. We saw several right behind the restaurant one evening.
   P.S. Save room for dessert.

PIREP FBO Comments: Very friendly folks have a hangar to put you up in for overnights if needed and 100LL/ Jet A on-site.

PIREP Flew into Big Sandy Regional with my partner and had a great lunch at The Cloud 9 Restaurant. Very friendly folks here and good cookin'! Kinda like a mini-Cracker Barrel! Highly recommend using this airport if you're transiting eastern Kentucky!

PIREP Nice and clean restaurant with an aviation-themed decor and menu. Pork chop was great. People very friendly.

Stanton, KY (Stanton—I50)

The Big Kahuna's 🍔🍔

Rest. Phone #: (606) 634-9627
Location: Short walk
Open: Daily: Breakfast, Lunch, and Dinner

PIREP Any lazy Saturday afternoon is a perfect day for a flight to Stanton, KY (I50) and lunch at The Big Kahuna's. Granted, there's no aviation theme, and you won't find Frankie Avalon and Annette Funicello surfing the waves, but you will find the Big Kahuna cheeseburger platter. If you're not up for grease, you can opt for a thick slab of bologna with a slice of cheese. Big Red and grape soda are readily available. During the week, delicious hot meals are cooked and ready for the hungry working man or woman.
   Kolb Aircraft ultralights are assembled at the airport, and usually you can get a quick tour of the assembly room.
   A definite two-burger rating, but the adventure is well worth the trip.

PIREP I'll second David Munday's report of the Airport Market (really a gas station). They've got some of the best pizza I've had, and you can get it by the slice for about a buck. No wait—they keep 'em hot under the heat lamps. Sounds dicey, but it's gooood!
   Also a full grocery store at the same corner, about 0.1 miles from the airport.

PIREP Walk a couple of hundred yards to the other end of airport road and you will find the Airport Market Convenient Store.
   There's nothing aviation-related about the place except the name. It's a convenience store with gas pumps, some tables, and a short-order cook. We had a breakfast of eggs and bacon, but their card says: "Pizza—Deli Sandwiches—Home-cooked Meals".
   I'd give it one burger.

The Hundred Dollar Hamburger

Louisiana

Abbeville, LA (Abbeville Chris Crusta Memorial—0R3)

Black's Oyster House 🍔🍔🍔🍔

Rest. Phone #: (337) 893-4266
Location: Long walk
Open: Daily: Lunch and Dinner

PIREP Just a couple of pireps to my previous report: during two visits in early 2005, I discovered Black's has changed hands. The ambiance has changed, and a lot of the wait-staff seem to have moved on along with the previous owner. The food is still okay, but the service was very bad on these visits and it seems to be going through a troubled phase right now.

Also, my estimate of "one-mile" walk into town is way off. It's nearer three (45 minutes). Sorry.

I can recommend the hotel that is adjacent to the airport. $55 is a reasonable price (AAA). Clean rooms. Ask for a room around #109 and you'll have an in-room view of the threshold of runway 33 at 0R3.

Failing Black's, check out Shucks (another oyster place), and I can heartily recommend Richard's Seafood Patio for the biggest and best crawfish around. (Not like certain crawfish places where they put a few big crawfish on top and underneath are all tiny ones—these are 100 percent big succulent crawfish. Priced accordingly—about $20 for five pounds.)

PIREP The food is still great. The crawfish at this time of year are big and succulent, and they were selling for about $2.95 a pound (two-pound minimum). Oysters just a little late in the season and getting milky. Still good. And the best seafood gumbo around. A couple of people said they prefer the alternate oyster house Shucks, so next time we'll try that for comparison and report back. On the way back to Tulsa, overheard a hapless 150 pilot busted by FTW Center for penetrating the Red River TFR around Texarkana while squawking the wrong code: "Do you have a pencil and paper handy? I need for you to make a phone call when you land."

There's NO courtesy car at the FBO. Taxi or Enterprise Rent-a-car will come pick you up. FBO is pretty basic and very small and understaffed. It may have the world's most difficult-to-use weather computer, which, as most people know, is quite a distinction to have. A line of quite-tall trees on the east side of the runway, so you can expect variations in winds/x-winds on short final and flare.

PIREP For those who don't know, Abbeville is an oyster, boudin, and seafood mecca. The airport is an easy in and out, although winds can be breezy from the Gulf. There is a long single runway, 15-33. It is a busy place with oilfield choppers and other business and even military aircraft. The last time I went, there was no courtesy car. You might check with the FBO in case that has changed: (337) 893-7128. But you can walk to downtown Abbeville (it's about a mile) if it's not too hot, or call a taxi or rent a car if you're staying longer than lunch.

The big draw is the Grand Isle oysters at Black's Oyster House.

There is an alternative place called Shucks, but I like the oysters at Black's the best, plus it is located in a historic building (a former dry goods store) with high, high wooden ceilings, wrought-iron decorative columns, and revolving fans. You really feel as though you are in an old settlement, which you are. Abbeville has a long history.

Oysters on the half shell come for about $3 per half dozen or $4 per dozen, so don't order by the half dozen, not that you'd want to. Three dozen is a healthy portion for the oyster lover. They have a salty succulence that makes you want to keep slurping them from their shells. For that matter, all the food at Black's is superb. If crawfish are in season (usually November through June), they have *the best* boiled crawfish in Acadiana (SW Louisiana). They also feature a wonderful mud-like seafood gumbo with lots of seafood. Po-boy sandwiches (fried oyster and shrimp). Also great people-watching. Black's is closed during the heat of the summer, July through September, when oyster houses traditionally close down.

If you have a car, you can ask for directions to Hebert's Meat Market (it's very near the airport but on a country road and impossible to find easily) and ask for a hot link of "red" (boudin)—or "white," for that matter. Both of these boudins are some of the best you'll find, but the red is special with its red-pepper spice and horror-movie appearance. It tastes superb, so don't be put off. Great cracklin's too, but ask for the "extra spice" to sprinkle on them.

The approach to 0R3 (that's ZERO–ROMEO–THREE) is over sugar cane fields, and the entire area is very scenic.

A five-burger landing destination, for sure.

Alexandria, LA (Alexandria Intl—AEX)

Cajun Landing 🍔🍔🍔

Rest. Phone #: (318) 487-4912
Location: Crew car
Open: Daily: Lunch and Dinner

PIREP FBO Comments: FBO seems to have just changed to Millionaire and was very helpful. They have a single courtesy car (van, actually) that is in good condition.

PIREP We arrived at the FBO (only one on the field), having flown in from Houston, and asked for where we might find some good Cajun food. We were strongly recommended to go and try out the Cajun Landing just ten minutes drive from the FBO (who provides you with a map, not that the directions are complex!).

Well, I have to say this was one of the best meals I have had for a while and well worth the stop. The blackened red snapper "Oscar" was to die for and the crawfish bisque was so delicious we went back for more from the salad bar (no extra charge). The service was also friendly, efficient, and prompt.

When the bill arrived, I was comfortably amazed by how small it was. A three-course meal for two was just $45 before the tip. And that was with us pigging out! Maybe I am getting used to high prices in Houston, but I assure you this was cheap for the quality we got. (I have had worse food at Gaido's in Galveston for nearly double this cost.)

I am already making plans to return and bring some more friends for a delightful experience.

PIREP If you've got the time to stick around for a while, try the Cajun Landing Restaurant. It's more of a "nice dinner" type of place, but the food is excellent. I've been living in Alexandria for about five months working as a CFI, so I know the value of a good meal on a budget. But this is one case where I would treat myself every once in a while. The restaurant is about five to ten minutes from the airport.

Baton Rouge, LA (Baton Rouge Metropolitan, Ryan Field—BTR)

Tony's Seafood 🍔🍔🍔🍔

Rest. Phone #: (225) 357-9669
Location: Crew car
Open: Daily: Lunch and Dinner

PIREP   FBO Comments: PAI Aero was excellent. Loaned us an Olds for several hours, no problem. Self-serve 100LL. We were recommended local eateries, and another gentleman (not sure if connected with the FBO) gave us an excellent history of Baton Rouge (that it was the only place apart from New England involved in the Revolutionary War) and surrounding plantations, not to mention some primers on pronunciation of place names.

PIREP   I have waxed lyrical about many places in the *Hundred Dollar* pages but I don't think I have ever come across quite such a remarkable place as Tony's Seafood Market (just three or four minutes by courtesy car from PAI Aero). The deal here is seafood. And the selection is just totally remarkable. Live, fresh, frozen, cooked. And all the fixin's. They have a cafeteria-style line where you can select plate lunches or whatever you fancy from a very appetizing selection of stuff that (no kidding) looks good, like your Mom made it. Very down home, but mmmm goo-oo-d. Don't miss the boudin balls! Everything here is to-go, so you load up with plate lunches and take them elsewhere (say, back to PAI) to consume them. Obviously, if you bring a cooler you can take treasures home. Did I mention they have an Olympic-size CATFISH POND (!) here? Chock-full of live catfish that they will scoop out and spread over the floor for you to choose the ones you like the look of. Now, that's fresh!

Their boiled seafood (we had some of the best boiled crawfish we'd ever eaten) is superb. They have a lot of stuff in the frozen cases that would ship back home well in the baggage compartment.

But most of all, you have to SEE this place. It is like a miniature Wonder of the World and always busy with wide-eyed (and hungry) patrons of all kinds, including, I guess, pilots. Everything is clean, fresh, and beautiful. Although crowded, you get served immediately and checked out in no time. One of Baton Rouge's prime tourist attractions—or should be, in my book!

Crowley, LA (Le Gros Memorial—3R2)

The Rice Palace 🍔🍔🍔

Rest. Phone #: (337) 783-3001
Location: Crew car
Open: Daily: Lunch and Dinner

PIREP   The Rice Palace is in Crowley, LA. The closest airport is Le Gros Memorial (3R2) in Estherwood, which is approximately seven miles from the restaurant.

The FBO (AJ Aviation) will provide transport to the restaurant. For those wishing to stay overnight, there are several good motels adjacent to the restaurant. The restaurant also contains a casino for those so inclined.

The Rice Palace is famous for its Cajun cuisine in that it is located in the heart of Acadia country. The menu is extensive; however, the chicken, sausage, and crawfish gumbo is hard to beat. Be sure to order a helping of corn nuggets, a house specialty. There is a wide variety of home-baked pies in the dessert cooler. Prices are moderate and service is attentive. Dress is casual.

De Ridder, LA (Beauregard Regional—DRI)

Ryan's Steak House 🍔🍔

Rest. Phone #: (337) 462-1122
Location: Crew car
Open: Daily: Breakfast, Lunch, and Dinner

PIREP   You can get a courtesy car. Ryan's Steak House is about three miles and has an excellent lunch buffet for about $6.10 with drink and taxes. Also site of very friendly FSS you may want to visit.

Franklinton, LA (Franklinton—2R7)

Franklinton Golf Club 🍔🍔🍔🍔

Rest. Phone #: (985) 839-4195
Location: On the field
Open: Daily: Breakfast and Lunch

PIREP   FBO Comments: I have never seen any fuel available.

PIREP   I fly down from Columbia on a regular basis and the quality of food and service is now better than ever. The establishment is under new management, and the price is even a bit lower. They had some fine ribs this last Sunday. The manager also runs a smoke house, so he knows his business with meats.

PIREP   I flew friends to Franklinton on a Sunday morning to have lunch at the country club based solely on the reviews from the $100 Hamburger website. We were not disappointed. I parked the Bonanza, and we walked across the runway and two fairways to get to the club. We enjoyed their nice buffet lunch, which included fried chicken, fried catfish, blackened catfish, lots of home-cooked vegetables, breads, and desserts. Sunday lunch is all you care to eat.

   Arrive before noon to beat the local after-church crowd. The airspace can be busy—six aircraft arrived and departed during our visit, so keep your eyes open and make those radio calls.

PIREP   Franklinton Golf Club is right across the runway from the terminal, within walking distance. The Sunday buffet is excellent—country cooking at its finest. Good view of golf course and dock ponds while dining. The price is very reasonable.

Houma, LA (Houma-Terrebonne—HUM)

Big Al's Seafood Restaurant & Market 🍔🍔🍔

Rest. Phone #: (985) 876-7942
Location: Crew car
Open: Daily: Lunch and Dinner

PIREP   FBO Comments: Butler is FIRST RATE! For $25 service fee, they loaned us their 12+-seater Dodge van for a day and a half. Like driving a big rig and good fun.

PIREP   The draw here is the raw oysters (pristine condition: fresh shucked) and boiled seafood. Please call to check hours. You might want to avoid the back "smoking" room, as the atmosphere is suitable for curing whole sides of beef and coats the tongue, detracting from the oysters a bit. Beer only in terms of adult beverages. If you can, scoot under the tunnel under the Intracoastal Waterway to Rob's 24-hour donuts and try one of their pecan-praline-éclair donuts—a masterpiece, and not too sweet.

   Houma KHUM is easy in-out with big military-style runways, but watch out for helicopters that are all over the place. Part-time Class D Tower. New Orleans handles App-Dep.

   Leave the airport. Take a left. At first stoplight another left. Drive straight ahead (one mile or so) to a stoplight, where you must turn either right or left. Take a left there and Big Al's is on the right almost immediately. So it's about as close as it gets, as there are not a lot of places close to the airport in Houma.

Jennings, LA (Jennings—3R7)
Holiday Inn 😊😊😊😊

Rest. Phone #: (337) 824-5280
Location: On the field
Open: Daily: Breakfast, Lunch, and Dinner

PIREP The Holiday Inn now has a Denny's Restaurant. There are several other choices about a block away including Shoney's and Burger King.

PIREP Landed at Jennings on my way back from Tennessee. What a great place! There is a hard surface tie-down area adjacent to the hotel. A 200-foot walk brought me to the restaurant. They serve a really nice buffet for only $8.25. I was told that they have the buffet everyday but Saturday and that it goes from 11:30 AM until 2 PM.
Jennings is well worth the stop!

PIREP There is a Holiday Inn right on the field, and you can park your plane within 100 yards of the hotel and restaurant. The food is reasonably priced and they have a decent selection.

Lafayette, LA (Lafayette Regional—LFT)
BUNS 😊😊

Rest. Phone #: (337) 232-3287
Location: Crew car
Open: Daily: Lunch and Dinner

PIREP Lafayette is known for its fine restaurants and Cajun hospitality. However, the best hamburger place is just off the airport. It is too far to walk, but the FBOs will normally lend you a car or give you a ride. The restaurant is called BUNS. It is across the street from a car dealership and considered one of the best hamburger joints in Lafayette. If you are interested in other restaurants, then you may want to try one of these: The Blair House, Cafe Vermillionville (a little pricey, but excellent food), Don's Seafood Hut, Don's Seafood and Steakhouse, Stroud's Steak House, Hub City Diner, LaFonda, Riverside Inn, and many others. Other good hamburger joints are the Judice Inn and Pete's, although these are not near the airport.

Lake Charles, LA (Lake Charles Regional—LCH)
Player's Casino 😊😊😊

Rest. Phone #: (337) 433-1645
Location: Call for pickup
Open: Daily: Breakfast, Lunch, and Dinner

PIREP Call Player's Casino from the terminal for courtesy transportation to and from the casino. They have a very nice, reasonably priced buffet. If you are playing blackjack, mention to the pit boss that you're a pilot who flew in—he/she will almost always comp you for the buffet.

Marksville, LA (Marksville Municipal—MKV)
Grand Casino Avoyelles 😊😊😊

Rest. Phone #: (318) 253-1946
Location: 1/2-mile-walk or call for pickup
Open: Daily: Breakfast, Lunch, and Dinner

PIREP   FBO Comments: Call Mr. Newman for fuel: (318) 253-9574.

PIREP   Been going to Marksville just to eat for the past five years. Always been great, and the shuttle drivers usually recognize the plane and are on their way to pick us up before we step out of the plane.

PIREP   They have a nice runway. I think you can get fuel with a phone call. The local attraction is the casino that is about two miles away. They will gladly pick you up if you call them. Many pilots fly there for their excellent buffet meal. There is a nice hotel if you would like to stay overnight.

PIREP   The Grand Casino Avoyelles is located about a half mile from the airport and has free shuttle service to the casino. They have a buffet for $12.95 that has over one hundred items on the menu and is one of the best I've ever had.

Minden, LA (Minden-Webster—F24)
Daddy O's Diner 🍔🍔

Rest. Phone #: (318) 377-7300
Location: Crew car
Open: Daily: Lunch and Dinner

PIREP   FBO Comments: Friendly, helpful staff; crew car available (first come, first served).

PIREP   Restaurant is downtown (a short drive in the crew car). The FBO will provide directions. The food is excellent. Also, check out the ice cream desserts. The service is friendly, cheerful, and prompt. The decor is '50s-based (around the movie *Grease*). The prices are appropriate.

Monroe, LA (Monroe Regional—MLU)
Airport Cafe 🍔🍔🍔

Rest. Phone #: (318) 323-1574
Location: On the field
Open: Daily: Breakfast and Lunch

PIREP   Not much in the way of burgers but some of the best home-cooked plate lunches around. Believe it or not, it is the airport's cafe. If you're in town overnight, even the airport lounge gets going.

Natchitoches, LA (Natchitoches Regional—IER)
The Landing 🍔🍔🍔

Rest. Phone #: (318) 352-1579
Location: Two miles, call for pickup
Open: Daily: Breakfast and Lunch

PIREP   Natchitoches (NAK-a-tish), KIER, is halfway between Shreveport and Alexandria. It's the oldest town in the Louisiana Purchase, very historic, and has a beautiful downtown area. The FBO at Natchitoches (Alpha Aviation: (318) 352-0994) has great service and 100LL is $2.21. They have only one courtesy car, so you may want to call ahead to reserve it.

Great restaurant downtown (only two miles) is The Landing. Louisiana-style food of all types. Reasonable lunch prices. If the courtesy car is out, call the restaurant. They may have someone who can come pick you up. Truth is, my brother owns the restaurant, but I eat there anyway! It's that good.

I base my Bonanza at IER.

New Orleans, LA (Lakefront—NEW)

Flight Deck Restaurant 🍔🍔🍔

Rest. Phone #: (504) 241-2561
Location: On the field
Open: Daily: Breakfast and Lunch

PIREP   The Flight Deck Restaurant, located by the old tower, taxi right up to the door. The whole front is glass and overlooks the field, and you can watch the aircraft take off and land while you dine. The food is excellent and the prices very reasonable.

PIREP   New Orleans Lakefront Airport is located on the south shore of Lake Pontchartrain on the north side of New Orleans. It's about a $10 cab ride to the French Quarter! What more needs to be said? Oh, there's a casino adjacent to the airport for those who don't have long to visit.

I've been well treated on each visit by the General Aviation FBO. There are warnings that the horizon may be obscure over the lake.

This is a wonderful getaway.

Castnet 🍔🍔🍔

Rest. Phone #: (504) 244-8446
Location: 10826 Haynes Blvd.
Open: Daily: Lunch

PIREP   New Orleans is still suffering and this area in particular. While restaurants may be hard to find, flying in to Lakefront Airport for dinner can still be a very memorable experience. When we were about to depart Lakefront Airport, we decided to go for dinner. The Millionaire FBO provided the crew car, a new Jaguar, making the drive unique. The Castnet restaurant offers a variety of inexpensive but quality seafood with flair and character. In spite of the difficulties facing this part of the city, a sense of warmth can be found at the Castnet. Visiting can only help.

Oakdale, LA (Allen Parish—ACP)

Grand Casino Coushatta 🍔🍔🍔🍔

Rest. Phone #: (800) 584-7263
Location: Call for pickup
Open: Daily: Breakfast, Lunch, and Dinner

PIREP   FBO Comments: Nice, well-maintained runway (no taxiway, so you have to back-taxi on the runway if you miss the cutoff at midfield). Friendly local family that runs the place; they were very helpful.

PIREP   Great spot to fly in if you want to visit the Grand Casino Coushatta. Call the casino ahead of time, and they will have a nice custom van there to take you to the casino free of charge. Also ask about the trailer they have at the airport; the driver usually has a key, and it is stocked with drinks (beer, soda, water, and so on) and is a cool place to relax before taking off or to get a cool drink before heading to the casino. The pilot's lounge is somewhat small and bare bones, but the A/C was working.

PIREP   Another one of my favorite destinations! *Very* nice pilot's lounge: complimentary sodas and comfortable furniture. The FBO offers attended fuel from the trailer home next to the pad. Call the Grand Casino Coushatta for courtesy transportation to the casino.

EXCELLENT buffet and fun!

Shreveport, LA (Shreveport Downtown—DTN)
Candy's Skyway Cafe 🍔🍔🍔

Rest. Phone #: (318) 221-7005
Location: On the field
Open: 6 AM–8 PM, 7 days a week

PIREP   FBO Report: Royal Air is a friendly FBO.

PIREP   The restaurant is in the old terminal building immediately adjacent to the FBOs.
Obvious aviation theme.
Breakfast through dinner. Varied and expanding menu. We had breakfast this past Sunday morning, and three people had a traditional eggs, meat, grits, and coffee meal for less than $11 total!

PIREP   The Shreveport Downtown Airport Restaurant in the old terminal building is named Candy's Skyway Cafe. You can fly right to the front door.
Cindy and I stopped in Sunday and had a great burger and homemade fries.

Tallulah/Vicksburg, MS, LA (Vicksburg Tallulah Rgnl—TVR)
Ameristar Casino 🍔🍔

Rest. Phone #: (601) 638-1000
Location: Crew car
Open: Daily: Breakfast, Lunch, and Dinner

PIREP   Fly into the TVR airport between 7 AM and 7 PM, ask for the courtesy car. Go to Highway 20, one mile south and eight miles east over the Mississippi River Bridge to the Ameristar Casino.
This is Vicksburg, the Gibraltar of the Confederacy.
Enjoy their buffet or the Blues Bar. Bring lot cash to help keep Mississippi green.
Enjoy.

Welsh, LA (Welsh—6R1)
Cajun Tales 🍔🍔🍔

Rest. Phone #: (337) 734-4772
Location: Very short walk
Open: Daily: Lunch and Dinner

PIREP   There is an excellent little Cajun restaurant in Welsh. Cajun Tales is on North Adams, about two blocks off I-10 and about two blocks away from the airport. The airport (Welsh, 6RI) is small and there is not much there. I flew in on a weekend and there was no one around. The taxiway is gravel, so take it slow. All the shortfalls of the airstrip are melted away by the excellent food.

Dairy Queen 🍔🍔

Rest. Phone #: (337) 734-3506
Location: Across the street
Open: Daily: Lunch and Dinner

PIREP   There is a Dairy Queen across the street. This is my $100 dipped cone when I'm not in the mood for a $100 hamburger.

The Hundred Dollar Hamburger

Maine

Auburn/Lewiston, ME (Auburn/Lewiston Muni—LEW)

The Landing Strip Cafe 🍔🍔🍔

Rest. Phone #: (207) 784-7881
Location: On the field
Open: Tues.–Sun. 6 AM–2:30 PM

PIREP Excellent restaurant in the main terminal with a view of the ramp. Many homemade soups and chowders. Good breakfast and lunch menu. Nice ambiance.

PIREP Second visit to the restaurant. They serve a lobster roll with fries and coleslaw for $5.95! Excellent food and service and very, very clean. Also some eye candy for pilots—there are two "connies" off the approach end of runway 22. According to the staff at the restaurant, it's an ongoing restoration project.

PIREP Nice airport, and we even left the plane at the fuel pump to walk into the restaurant located a few steps away. The restaurant is now called The Landing Strip Cafe, and it had a trio playing keyboard and guitars and singing to the packed house. The server was over-burdened, having all tables filled, but she was doing her best and maintained a good attitude. The omelet delivered was not quite the one ordered so we can't tell whether a fully loaded version is amazing. The blueberry pancakes with syrup were very good. Nice place.

Augusta, ME (Augusta State—AUG)

Thai Restaurant 🍔🍔🍔

Rest. Phone #: (207) 621-8575
Location: On the field
Open: Daily: Lunch and Dinner

PIREP FBO Comments: Maine Instrument Flight, very friendly folks.

PIREP Thai food? At a Maine airport restaurant? You betcha, and it was great. If you like spice you must specifically request it, as they try to accommodate the less tolerant American palate.

PIREP   A new Thai restaurant has opened on the field in the same building as the old restaurant. I haven't made it up there myself but have heard multiple good reviews. They're open Monday through Saturday 11 AM–8 PM, and Sunday 1 PM–8 PM.

Bangor, ME (Bangor Intl—BGR)

Red Baron 🍔🍔🍔

Rest. Phone #: (207) 947-4375
Location: On the field
Open: Daily: Lunch and Dinner

PIREP   Who says that big airports are unaccommodating to smaller general aviation pilots—especially hungry ones? Head for Bangor International Airport and segue in between the British tour jet traffic and head for the GA ramp once on the ground. Staff will be happy to give you a lift to the domestic terminal, where you can take refuge in the Red Baron Lounge for some inexpensive and delicious fare. Shop for some souvenirs before departing on 11,439 feet of runway. Fuel prices are the lowest around, so tank up before you go!

Bar Harbor, ME (Hancock County–Bar Harbor—BHB)

Lompoc Cafe 🍔🍔🍔

Rest. Phone #: (207) 288-9392
Location: Short drive
Open: Daily: Lunch and Dinner

PIREP   FBO Comments: Columbia was decent to us. No complaints. They seem to run a very professional organization.

PIREP   This was a trip to Bar Harbor to see an old friend. It was Sunday and off season, so options for food were few. This place was excellent. It is something of a local hangout. We took a rental car as it is too far to walk (probably a short cab ride), and we also had to go to Ellisworth after dinner. But the food was perfect, the service was great, the ambiance was local Maine, and the price for the lobster special (all four of us had it) was quite decent, thank you.
   This is a town that I would want to check out more, closer to summer season.

PIREP   BHB lies on the mainland, just before the little causeway that takes motorists to Bar Harbor. The road outside the airport (half-mile walk) is lined with lobster pounds, where steamed lobsters, fish, clam chowder, and seafood of all descriptions is available fresh. The restaurant right at the airport entrance is excellent; the others look good too.

PIREP   Bar Harbor has a lot to offer for the fly-in types. There is an excellent shuttle service that covers the entire Mt. Desert Island (including downtown Bar Harbor) for free. The shuttle is on a 20-minute schedule and you can get on or off as you wish; they do go into most major campgrounds as a scheduled stop. I recommend Rupunini's as a great place for a prime rib steak. The Lompoc Cafe is a little weird but offers a nice variety of microbrew beer. The Atlantic Brew Co. is a must, and tours are free (the shuttle will take you there) for some of their great beers. The staff at the airport is really nice, and there are some funky planes to see. We waited a few moments prior to departure for a privately owned L-39 on final to land and taxi back to home base after some aerobatics.
   Really a nice spot!!

Belfast Soup and Sandwich 🍔🍔

Rest. Phone #: (207) 338-6277
Location: Short bike ride. Loaner bikes available.
Open: 11 AM–7 PM Mon.–Fri., 11 AM–2:30 Sat., closed Sun.

PIREP   Fresh bread made daily, outstanding hearty fare, 2 mile or less bike ride into town (airport provides bicycles). Sandwiches $6.50 or so (lobster rolls higher), soups $3.50.

Belfast, ME (Belfast Muni—BST)
Young's Lobster Pound 🍔🍔🍔

Rest. Phone #: (207) 338-1160
Location: 1-mile walk
Open: Daily: Lunch

PIREP   Belfast is a quaint harbor town that seems to be overlooked by most tourists. It doesn't get the attention of a Camden or Bar Harbor, and it doesn't get the crowds either. Belfast has a fine airport that is about a one-mile walk from downtown. Walk out the airport road, turn left at the "T", cross Route 1 (a main road), and continue into town. It will be obvious where the center of Belfast is. About one mile down you have to take a right and go downhill a few blocks. There are a number of restaurants that have fine food. These are our four favorites.

90 Main. Address is the same. They serve down-east regional foods prepared in a eclectic, gourmet style. They have their own bakery and serve unique and downright decadent sweets for breakfast (bakery open in the morning) or after meals.

Darby's, on High Street next to the movie theater, is within view of 90 Main. It is more of a pub atmosphere with a tin ceiling and an old bar. The good food is more bistro type. The mood is friendly and comfortable. A favorite of locals.

Weathervane, on lower Main Street at the public landing, has the best views and is very popular. Even though this is a seafood restaurant chain, it serves a wide variety of excellent food at reasonable prices. They have outside seating and large windows that allow you to feel that you are eating at the banks of the harbor (and you are).

Young's Lobster Pound is across the harbor from the Weathervane. It is a long walk from the airport. There are taxis in Belfast, and Young's is worth the trip. At Young's, you can order lobster, steamers, and corn and have it cooked in salt water. You can then eat it inside or outside on tables on the banks of the north shore of Belfast harbor. The prices are right, the food fresh, and the view serene.

Young's has a year-round seafood market on the airport road that will pack lobster and other seafood for your trip out. The market is just a couple hundred yards from the airport.

Bethel, ME (Bethel Regional—0B1)
The Sunday River Brew Pub 🍔🍔

Rest. Phone #: (207) 824-4253
Location: 1 mile
Open: Daily: Lunch and Dinner

PIREP   The Sunday River Brew Pub is located approximately one mile NE of the Bethel Maine (Col. Dyke Field) airport. The restaurant serves food typical of generic brew pubs—good burgers, ribs, salads, and so on. Of course, they also have the required excellent selection of various beers brewed onsite (remember: not while flying). It can be walked, but I would recommend calling unicom five to ten miles out and asking them to call a cab. The cab/van will be waiting for you when you land, and the cab fare is only $2–3 per person. If you want to sample/enjoy the beer, stay at any of the local B&Bs or hotels. For the downhill adrenaline junkies, Sunday River ski area is a cab ride away.

L'Auberge 🍔🍔🍔🍔

Rest. Phone #: (800) 760-2774
Location: Call for pickup
Open: Daily: Breakfast, Lunch, and Dinner

PIREP   FBO Comments: Nice airport in a valley. Tie-downs and hangar rental available. Say hello to Harold when you fly in.

PIREP   L'Auberge is a B&B with a very, very fine restaurant. Although burgers aren't on the menu, the chef/owner will be happy to make one up. Better choices are the New York strip and the Lobster L'Auberge. The food is excellent and the B&B is very comfortable. The accommodating owner or his wife will pick you up at the airport if you give them a call. Reservations are recommended, particularly during the busy seasons.

The Bethel Inn 🍔🍔🍔🍔

Rest. Phone #: (207) 824-2175
Location: Call for pickup
Open: Daily: Dinner

PIREP   The Bethel Inn (free pickup for guests, not sure for dinner) has A+ and reasonably priced accommodations, with a shuttle for skiing and a nice golf course.

    The dinner in their main restaurant is exceptional, with a nice dining room and lots of charm. You will need reservations for dinner.

    Heated outdoor pool for guests, even during ski season.

Biddeford, ME (Biddeford Muni—B19)

The Ready Room Cafe 🍔🍔🍔

Rest. Phone #: (207) 286-1880
Location: On the field
Open: Daily: Breakfast and Lunch

PIREP   Biddeford has a cafe at the airport now serving breakfast and lunch. The Ready Room Cafe serves up a mean sandwich and all the hangar flying you can handle.

    Denise Sullivan runs the cafe—her hours are 6:30 AM–2 PM Monday through Friday and 7:30 AM–2 PM on Saturday and Sunday.

The Dry Dock 🍔🍔

Rest. Phone #: (207) 282-3775
Location: Crew car
Open: Daily: Breakfast and Lunch

PIREP   Biddeford, ME is less than ten miles south of Portland. The Dry Dock is a five-minute ride in a Lincoln Continental, provided as the free courtesy car by Biddeford Muni. The specialty is fresh seafood. Try a cup of chowder. The seafood basket includes a generous helping of several types of seafood. The largest strawberry shortcake I've ever had makes a good dessert. I suggest you divide this with a partner, as the portions are too large for an individual.

Millinocket, ME (Millinocket Muni—MLT)

The Hotel Terrace 🍔🍔🍔

Rest. Phone #: (207) 723-4525
Location: Short walk or crew car ride
Open: Daily: Breakfast, Lunch, and Dinner

PIREP   Head north towards Mt. Katahdin and land at Millinocket, ME (MLT). Although the airport has a grill that occasionally functions, hoof your way a short distance—maybe a half mile, or somebody hanging out at the airport will probably give you a ride—to the Hotel Terrace for lunch. Be sure to say hi to the proprietors, the Bisques. Cheap, plentiful, and if you get weathered in, you are at the hotel.

PIREP   Just a note to say how good the Hotel Terrace is at Millinocket, ME. Last Friday their menu included a buffet including a great fish chowder, green salad, mashed potatoes, fried fish, meatloaf, veggie, and dessert.

Atmosphere is great too: a big stone fireplace and clean, spacious surroundings.

PIREP   We flew from 08B Merrymeeting Field into Millinocket (MLT) for breakfast and were directed to the courtesy car for the half-mile (should have walked) drive to the Hotel Terrace for the breakfast buffet.

All you can eat, and it was terrific. We had to check the weight and balance after the big breakfast, but fortunately Millinocket has long runways.

Old Town, ME (Dewitt Field, Old Town Muni—OLD)

Johnny's Restaurant 🍔🍔🍔

Rest. Phone #: (207) 827-3848
Location: Crew car
Open: Daily: Lunch and Dinner

PIREP   FBO Comments: GREAT FBO! Old Town Aviation. They had already loaned out the courtesy jeep, but the lineman allowed the use of his personal vehicle for the two-mile trip into town. At the FBO, there was free haddock chowder, sandwiches, chips/crackers, and dip—apparently they always have refreshments available!

PIREP   We opted for a drive into Johnny's Restaurant. Nothing fancy—serves the college-age crowd with paper placemats and plastic utensils (University of Maine at Orono is one town below Old Town). The food quantity and quality is very good and the prices are amazingly low—a 12" all-meat pizza (sausage, hamburger, bacon, pepperoni) was under $8 and I could only eat three slices. Brenda's soup was obviously homemade, with big chunks of vegetables and meat.

There's a place called Chocolate Grill nearby, which we were told was a bit more expensive but had a nicer view of the river—we'll try that one next time.

Portland, ME (Portland Intl Jetport—PWM)

Dimillo's Floating Restaurant 🍔🍔🍔

Rest. Phone #: (207) 772-2216
Location: 10-minute drive
Open: Daily: Dinner

PIREP   FBO Comments: Northeast was very professional. They had our rental cars waiting for us with no hassles. Self-serve fuel available for 50-cent discount.

PIREP   Flew to Portland from NJ with two other planes. Flew up for a lobster dinner. Found Dimillo's Floating Restaurant on the Web. What an excellent choice. Service, food, and atmosphere were great. Would definitely recommend to anyone flying to Maine for a very good dinner. Was just a short ten-minute ride from the airport. Definitely five stars.

Gilbert's Chowder House 🍔🍔🍔

Rest. Phone #: (207) 871-5636
Location: Crew car
Open: Daily: Lunch and Dinner

PIREP   Northeast Air has crew cars that they will loan visiting pilots to go into town and grab a bite to eat. Don't let the Formica tables and the interior that hasn't been redecorated in decades fool you. Located on Commercial Street in the Old Port (the historic district of downtown Portland), Gilbert's Chowder House offers the best chowder on the face of the earth, the kind that hangs off the bottom of the spoon. For the true seafood lover, you can order the Super Seafood with twice the seafood in the chowder. Also, if the weather is nice and you don't care much for Formica, there is a deck in the back that looks out onto the bay.

Ricetta's Brick Oven Pizzeria 😋😋😋

Rest. Phone #: (207) 775-7400
Location: Crew car
Open: Daily: Lunch and Dinner

PIREP   This is the best gourmet pizza we have ever eaten. It's better than the $100 hamburger. Portland has an FBO (Northeast Air at (207) 744-6318) who has a courtesy van available and a free map of the area. Ricetta's is a 5–10-minute ride from the FBO by the Maine Mall at the junction of Gorham Road and Philbrook. Lunchtime is an all-you-can-eat buffet of soup, salad, and gourmet pizza, including a dessert pizza. I promise you'll love it! I suggest you arrive at Ricetta's before noon to beat the local rush.

   Mangiamo!

Presque Isle, ME (Northern Maine Regional Airport At Presque—PQI)

Winnie's 😋😋

Rest. Phone #: (207) 769-4971
Location: 1.6 miles
Open: Daily: Lunch and Dinner

PIREP   FBO Comments: Very friendly, but very expensive fuel!

PIREP   Given a ride to Winnie's, 1.6 miles from airport. Kind of a combination of Seafood Shack, burger and fries joint, and ice cream stand, but extensive menu selection, including fresh salads and low carb items. The fresh french fries (from Maine potatoes—I mean, it is the Heart of Maine's potato farming area) were great!

Rockland, ME (Knox County Regional—RKD)

Owl's Head Transportation Museum 😋😋😋

Rest. Phone #: (207) 594-4418
Location: On the field
Open: Daily: Lunch

PIREP   They have a concession stand but also have various air shows throughout the summer.

Sanford, ME (Sanford Regional—SFM)

Rudy's Cockpit Cafe 😋😋😋

Rest. Phone #: (207) 324-7332
Location: On the field
Open: Daily: Breakfast and Lunch

PIREP   FBO Comments: Fuel price at Presidential.

PIREP   I have eaten at this restaurant many, many times, and they must recognize my face. I have always been very satisfied with the food and service until today. They close at 12 noon on Saturday. I arrived at exactly 11:57 AM today and was turned away because their clock said 12:05.

I'll never return! There are too many good airport places to eat to accept that kind of behavior. Service in the past has always been at least average.

PIREP   Great little restaurant located on the north side of the field in the same building as Sanford Air. Plenty of parking at the ramp and out front for cars. The folks at Sanford Air were very helpful. The restaurant only serves breakfast and lunch. The breakfast hour can get a little busy especially during the weekends. The food is very good and reasonably priced. Service is very good and the wait-staff is very attentive. The atmosphere is good. Excellent recommendation.

PIREP   Went to Sanford the other day and ate at the Cockpit Cafe. Pretty cool place. The airport is huge, a former naval base with a 6000' runway. The old Navy control tower remains on the north side of runway 25. The cafe itself was nice too. The food is diner-style fare and pretty good. I had the chicken and biscuits, which the waitress said is one of their most popular dishes. Everything was good, and the french fries were among the best I've had. The prices at this place are so low, it's a wonder they can stay in business! The previous pireps about the place being busy seem to be true. I left at around noon, and the place was filling up fast.

Turner, ME (Twitchell—3B5)
Lobster at the Mini Mart ♨♨♨

Rest. Phone #: (800) 727-4645
Location: Very short walk
Open: Daily: Lunch

PIREP   If you're going up through central Maine and need a gas stop, you can land just north of Lewiston at Twitchel's. This is also a seaplane base. A two-minute walk across the road takes you to a mini-mart that has good coffee and lobster sandwiches.

The 2300' strip has trees at both ends, so drag it in a little. This is kind of an old-school family airport. Not much in atmosphere, but okay food.

Also located just across the highway from Twitchell's is the Chickadee Restaurant. Excellent local dining with many seafood specials. Watch your gross weight on takeoff after a meal here!

Waterville, ME (Waterville Robert Lafleur—WVL)
Riverside Farm Market ♨♨♨

Rest. Phone #: (207) 465-4439
Location: Cab ride
Open: Daily: Lunch

PIREP   Place started as a farm stand, has now progressed to a farm stand/deli/restaurant with nice selections of sandwiches and desserts to die for. It's actually in a town next to Waterville but well worth (in our opinion) a cab ride from the airport—it's not too far, but outside of walking distance.

We had lunch for four (sandwiches and drinks) for under $35.

Very friendly folks.

Wiscasset, ME (Wiscasset—IWI)
Sarah's Cafe ♨♨♨

Rest. Phone #: (207) 882-7504
Location: Crew car
Open: Daily: Lunch and Dinner

PIREP Sarah's Cafe in Wiscasset, ME is the best! They have everything from pizza, pasta, seafood, and even Mexican. Call ahead to the FBO (Wicked Good Aviation (207) 882-5475) to check on courtesy car availability. For longer stays, Enterprise Rent-A-Car has an office in town, (207) 882-8393. I usually go home with a doggie bag! There are also other restaurants and cafes in town.

The Hundred Dollar Hamburger

Maryland

Annapolis, MD (Lee—ANP)
Annapolis Seafood Market 🍔🍔🍔🍔

Rest. Phone #: (410) 798-9877
Location: Very short walk
Open: Daily: Lunch and Dinner

PIREP Hayman's is no longer Hayman's. It is now the Annapolis Seafood Market. This is the first time I went there, so I don't know what it was like before. It is a seafood market with a small restaurant. There is no ambiance, but the food was great. We had cream of crab soup and my wife and I split a dozen steamed crabs. It was delicious. The walk from ANP is nothing: 1–2 short blocks. We will definitely be back. In fact, my wife accidentally (?) left her jacket, so we'll be back soon.

PIREP Hayman's is a short walk, about two or three blocks south along Route 2. Their claim to fame is their delicious crab dishes. Try a soft crab sandwich or eat 'em by the dozen. If you want to sample the finest that the Chesapeake Bay has to offer, give Hayman's a try.

When you land, walk over to Chesapeake Air, the repair facility. Ask for directions.

PIREP Hayman's Crab House. A popular seafood restaurant about one block south on Route 2.

Baltimore, MD (Martin State—MTN)
Decoy's Restaurant 🍔🍔🍔🍔

Rest. Phone #: (410) 391-5798
Location: 5-minute ride, crew car or shuttle van
Open: Daily: Lunch and Dinner

PIREP If you fly into Martin State Airport, get a ride in the airport courtesy van to Decoy's Restaurant. It's about a five-minute ride from the General Aviation parking area. They serve burgers and Maryland-style seafood. Good lunch or dinner stop!

PIREP The Decoy is waterfront and you can eat on a deck over the water. Boats tie up at the pier!!! Very good atmosphere.

Cambridge, MD (Cambridge-Dorchester—CGE)
Signature Cafe 😋 😋 😋

Rest. Phone #: (410) 228-1121
Location: On the field
Open: 6 AM–2 PM; closed Mondays

PIREP My girlfriend and I ate lunch at the Signature Cafe. As usual, the food was outstanding—great burgers, hot and fresh—but there was one standout. At the suggestion of another pilot, we tried the cream of crab soup and it was heavenly! Made with all back-fin meat, large chunks of crab and plenty of them in a well-seasoned white base. You gotta try this; it is outstanding. The people that work there are friendly, courteous, and more than willing to carry on a conversation. Many locals eat there and I cannot say more! It is also great to park your steed not 100 feet away from the entrance where you can admire it and the other planes while you eat!

PIREP I second that emotion about the cream of crab soup. Yummmmmm. I highly recommend this place. Just so that you're not surprised, they don't accept credit cards.

College Park, MD (College Park—CGS)
The 94th Aero Squadron 😋 😋 😋 😋

Rest. Phone #: (301) 699-9400
Location: On the field
Open: Daily: Lunch and Dinner

PIREP The manager was rude to everybody and hard on the very limited staff working on this busy day. The door to the deck was kept open so that the busboys could clear the outside tables despite our repeated complaints that this made our table cold and unpleasant. Dishes were not cleared nor drinks refilled in a timely manner. And while cocktails, juices, tea, and coffee were included in the price, soda was extra. Gimme a break. I'm not sure I'll go back.

I called the restaurant directly to complain, and they said they will get back to me and probably offer me a complimentary brunch. In the past I have really liked the place.

PIREP What a wonderful experience.

I read with interest and, I have to say, fright all the negative reports about the 94th Aero Squadron, College Park, MD. Based on them I wouldn't have gone to this venue—had it not been for my brother (who was paying), I wouldn't have realized what I missed.

This restaurant has recently undergone a drastic change in management, so the waitress informed me. The ambiance was the first thing that caught my attention (first class): soft lighting and candles with lit fireplaces and exposed beams were the perfect mood setter for what turned out to be a delightful experience. The wait-staff was very polite and attentive, the food was not only served quickly and hot, but also with a zing of flavors.

Although my brother was paying, it seemed very reasonable, with a steak costing around $20. I found no rudeness that previous people mentioned. The staff was very polite and professional. I would love to go back and experience all that this restaurant has to offer again. And I would encourage everyone that read the previous comments to go out and find out for yourself—in my opinion. this really is a jewel.

Crisfield, MD (Crisfield Muni—W41)
Captain's Galley 😋

Rest. Phone #: (410) 968-3313
Location: Long walk
Open: Daily: Lunch and Dinner

PIREP   I am sorry to say that Max is gone from Crisfield Airport. We have always looked forward to the friendly FBO and even Max's predecessor. It is still friendly, but no longer is there transportation into town to the fantastic restaurants. The present FBO folks say they had to stop that. All there is available is a taxi that charged $5 each way per person ($20 plus tip for 2). When we arrived on a Sunday afternoon, when things were usually hopping, we were the only plane on the ramp.

The Captain's Galley's crab cakes and crab imperial were as tasty as ever, but I missed the homemade breads that accompanied the salad bar. As regular patrons of Crisfield Airport and The Captain's Galley, we were a bit disappointed.

PIREP   Homemade chowder, sandwiches, and baked goods still greet pilots traveling into W41 during the winter. Winter is a special time on the Eastern Shore, and the ice in the bay made the scenery even more spectacular. Tangier Island (TGI) was iced in, and the only transport was by air into Crisfield. We all need to support the airport at Crisfield. The town and county showed their support of the airport and the fine work done by Max by taking away the Pax Van that Max used to transport folks into town. Hopefully, they will see the light before the busy summer season, or the local restaurants will have to help out.

Stop in and see Max, and when you get back home, drop a note off to the town manager (who doesn't seem to have a clue about tourism, business development, or the importance of GA airports like W41). The hospitality of the FBO rates five burgers! The lack of a crew car rates a big –2.

Cumberland, MD (Greater Cumberland Regional—CBE)

Emma's Restaurant 🍔🍔🍔

Rest. Phone #: (304) 738-8700
Location: In the terminal
Open: Daily: Breakfast and Lunch

PIREP   My girlfriend and I ate at Emma's in the old terminal building. The food was fresh and good, and there was plenty of it at a very reasonable price. You can sit and watch your steed while you eat, which is nice. Very friendly people and excellent food.

PIREP   A friend and I flew to CBE pretty much on a whim (the close-your-eyes-and-put-a-finger-somewhere-on-the-chart kind, only with a little more planning involved). The flight from Leesburg was short and scenic, with the mountains and farms and so on. Surmounting the final ridge, if you look closely, you might see on top of the ridge the (surprisingly long) grass strip owned by a local doctor. The approach to the field was also interesting, as it involves a rather steep drop-off on nearly all sides, plus several rather large hills (or mini-mountains, if you will) where there are not drop-offs. However, we had no problems getting in or out.

Upon landing, we were able to taxi right up to the restaurant (which was next to Dirty Bird's Airplane Detailing—the sign itself is almost worth the trip!). There are two levels inside, both with comfy booths and a smallish bar upstairs, along with some sort of video gambling room.

The food was very good, not too expensive, and there was quite a lot of it! The burger (local beef) was accompanied by slaw and fries. I didn't finish mine, if that's any indication, though it certainly wasn't from lack of trying! After lunch we explored the new terminal with its lovely observation deck upstairs, mural of the Wright Flyer, a nice pilot's lounge, and even a quiet room for transient pilots' rest.

Very nice place!!!

Easton, MD (Easton/Newnam Field—ESN)

Hangar Cafe 🍔🍔🍔🍔

Rest. Phone #: (410) 822-8560
Location: On the field
Open: Daily: Breakfast and Lunch

PIREP Just came back from a $100 (actually $240 now) brunch flight from Manassas to Easton. We arrived at Easton before 10 AM to find plenty of parking and only two other planes parked at the terminal. However, when we got in, the cafe was packed, and there was a waiting list to get in. We have experienced this before. Apparently this place is popular with the local nonflying public as well as just pilots. The breakfast was hearty and satisfying, and the tab for two adults and one kid was less than $15 including tip.

PIREP The Hangar Cafe never fails to satisfy. I've eaten breakfasts and lunches there three times in the last month and the food and service is above average. The crab soup is spicy and good to the last drop. The burgers are cooked to order and of good quality. Be careful—the breakfasts are filling enough to make you drowsy on the flight home! Easton's Hangar Cafe will always be one of my fail-safe alternates when I fly to the Eastern Shore.

PIREP FBO Comments: Folks at Maryland Airlines are great, very friendly and helpful. Gave us a courtesy car for the night when we got weathered in, and told us about historic Easton. We had dinner at the Pub on Washington Street; the food, atmosphere, AND prices were top notch. Highly recommend.

PIREP Hangar Cafe was great! Food was great (I had the "Pitts" burger and my friend had the barbecue), service was friendly and quick. Gotta make the trip!

Elkton, MD (Cecil County—58M)

Walsh's Dockside 🍔🍔🍔

Rest. Phone #: (410) 392-6859
Location: Short walk
Open: Daily: Lunch and Dinner

PIREP Walsh's Dockside is fine, affordable, year-round waterfront dining within walking distance of the airport.

Frederick, MD (Frederick Muni—FDK)

Airways Inn 🍔🍔🍔🍔

Rest. Phone #: (301) 662-3077
Location: On the field
Open: Daily: Lunch and Dinner

PIREP FBO Comments: Clean, friendly FBO.

PIREP Restaurant was clean, food priced fairly for quality and service. Staff was friendly, view of the field is excellent.
Would go again.

PIREP What a great find!
Outstanding food, friendly service, and reasonable prices at an airport cafe. Highly recommended are the Eggs Chesapeake (with Maryland crab), fried oysters, fried shrimp, and great crab cakes. Try their Mariner's salad for a light summertime meal. A very popular lunch place, get there early or expect to stand in line. Located at FDK, practice your instrument approaches, visit AOPA and AVEMCO, have a great lunch, meet some wonderful people—all in one modestly decorated place. A must-visit site.
Give it four and a half hamburgers! Open for lunch and dinner.

Gaithersburg, MD (Montgomery County Airpark—GAI)

Montgomery County Airport Cafe 🍔🍔🍔

Rest. Phone #: (301) 330-2222
Location: On the field
Open: Daily: Breakfast and Lunch

PIREP   Good prices, small portions, nice view, friendly people, and atmosphere.

PIREP   The Airport Cafe is located in the terminal building. Open daily from 6:30 AM to 8 PM for breakfast, lunch, and light dinners. The food is good and prices reasonable. It is a popular place. The owner plans to add a deck on the flight-line side in the future for additional seating in summer months.

Hagerstown, MD (Hagerstown Regional-Richard A Henson Field—HGR)

Nick's Airport Inn 😋😋😋😋😋

Rest. Phone #: (301) 797-8001
Location: On the field
Open: Daily: Lunch and Dinner, closed Sundays

PIREP   Flew up for typical hamburger and was quite impressed. Park at the FBO, walk through it, and the restaurant is immediately in front of you (look for the solarium thing). Nick's is open Monday through Friday 11 AM–2 PM for lunch and 5–10 PM for dinner, Saturday only open 5–10 PM for dinner, closed Sunday.

Suggest you call before launching to check the hours. We were quite underdressed for dinner, but the staff was accommodating. I got the idea that shorts and an old T-shirt were okay if you were a pilot.

They have a cheese spread thing with crackers to start you out that I would swear was "spray" cheese, if it wasn't that we were sitting at a table with fine linen and mood light with fancy waiters around us.

Bow-tie pasta and shrimp with sun-dried tomatoes was excellent, and the baked crab imperial was very tasty. I did find a few pieces of crab shell though, and the waitress confirmed that several people have had similar experiences. Dessert was great.

PIREP   One of the very best airport accessible restaurants I have found in 20-plus years of flying 1550W. Just a few yards walk from the FBO and ramp. Excellent in all respects, reasonably priced. Great fried oysters.

Leonardtown, MD (Duke Regional—2W6)

Rick's Cheesesteak Factory & Italian Eatery 😋😋😋

Rest. Phone #: (301) 862-1181
Location: 1-mile walk
Open: Daily: Lunch and Dinner

PIREP   Went to 2W6 on a summer weeknight after work. Paid close attention to the ADIZ and the restricted areas associated with NAS Patuxent but controllers were friendly and helpful with VFR flight following. Airport was quiet after 5 PM with no one maintaining unicom and no apparent fuel available after hours. No courtesy car available but enjoyed an easy 1-mile walk (past Outback Steakhouse 0.9 mi) to Rick's for great hoagies and burgers. If you're not up for the walk, call 'em and they'll deliver to 2W6. VERY NICE terminal built for commercial service.

Ocean City, MD (Ocean City Muni—OXB)

Captain's Galley 😋😋😋

Rest. Phone #: (410) 213-2525
Location: Short ride into town
Open: Daily: Lunch and Dinner

PIREP Ocean City MD (OXB) is, as we all know, a great place for all types of food. One of my favorites is Captain's Galley located at the harbor. It features one of the best crab cakes around. Or you could go into town to the Ocean Club on 49th Street for a great prime rib. Huge hunk of meat—even the ladies' portion should be plenty enough for anyone. Take the shuttle from the airport into town for a couple of bucks or share a cab with others. Once in town, the O.C. bus is only $1.25 anywhere.

Stevensville, MD (Bay Bridge—W29)

Hemingway's 😑😑😑😑😑

Rest. Phone #: (410) 643-8825
Location: Short walk
Open: Daily: Lunch and Dinner

PIREP Easy walking distance. There is a worn path through the high grass behind the tetrahedron on the north side of the runway that will lead you to the marina maintenance and storage area. Follow the gravel road to the water and then the walkway along the water to the last building on the path. Casual dining on the banks of the Chesapeake Bay with an unobstructed view of the sunset near the base of the Chesapeake Bay Bridge on Kent Island. Seating inside or on the deck outside. Prices are moderate and the food is excellent. Their specialty is crab. Reservations are suggested

PIREP FBO Comments: Handy transient tie-downs, friendly folks.

PIREP Flew two friends to Hemingway's for a romantic anniversary dinner. I sat separately and had a shrimp appetizer and their Chesapeake Burger (a third-pound burger topped with crab imperial). That was probably one of the best burgers I have ever had! The crab topping was absolutely superb and very flavorful.

Shrimp (so-so flavor): $9. Chesapeake Burger: $9

Ate on the balcony and watched the sun set. Gorgeous view, great service. Could hear the live music from the waterfront bar downstairs, but it wasn't loud or intrusive.

Stevensville, MD (Kentmorr Airpark—3W3)

Kentmorr Harbour Restaurant 😑😑😑😑

Rest. Phone #: (410) 643-2263
Location: Short walk
Open: Mon.–Fri. 10 AM–4 PM

PIREP How I've missed the Chesapeake. Haven't been down since before 9/11. We had a great time! The food, friendly service, and ambiance at the Kentmorr Marina are top notch! You can also get pool and beach access for $2 a person. If you haven't been there, what are you waiting for?

PIREP Kentmorr Airpark is a grass strip, public-use residential airpark located on Kent Island, about five miles south of Bay Bridge Airport. Great crab cakes and black angus beef sandwiches at nearby Kentmorr Harbour Restaurant. Just park on the north side of the runway at the Chesapeake Bay end, opposite the homes/hangars, and walk a block to the marina. Food is excellent. Makes a good fly-in lunch spot, with a great view of the water and all the marine traffic in the bay.

Massachusetts

Bedford, MA (Laurence G Hanscom Field—BED)

A.P. Pizza ☕☕

Rest. Phone #: (617) 274-0133
Location: Short walk
Open: Daily: Lunch and Dinner

PIREP The restaurant is a short walk from the east ramp in bldg 1534. They have great pizza, calzone, subs, spaghetti, salads, and soft drinks for a very reasonable price.

Beverly, MA (Beverly Muni—BVY)

Something Different ☕☕☕

Rest. Phone #: (978) 927-0070
Location: On the field
Open: Daily: Breakfast and Lunch

PIREP It appears that the East Side Cafe is once again opened at the Beverly Airport. It is open on weekends.

PIREP The East Side Cafe is now open! The food is excellent. Old WWII and later memorabilia is scattered throughout the restaurant. It is brightly lighted with a west-facing view toward the ramp and runways. The transient ramp is located nearby.

Northern Grind ☕☕☕

Rest. Phone #: (978) 922-9288
Location: FBO takes you and brings you back
Open: Daily: Breakfast and Lunch

PIREP FBO Comments: Aviators of New England. A real old-fashioned family run business. They are happy to give you a lift into town. There is also a restaurant on the field, but it was closed Mondays.

228

PIREP   We have found Beverly to be a very nice way to get to the Boston area. There is a commuter rail line, and the folks at Aviators are very nice to take you there and pick you up.

While waiting for the train we found a neat little coffee-house style restaurant called the Northern Grind. I had the excellent clam chowder. They also have a good breakfast menu and a selection of sandwiches.

Chatham, MA (Chatham Muni—CQX)

Cloud 9 🍔 🍔 🍔

Rest. Phone #: (508) 945-1144
Location: On airport
Open: Closed Mon., Sun. 6 AM–12 noon, otherwise 6 AM–2 PM

PIREP   Recent breakfast visit was a pleasant surprise. Years ago I had been turned off, but the restaurant has been under new management for quite a while now as I understand it. The food is outstanding. Homemade breads, original and terrific entrees. Nice people, too. The Crab Cakes Bene are worth the flight alone. Highly recommended.

Edgartown, MA (Katama Airpark—1B2)

Plane View 🍔 🍔 🍔

Rest. Phone #: (508) 693-1886
Location: On the field
Open: Daily: Breakfast and Lunch

PIREP   Katama Airpark is a beautiful, well-maintained turf airport with three runways. Two parking areas, one by the restaurant and one by the beach where you park and walk a path crossing one street to the Atlantic. We were about number ten at the beach parking, and about an hour later I counted 29 planes (with more coming in) and more at the restaurant end.

We stayed a few hours at the sandy beach with dunes (the name is South Beach), visited some friends, and took off for another sunset tour around the island to the Gay Head Cliffs (town of Aquinnah) end.

The small fee you pay at the restaurant is used to maintain the fields. Katama unicom gave a "thanks for visiting" call as I was leaving. Nice touch.

PIREP   I used to teach flying at Plymouth Airport, which itself is a great spot to dine. But for the more adventuresome folk, go down to Katama Field on Martha's Vineyard and set down on the largest turf airport in the country!!!!! Yes!!!! They have three long strips maintained very well that sit next to a gorgeous beach that you can taxi to! Seasonal, yes, but there's a cute little restaurant with good food and service, and you'll probably see rides being given in a Staggerwing or see a DC-3 occasionally set down. Off the beaten path and makes for a fun afternoon.

Fitchburg, MA (Fitchburg Muni—FIT)

Gene Collette's Airport Restaurant 🍔 🍔 🍔

Rest. Phone #: (978) 771-8133
Location: On the field
Open: Daily: Breakfast and Lunch

PIREP   I stopped in for lunch the other day at FIT for lunch. First, some useful information: the restaurant is located in the main terminal right beside the AC transient parking.

The food was not fancy, but it was good—some of it was even homemade (good pies). There appeared to be plenty of seating. The service was excellent and prompt and catered to all requests.

Enjoy yourself.

PIREP   Another local fly-in with a basic menu highlighted, with great homemade pies for dessert. A good view of the active runway is available from the restaurant.

100LL is discounted on weekends. Fly in, request a top off, and your refueled airplane is ready to go when you are.

Great Barrington, MA (Walter J. Koladza—GBR)

MOM'S 🍔🍔

Rest. Phone #: (413) 528-2414
Location: 3 miles away
Open: Daily: Breakfast and Lunch

PIREP   We flew into GBR this morning (Great Barrington, MA), out in the Berkshires. Landing was on a 2500' strip with crosswinds. It was interesting. The FBO let us use their car to go to Mom's restaurant, about three miles away.

Just a quaint little place. Food good and moderately priced for that area of the world. The FBO wouldn't take any money for the use of the car, so I gave him some pesos and bought some fuel.

I would imagine that the area is breathtaking with the change of the leaves.

Hopedale, MA (Hopedale Industrial Park—1B6)

Landing Strip Cafe 🍔🍔

Rest. Phone #: (508) 422-9297
Location: On the field
Open: Daily: Breakfast and Lunch

PIREP   Airport Comments: Approach to runway 18, a good lesson in high, short field. Get rid of those trees! Not a soul on the field.

PIREP   Restaurant on field was empty. Poor service, all TVs were on, tuned to different soap operas, with the volume turned up. One of which the so-called waitress (I'm sorry—server!!!) was fixated on. Took half an hour to prepare a meatball sub—the pickle was good.

Had to fly here just because I hadn't been there before, but guess what? Not again.

PIREP   I just went to the restaurant at Hopedale-Draper. On the runway next to the terminal ramp. Excellent inexpensive lunch and dinner. Opens around noon. Also has a bar. In the same building is a nice family billiard hall—bring your sticks.

Hyannis, MA (Barnstable Muni-Boardman/ Polando Field—HYA)

DJ's Wings 🍔🍔🍔🍔

Rest. Phone #: (508) 775-9464
Location: Short walk
Open: Daily: Lunch and Dinner

PIREP   Park at Air Cape Cod east ramp, walk down Mary Dunn Way diagonally across Main Street. Best wings and ribs in the area. Family sports pub with a roof deck to watch airplane ops when 13 is in use.

Lawrence, MA (Lawrence Muni—LWM)

Joe's Landing 🍔🍔🍔

Rest. Phone #: (978) 682-8822
Location: On the field
Open: Daily: Lunch and Dinner

PIREP   You should give this restaurant a couple more hamburgers. It's super.

PIREP   I visited LWM (Lawrence, MA) on business today and stayed for lunch at Joe's Landing Strip Cafe. Reasonably priced at less than $20 for two people. I found the food to be exceptional. The service made us feel like we were at home. I would recommend the Kafta Wrap very highly.

PIREP   Joe's Landing is a Greek-oriented sandwich place and was not too bad when I tried it yesterday. It is located in the terminal building.
    LWM has two long, hard surface runways, tower control, and lots of parking.

Lobster Claw Restaurant 🍔🍔🍔

Rest. Phone #: (978) 664-6349
Location: 10-minute walk
Open: Daily: Lunch and Dinner

PIREP   The restaurant in the terminal building is okay but somewhat of a greasy spoon.
    There are several good restaurants a short (five- to ten-minute) walk from the airport, although the roads are not exactly pedestrian friendly. Ask for directions to the Butcher Boy Mall.
    You'll find the Lobster Claw, a fast seafood place; Orza, an Italian restaurant; and a cafe (name?) on the far end serving cappuccino, bagels, and so on.
    Across from Butcher Boy is Joe's Fish, a full-service seafood restaurant, and past that is the Loft, a full-service American restaurant. I'd recommend any of these places.

Treadwell's Ice Cream 🍔🍔🍔🍔

Rest. Phone #: (978) 686-1850
Location: Short walk across the street
Open: Daily

PIREP   Anybody flying in to Lawrence Airport (LWM) MUST walk over to Treadwell's Ice Cream, right at the end of the runway. The only year-round ice cream stand I know of where you even have to wait in line in January.

Marshfield, MA (Marshfield Municipal— George Harlow Field—3B2)

Santoro's Pizza and Subs 🍔🍔

Rest. Phone #: (508) 866-7334
Location: Short walk
Open: Daily: Lunch and Dinner

PIREP   FBO Comments: Wonderful FBO, very service oriented. Extremely nice building, full pilot shop, reasonable prices.

PIREP   Santoro's Pizza and Subs is a deli-style restaurant with seating for about 20. Nothing too fancy but good food within a four-minute walk of the airport. We got subs to go and took them back to the airport and ate on the lawn overlooking the runway.

Good little place to have fair-priced meal.

Other restaurants available, but you need to use the FBO's courtesy van to get to them, which is, naturally, on a first come, first served basis.

Marston Mills, MA (Cape Cod—2B1)
Golf Course Restaurant 🍔

Rest. Phone #: (508) 420-1143
Location: Across the street
Open: Daily: Breakfast, Lunch, and Dinner

PIREP   There is a golf course across the street with a restaurant on it. They open at 7 AM. You can park your plane and walk across the street. The airport is a turf strip with long, lumpy runways.

This restaurant was a double bogie.

It seemed that they were almost surprised to see people coming to eat. We had a hard time getting hot coffee, and when we finally got it there were no utensils—you kind of had to swirl your coffee around and drink it while it was hot. The wait-staff couldn't remember to refill. They were out of orange juice and apple juice; they were also out of waffles—must use the frozen kind. We were in the middle of cranberry country (Cape Cod), so they must have been scared to say they were out of that—instead they watered down what little they had left so it tasted like bad water. Oh yes, almost forgot: they were out of sausage!!! This was 8:30 in the morning. When we did finally get all the food, it wasn't bad.

There were two more coming when we were leaving. We sent them to Plymouth, where you always get a decent breakfast. Give it a try and see for yourself.

We may have come at a bad time. "Breakfast"!?

Montague, MA (Turners Falls—0B5)
Country Creemee 🍔🍔

Rest. Phone #: (413) 863-3529
Location: Across the road
Open: Daily

PIREP   There's a small ice cream/hot dog/hamburger stand just across the street from the midfield parking area. Large portions and good ice cream with six picnic tables under shade trees.

Nantucket, MA (Nantucket Memorial—ACK)
Hutch's 🍔🍔🍔

Rest. Phone #: (508) 228-5550
Location: On the field
Open: Daily: Breakfast and Lunch

PIREP The terminal restaurant at ACK is Hutch's. The food is decent and cheap (at least by Nantucket standards), and the service was quick and efficient. Overall though, I'd say a three-burger rating is about right.

One other thing, this place is quite busy in the summer and on weekends, so be alert for other traffic. And there is a modest parking/landing fee ($6). I also congratulate the ACK ops department for running the most efficient and friendly airport I have yet visited. Other airports should follow ACK's example.

PIREP Unicom summons a yellow Follow-Me van that leads you to parking and gives you and your baggage a lift to the terminal. The $10+ cab fare downtown gives access to a plethora of restaurants. Try Arnos on Main Street or the Tap Room at the Jared Coffin House on Broad Street. American Seasons on Center is exclusive and $$$$, a bit noisy, but quite interesting.

Nantucket is simply the best. It has good food, a nice staff and, of course, lots of the tourists coming though.

Newburyport, MA (Plum Island—2B2)
Plum Island Grille 🍔🍔

Rest. Phone #: (978) 463-2290
Location: Reasonable walk
Open: Sunday Brunch, Dinner other days

PIREP FBO Comments: The FBO at Plum Island Airport is run by Eagle East Aviation headquartered at Lawrence Airport (LWM).

PIREP The Plum Island Grille is a 20–25-minute walk to the east of the airport. The restaurant is small and cozy. The menu has quite a range of gourmet choices. The Sunday brunch menu is available from 12 noon until about 3 PM. The dinner menu is superb. The restaurant is open Thursdays 5–9 PM, Fridays and Saturdays 5–10 PM, and Sunday 12 noon until 8 PM. The food was excellent.

Reservations are recommended.

Bob's Lobster 🍔🍔🍔🍔

Rest. Phone #: (978) 465-7100
Location: 1/4-mile walk near the beach
Open: Daily: Lunch and Dinner

PIREP If you're in a hurry, the closest food is at Bob's Lobster (less than .25 mi. east). They have fresh seafood, hot dogs, chicken fingers, and ice cream. If you have a while, though, you can explore the many other places to eat within walking distance. However, I recommend making it a day trip if you can; it's a really fun little place.

There are a number of different things to do in the small community, some of which include visiting the restaurants. There is a small road right at the airfield. Go one mile east and you get to Plum Island, on which there is a National Wildlife Refuge, restaurants and, of course, the beach! Go one mile west from the field, and you get to downtown Newburyport, a quaint little main street with many shops and things like that.

This airfield is very small and fun to get in and out of with a small plane.

New Bedford, MA (New Bedford Regional—EWB)
The Airport Grille 🍔🍔🍔

Rest. Phone #: (508) 994-1600
Location: On the field
Open: Daily: Lunch and Dinner

PIREP   A great place to sample the local seafood! Very tasty lobster bisque, mussels, and swordfish. There was a short list of dinner specials, all reasonably priced. There was great attention to the details and even the side dish veggies were crisp and nicely done. Full bar available (for the nonpilots, of course).

The restaurant is directly adjacent to transient parking with an outdoor deck that overlooks the tarmac. The service was first rate, and they squeezed in our group with only a very short wait on a busy evening. The indoor dining area is on the small side, and it was quite crowded with both locals and aviators, a testament to the chef's skills. Cooking is done in a semi-open kitchen with lots of flair. The ambiance is average, but the cooking is first rate, a rare find in on-airport restaurants.

Northampton, MA (Northampton—7B2)

Witzwilly's 🍔🍔🍔

Rest. Phone #: (413) 584-8666
Location: Reasonable walk
Open: Daily: Lunch and Dinner

PIREP   FBO Comments: $5 ramp fee, even with gas.

PIREP   Great destination! The town is delightful; lots of nice shops and restaurants offering a wide choice of cuisine; Thai, Japanese, Chinese, Mexican, Moroccan.

Witzwilly's is very nice, and my shrimp and chevre salad was perfect. Service was swift once I told the waiter we were a bit pressed. The girl at the FBO said the walk was about 15 minutes, but it's a bit more than that. Still, a great place for lunch and a stroll.

Norwood, MA (Norwood Memorial—OWD)

The Runway Cafe 🍔🍔🍔

Rest. Phone #: (781) 769-3550
Location: On the field
Open: Daily: Breakfast and Lunch

PIREP   The Runway Cafe is now opened in the location of the former Prop Stop Restaurant. Initial comments for returning customers have been favorable. Open for Breakfast and Lunch, closes mid-afternoon.

Orange, MA (Orange Muni—ORE)

Amy's White Cloud Diner 🍔🍔🍔

Rest. Phone #: (978) 544-6821
Location: On the field
Open: Daily: Breakfast and Lunch

PIREP   FBO Comments: Airfield in very good condition. Had a chat with the airport manager, and this airport is getting to be a popular stop. Camping still allowed on field.

PIREP   Visited for the second time with time to check out the restaurant. A beautiful woman by the name of Amy bought the place and has already made a difference. A very short walk across the street from the terminal makes this one of my favorite stops now. The food was good to very good for a breakfast and average in cost. Two eggs, toast, hash, home fries, sausage, coffee, and O.J. for, like, 8 bucks. It almost felt homey, with the decor and a fireplace.

Will definitely head back soon for lunch and conversation. The diner closes at 2 PM. Without hesitation, Amy has this place up to a four-star rating.

PIREP   Easiest access is the White Cloud (breakfast and lunch), across the street from Jumptown, the skydiving club. It's under relatively new management and said to be much better than in the past.

Plymouth, MA (Plymouth Muni—PYM)
Plane Jane's Cafe ☺☺☺☺

Rest. Phone #: (508) 747-9396
Location: On the field
Open: Daily: Breakfast and Lunch

PIREP   Plane Jane's Cafe, at Plymouth airport (PYM), is one of the finest airport cafes in the Northeast. My flying buddy and I have been to just about every $100 breakfast cafe in New England and New York and the continued reliability, friendly service, never-ending coffee refills, and good and plentiful food is always served with a smile.

Plymouth airport has wide and long runways for those iffy winter days when the wind is howling and you are wondering if the snow is off the blacktop.

Give Plane Jane's a try, you will not be dissatisfied.

PIREP   Plymouth, MA (PYM) open from around 7 AM, closes at 2 PM. Great lunches and fast service. Second floor in the administration building. Plenty of parking right in front. Prices are reasonable.

Provincetown, MA (Provincetown Muni—PVC)
Picnic on the Beach ☺☺☺☺

Rest. Phone #: (508) 487-0241
Location: Short walk
Open: Daily

PIREP   Provincetown is out on the very tip of Cape Cod. The airport is essentially right on the beach. When you land, a "follow me" golf cart will guide you to a parking spot and take you up to the terminal building. The driver will collect your $5 landing fee. It helps to have the right bill.

The beach is only a five-minute walk from the airport building. There are restrooms, showers, and changing facilities.

If you'd rather go into town, there are usually taxis waiting at the terminal building. The ride into town is $5/person each way. It's a good idea to copy down all the cab company phone numbers posted at the airport because you may not be able to get the same cab back. A cell phone does come in handy.

Southbridge, MA (Southbridge Muni—3B0)
Jim's Fly-in Diner ☺☺☺

Rest. Phone #: (508) 765-7100
Location: On the field
Open: Daily: Breakfast and Lunch

PIREP   Closed this winter. Call first to check when they reopen.

PIREP   My wife and I visited Southbridge Municipal (3B0) and ate at Jim's Fly-in Diner. The food was decent and the diner is directly off the ramp. You can sit outside in good weather, like the day of our visit, and watch takeoff and landing activity.

PIREP   Airport Comments: runway 02-20 is 3500 X 75' in good condition. Plenty of parking.

PIREP   A very pleasant airport and old-style diner. Aircraft parking is as close as ten feet away with plenty of window and deck seating.

Hours are Tuesday and Wednesday 7 AM–2 PM (breakfast/lunch), Thursday and Friday 7 AM–8 PM (breakfast, lunch, and dinner), Saturday 6 AM–2 PM (breakfast/lunch), Sunday 7 AM–2 PM (breakfast only). Fridays they have their seafood specials with haddock, scallops, whole clams, and chowder.

You won't leave hungry, there is plenty to choose from, and the prices are very reasonable. It's a pretty area to fly into.

At least 20 people were in the diner, with a birthday party in the back. At least six planes had flown in.

Stow, MA (Minute Man Air Field—6B6)
Nancy's Air Field Cafe 😋😋😋😋😋

Rest. Phone #: (978) 897-3934
Location: On the field
Open: Daily: Breakfast and Lunch

PIREP   Great restaurant, one of the best in MA!

PIREP   Three friends and I flew up to Minuteman in a Skyhawk and Cherokee 180, late morning. The staff at the FBO were friendly and helpful. We dined at Nancy's for lunch and had quite the tasty assortment of designer sandwiches. Our waitress was a hoot—good sense of humor and attentive. We're sorry that we didn't try the chili, but we'll be back.

Vineyard Haven, MA (Martha's Vineyard—MVY)
Airport Cafe 😋😋😋😋

Rest. Phone #: (508) 693-1886
Location: On the field
Open: Daily: Breakfast and Lunch

PIREP   I went to Martha's Vineyard this morning. There were four planes and nine people in my group. The food was good, as it has been before. The waitress has a funny way about her, never know if she is joking or not.

Anyway, food service was timely and well prepared. Parked in front of the restaurant, which is located in the terminal. Gets much busier in the summer months. Try the French toast made with Portuguese sweet bread—maybe it should be called Portuguese Toast or Masa Toast ("masa" is Portuguese for "sweet bread"). I recommend going here for breakfast, but be prepared for a longer wait in season.

PIREP   The restaurant in the new terminal is now open. Most tables have a nice view of the field when there isn't a Cape Air plane parked in front loading or unloading. Food is reasonably priced and varies from hot and cold subs to burgers to chowder, as well as salads and dessert.

There is taxi and shuttle service into Edgartown if you want to look for something more authentically MV-ish.

PIREP   Had breakfast at MVY. Food was okay, service was s-l-o-o-o-o-w. It was busy, and it seemed that the kitchen was overwhelmed.

Westfield/Springfield, MA (Barnes Muni—BAF)

The Flight Deck 🍔🍔🍔

Rest. Phone #: (413) 568-2483
Location: On the field
Open: Daily: Breakfast and Lunch

PIREP The Flight Deck is still open (closes at 2 PM) and still good and friendly. The A-10's still provide lunch-time entertainment, and we flew over a C-5 doing touch-and-go's at Westover (CEF) next door.

PIREP Two long runways, interesting National Guard traffic (currently A-10's), and an excellent classic home-cooking breakfast and lunch restaurant on the field with a view of it all. Soups and burgers are recommended, atmosphere very pleasant. Talk to Bradley on the way in and out as there is a lot of traffic overhead.

Allegan, MI (Padgham Field—35D)

The Grill House 🍔🍔🍔

Rest. Phone #: (269) 686-9192
Location: They will pick you up for free
Open: Daily: Lunch and Dinner; closed Sundays

PIREP FBO Comments: 24-hour fuel with credit card. FBO is Dodgen Aircraft; open 7 days a week, 8 AM to dusk; courtesy car available during open hours, restaurant will provide transportation to/from restaurant.

PIREP The Grill House Restaurant will pick you up and drop you back at the airport. The food is very good—there are two venues—cook your own steaks, chicken, fish, or shrimp on the upper level, and sandwich fare on the lower level. Restaurant is open Tuesday through Sunday (closed on Monday). They have excellent food and service. Cocktails are available.

Cadillac, MI (Wexford County—CAD)

Hermann's European Cafe 🍔🍔🍔

Rest. Phone #: (231) 775-9563
Location: Free pickup
Open: Daily: Lunch and Dinner

PIREP FBO Comments: Fuel onsite, very friendly and helpful.

PIREP A great destination to stop. They will pick you up from KCAD and bring you back. Menu is upscale, and pricing is reasonable—food is terrific!! A very nice place to hang out, and Lake Cadillac is half a mile away if you're interested, within walking distance.

Chesaning, MI (Howard Nixon Memorial—50G)

Burger King 🍔🍔

Rest. Phone #: (989) 723-8468
Location: 1/2-mile walk
Open: Daily: Breakfast, Lunch, and Dinner

PIREP Burger King, Little Caesar's Pizza, and an ice cream stand are all within a half-mile walk.

Clinton, MI (Honey Acres—7N4)
McDonald's 🍔🍔

Rest. Phone #: (517) 456-8700
Location: Short walk
Open: Daily: Breakfast, Lunch, and Dinner

PIREP Airport Comments: Airport is sod, long and smooth. The runway can be soft after a heavy rain; there is no snow removal.

PIREP The only airport around with a McDonald's. Like all McDonald's, it's pretty good.
Park at the south end and it's a short walk to the east.

Race Track Inn 🍔🍔🍔

Rest. Phone #: (517) 456-7768
Location: Short walk
Open: Daily: Breakfast, Lunch, and Dinner

PIREP Honey Acres Airport (7N4) in Clinton, MI has a great little restaurant called the Race Track Inn. Park on the south end of the runway and it's about 300 yards west. Dim ambiance, average prices, and good food and drink (for the nonflyer); it's full of old horse-racing decor (hence "Race Track Inn").

Detroit/Grosse Ile, MI (Grosse Ile Muni—ONZ)
Airport Inn 🍔🍔🍔

Rest. Phone #: (734) 675-4200
Location: Across the street
Open: Daily: Lunch and Dinner

PIREP The Airport Inn is still going strong and is an easy walk right across the street from the FBO. Excellent hand-tossed pizza, and the burgers are good too!

PIREP The Airport Inn is a comfortable neighborhood restaurant/bar with good pizza and moderate prices. It is about 100 yards east of the terminal on the road that runs past the ramp on the north side of the field. Ask to see the airplane of Dairy-Aire, good fun! Be warned: the FBO tends to close early, so have full tanks if you go in late.
This is a great night currency flight if you live around Detroit or Toledo. You can cap a good evening for your guests with a flight up the Detroit River over the spectacular Detroit and Windsor riverfronts, then turn back west at Detroit City.

Dowagiac, MI (Dowagiac Municipal—C91)
The Round Oak Restaurant 🍔🍔🍔

Rest. Phone #: (269) 782-5128
Location: Call for free pickup
Open: Daily: Lunch and Dinner; closed Sundays

PIREP   We flew into the Dowagiac Municipal Airport (C91) and telephoned the famous Round Oak Restaurant. The owner, Mr. Douglas McKay was quick to offer to pick us up at the airport. We had a pleasant meal, well served and at a fair price. Steaks, fish, prime rib, salad bar, and evening entertainment.

Wahoo's Eatery 🍔🍔🍔

Rest. Phone #: (269) 782-0601
Location: Short walk
Open: Daily: Lunch and Dinner

PIREP   FBO Comments: Airport was unattended on Sunday.

PIREP   Should have called ahead. Found out the Round Oak is not open on Sunday. However, a short four-block walk to the north from the airport took us to Mr. Wahoo's, a family dining spot with good food, generous portions, and a very reasonable price. You will find it just behind the Burger King.

Flushing, MI (Dalton—3DA)

Gabby's 🍔🍔🍔🍔

Rest. Phone #: (810) 732-4650
Location: On the field
Open: Daily: Breakfast, Lunch, and Dinner

PIREP   Happened to stop into Dalton airport today, Flushing, MI, which sits under the Flint, MI Class C. What a great little strip, with parking in the grass right in front of the on-field restaurant. I don't know what the story is, but there are a slew of new hangars and a good number of homes with hangars—some very nice and one a genuine single-wide mobile home, with hangar. Truly an "everyman's airpark."

Flint approach and tower were great about getting us through the charlie airspace, and the food was good. Watching the local drop 'em in over the trees on the 36 approach was entertaining. There seemed to be an unspoken contest for "just how little of this short strip can you use?!" Replete with friendly folks in lawn chairs sitting out in front of their hangars giving advice, comments and ratings on landings, takeoff, and so on. Always good natured. Got some nice comments on 43W.

Had a couple of decent omelets, and the lunch and dinner menu looked pretty extensive. Very nice waitress, although service was a bit slow from the kitchen.

It's a shame that they allow two rows of vehicles to park between the restaurant and the runway, though. Kills the view. Other than that, worth the stop.

PIREP   FBO Comments: Self-serve, pay at the pump with credit card. Lightweight fuel hose, ladder.

PIREP   Just park your airplane on the grass, north end of 18/36, and walk over. Very convenient. Family style, good food, excellent and friendly service. No complaints.

Frankenmuth, MI (Wm 'Tiny' Zehnder Field—66G)

Zehnder's 🍔🍔🍔🍔🍔

Rest. Phone #: (800) 863-7999
Location: Call for pickup
Open: Daily: Lunch and Dinner

PIREP   Frankenmuth, MI is about ten miles north of Flint, MI. The whole village is done in a Bavarian motif and has lots of shops and the world's biggest Christmas store. Beautiful grass strip about two miles from town. Not super long, but my brother gets in and out with four people in his Bonanza. The BIG restaurant in town is well known for their family chicken dinners. They will come out and pick you up and drop you off.

Gaylord, MI (Lakes of The North—4Y4)
Settings 😊😊😊😊

Rest. Phone #: (231) 585-6000
Location: On the field
Open: Daily: Lunch and Dinner

PIREP   Great flying destination. Settings restaurant on the field. Deer Run Golf Course also on field.
Bring your clubs and play a round of golf, then enjoy a sandwich at Settings.

Gladwin, MI (Sugar Springs—0MI1)
The Hearth Restaurant & Pub 😊😊😊😊😊

Rest. Phone #: (989) 426-9203
Location: On the field
Open: Mon.–Sat: Dinner from 4:30 PM; Sunday: Breakfast, Lunch, and Dinner

PIREP   This is a terrific place to slip away to for a round of golf or a good meal. Remember, this is a PRIVATE airstrip and permission MUST be granted before you land. It's easy to obtain. Simply call the Property Owners Association Office at (989) 426-4111, the Golf Pro Shop at (989) 426-4391, or the Hearth Restaurant at (989) 426-9203.

Park at the west or east end (the west end is easier and has an access road). Bring your own tie-downs and have a member or guest pass with you.

The Hearth Restaurant & Pub is relaxing and the food is excellent. They may well add daily breakfast service during the summer months IF there is enough interest. Call before you make the trip just to be sure.

The 6,737-yard, par 72, 18-hole championship course, designed by Jerry Matthews, weaves through a wooded front nine and a more open back nine with subtle elevation changes, water in play on five holes, and greens guarded with bunkers, which create an enjoyable challenge for golfers of all skill levels. The 18-hole green fee is only $32!

This is a family-oriented resort. It includes a wonderful campground, an Olympic-sized swimming pool, archery range, and tennis courts.

I plan on camping out here over the July 4th weekend. They put on a VERY nice fireworks display.

For complete information visit the website at www.sugarsprings.net/.

Gladwin, MI (Gladwin Zettel Memorial—GDW)
The Peppermill Restaurant 😊😊

Rest. Phone #: (989) 426-8922
Location: Short walk
Open: Daily: Breakfast, Lunch, and Dinner

PIREP A good place to go for a good, reasonable breakfast is the Peppermill Restaurant in Gladwin, MI (GDW). A varied menu of eggs, omelets, and pancakes has always satisfied our appetite. It's a typical small-town restaurant that gets crowded after church on Sunday mornings. The airport has a courtesy car, although it has seen better days! It is a short walk (approximately half a mile) if the car is not available.

Harrison, MI (Clare County—80D)
Yvonne's Aero-Port Restaurant 😋😋😋

Rest. Phone #: (989) 539-1736
Location: On the field
Open: Daily: Breakfast, Lunch, and Dinner

PIREP One of my favorite breakfast places. Fresh, homemade bread and baked desserts. Great omelets. Very small restaurant. Waitresses who call you "Hon." Beware: runway 18-36 is paved but often not plowed in winter. Quarter-mile taxiway from runway to restaurant is grass.

PIREP Nice little restaurant. Friendly service, great chili. Taxi up to the back door at north east corner of airport. Watch out for very long displaced threshold for 36.

PIREP Open seven days per week. Home-style cooking and homemade baked goods. Good, inexpensive breakfast. Great wet burrito.
Taxi right to the back of the building located at the northeast side of the airport.

Hessel, MI (Albert J Lindberg—5Y1)
Hessel Bay Inn 😋😋😋😋

Rest. Phone #: (906) 484-2460
Location: 1.5 miles, call for pickup
Open: Daily: Breakfast, Lunch, and Dinner

PIREP Airport Comments: No fuel. Get it in St. Ignace. No instrument approaches.

PIREP Approximately 20 nm east of St. Ignace/Mackinaw Island. This runway in the woods is approximately 1.5 mi. from the village of Hessel, MI. Hessel is the gateway to Les Cheneaux Islands, the most beautiful area in Michigan.
Nice deck overlooking the marina and Hessel Bay. My family has had a place here since 1901. They say that God's area code is 906. Once you've visited, you'll know why.

Houghton Lake, MI (Roscommon County—HTL)
Spikehorn Bar 😋😋

Rest. Phone #: (989) 366-9698
Location: Short walk
Open: Daily: Lunch and Dinner

PIREP The Spikehorn is a bar and grill within walking distance of the Houghton Lake Airport. They serve a mean burger.

Howell, MI (Livingston County—OZW)
Tomato Brothers 😋😋😋

Rest. Phone #: (517) 546-9221
Location: Short walk
Open: Daily: Lunch and Dinner

PIREP FBO Comments: From the fuel truck, I think there is cheaper self-serve available.

PIREP Excellent, mainly Italian restaurant about one-quarter mile from the airport.

Iron Mountain Kingsford, MI (Ford—IMT)
The Blind Duck Inn 🍔🍔🍔🍔

Rest. Phone #: (906) 774-0037
Location: Very short walk
Open: Daily: Lunch and Dinner

PIREP The Blind Duck Inn is adjacent to the airport, 600 feet across the road with a lake view. It opens at 11 AM Monday through Saturday (CST) and closes at 10 PM. I enjoyed their North Woods atmosphere. The menu offers Mexican, American, and Italian dishes. I like the Italian the best.

I rate it as one of the best in Michigan's U.P.!

Jackson, MI (Jackson County-Reynolds Field—JXN)
The Airport Restaurant & Spirits 🍔🍔🍔

Rest. Phone #: (517) 783-3616
Location: On the field
Open: Daily: Breakfast, Lunch, and Dinner

PIREP Just a great airport joint. Very welcoming and great menu variety. Good burgers, tuna salad, breakfast fare, and so on. I recommend the fish sandwich. Come to think of it, I've never had a bad sandwich or anything else here. Friendly waitresses, reasonable prices. Very clean restrooms.

Lots of planes coming in and out and many friendly pilots to swap lies with. A towered field—most of the time—and the controllers all around the area (Detroit, Toledo, Lansing, Kalamazoo, South Bend) are, from my experience, about the best in the U.S. They are always willing to help and always provide handoffs.

Area to the south of Jackson is called "The Irish Hills." A nice moraine lakes area, very scenic for slow flight. Watch for lazy traffic above the various auto company proving grounds and the Michigan International Speedway—and a couple of monster towers between Jackson and Detroit. Jackson is always worth a stop.

PIREP Four stars, er, burgers. Excellent food great prices. Extremely popular stop. And the gas is reasonable too.

Kalamazoo, MI (Kalamazoo/Battle Creek International—AZO)
The Air Zoo "Kitty Hawk Cafe" 🍔🍔🍔🍔

Rest. Phone #: (866) 524-7966
Location: On the field
Open: Daily: Lunch

PIREP FBO Comments: Duncan Aviation.

PIREP The Air Zoo should be every pilot's destination when they visit Kalamazoo. The Kitty Hawk Cafe is upstairs from the main lobby at the new Air Zoo overlooking the main floor of the museum.

The recently NEW Air Zoo is a $19.50 multimedia and aviation experience. The OLD Air Zoo with its numerous WWII warbirds is still open and is right on the airport; the NEW Air Zoo is a short walk away.

Lewiston, MI (Garland—8M8)

Garland Resort 🍔🍔🍔🍔🍔

Rest. Phone #: (989) 786-2211
Location: Call for pickup
Open: Daily: Breakfast, Lunch, and Dinner

PIREP FBO Comments: No fuel available. Small office in back of pole building has sign-in sheet. No telephone available. Must contact Garland Lodge for shuttle ride on unicom 122.8 or by cell phone at (989) 786-2211. About a ten-minute ride each way. Verizon has coverage in this area, Nextel does not. Garland no longer owns the airport—it was sold to a local company. Airport is fenced to keep deer away. Still plenty of wildlife on and in the vicinity of the runway.

PIREP Absolutely gorgeous facility. AAA four-star rating. Food and service is always excellent. We had lunch in the bar while watching the British Open. I had Hermann's Perch Sandwich with onion rings (highly recommended). The grounds and the four golf courses (soon to be five) are groomed and manicured to perfection.
   Bringing my golf clubs next time.

Ludington, MI (Mason County—LDM)

PM Steamers 🍔🍔

Rest. Phone #: (231) 843-9555
Location: Crew car
Open: Daily: Dinner

PIREP I took a Sunday trip up to Ludington for dinner. PM Steamers is an excellent choice. Be careful, as they don't open until 5 for dinner. Great view of the harbor and the ferry as it arrives around 6 PM. The city park beach is beautiful as well.
   Good service and a nice terminal.

PIREP Try PM Steamers. They are excellent. There is a courtesy car at the LDM terminal or call a cab. Well worth the trip!!! I give the food a five-hamburger rating!! If the courtesy car is gone, cab transportation is easy as well.

Luzerne, MI (Lost Creek—5Y4)

Lost Creek Sky Ranch🍔🍔🍔

Rest. Phone #: (517) 826-9901
Location: On the airport
Open: Daily: Lunch and Dinner

PIREP The Lost Creek Sky Ranch (5Y4) located in Luzerne, Michigan is an excellent choice for just a meal or a weekend destination. The airport is public use with two grass strips 2200' and 2600' in good shape. The flying "R" ranch offers two restaurants, one upstairs serving anything from burgers to steaks, prime rib, BBQ ribs and seafood. The lower restaurant is a pizza joint. I have eaten there on several occasions; the prices are reasonable ($8–20) the food and service has always been very good! They also serve drinks and have a band on the weekends. Overnight accommodations are available at a very reasonable price! Rooms go for $25–30; although

I have not seen the rooms the rest of the place is kept up very well! I believe tent camping is still available. The ranch also caters to the horse crowd and horse rental is available for anyone crazy enough to ride one of those things! From a quick burger to a weekend getaway, it is certainly a great destination. No services are offered so plan your fuel and bring your own tie-downs with stakes.

Mackinac Island, MI (Mackinac Island—MCD)

Pub & Oyster Bar 🍔🍔🍔🍔

Rest. Phone #: (906) 847-9901
Location: Horse-drawn wagon ride to town
Open: Daily: Lunch and Dinner

PIREP   FBO Comments: Very nice, quaint airport with limited services. Tends to get busy in the summer.

PIREP   Went with several friends up to Mackinac Island in mid-July for a late lunch. Everyone had a fantastic time. The only transportation available from the airport is a horse and buggy taxi, and I recommend you call ahead to request one. The ride into town takes approximately 25 or 30 minutes, but it's very scenic and interesting. The food and service at the Pub & Oyster Bar was excellent!

A highly recommended trip!

The Grand Hotel 🍔🍔🍔🍔🍔

Rest. Phone #: (906) 847-3331
Location: Horse-drawn carriage
Open: Daily: Breakfast, Lunch, and Dinner

PIREP   Just returned from three nights at the Grand Hotel on Mackinac Island. A neat place to fly to and relax. This was the fourteenth or fifteenth time we have done this since I started flying.

PIREP   A MUST trip on a midsummer night. The horse-drawn taxi is required for arrival at the hotel. Suit coats are required for dinner. The ambiance is grand dining, complete with dance orchestra.

For night departures, be prepared for immediate flight by instruments as there will be no horizon with liftoff and flight over water.

Manistique, MI (Schoolcraft County—ISQ)

Kewadin Casino 🍔🍔🍔

Rest. Phone #: (906) 341-5510
Location: Short walk
Open: Daily: Breakfast, Lunch, and Dinner

PIREP   Just a short walk to the casino, with a nice restaurant inside at reasonable rates (read: more $ for fuel) and reasonable prices for fuel.

Open 10 to 2 AM.

Marquette, MI (Sawyer International—SAW)

Tailwinds Grill & Bar 🍔🍔🍔

Rest. Phone #: (906) 346-3840
Location: On the field, but it's a mile away
Open: Daily: Lunch and Dinner

PIREP   FBO Comments: Boreal Aviation, fuel after hours by calling a pager number.

PIREP   This place is only about a mile by crew car. They have a full menu with sandwiches, steaks, ribs, and fish, as well as much more.

Kitchen open Sunday through Thursday 11 AM–9 PM, Friday through Saturday 11 AM–10 PM. Extended bar hours every night—oh yeah, there's a full bar. The building used to be the Officers' Club when the base was open, so it looks a little fancy but it's a family restaurant.

There is a hotel right next door called the Red Fox Inn: (906) 346-3355.

Marshall, MI (Brooks Field—RMY)

Schuler's Restaurant 😋😋😋😋😋

Rest. Phone #: (269) 781-0600
Location: Call for pickup
Open: Daily: Lunch and Dinner

PIREP   This is a real delight, located two miles north of Brooks Field in Marshall, MI. Just use the phone ((269) 781-0600) at the friendly FBO, and a courtesy van will pick you up on the spot! Real dining pleasure from a full-featured menu at this world-renowned restaurant.

After dining, take a leisurely stroll through Fountain Park just across the street. For a glimpse of real Victorian architecture, take the Historic Home Tour; you won't be disappointed.

PIREP   My wife and I recently took a flight to Marshall, MI to visit their very historic homes and have lunch at Schuler's Restaurant. The restaurant came highly recommended from friends of ours.

We called the phone number ((269) 781-0600) at the FBO, and within five minutes one of their assistant managers was there to take us the short trip (two miles) to the restaurant.

The food was very good and fresh tasting. We enjoyed everything from their home-style soups and great main dishes to the delicious desserts.

The staff was very friendly and, when we were finished, they even told us to walk around the small town and visit their many local sites (they have town maps at the front desk). They offered to drive us back to the airport whenever we got back.

I very highly recommend flying into Marshall, MI and eating at Schuler's Restaurant (and walking around this historic little town).

Mecosta, MI (Mecosta Morton—27C)

Country Lake Inn 😋😋😋😋

Rest. Phone #: (231) 972-3165
Location: Across the street
Open: Daily: Lunch

PIREP   The Country Lake Inn is just across the street from the tie-downs of this short, 2010' turf strip. The food is good, featuring big hamburgers and omelets worth checking out. The prices are better than fair. I recommend it.

Midland, MI (Jack Barstow—3BS)

Shirlene's Cuisine 😋😋😋😋

Rest. Phone #: (989) 631-8750
Location: Call for pickup, it's very close
Open: Daily: Lunch and Dinner

PIREP I recommend Shirlene's Cuisine, adjacent to the Midland Barstow (3BS) airport. Fly in and give them a call from the terminal and they will send a car. They have a varied menu and great salad bar! Meals are reasonably priced and pilots made to feel welcome.

Monroe, MI (Custer—TTF)
Cabela's - Tamarack Cafe 🍔🍔🍔🍔

Rest. Phone #: (734) 384-9616
Location: Call for pickup
Open: Daily: Lunch

PIREP The restaurant, Tamarack Cafe, is located on the premises of Cabela's giant sporting goods store.
Plan to visit Cabela's massive and spectacular new omniplex dedicated to wildlife conservation and outdoor sportsmen and women worldwide! 225,000 square feet of the World's Foremost Outfitter's quality merchandise. If you are an outdoorsman, expect your lunch to include some equipment purchases.
Monroe Aviation-FBO has a courtesy car that you may use to get to and from Cabela's and the Tamarack Cafe.

Mount Pleasant, MI (Mount Pleasant Muni—MOP)
The Embers 🍔🍔🍔

Rest. Phone #: (989) 773-5007
Location: About 1 mile
Open: Daily: Lunch and Dinner

PIREP The Embers, "Home of the One-Pound Pork Chop," is located about a mile from the airport at 1217 S. Mission Street. On a hot day, call a cab, (989) 772-9441. It's definitely worth the trip!!!

PIREP Regarding the Embers Restaurant in Mount Pleasant, MI (MOP): you need not call a cab, the restaurant will pick you up. I've been going there for years and consider it a place to take that "special person." A bit pricey; my only advice it to call ahead to confirm their serving hours.

Muskegon, MI (Muskegon County—MKG)
The Brownstone Restaurant 🍔🍔🍔🍔

Rest. Phone #: (231) 798-2273
Location: In the main terminal
Open: Daily: Lunch and Dinner

PIREP The Brownstone is clean and friendly, and the food is delicious. They offer an awesome variety of unique homemade soups. Also sandwiches/baguettes, appetizers, pastas, salads, pizza, and entrees. MKG itself is very parklike—very well landscaped, and the architecture is wonderful. The Brownstone does not seem to have their own website! But I think they are open everyday and well worth the effort to visit. I don't think you need to dress up—we sure didn't!

PIREP FBO Comments: Executive Air, fast and efficient fuel service. Nice pilot briefing facility in the building, great large screen TV, and a couch you can fall asleep on if you're not careful. Fuel discount if you pay cash. Fairly decent chart availability if you need them, but be advised the Michigan ones as usual are not free as in most states. Fairly decent fleet of rental airplanes on the line, inclusive of an Arrow IV (N604EA).

PIREP The Brownstone in the main terminal has excellent food. The soup of the day, whatever it may be, is typically more than a meal in itself. If you don't try the veggie wrap you are missing out—it's quite good! Excellent waitress staff, very polite and friendly. Nice respite from a trip either way across that lake, for sure. Apch./Twr. Controllers at Muskegon are top-notch and are very accommodating and professional. Makes it a breeze to fly in there and nab a meal on the way home wherever your final destination.

Napoleon, MI (Napoleon—3NP)

Napoleon Cafe 🍔🍔

Rest. Phone #: (517) 536-4244
Location: Short walk
Open: Daily: Breakfast and Lunch

PIREP Just reviewing the site with my CFI/cargo pilot pal Skip—when he saw the two-burger rating for the Napoleon Cafe, he was highly offended! He says it is "a great greasy spoon—at least three, if not four, burgers!"
Skip says go for it—you won't be disappointed.

PIREP Lunch was average. Kids received hats to color and wear. The cafe was just a ten-minute walk from the airport. Service was average. The Napoleon Burger was huge. Four hamburger patties, American and Swiss cheese, two layers of bacon strips, and chips. A kids menu was also available.

Newberry, MI (Luce County—ERY)

JJ's 🍔🍔🍔

Rest. Phone #: (906) 293-8281
Location: 1/2-mile walk
Open: Daily: Breakfast and Lunch

PIREP JJ's has a sign above the door that states, "Good Food." No argument here, it's exactly what you're looking for when you think $100 hamburger. It's only about a half-mile from the airport to the restaurant, but if the weather is not ideal (this is U.P. Michigan, after all), Wayne will probably give you a lift.

New Hudson, MI (Oakland Southwest—Y47)

New Hudson Inn 🍔🍔

Rest. Phone #: (248) 437-6383
Location: Short walk
Open: Daily: Lunch and Dinner

PIREP The New Hudson Inn has great burgers and a great atmosphere. It's the king of dive bars, but since you're going to be flying home, you'll be drinking iced tea, right? They've got great burgers, bring you all the fixin's so you can do it up your own way, and they've got a pool table in the back. It's across the street from Putters. If I'm working on my plane on Saturday and want lunch without getting sideways stares from the elderly couples because of the oil on my shirt or the grease under my nails, I'll go to the Inn. So do most of the other locals based at New Hudson International. Meet me there and I'll play you a game of pool.

PIREP The airport is 3000', has a VASI at both ends, and is lighted dusk to dawn. 07/25 runway has a taxiway, and in the middle of this is a taxiway that leads to a generous-sized ramp that is usually crowded. The airport is served by an FBO and has a self-serve fueling station with a credit card reader (all major types accepted).

If you arrive at night, be sure to follow the VASI as there are trees at both ends and a wire on 25 approach. The nonprecision VOR-A approach is very accurate as it is only 4.5 miles to the Salem VOR. On the approach you will see the hangars first as there are many white buildings. The controllers usually bring you on to the VOR very tight in, so be on your toes.

You'll love the Inn!

Leo's Coney Island ☕☕☕

Rest. Phone #: (248) 446-1008
Location: Short walk
Open: Daily: Breakfast, Lunch and Dinner

PIREP Leo's is a typical Michigan Coney Island restaurant. They serve great breakfast for a reasonable price. The restaurant is a short walk from the airport—ask at the FBO for directions.

Owosso, MI (Owosso Community—RNP)

Bob Evans ☕☕☕

Rest. Phone #: (989) 723-9770
Location: On the field
Open: Daily: Breakfast, Lunch, and Dinner

PIREP If you park on the grass at the west end of runway 10/28 there is a path leading north that takes you almost to the back of the Bob Evans—take a short detour through the bus lot next door to get there. Alternatively, the airport has loaner bicycles and you can ride there.

Bob Evans is a typical chain diner-style restaurant. Food is good and predictable but not exotic—prices are reasonable.

PIREP Several restaurants in the area are within walking distance, including Arby's, Burger King, Big Boy, Ponderosa, Bob Evans, Pizza Hut, Golden Coral, Subway. Besides these, we have BJ's home-cooked food, and in the winter months the local pilots' association puts on a breakfast every third Sunday starting at 9 AM, including eggs, sausage, bacon, toast, hash browns, orange juice, and coffee, and you may also find homemade breakfast items such as biscuits and gravy or a cinnamon roll or two.

Crosswind Cafe ☕☕

Rest. Phone #: (989) 725-1969
Location: On the field
Open: Saturday and Sunday, Breakfast and Lunch

PIREP We went to Crosswinds for breakfast yesterday—it is great! Home-style, serve yourself coffee, and a wonderful pricing system: "The menu is on the board—you can have anything you want, it is $5."

Good food cooked to order, and amazingly good sticky buns seem to go with everything!

A great, fun breakfast destination—we will be back!

PIREP We serve breakfast and lunch on Saturdays and Sundays from 8:30 AM to 2 PM, for now. We hope to increase our hours and days in the future to offer outside patio seating in the summer. It is the old terminal building at the airport, so it is not much of a walk.

It is not about fancy or anything, but about food, fun, and friends. Customers can heckle the help, and sometimes, depending on the attitude, the help heckles the customers.

We do have two resident musicians who provide music for us, and singing along is encouraged.

PIREP   The Owosso Airport Association has now opened the Crosswinds Cafe in the old terminal building. Locals say, "It is a cafe with an attitude." The building was remodeled by volunteers and is a great gathering spot, with an impressive wall-sized C-172 mural painted by a local high school student. We've enjoyed many full breakfasts over the past few months. Hours may expand but currently are 8:30 AM–11:30 AM Saturdays and Sundays. Come before 3:30 and they may have lunch! Proceeds support area aviation education for teachers and students. It's worth a stop! And they say RNP stands for "real nice people." We've found it so.

Pellston, MI (Pellston Regional Airport of Emmet County—PLN)

Brass Rail 😋😋

Rest. Phone #: (231) 539-8212
Location: Close by
Open: Daily: Lunch and Dinner

PIREP   Fly in to Pellston and visit the Brass Rail—great food, and Friday's all-you-can-eat fish dinners. Check it out, instrument approach, fuel all the good stuff.

Plainwell, MI (Plainwell Municipal—61D)

Fly Inn Again 😋😋😋

Rest. Phone #: (269) 685-1554
Location: On the field
Open: Daily: Breakfast and Lunch; closed Sundays

PIREP   Fly Inn is my all-time favorite place to fly in to eat. The restaurant is right on the ramp, prices are very reasonable, and the food is excellent—especially breakfast. Unbelievable omelets. The service is fast and friendly, and they even serve Diet Mountain Dew, which is very rare.

The dining room is very small, but the clientele is friendly—you can share a table and meet other pilots and their friends.

They are closed Sundays and mostly open for breakfast and lunch with some hours on Friday and Saturday night, I think—call (269) 685-1554 for current hours.

PIREP   FBO Comments: 100LL and mechanic on field

PIREP   Great burgers (get a 6-ounce), as well as a large menu selection. Great service and prices too. Will be back again and again.

Plymouth, MI (Canton-Plymouth-Mettetal—1D2)

Canton Coney Island 😋😋

Rest. Phone #: (734) 414-0890
Location: On the field
Open: Daily: Lunch

PIREP   The Coney has reopened and is serving good Greek food. And it is about 300 feet from the FBO.

Rothbury, MI (Double JJ Resort Ranch—42N)
Sundance Bar and Steakhouse 😋😋

Rest. Phone #: (616) 894-4444
Location: On the field
Open: Daily: Lunch and Dinner

PIREP We found an outstanding place in Rothbury, MI called the Double JJ Resort Ranch. It is a 1000-acre resort that has a 3600' private grass airstrip.

If you call ahead, a shuttle will pick you up and take you a mile down the road to their restaurant called the Sundance Bar and Steakhouse. The food was wonderful.

We had 9 planes and 22 people. They were very friendly and took us on a tour of the grounds. The service was great also. The man who gave us a ride back told us they close the strip for the winter in November.

Saginaw, MI (Mbs Intl—MBS)
MBS Grill 😋

Rest. Phone #: (989) 695-5555
Location: On the field
Open: Daily: Lunch

PIREP The MBS Grill is inside the terminal and the food is pretty good; however, the price is pretty high. If you can get a ride into town, Freeland has a good, friendly staffed, reasonably priced restaurant called Antonio's.

St Ignace, MI (Mackinac County—83D)
North Bay Inn 😋😋😋

Rest. Phone #: (906) 643-8304
Location: On the field
Open: Daily: Lunch and Dinner

PIREP The North Bay Inn offers family dining, buffet style or off the menu. SE corner of airport, aircraft parking on slope off runway.

Three Rivers, MI (Three Rivers Muni Dr Haines—HAI)
Fisher Lake Inn 😋😋

Rest. Phone #: (269) 279-7984
Location: 2.5 miles, call a cab
Open: Daily: Lunch and Dinner

PIREP Their sandwiches are huge!! Best Reuben I've ever had.

PIREP FBO Comments: Very friendly and helpful FBO. No courtesy car, but there is a great new cab service in the area. Shorty's Cab Service is located right in Three Rivers at (269) 273-7333.

PIREP Excellent restaurant right on Fisher Lake, with nice sunset views and great food. Lots of seafood, steaks, prime rib, and a soup and salad bar. If you can afford it after topping off the plane, try their surf and turf with the large lobster tail (you can also order it with a smaller tail). Great steak and not only one of the best tasting lobster tails, but certainly the largest I have ever eaten. The rest of the menu is priced reasonably at about what you would expect for the area and this type of restaurant.

The decor and ambiance is that of a lake lodge that has been around a while but is kept up and clean. They also have an outdoor deck right on the water with their own dock for boats to tie up to. They have a big fireplace inside when it is cold out. I have eaten there several times when in the area, and so far we have had excellent meals. I would recommend it highly.

Nice location on the lake, but since the restaurant is 2.5 miles down the road from the airport, you need to hike it in or call Shorty's Cab Service. Definitely worth the short cab ride.

Traverse City, MI (Cherry Capital—TVC)
The Grand Traverse Resort 🍔🍔🍔🍔🍔

Rest. Phone #: (616) 938-2100
Location: Call for pickup
Open: Daily: Breakfast, Lunch, and Dinner

PIREP Though it is not located on the field, the Grand Traverse Resort will dispatch a car to meet you at the General Aviation Terminal. Call a few hours in advance to make arrangements with the resort. After landing, ask the FBO to call Grand Traverse. By the time you're tied down and have your fuel order placed, the van should be there.

The Trillium Restaurant is on the top floor and has a terrific view of Grand Traverse Bay. We recommend going on a night when the full moon rises about sunset. A table at the west window will give you the treat of a spectacular moonrise while the sun sets.

Mabel's Restaurant 🍔🍔

Rest. Phone #: (231) 947-0252
Location: 1/2 mile
Open: Daily: Breakfast, Lunch, and Dinner

PIREP FBO Comments: Price is above average for the area. Harbor Air's service was outstanding! Prompt and friendly attention, and they unhesitatingly offered us a courtesy car to get to the restaurant.

PIREP The bad news: The restaurant in the TVC terminal is closed permanently.

The good news: There are a couple of decent and inexpensive restaurants within a half-mile drive. We had lunch at Scheile's. Excellent service and food (try the Leelenau chicken sandwich) and not too crowded on a Sunday afternoon.

The other place is Mabel's, just a couple of blocks north of Scheile's. Mabel's was the unanimous choice of the FBO staff, but there was a 30-minute wait for a table, and this isn't a small restaurant. The food looked and smelled good though.

White Cloud, MI (White Cloud—42C)
Charlie's Pub 🍔🍔🍔

Rest. Phone #: (616) 689-6143
Location: Close by
Open: Daily: Lunch and Dinner

PIREP We received *The $100 Hamburger* travel guide as a gift from our daughters and use it often. We just had a great burger at a little town pub while flying on Sunday. The airport was White Cloud Airport (42C).

Charlie's Pub in downtown White Cloud, MI was open on Sunday during game time, so I think they are always open. We had HUGE burgers and they were under $4. The atmosphere was great, small town with a friendly staff.

Charlie's Pub is an easy short walk to downtown (1/4 mile at most).

Sally's 🍔🍔🍔

Rest. Phone #: (231) 689-6560
Location: Close by
Open: Daily: Breakfast, Lunch, and Dinner

PIREP Sally's Restaurant is just a quarter-mile downhill from the White Cloud, MI (42C) airport. They are open seven days a week for breakfast, lunch, and dinner. On weekends they have a breakfast buffet. Their cinnamon rolls are huge—a meal by themselves.

PIREP All was as previously described: nice airport, friendly folks in the office/pilot lounge, short walk to Sally's, and good down-home food. A very nice breakfast destination!

Minnesota

Alexandria, MN (Chandler Field—AXN)
Arrowwood 🍔🍔🍔🍔🍔

Rest. Phone #: (320) 762-1124
Location: Call for pickup, 4 miles away
Open: Daily: Breakfast, Lunch, and Dinner

PIREP Alexandria, MN (AXN) is located in the far west central part of Minnesota. It is the "jumping off" spot for much of the Minnesota Lake country. Named one of America's top family resorts, Arrowwood, a Radisson Resort, has 450 acres on Lake Darling with horseback riding, indoor and outdoor tennis, golf, marina, indoor and outdoor swimming pools, sauna, cross-country skiing, snowmobiling, skating, and over 15,000 square feet of meeting space.

Just four miles from the airport, they have a beautiful dinning room and deck that looks out over the lake. They will pick you up and drop you off at the airport. Arrangements can be made by calling (320) 762-1124.

Alexandria is also the manufacturing site for Bellanca Aircraft, who are on the airport premises.

Appleton, MN (Appleton Muni—AQP)
Shooters Bar & Grill 🍔🍔

Rest. Phone #: (320) 289-1100
Location: 1-mile walk or crew car
Open: Daily: Lunch and Dinner

PIREP Appleton is home to Shooters Bar & Grill. It is a reasonable distance from the tie-downs, only one mile west of airport. They serve a good burger basket and more. Call ahead to verify availability of transportation ((320) 289-1100) if you're not up for the hike!

Backus, MN (Backus Muni—7Y3)
The Corner Store 🍔🍔

Rest. Phone #: (218) 947-4115
Location: Across the street
Open: Daily

PIREP   The Corner Store has good food and is right across the highway from the airport.

Bemidji, MN (Bemidji Regional—BJI)
Gangelhoff's Restaurant & Lounge 🍔🍔🍔

Rest. Phone #: (218) 444-9500
Location: Short walk
Open: Daily: Dinner

PIREP   On the way up to Canada from the Twin Cities this last summer, my friend and I, both pilots, stopped for fuel for both ourselves and the plane in Bemidji, MN. It is a quite large airport with both a GPS/NDB, and an ILS approach. There is one large FBO on the field. To our delight, the Northern Inn Hotel is only about one to two blocks away from the airport. The Gangelhoff Restaurant and Lounge is in the hotel. Their food is terrific. They have a full menu ranging from shrimp cocktail for starters to N.Y. strip steak for your main course. You can also get a array of different burgers. You can expect to pay between $10–20 per person, but it's good food and it's only a short walk from the airport.

Brainerd, MN (Brainerd Lakes Regional—BRD)
The Brainerd Cafe 🍔🍔🍔🍔

Rest. Phone #: (218) 829-3398
Location: On the field
Open: Daily: Breakfast, Lunch, and Dinner

PIREP   The airport cafe has been completely remodeled and the menu is new. I highly recommend the bacon cheeseburger or the champ tuna fish sandwich. If you are on floats, there is Maddens on Gull Lake that has a small seaplane base. Brainerd's Cafe has a burger that is hard to beat. I give it two thumbs up.

PIREP   While doing my instrument training earlier this year, I came across this nice local friendly eatery. While the food/ambiance/service will never be confused with four-star restaurants, I had a nice fresh burger, served by a friendly waitress in the FBO building, providing a view of the goings on at the airport. Hangar flying was, of course, rampant at the tables.

Duluth, MN (Duluth Intl—DLH)
The Afterburner 🍔🍔🍔

Rest. Phone #: (218) 727-1152
Location: On the field
Open: Daily: Breakfast, Lunch, and Dinner

PIREP   At Duluth International there is the Afterburner; it's in the main terminal. I've heard that the food is excellent and at a reasonable price, but it doesn't open until noon. For the Dawn Patrol, the Coffee Shop is right next door, and despite its name is really a complete cafe. I had breakfast there, and the food and service were both good.

You can get a ride to the terminal from North Country Aviation.

PIREP   FBO gives rides to terminal building for fancier grub at good prices. Nice place for dinner.

Duluth, MN (Sky Harbor—DYT)
Grandma's Saloon and Deli 😋😋

Rest. Phone #: (218) 722-9313
Location: In town
Open: Daily: Breakfast, Lunch, and Dinner

PIREP   Sky Harbor is located on a narrow sand peninsula in the southwest end of Lake Superior, where the St. Louis River meets the lake. This destination is beautiful and relaxing in the warmer months of summer. The airport features both a paved runway and a seaplane area in the bay. Either way, taxi to the dock or FBO, walk behind the FBO and over the sand dune, and there before you stretches the beautiful lake and miles of clean sand. Though the water remains a chilly 50 degrees even in summer, hearty souls and some of the locals venture in on hot days. The airport is a about a three-mile cab ride from the Duluth, MN harbor, which is a seaport for ocean-going vessels and also features shopping and restaurants, including Grandma's Saloon and Deli. Great hamburgers and many other tempting items. A couple more miles across the lift bridge and you're in downtown. From Duluth north, the landscape rises from the lake and is very scenic. A great day trip or weekend visit.

PIREP   Duluth Sky Harbor is still there and still a fun destination. There is usually a courtesy car available to tool out for a burger. I disagree with the previous poster—Grandma's is not a very good place for food (but it is a good place to sit on the deck and watch ships). There are a bunch of other restaurants in the area, too—mostly pretty corporate. Now back in 1976 when it opened, Grandma's was terrific (and you could probably rent a 150 for $10/hour).

The two stars rating is probably pretty accurate.

If you want edible chow without driving, see the listing for Superior Wisconsin, which is about four miles away.

East Gull Lake, MN (East Gull Lake—9Y2)
Madden's on Gull Lake 😋😋😋😋

Rest. Phone #: (218) 829-2811
Location: On the airport
Open: Daily: Breakfast, Lunch, and Dinner

PIREP   Madden's Resort is one of the best aviation destinations in the country. This is a very large first class resort that caters to the pilot and aircraft owner.

This resort is a golfer's dream come true: four excellent courses, including the 33rd top-rated course (The Classic) in the United States. Their culinary offerings are excellent; this is a resort destination for the whole family. They just completed the new spa, and there's definitely something for every family member to do and enjoy at this resort. They even offer flight training packages for seaplane ratings and tail-wheel endorsements!

PIREP   FBO Comments: Fuel is available at Brainerd Airport, a ten-minute flight away.

PIREP   A great place to fly into for the weekend or for lunch is Madden's on Gull Lake. There is plenty of dock space for floats as well. There are 63 holes of golf. The Classic is rated the 33rd best public golf course in the U.S. by *Golf Digest*.

East Gull Lake Airport is an excellent grass strip and is a blast.

Eveleth, MN (Eveleth-Virginia Muni—EVM)

K+B Drive In 🍔🍔

Rest. Phone #: (218) 744-2772
Location: Crew car
Open: Daily: Lunch and Dinner

PIREP   FBO Comments: Very friendly FBO; they have a courtesy car available. They can also give you a nice little map of the town, including how to get to K+B Drive-in (approximately 2.5 miles away).

PIREP   Heard about K+B Drive-in from another pilot and decided to go check it out. Wow, great service, good food, and a relaxing atmosphere. You can either sit outside, or you can sit in your car and order the food to take back to the very comfy couches at the FBO. Certainly worth the visit if you're up in that neck of the woods.

Faribault, MN (Faribault Muni—FBL)

The Depot Bar and Grill 🍔🍔

Rest. Phone #: (507) 332-2825
Location: Crew car
Open: Daily: Lunch and Dinner

PIREP   The food at the Depot Bar and Grill is really good. It's quite a trek across town (take the free courtesy car), but it's a great place for eating. It was built in an old railroad depot (imagine that) and has railroad decor all over the place. It's got great outdoor seating in summer, and if a train goes by when you're there you get a free drink.

Grand Rapids, MN (Grand Rapids/ Itasca County-Gordon Newstrom Field—GPZ)

The Blue Loon Cafe 🍔🍔🍔

Rest. Phone #: (218) 326-1226
Location: On the field
Open: Weekdays: Breakfast, Lunch, and Dinner

PIREP   The Blue Loon Cafe is on the field. I was there on a Saturday in January and it was closed, so it may only be open weekdays and/or during the summer.

Hinckley, MN (Field of Dreams—04W)

Grand Casino, Hinckley 🍔🍔🍔🍔

Rest. Phone #: (320) 384-7777
Location: 1 mile N of airfield
Open: 24 hours

PIREP   Hard to believe this place hasn't been added yet. Field of Dreams is a (virtually) brand new airport located just outside of Hinckley, MN and approximately one mile north of Grand Casino Hinckley. The casino will provide shuttle service to and from the airport. Contact them by phone from the phone box located just outside the door of the FBO. The numbers for the casino, a limo service, and several other contact numbers are on the inside cover of the phone box. There's an FBO on the field.

The airport is privately owned but open to the public. It's on both the Green Bay and the Minneapolis sectionals. Airport ID 04W. Approx 2700 × 75-foot asphalt runway. Nice facility. Self-service gas at a good price. There's a $10 parking fee that's waived if you buy 50 bucks worth of gas or more.

Grand Casino is a 24-hour operation with several restaurants and a 300-plus room hotel. They have regularly scheduled big-name entertainment. Numerous other businesses and attractions in the area as well.

International Falls, MN (Falls Intl—INL)
Chocolate Moose 🍔🍔🍔

Rest. Phone #: (218) 283-8888
Location: 1/2-mile walk or use crew car
Open: Daily: Lunch and Dinner

PIREP The Falls International Airport (INL) in Minnesota is always fun to fly into—it's about as far north as you can get before you're in Canada. The weather tends to be a bit difficult, and I think the Falls served as the model for Rocky and Bullwinkle's Frostbite Falls. Whenever I've wanted to fly there, conditions tended to be MVFR, as it was yesterday. But I got in, and it was worth it. Stopped at the Emerson Brothers FBO for refueling. Offered a courtesy car and a map of the area, I went to the Chocolate Moose to have supper (three burgers).

Even without the courtesy car, the restaurant is within walking distance (about a half mile), as are motels. The Canadian border is not too far either—just cross the toll bridge over the Rainy River to visit Fort Frances. More shopping and restaurants both sides of the border.

PIREP I took a student there on a long cross country. We took the courtesy car and ate at the Chocolate Moose. I had a walleye and steak dinner. It was very good. The people at the FBO were very helpful, and you get a free jar of jam if you buy fuel.

Litchfield, MN (Litchfield Muni—LJF)
Peter's On Lake Ripley 🍔🍔🍔🍔🍔

Rest. Phone #: (320) 693-6425
Location: Short walk
Open: Daily: Lunch and Dinner

PIREP FBO Comments: Great pilot hangout, clean comfortable, 32" TV with VCR.

PIREP The food is great, try the buffet; besides the ribs, there is chicken, shrimp, fish, and so on, all wonderful.
The courtesy car at the airport was already at Peter's, so we called and Peter himself came out to get us.
After we finished with dinner, he drove us back to the airport. We chatted airplanes and things. He flies a Pacer, if I remember correctly.
All in all, absolutely outstanding.
My rating: 8

PIREP A friend and I flew in to LJF to play golf and visit Peter's restaurant. The Litchfield Golf Club is a nice little municipal course, in good condition, with friendly people in the pro shop.
Peter's restaurant was everything we had hoped for: nice atmosphere, friendly staff, and excellent food. The ribs were great—better than Famous Dave's (which aren't bad, either). A half rack, plus potato and salad bar, made for just the right size dinner. I heartily recommend Peter's on Lake Ripley. It's worth the trip.

PIREP Peter's may have great ribs, but they also have excellent prime rib. They have a dinner buffet and salad bar that is extensive, and the prices are very reasonable.

Peter's is located on Lake Ripley and it also serves as the golf course club house.

Peter's is a one-mile walk from the airport, or there is a former Litchfield cop car at the airport with the keys stored in the pilot lounge.

PIREP This is truly a pilot-to-pilot e-mail, as I have been in the cockpit since 1970. I hold a private ticket with an inst. rating and am currently enrolled in aerobatic training. I am a member of EAA, The Short Wing Club, and Aopa. My ride is a 1958 fully restored Pacer 160. My base is on 16 acres adjacent to LJF.

My restaurant is one mile from the field. There is a courtesy car, and we offer a free ride to anyone who calls ahead. We offer a complete menu from burgers to steak and seafood. We are renowned for our barbecued ribs. We also ship these ribs all over the country. Hours are Mon.–Fri. lunch from 11 AM–1:30 PM. Evenings from 5–10 PM Mon.–Sat. We also have a Sunday brunch next to none: 28 feet of food.

We are located on a 500-acre lake with floatplane parking within a block. We are also set up to land helicopters. Here is something of interest for you: we have 18 holes of great golf and boast the lowest green fees in the state.

PIREP The Sunday brunch has been discontinued in an effort to free up some time for Peter and Luann. Thankfully, light summer luncheons are still available. Peter's has become a regular stop for many, including choppers that land on the front yard. Patrons get a real kick out of that!

This really is THE place for ribs! I know that it would be a long flight from California in a 172. Peter has solved that problem by offering mail-order ribs! I didn't believe that ribs from a UPS truck could be any good. Peter sent me some. Awesome—this mail order rib idea works. Get some TODAY. Suffer no longer!

ORDER BIG! http://www.petersribs.com.

Longville, MN (Longville Muni—XVG)
Patrick's Fine Dining & Lounge 🍔🍔🍔🍔

Rest. Phone #: (218) 363-2995
Location: Short walk
Open: Daily: Lunch and Dinner

PIREP A short walk in from the Longville, MN airport. A resort/dinner club kind of place—high ceilings—very nice.

Maple Lake, MN (Maple Lake Muni—MGG)
Maple Lake Cafe 🍔🍔🍔

Rest. Phone #: (320) 963-3907
Location: 3/4 mile
Open: Daily: Breakfast, Lunch, and Dinner

PIREP Maple Lake Cafe is three-quarters of a mile from airport—good food and a nice hometown setting. Very reasonable pricing: $5 dinner. Special transportation can usually be arranged onsite (read: bum a ride!).

Minneapolis, MN (Flying Cloud—FCM)
The Lion's Tap 🍔🍔

Rest. Phone #: (952) 934-5299
Location: Close by
Open: Daily: Lunch and Dinner

PIREP   Just five minutes south of Flying Cloud Airport in Eden Prairie, MN is a wonderful burger place called The Lion's Tap, where you will experience great service, fun atmosphere, and most importantly, excellent burgers!! They have a limited menu of entirely burgers, but they make one heck of a burger!!!! Please take the time to drop into Flying Cloud (FCM) and ask anyone there how to get to Lion's Tap—trust me, they'll know!

Motley, MN (Morey's—22Y)
The Countryside Cafe ⊜⊜⊜

Rest. Phone #: (218) 352-6777
Location: Very short walk
Open: Weekdays: Dinner; Weekends: Lunch and Dinner

PIREP   There is an excellent restaurant, The Countryside Cafe, just up the road toward town about three blocks walking distance. It is open evenings and weekends. On the way back to your plane, you can stop and pick up some delicious smoked salmon and so on at the famous Morey's Fish, right on the runway's edge.

Olivia, MN (Olivia Regional—OVL)
Sheep Shedde ⊜⊜⊜

Rest. Phone #: (320) 523-5000
Location: 2 blocks
Open: Daily: Lunch and Dinner; Sunday: Brunch

PIREP   The Sheep Shedde is an Old English restaurant about two blocks from the airport. Good lunch buffet on weekdays, Sunday brunch from 10 AM to 2 PM. Walk or call (320) 523-5000 for a ride. Shuttle to Firefly Casino.

The Chatterbox Cafe ⊜⊜⊜

Rest. Phone #: (320) 523-5384
Location: 1 block
Open: Daily: Breakfast, Lunch, and Dinner

PIREP   The Chatterbox Cafe, located between the airport and the Sheep Shedde, is a better value with more local flavor. Good food, friendly staff, and a partial view of the runway. It's hidden in the Ashland gas station building.
   One of the more frequent fly-outs for the Faribault (MN) Area Pilots' Association.

Ortonville, MN (Ortonville Muni-Martinson Field—VVV)
The Matador ⊜⊜⊜⊜

Rest. Phone #: (320) 839-9981
Location: On the field
Open: Daily: Lunch and Dinner

PIREP   The Matador is one of our favorites. You can park by the FBO or at the west end of the grass runway and walk less than 50 yards and be in their back parking lot. It's the big brown building. Besides being a nice supper club, they offer a full salad bar with soup. They are also open Sunday evenings.
   Ortonville Airport also offers self-service fuel with a credit card at a very reasonable price.

Owatonna, MN (Owatonna Degner Regional—OWA)
Cabela's - Northwood's Cache 🍔🍔🍔🍔

Rest. Phone #: (507) 451-4545
Location: Call for free pickup
Open: Daily: Lunch

PIREP   FBO Comments: Rare Aircraft Ltd. (507) 451-6611.

PIREP   Landed here with my CFI, practicing some icy field landings, taxiing, and take-offs. We parked, went inside, and used a courtesy phone to call Cabela's, who sent a shuttle van over for the five-minute ride to the store. Cabela's is known for their quality products and amazing displays in their huge retail stores; this one is no exception. Their cafeteria serves a wide variety of good food and some unique items like buffalo burgers and venison brats. FBO service was prompt and friendly, too. Great getaway stop just 50 miles south of the Twin Cities along I-35.

PIREP   The restaurant, Northwood's Cache, is located on the premises of Cabela's giant sporting goods store.
   If you are an outdoorsman, expect your lunch to include some equipment purchases.
   The airport provides a shuttle to and from Cabela's and the Northwood Cache. Call Dave Beaver, the airport manager, in advance of your arrival to make arrangements. Airport phone: (507) 451-6611.

Princeton, MN (Princeton Muni—PNM)
Merlin's Family Restaurant 🍔🍔🍔

Rest. Phone #: (763) 389-5170
Location: Short walk
Open: Daily: Lunch and Dinner

PIREP   Merlin's Family Restaurant is about two blocks further north and east of the Pine Loft. Merlin's is open for breakfast and has great pancakes.
   Taxi to the northeast corner of the airport. There are several hangars east of the taxiway. I usually park my plane in the grass just west of the taxiway near the white wooden fence. You can walk north through the fence to the Pine Loft and Merlin's on the gravel road that runs north.

Pine Loft Restaurant 🍔🍔🍔

Rest. Phone #: (763) 389-4762
Location: Very short walk
Open: Daily: Lunch and Dinner

PIREP   The Pine Loft Restaurant is about a quarter-mile from the north end of the old runway; open from 11 AM every day. Other restaurants close by.
   Tour FSS facility on airport. They're always glad to see you (not too late—keeps 'em awake).

Red Wing, MN (Red Wing Regional—RGK)
Lavender Rose 🍔🍔🍔🍔

Rest. Phone #: (715) 792-2464
Location: Across the street
Open: Thurs.–Sun.: Dinner

PIREP You can walk across the field to the Lavender Rose (open Thurs.–Sun.). They have a somewhat limited (read: "trendy") menu and have food to die for. Prices are quite reasonable. Their desserts are excellent.

St. James Hotel 🍔🍔🍔🍔

Rest. Phone #: (651) 227-1800
Location: Crew car or call for pickup
Open: Daily: Breakfast, Lunch, and Dinner; Sunday: Brunch

PIREP Fly into Red Wing, MN (actually in WI), airport designation (RGK). Sunday brunch at the historic St. James Hotel is a treat. Overlooks the Mississippi River and an old 1880s hotel. Prices are reasonable.
Red Wing provides a courtesy car or St. James will pick you up. It's four–five miles to town.

Redwood Falls, MN (Redwood Falls Muni—RWF)
Jackpot Junction Casino 🍔🍔🍔

Rest. Phone #: (507) 644-3000
Location: Call for pickup
Open: Daily: Breakfast, Lunch, and Dinner

PIREP Redwood Falls has an excellent fly-in place to eat. And you can have a little fun, too! Jackpot Junction Casino will zip right over to the airport and pick you up in their courtesy van. It's only a four-minute ride to the casino. They have an excellent Sunday brunch, daily buffets for breakfast, lunch, and dinner. They also have the Dakota Restaurant. When you get there, the hospitality desk will give you $5 in free silver. The food is excellent, and it's an enjoyable flight to the Minnesota prairie country.

Rochester, MN (Rochester International—RST)
Hangar Bar & Grille 🍔🍔🍔🍔

Rest. Phone #: (507) 288-8444
Location: 3/4-mile walk or transportation provided
Open: Daily: Lunch and Dinner; Sunday: Brunch

PIREP I also have been to the Hangar. Great food, great prices, free transportation, and a pilot discount. It's a nice clean place with aviation decor all around. The people in the control tower are great folks. When I was a student pilot, they had patience to deal with me, and I am happy to say they are some of the best tower controllers in Minnesota. This restaurant is awesome.

PIREP There is a wide variety of food and drink from a creative owner and chef. Since its location is in Olmsted County, it is entirely smoke-free. However, it has a very nice outdoor patio facing the airport for smokers and non-smokers alike. A very nice brunch is available on Sunday mornings from 9 AM to noon, for a change of pace from the traditional pancake breakfast. The ambiance is a perfect fit for pilots and aviation enthusiasts, as it is an aviation theme. The personal plane of a local aviation celebrity hangs from the ceiling, and the history of the individual and his accomplishments is mounted on the wall at the entrance. Pilots get a 10 percent discount. Transportation is provided by a crew car and Rochester Aviation employees. This FBO just opened in a brand new building that contains an excellent pilot's lounge, showers, pool table, planning and briefing rooms, and separate vending area. If they are busy, you may have to wait a bit for your ride. However, they are an extremely friendly and enthusiastic bunch and will make every attempt to accommodate you. If you are hardy, the restaurant is walkable; I'd guess about three-quarters of a mile from the general aviation ramp. For those of you who do not normally fly towered airports, do not be intimidated. It is one of the most laid-back and friendly towers I have flown. Simply listen to ATIS, then call the tower from 10 nm out with your location and intent to land. They almost always give you the shortest pattern based on your location.

This is a special place and you will not be disappointed!

St Paul, MN (Lake Elmo—21D)

Lake Elmo Inn 🍔🍔🍔🍔🍔

Rest. Phone #: (651) 777-8495
Location: 1-mile walk
Open: Daily: Lunch and Dinner

PIREP If you're willing to walk a mile for great food, park your plane at Mayer Aviation and head west along the railroad tracks for one mile to the Lake Elmo Inn. It is one of the better restaurants in the Twin Cities area. You can get into lunch without a reservation (dinner is a different story). The prices are not cheap, but the food is well worth the price and the hike.

St Paul, MN (St Paul Downtown Holman Field—STP)

El Burrito Mercado 🍔🍔🍔

Rest. Phone #: (651) 227-2192
Location: Medium walk
Open: Daily: Lunch and Dinner

PIREP There is a truly terrific Mexican place a couple miles away at Concord and State Streets across from Concord Drug called El Burrito Mercado.

This is a market/cafeteria-style place and is actually run by Mexican Minnesotans. I don't always know what I'm pointing at, but it is invariably good, even when it turns out to be stewed pork rinds and cactus. The fresh tortillas are swell. Expect to pay about $7 for a big meal. It is a full-service grocery store too, and they take credit cards. There are a bunch of other Mexican places in the area that I can't report on, since once I found this place I quit trying new ones.

If you don't feel up to the walk, Regent will give you a ride. I think there is a $25 handling fee for single-engines, or they waive it if you top up. They've waived it in the past for me if I bought something at the pilot shop (which is the best one in the Twin Cities).

Starbuck, MN (Starbuck Muni—D32)

Water's Edge 🍔🍔🍔

Rest. Phone #: (320) 239-9117
Location: Very short walk
Open: Daily: Lunch and Dinner

PIREP It's grass and at 2512' in length not the longest strip in the state, but Starbuck Airport (D32), hard on the shore of Lake Minnewaska, is a very short stroll to a lakeside eatery called, appropriately, the Water's Edge.

Family type of a restaurant, but with a full bar, the place serves good food at reasonable prices. Make sure to practice your short field, crosswind landings and takeoffs before venturing there.

Mississippi

Brookhaven, MS (Brookhaven-Lincoln County—1R7)

The Fish Fry 🍔🍔

Rest. Phone #: (601) 835-2104
Location: Crew car
Open: Daily: Lunch

PIREP On a recent cross-country trip from Texas to Virginia and back again we stopped at Brookhaven, MS (1R7) both going up and coming back. If I was ever in doubt about the meaning of Southern hospitality, I am not anymore. On the way to Virginia we just stopped for gas at Brookhaven. The FBO, Al Jordan, directed us to the gas pump, filled up our plane for us, then treated us to soft drinks and snacks. The airport terminal building and ramp were extremely well kept; everything was spit polished. On the way back from Virginia we stopped for lunch and gas. Al Jordan handed us the keys to the courtesy car, and directed us to a place not far from the airport called the Fish Fry.

The Fish Fry has, as the name implies, fried everything, including DILL PICKLES! I had never had fried dill pickles before, but as it turned out they were very good and went well with the fried catfish and Cajun-fried potatoes. (They did not fry the coleslaw!) The Fish Fry is a plain, small-town restaurant that serves good food for a reasonable price. Lunch was about $5 per person.

Back at the airport after lunch we decided the weather was not good enough to continue our trip and that we should stay in Brookhaven. Al made us a reservation at the local Best Western and said keep the courtesy car for as long as we needed it. Al and his wife, Helen, made our stay in Brookhaven a pleasant and memorable occasion. If you ever have occasion to fly in that part of Mississippi, stop by, you won't be disappointed. (And if you want a good conversation on fishing the Mississippi River, have Mrs. Jordan introduce you to her brother Earl!)

Cleveland, MS (Cleveland Muni—RNV)

Airport Grocery Restaurant 🍔🍔🍔

Rest. Phone #: (662) 843-4817
Location: 200 yards
Open: Tues.–Sat. 11 AM–9:30 PM

PIREP   Restaurant is about 200 yards from the airport—nice way to stretch your legs after a short or long flight. Excellent food: barbecue sandwiches, hamburgers, grilled corn on the cob, and so on. Several tasty appetizers—fried mushrooms and fried pickles are a few of the less common that you will find at this restaurant.

Prices are reasonable—$5–10 per meal.

PIREP   The Airport Grocery Restaurant is a short walk from the Air Repair, Inc., ramp on the south side of the field (1 burger). Food is excellent; I'd give it 1-1/2 on your food scale. The setting is an old country grocery store. For 1 on ambiance, a total of 3.5 burgers, if you allow half burgers.

I'd say this is a definite repeat stop.

Columbus, MS (Columbus-Lowndes County—UBS)
Kountry Kitchen 🍔🍔

Rest. Phone #: (662) 327-9207
Location: Very close
Open: Daily: Breakfast and Lunch

PIREP   FBO Comments: Great friendly service, will get you whatever you need. Taloney Air Service has a limo and a courtesy car to cruise in.

PIREP   Home-town cooking, Southern style. Very close to the airport, open for breakfast and lunch only, and very affordable. Have more than enough food to eat, plus unlimited sweet tea.

Greenville, MS (Mid Delta Regional—GLH)
"Original" Doe's Eat Place 🍔🍔🍔🍔

Rest. Phone #: (662) 334-3315
Location: Crew car
Open: Mon.–Sat. 5:50 PM–10 PM

PIREP   FBO Comments: General Aviation Services, very friendly, courtesy car available.

PIREP   Doe's Eat Place is not located on the airport, but the FBO will be glad to loan you the courtesy car to go there. There are Doe's Eat Place restaurants in several cities, but there is only one ORIGINAL Doe's Eat Place and it is this one at 502 Nelson Street in Greenville. It is still in the same old wooden store building where it was started in the early 1940s and is still run by the same family.

There is no written menu and their specialties are steaks (they are huge and you pick your own steak before they cook it), fried shrimp (we have tried fried shrimp all over the United States, the Caribbean, and Mexico and have never found any as good as Doe's), and hot tamales (some folks send their corporate jets to Greenville just to pick up Doe's hot tamales). Don't let the neighborhood deter you, and I guarantee you will have one of the finest meals in one of the most unique atmospheres you have ever experienced.

It is not inexpensive and reservations are recommended on Friday and Saturday nights.

Grenada, MS (Grenada Muni—GNF)
Williams Aviation, Inc 🍔🍔🍔🍔

Rest. Phone #: (662) 227-8402
Location: On the field
Open: Daily: Lunch

PIREP   FBO Comments: SHELL -100LL. Hours: 7 AM to 5 PM weekdays, 8 AM to 5 PM weekends, after on request. SHELL jet premix. Catering, computerized weather, pilot lounge, restrooms, showers, snooze room.

PIREP   Mr. Wayne is the operator at Grenada Muni. He is a super nice guy. He does not serve hamburgers, but he does have hot dogs and the fixings. Best of all, he does not charge anything for the hot dogs.
  That's right! Lunch is FREE at GNF and, best of all, he enjoys your company.
  Go and see Wayne—you will enjoy the stop. I know I do every time.

Gulfport, MS (Gulfport-Biloxi Intl—GPT)
Montana's BBQ 🍔🍔

Rest. Phone #: (228) 864-1113
Location: Crew car, short drive
Open: Daily: Lunch and Dinner

PIREP   FBO provided transportation, three minutes from ramp. Food is buffet style, all you can eat. The BEST barbecue for $8.95 before 3 PM, $10.95(?) after 3 PM.
  FBO requires fuel purchase to use courtesy vehicle.

Hattiesburg, MS (Bobby L Chain Muni—HBG)
Dan's Cafe 🍔🍔

Rest. Phone #: (601) 582-7055
Location: Crew car
Open: Daily: Breakfast and Lunch

PIREP   Flew into Craig Muni on way to the Gulf Coast. Was lunch time and needed fuel. Was greeted by friendly, helpful staff. Crew car offered at no charge and was given directions to Dan's Cafe (about a five-minute drive). Above average meat and two for $6.50.
  Fuel reasonably priced and service was excellent.

McComb, MS (McComb/Pike County/John E Lewis Field—MCB)
The Dinner Bell 🍔🍔

Rest. Phone #: (601) 684-4883
Location: Call for pickup
Open: Daily: Lunch and Dinner

PIREP   It's a little tough to find, but I'm sure if you call the Dinner Bell in McComb they would be more than happy to come pick you up. There is also a courtesy car on the field. Everything is served on a lazy Susan around a huge table. Great food, low price.

Meridian, MS (Key Field—MEI)
Magnolia Restaurant 🍔🍔🍔

Rest. Phone #: 601-482-0233
Location: Crew car
Open: Daily: Lunch and Dinner

PIREP I work at Meridian Aviation in Meridian, MS. Many times I chauffeur pilots and passengers. Most people like to go to a place that they can sit down and eat in a hurry but get a good selection and atmosphere. Two of the really popular places in Meridian are Magnolia Restaurant (great lunch specials) and the Hungry Heifer (enjoyable dinner atmosphere).

Natchez, MS (Hardy-Anders Field Natchez-Adams County—HEZ)

The Lady Luck Casino 🍔🍔🍔

Rest. Phone #: (800) 722-LUCK
Location: Call for free pickup
Open: Daily: Breakfast, Lunch, and Dinner

PIREP The Lady Luck Casino, a riverboat on the Mississippi River, is a great place to go. Land at HEZ and the FBO will call the casino to have you picked up FREE of charge. They have a large buffet and lots of gaming. With a little luck you can go home with full stomachs AND full pockets.

Oxford, MS (University-Oxford—UOX)

Old Venice Pizza Company 🍔🍔

Rest. Phone #: (662) 236-6872
Location: Crew car
Open: Daily: Lunch and Dinner

PIREP We flew our Cherokee to Oxford, MS (UOX) at Ole Miss. The FBO will give you a courtesy car and directions to the square at Ole Miss. Lots of little sandwich shops and pizza. Old Venice Pizza Company makes one of the best calzones I've ever had. There is a white tablecloth restaurant right next to it if you prefer "artsy fartsy" food.

No fees at the FBO for anything. I wish every airport was like this one.

Picayune, MS (Picayune Muni—MJD)

Dockside Seafood 🍔🍔

Rest. Phone #: (601) 749-0400
Location: Crew car
Open: Daily: Lunch and Dinner

PIREP You can get a nice diner-type of seafood meal down the road about a mile. The FBO operator will take you there and pick you up again when done eating. Good seafood, po-boys, and so on at very reasonable prices. I make this a regular stop when I'm heading toward Florida.

Tunica, MS (Tunica Municipal—UTA)

The Grand Casino 🍔🍔🍔🍔🍔

Rest. Phone #: (662) 363-2788
Location: Crew car
Open: Daily: Breakfast, Lunch, and Dinner

PIREP   Stay at one of the casinos (recommend the Horseshoe but they are all are supposed to be good), gamble, golf, dance, dine, catch a show, swim, sunbathe, visit Graceland, you name it. Make reservations at least two weeks in advance unless you are a gambler with a track record there. Rooms are very nice, golf is wonderful, food is great, and the entertainment is the best outside of Vegas.

(Gambling is better than Vegas.)

PIREP   FBO Comments: Nearly new airport built for gamblers going to the casinos north of Tunica. Very nice people. No courtesy cars but rentals are $36 per day. Red carpet service for a guy in a Bonanza. Airport is already under expansion. No instrument approaches—yet. Runway is 5100' and pilot remote lighting on 123.00 has just been made operational. GPS approaches coming. Linemen go out of their way to help. Tipped my guy $5 because he held an umbrella to keep the rain off us after spotting us right in front of the FBO door. He insisted on carrying our bags. This is before I tipped him. When we returned two days later, he not only recognized us but went out of his way to make sure we had everything we wanted.

PIREP   The Tunica Airport is in the middle of cotton fields in the Mississippi delta. It is surprisingly far from the casinos, so you will need to rent a car to get anywhere. The best casino in Tunica is the Grand Casino, though they are all good. It has a great golf course and sporting clays course.

The Hundred Dollar Hamburger

Missouri

Bolivar, MO (Bolivar Municipal—M17)

The Plane Cafe 🍔🍔🍔

Rest. Phone #: (417) 777-6080
Location: On the field
Open: Daily: Breakfast and Lunch

PIREP Flew in for a Sunday lunch. Taxi in, shut down, and walk in the door for lunch. Very friendly staff serving good food at a decent price. Great facility with little to no traffic.

PIREP A P.S. to the Bolivar, MO, info: on weekends they have a nice breakfast buffet in the mornings, with fruit, pancakes, two kinds of scrambled eggs, hash browns, bacon, sausage, biscuits, gravy, and a couple of other things. Worth the few bucks they charge for it.

PIREP The Bolivar Airport (Bolivar, MO) has a great little grill located in the terminal. It's pretty well packed on nice Sundays. Good food, reasonable. The 18/36 asphalt runway is smooth.

Camdenton, MO (Camdenton Memorial—H21)

CJ's 🍔🍔

Rest. Phone #: (573) 346-6133
Location: Crew car
Open: Daily: Breakfast and Lunch

PIREP The outstanding point of interest here is not the restaurant, but the castle! Call Bob at (573) 346-0300 for availability of courtesy car (ex-police car) and directions to the castle (ten miles). We ate at CJ's because it was on the way, but there are better restaurants close by.

The castle was built in 1905, the owner died in a car accident in 1906 (you gotta ask yourself, how many cars were there in Missouri in 1906?). His sons finished construction in 1924, then it burned in 1946. Today the walls stand in Ha Ha Tonka State Park.

Great for a walk-through; fantastic scenery.

Cape Girardeau, MO (Cape Girardeau Regional—CGI)
The Landing Place 🍔🍔🍔

Rest. Phone #: (573) 651-3663
Location: On the field
Open: Daily: Breakfast and Lunch

PIREP We stopped at Cape Girardeau, MO, for fuel for our plane and ourselves on a trip from Dallas to Chicago—good choice. The FBO suggested the Landing Place Restaurant right next door in the terminal building. Good selection of sandwich and salad items. Our family had chicken strips, hamburger, BLT—all tasty.

Great view of the ramp.

Columbia, MO (Columbia Regional—COU)
Anita's 🍔🍔🍔

Rest. Phone #: (573) 814-2100
Location: On the field
Open: Daily: Breakfast and Lunch

PIREP Anita's took over the space that used to be Dessa's. They have a small and inexpensive menu, but the steak quesadilla and salad that I had were delicious. Nice view of the runway and pleasant staff. They are located in the terminal building a short walk away from the FBO.

Dexter, MO (Dexter Muni—DXE)
Airways Cafe 🍔🍔🍔

Rest. Phone #: (573) 624-4377
Location: On the field
Open: Daily: Breakfast and Lunch; closed Sunday

PIREP The Airways Cafe at Dexter Municipal Airport is a great little hometown cafe with very friendly people! As friendly as everyone was, you would never have guessed it was my first time walking through their door. Everyone had time to stop, talk, visit, and even offer us a tour of the older aircraft hidden within the airports hangars.

My sons and I enjoyed the breakfast and hospitality so much we came back the next day with my daughter since it was her day off. The second day we got there after the lunch menu had been posted, but they still let us order from the breakfast menu. I can't speak for their lunches since I had breakfast both days; but if lunch is anything like breakfast, and from the looks and smells of the lunches the other guests were getting, you won't be disappointed getting either.

Their hours are 0630–1400 Monday through Friday, 0630–1300 on Saturday, and closed on Sunday. They're located right on the airport with the ramp on one side and the parking lot on the other side. This place is well worth the trip; you WON'T be disappointed.

Hickory Log Restaurant 🍔🍔🍔

Rest. Phone #: (573) 624-4950
Location: Call for pickup
Open: Daily: Lunch and Dinner

PIREP   What a treat to be able to fly into such a nice place as Dexter (everyone was very friendly) and get such great BBQ. The service was wonderful and the food was outstanding. We had no wait at all to return to the airport and they even bagged us up some ribs to go. The Hickory Log is a must visit if you're hungry for BBQ.

PIREP   Being 70 miles south of Dexter, MO in northeast Arkansas give me a advantage of being able to make Dexter a short little supper run. The wife and I fly into Dexter and use the phone on the airport grounds to phone the Hickory Log Restaurant......I can't say enough about the food and the people. Either the owner or his son have always seemed cheerful when picking us up at the airport no matter the time of night and they will not accept a tip for the trip.

Once supper is over there is hardly any wait to get back to the airport as I've seen in some locations; instead again the owner or his son is ready and willing to fit their transportation into your schedule. All in all I'd recommend the Hickory Log for any time you're near southeast Missouri. Dexter is the place to stop.

Excelsior Springs, MO (Excelsior Springs Memorial—3EX)
The Wabash 😋😋

Rest. Phone #: (816) 630-7700
Location: Crew car
Open: Daily: Lunch and Dinner

PIREP   My wife, daughter, and I happened into the Excelsior Springs Airport sometime in June. The airport manager met us and told us of a great barbecue place, The Wabash, just in town. He actually loaned us his own truck for the five-minute ride into town. Upon our return, he gave us a quick tour of the facility including the fledgling biplane exhibit they have. I would HIGHLY recommend this airport and the Wabash.

Farmington, MO (Farmington Regional—FAM)
The Warehouse Bar-B-Q Co. 😋😋😋

Rest. Phone #: (573) 760-1600
Location: Call for pickup
Open: Daily: Lunch and Dinner

PIREP   FBO Comments: Friendly staff with mechanic on duty. Courtesy car.

PIREP   The Warehouse Bar-B-Q Co. sent a car over to pick us up . They have the finest barbecue around. It was a short drive over to the restaurant. The staff was very friendly and the food outstanding. Homemade potato salad and, of course, the barbecue was fabulous. Ruth Berry, the owner, couldn't have been nicer to us. Great food stop in the area of Farmington, MO.

Hayti, MO (Mid Continent—M28)
Chubby's 😋😋😋

Rest. Phone #: (800) 553-0415
Location: Short walk
Open: Daily: Lunch and Dinner

PIREP   Airport Comments: Right in the boot heel. Hayti is a 3400' 18/36 grass strip and is the home to Mid Continent Aircraft, which is a top-notch Stearman restoration shop, as well as being an Ag Cat, Dromadier, and Cessna Single Engine dealer.

PIREP   Chubby's is owned by Chubby, the local sheriff, and located about one mile west of the airfield on Hwy 412. It is absolutely the best barbecue I have ever eaten. The pulled pork is unreal, and even the sides are phenomenal. I've only gone for lunch, but the place is always packed with locals. The motif is exactly what all those chain places try to replicate, but this is the real thing, right down to the plywood walls and the bulletin board.

Jefferson City, MO (Jefferson City Memorial—JEF)
Nick's Family Restaurant ☺☺☺☺

Rest. Phone #: (573) 634-7050
Location: On the field
Open: Daily: Lunch and Dinner

PIREP   Nick's Family Restaurant will be open in the terminal building at Jefferson City Memorial Airport (JEF), Jefferson City, MO. Nick's has a long-standing reputation in Central Missouri for their excellent fried chicken and country ham meals.

Kaiser Lake Ozark, MO (Lee C Fine Memorial—AIZ)
J.Bruners ☺☺☺

Rest. Phone #: (573) 348-2966
Location: Crew car
Open: Daily: Lunch and Dinner

PIREP   If you are going to Lake of the Ozarks, do yourself a favor and try out the best steak place on the lake. It is called J.Bruners and is located just north of the bridge near Lynn Creek. The steak for two is wonderful, the salad dressings are all homemade and delicious, and the onion rings are made from Vidalia onions and are the lightest and tastiest I have ever had.

Unlike the closer airport (Lynn Creek/Grand Glaize), Lee C. Fine Memorial has clear approaches, a long runway, a friendly FBO and, most importantly, a free courtesy car. I flew in on a cross-country from California and found the entire stop very enjoyable.

Give 'em a try.

Kansas City, MO (Charles B. Wheeler Downtown—MKC)
Hereford House ☺☺☺☺

Rest. Phone #: (816) 842-1080
Location: Short drive, crew car
Open: Daily: Lunch and Dinner

PIREP   FBO Comments: Excellent staff and service. Provide nice, clean, and fairly new courtesy cars. A first class operation.

PIREP   On the way home to Colorado from Illinois, stopped in KC for a steak dinner. The nice lady at the FBO suggested the Hereford Steak House and we were not disappointed. Located a short distance from the airport, we enjoyed a very nice dinner. The wait-staff couldn't have been better, the food was excellent, and it had a nice ambiance. You can get a 6–ounce to 12–ounce steak with salad and great sides to choose from. They have an extensive wine list, but of course we couldn't take advantage of it at the time. We arrived home a little later than we had originally planned, but it was worth it.

Kennett, MO (Kennett Memorial—TKX)

Mi Ranchito 🍔🍔🍔

Rest. Phone #: (573) 717-7070
Location: Short walk
Open: Daily: Lunch and Dinner

PIREP Mi Ranchito serves above average Mexican food at a reasonable price. Try the Diablo shrimp—they call this something else but I don't remember the name, just the flavor. Peeled and deveined shrimp in an unusual red spicy sauce. Our ticket was about average for this part of the world. The service was somewhat above average but not over the top.

They earned a 20 percent tip.

They are 200 yards from the approach end of runway 18.

Lincoln, MO (Lincoln Muni—0R2)

Papa Joe's 🍔🍔🍔

Rest. Phone #: (660) 547-2706
Location: Short walk
Open: Daily: Breakfast, Lunch, and Dinner

PIREP A gorgeous grass strip next to the tiny Midwest town of Lincoln. This place will take you back to the days of tail-draggers. A short walk will take you to a bunch of diners and a Pizza Hut. I've flown here quite a few times in a 150, and I must say, I love this little 'port! Five burgers for sure.

PIREP Landing at Lincoln's grass strip is worth the trip alone, but the short walk to Papa Joe's eatery for breakfast or dinner is a real treat. Great food, very friendly folks, and no calorie counting allowed!

PIREP My friend and I flew "the Jewel" in today and had a piece of pie at Papa Joe's. It was great food and good service. Five burgers. They have a great menu and reasonable prices. We'd recommend it.

Osage Beach, MO (Grand Glaize—Osage Beach—K15)

The Kenilworth Cafe 🍔🍔🍔🍔

Rest. Phone #: (573) 348-5959
Location: Short walk
Open: Daily: Breakfast, Lunch, and Dinner

PIREP The Kenilworth Cafe (the departure end of runway 14 and a two-minute walk from the ramp) serves excellent homemade food, including soups and sandwiches. For days when you are not flying (or for your passengers), the Osage Brewing Company, with many microbrew selections, is located in the Kenilworth.

PIREP No trip in the Midwest is complete without a stop at Bagnel Dam or The Lake of The Ozarks area in Central MO. Lake Ozark is the second largest manmade lake in the world, boasting a snake-like length of some 120 miles and 1,200 miles of shoreline. It features some of the most fantastic upscale waterfront resorts and homes anywhere! The area features 3+ airports: Camdendon, Osage Beach (K15), and Lee Fine Memorial. Ocean-going yachts, world-class "go-fasters," PWC's, and rental houseboats abound, often stopping at the countless waterfront restaurants and dockside attractions and discos. The 3500 ft+ Lynn Creek/Osage Beach (K15) Airport is in the heart of things, including major dinner theatres, great restaurants and amusement parks, a major regional outlet mall, a six-screen theater, and (of course) one of the most beautiful deep clearwater lakes in the U.S. Most native

Midwesterners call this their "summer playground" and one of the lake's countless inlets or coves makes national news as "The Party Cove," with bikini clads abounding in such great quantity that you can almost walk across the large cove from boat to boat w/out getting wet! It is also featured in many commercials, such as the famous ESPN/Budweiser spots! Numerous ESPN water ski tourneys dot the summer calendar.

Rental boats, including houseboats and PWCs, make this a "must do" for any pilot seeking (what must be) one of the top ten fly-in vacation destinations in the Midwest. The area is family oriented, and several major water parks and/or minor theme parks abound. The scenery is native Ozark mountain with upscale summer places thrown in.

Our family has been going there since I was little, and most all folks from St. Louis, Memphis, Chicago, and Kansas City visit regularly and/or have summer homes/condos there. I can't recommend it enough and can't imagine how any fly-in visitor could complain. Most resorts require reservations and have free airport pick-up, although a rental car is a must to enjoy the nonwater activities along Hwy 54. Holiday weekends are crowded and the water can be rough, but we bet you will visit more than once!

Ozark, MO (Air Park South—2K2)

Lambert's Cafe 🍔🍔🍔

Rest. Phone #: (417) 581-7655
Location: 1-1/2 miles, call for pickup
Open: Daily: Lunch and Dinner

PIREP There is another Lambert's Cafe here now. Air Park South Airport (2K2) is very close; 1 to 1-1/4 miles. There are no services at the airport and no courtesy car. The runway is not in real good shape but it is smoother than I thought it would be. The walk to the restaurant is well worth it. Now those of us who live too far west to get to Sykeston, MO on a regular basis will be able to enjoy Lambert's more often.

PIREP It's a 1-1/2 mile walk to the restaurant.

If you call the restaurant and ask the manager to pick you up (bring your cell phone, because there isn't one at 2K2), they will pick you up. Make sure they know it's at Air Park South and not Springfield (if they don't know the airport, tell 'em to cross US 65 and turn left at the Conoco station. It's about a mile down on the right). They have been known to drive to Springfield to pick up folks (they probably had bigger airplanes than I do.).

2K2's runway is in acceptable condition. 2500' × 40' wide, no potholes, no loose gravel on the runway. The runway slopes uphill to the south. With a little care, a Cherokee 235 or C182 can land and still make the taxiway turnoff about a third down the runway. Take care on the taxiway as it is tar and gravel (some minor loose gravel).

PIREP If you call the restaurant after flying into 2K2, they will pick you up. Restaurant, service, and so on is much nicer in my humble opinion than the Sikeston location.

St Charles, MO (St Charles County Smartt—SET)

Kilroy's 🍔🍔🍔

Rest. Phone #: (636) 250-3303
Location: On the field
Open: Mon. 10 AM–1 PM, Tues.–Thurs. 10 AM–6 PM, Fri. 10 AM–8 PM, Sat./Sun. 8 AM–4 PM

PIREP Kilroy's is now open at the south end of the field, near the Confederate Air Force self-serve pumps (where Mary's then JB's used to be)

Home cooking, including spicy juicy beef on rye, pies and cobblers, and daily specials (when we were there it was fried chicken). I would rate them better than JB's or Mary's (their predecessors).

Clean restaurant, widescreen TV, hand-painted airplane murals.

Note: there is currently no sign identifying the restaurant from the runway. You need to know it's at the south end of the airport building, just north of the CAF hangar.

St Joseph, MO (Rosecrans Memorial—STJ)

The Airport Cafe 🍔🍔🍔🍔

Rest. Phone #: (816) 364-6211
Location: On the field
Open: Daily: Breakfast, Lunch, and Dinner

PIREP My wife and I stopped here for lunch. The food and service was very good. I love a restaurant where I can taxi right up to it and enjoy my lunch while having a great view of the runway.

PIREP I flew into the St. Joseph, MO airport for lunch today. The airport restaurant is located right under the control tower. Ground control instructed me to park in front of the tower, which parks you in full view of the restaurant just a few feet away. This is my kind of airport restaurant. The service was really good.

PIREP FBO Comments: All of the people at the St. Joseph airport were very helpful when we were stuck there due to bad WX; they assisted in finding us lodging and rental cars and tied down our plane until the storms had passed. We sure appreciate all of their help during this weather setback. If we ever get stuck again, we hope it is at St. Joseph.

PIREP While flying from Ohio to Colorado, bad weather caused us to divert to KSTJ and stay overnight. The airport cafe at the field provided us with outstanding service and food and friendly people. I plan on making this airport a regular stop on my trips.

St Louis, MO (Spirit of St Louis—SUS)

Wente's O&W Restaurant & Saloon 🍔🍔🍔

Rest. Phone #: (636) 530-9994
Location: Nearby easy walk
Open: Daily: Lunch and Dinner

PIREP Just off the airport proper, Wente's O&W Restaurant & Saloon is a great little bar and grill. Good American food at a good price. A bit of a long walk from the FBO's (10-plus minutes), but a real quick drive with any of the courtesy cars. The closest FBO is JetCorp, but any of the four will do fine. Call ahead for hours of operation.
  If you are a fan of racing, in particular the open wheel kind, this is the place for you.

Sikeston, MO (Sikeston Memorial Muni—SIK)

Lambert's 🍔🍔🍔🍔

Rest. Phone #: (573) 471-4261
Location: Call for pickup
Open: Daily: Lunch and Dinner

PIREP This is an amazing place that I found by accident when stopping off my normal route from Key West to St. Louis for fuel. I was tying down the plane to top off my tanks, take a break, and check the weather when the friendly FBO drove up and said, "I already called the courtesy van for Lambert's and they said they would be here in five minutes." When I asked what Lambert's was he told me that the only two reasons anyone ever comes to

Sikeston is to buy a John Deere tractor or to eat at Lambert's. He also said that on a weekend, there would be as many as 40–50 planes that fly in there just for lunch. He assumed I was here to eat as I was tying down my plane. I found out that the fuel was self-serve and very reasonable.

A quick van ride, and I was escorted in the back door, past the very long line of people waiting to get in VIP style, and seated immediately. The dining room is about the size of K-Mart but sectioned off. The tables are picnic style, and the atmosphere is definitely down-home country. There was even a live piano player (in the middle of the day at lunch time, no less). The drinks are brought out in a variety of giant cheesy plastic mugs, and the food is very good and plentiful. The main attractions are the rolls and pots. The restaurant is famous for "throwing your rolls." They wheel out this big cart with fist-sized rolls steaming and fresh from the oven. The guy starts calling out who wants rolls, and you raise your hand and get it tossed to you from clear across the room! Another hillbilly-dressed server comes along behind the cart and offers "honey for your rolls!" which is dolloped out (but fortunately not thrown). Every five minutes or so another server comes around with a huge pot straight from the kitchen with fried okra and other country dishes, dishing it out for whoever wants it. Just tell your server when you are finished and they will take you straight back to the airport (about a five-minute ride). It is now on my permanent fuel stop list, whether I need fuel or not.

PIREP   FBO Comments: Very nice airport. FBO attendant very helpful.

PIREP   Lambert's is the place to go. Opens 10:30, and a van will take you from the airport to the restaurant. You get seated ahead of the never-ending crowd. It is not unusual to have over one-hour waits on the weekend. We had no wait. The food is the best. Something to meet all tastes.

Make it your next trip.

PIREP   What a great place to go!! You are treated like a celebrity!!

I called the FBO about 12 miles out and let him know our intentions and asked if he could give Lambert's a call. By the time we shut down and walked in the FBO, the shuttle was there ready to go.

When we pulled into the Lambert's parking lot, a five-minute ride max, there must have been 100 people waiting to be seated—people come by the busload. Our shuttle driver took us through the exit door and told us to wait there. Five minutes later, we were seated!!

Our food and service was excellent. They kept telling us, "Everything is all you can eat so just let us know when you are ready!" With the huge portions, I don't know how you could order seconds. The servers walking around slinging rolls through the air was extremely entertaining and unique.

After we were finished, we let the cashier know that we were ready to go and within five minutes the shuttle whisked us back to the airport.

Sikeston, MO and Lambert's cafe is a great place to go whether on your way to another destination and need a great meal or if you need someplace new to go on a weekend getaway.

Springfield, MO (Springfield-Branson Regional—SGF)
Hemingway's Restaurant 🍔🍔🍔🍔

Rest. Phone #: (417) 887-7334
Location: Crew car
Open: Daily: Lunch

PIREP   FBO Comments: FBO = Springfield-Branson Regional (SGF) Airport. Three courtesy cars available for max three hours.

PIREP   Missouri's #1 tourist attraction is a sporting goods store, Bass Pro Shops, in which you will find Hemingway's Restaurant. Just 20 minutes by courtesy car from SGF, the restaurant features an awesome menu as well as a great brunch for $13 on Sundays. Afterward, you'll need at least an hour to wander all around the store viewing numerous ponds, aquariums, and demos.

Warsaw, MO (Warsaw Muni—RAW)

Bradley's Restaurant 🍔🍔🍔

Rest. Phone #: (660) 438-5522
Location: Short walk
Open: Daily: Lunch and Dinner

PIREP   I've been landing here ever since the airport opened a few years ago. Bradley's Restaurant is just a 15-minute walk away (or five minutes if you want to climb some cattle fences), but they also have a courtesy car at the airport now, or Bradley's wife Theresa will pick you up if you call them.

Their buffet is delicious and reasonable, and you can experience live country music on weekends after 8:30 PM. They also have a motel next door if you really want to make it a night out.

Bozeman, MT (Gallatin Field—BZN)

Overland Express 🍔🍔🍔

Rest. Phone #: (406) 388-4565
Location: In the terminal
Open: Daily: Breakfast and Lunch

PIREP Bozeman has a very nice airport, and the restaurant in the public terminal has reasonably good food and a nice view of the runways.

Sir Scott's Oasis 🍔🍔🍔🍔

Rest. Phone #: (406) 284-6443
Location: Short drive, use the crew car
Open: Daily: Lunch and Dinner

PIREP FBO Comments: We parked at Arlin's, nice new clean place, high-speed Internet, and so on, and since the courtesy van was being serviced, Arlin loaned us his pickup, deer horns included in the back. No one suspected we were from out of town.

PIREP We'd heard some Montana natives rave about this place and we were more curious than hungry. Some of these people don't usually pass out compliments on a daily basis. How could a roadside bar and grill in a two-horse town require dinner reservations?

We were not disappointed. We went for lunch even though their claim to fame is supper. The prime rib sandwich was the best I've ever had, even the bread was homemade. I paid an extra quarter for the Cajun version, the half-inch slab of prime rib was an honest three-quarter pound. They have a daily lunch special; burgers are on the menu as well.

We usually aren't crazy about bar food because of the smoke and drunks, but this place is separated nicely and you don't even have to go into the bar at all. It's about eight miles south of the Bozeman Airport (BZN) on the interstate or eight miles north of the Three Forks Airport (9S5). The town is tiny, one exit—drive to the four-way stop, then left along RR tracks one block. We fueled Arlin's truck, then burned most of the gas driving to Bozeman and back because we missed our exit to Belgrade where the airport is. Keep in mind Belgrade is also a one-exit town; don't wait for the next closer one.

Dell, MT (Dell Flight Strip—4U9)

Yesterday's Cafe 😋😋😋

Rest. Phone #: (406) 276-3308
Location: 1/2 mile
Open: Daily: Breakfast and Lunch

PIREP   Dell is located about 40 miles south of Dillon. There is a paved strip about half a mile away, or you can land on the gravel road in front of the cafe and park in the large parking lot.

The food is all homemade, right down to the hamburger buns. If your group wants to all eat the same thing, like chicken dinner, it is served just like at home—help yourself to how much you want. If you are not already stuffed, you can get a large slice of pie, homemade of course, and about ten types to choose from.

If you did not park "out front," just let the waitress know and she will arrange a ride back to your plane.

Ennis, MT (Ennis–Big Sky—5U3)

Sportsman's Lodge 😋😋😋

Rest. Phone #: (406) 682-4242
Location: A drive away, or land at the lodge
Open: Daily: Breakfast, Lunch, and Dinner

PIREP   Sportsman's Lodge, downtown Ennis, MT (*not* at the airport! 5U3). Land behind the lodge, taxi up to restaurant. Get permission to land and check conditions first.

The strip was mowed and cleared. I came in from the north over the gas station last time. It's right in the center of a very small town, near the main intersection.

The Continental Divide 😋😋😋😋

Rest. Phone #: (406) 682-7600
Location: Short walk
Open: Daily: Lunch and Dinner

PIREP   The Continental Divide is a five-star continental (French and so on) restaurant. It has a superb reputation; you can walk a couple hundred yards to it from the strip.

Hamilton, MT (Ravalli County—6S5)

Hangar Cafe 😋😋😋

Rest. Phone #: (406) 363-4317
Location: On the field
Open: 6:30 AM–2 PM, 7 days a week

PIREP   The Hangar Cafe has opened at the Hamilton Ravalli County Airport. It is located on the west side of runway 16/34, at the midfield intersection taxiway. You can park in front or to the south side in a large parking area. They even offer free chocks to use.

This restaurant is also a pilot lounge and offers good food and friendly service. A fun and scenic place to fly to.

Helena, MT (Helena Regional—HLN)

Carol's Fly Away Cafe 😋😋😋

Rest. Phone #: (406) 443-6472
Location: On the field
Open: Daily: Breakfast, Lunch, and Dinner

PIREP Carol's Fly Away Cafe is located at Helena, MT (HLN) on the south side of the airport and is available by road or runway. It is in the old airline terminal underneath what used to be the shortest operating control tower in the U.S. (It now is empty since it was replaced with a new tower about two years ago.) The building also houses a small charter and freight business.

The service is never without a smile…and a humorous or sarcastic remark, which is all part of the Fly Away fun. The better your sense of humor, the more fun you'll have. The food is about average. They serve a good breakfast and lunch. Their burgers are pretty good, and I recommend the French toast for breakfast, but there is also a variety of other items to choose from in case you're in the mood for something else.

A weather briefing station allows you to see how friendly the skies are and to file your flight plan out.

They have their weekend regulars who keep coming back for the fun atmosphere and servers. All the wait-staff and cooks are really quite a hoot and I'm sure you'd enjoy the Fly Away experience.

Kalispell, MT (Kalispell City—S27)
Outlaw Inn 🍔🍔🍔

Rest. Phone #: (406) 755-6100
Location: Across the street from the airport
Open: Daily: Breakfast, Lunch, and Dinner

PIREP The Outlaw Inn is across the highway from the airport with an excellent restaurant.

In addition to a good menu with healthy food, they have a wonderful Sunday brunch! Last time I flew over, they had salmon, red snapper, shrimp, roast beef, a great pork roast, some delicacy similar to a waffle, lots of pastries, perfect scrambled eggs, salads, fruits, apple crisp for dessert, and a bunch of stuff I did not get to!

The brunch alone is good enough to justify a trip, and the adjacent Glacier National Park offers some of the most spectacular scenery I have seen between Arizona and Alaska!

(Try to honor the 2000' altitude minimum over the park.)

Seeley Lake, MT (Seeley Lake—23S)
Lindey's Prime Steak 🍔🍔🍔🍔

Rest. Phone #: (406) 677-9229
Location: They leave a truck at the field for you
Open: Daily: Dinner

PIREP FBO Comments: 4500' gravel/turf airport owned by Montana Department of Aeronautics. Mountainous terrain. Runway 34/16 land north and take off south. Easy to get into and out of. No fuel on airport. Missoula is 31nm to the southwest for fuel.

PIREP Park at tie-down area on west side of the airport. Walk less than 75 yards from your airplane to the Lindey's truck, a.k.a. "Lindey's Mobile." The keys are above the visor, then drive 1-1/2 miles into town for one of the best steaks you'll ever have.

However, they only have a special sirloin, a prime sirloin, and a prime chopped steak on the menu. Hours are 5–10 PM, 7 days a week in the summer, and 5–9 PM, five days a week the rest of the year. Reservations not required.

This is an old, established family restaurant, with other restaurants in Minnesota; the owners are Jenny and Mike Lindemer.

Three Forks, MT (Three Forks—9S5)

Custer's Last Stand 🍔🍔🍔

Rest. Phone #: (406) 285-6713
Location: 1/2-mile walk
Open: Daily: Lunch and Dinner

PIREP Three Forks, MT offers a great little burger place, about half-mile away from the airport, called Custer's Last Stand. It is a wonderful place to get a burger and an ice-cold root beer in a mug. They also have a pretty good pizza.

The first week in August, the airport fills with activity as the Montana Antique Aircraft Association holds their annual fly-in. There are usually a dozen or so pre-WWII aircrafts, along with better than a hundred other private planes. On that Saturday morning, a good part of the group will usually fly for breakfast to some other nearby airstrip. For the weekend, a local restaurant sets up on the field and serves breakfast, lunch, and one main course for dinner. When it gets dark, there is a hangar dance with a good C&W band. Hotels and motels are available in town; however, plan on camping under your wing, to experience it all. There is not a better alarm clock than a some radial-engined aircraft flying 30' over the top of your tent!

West Yellowstone, MT (Yellowstone—WYS)

TJ's Bettola 🍔🍔🍔

Rest. Phone #: (406) 646-4700
Location: On the field
Open: Daily: Lunch and Dinner

PIREP It's still a great little FBO but the restaurant is now an Italian place called TJ's Bettola. Super food but not cheap, we go for dinner and then fly back over the park on the way home at sunset. TJ's # is (406) 646-4700. You usually need reservations.

Nebraska

Alliance, NE (Alliance Muni—AIA)
The Elms 🍔🍔

Rest. Phone #: (308) 762-3425
Location: Crew car
Open: Daily: Breakfast, Lunch, and Dinner

PIREP In Alliance, NE we ate at The Elms. It's a homey, Midwestern cafe. The service was friendly, the hamburgers tasty, and the pies looked tantalizing (but we would have had to do another weight and balance if we had eaten any pie!).

While you're there, check out Carhenge, a unique car sculpture! The FBO at AIA provides a courtesy car—just ask for directions.

Broken Bow, NE (Broken Bow Muni—BBW)
Uncle Ed's Steakhouse 🍔🍔

Rest. Phone #: (308) 872-3363
Location: Crew car
Open: Daily: Lunch and Dinner

PIREP Take the courtesy car to the Arrow Hotel, downtown. Outstanding steaks! Neat atmosphere; reasonable prices.

Cozad, NE (Cozad Muni—CZD)
Plainsman 🍔🍔

Rest. Phone #: (308) 784-2080
Location: Crew car
Open: Daily: Breakfast, Lunch, and Dinner

PIREP   A fine old Chevy sedan is available. I believe the air conditioner works and it has a CB! Eat downtown at the Plainsman.

Grant, NE (Grant Muni—GGF)
Jenny's 🍔🍔🍔

Rest. Phone #: (308) 352-4178
Location: 1-mile walk
Open: Wed.–Sat.

PIREP   There is a nice family-type restaurant called Jenny's in Grant, NE that is open Wednesday through Saturday. It's very popular with the locals as well. They have homemade pies and a very nice salad bar. It's about a mile to town from the airport but there is a nice courtesy car at the airport.

Hastings, NE (Hastings Muni—HSI)
The Village Inn Pancake House 🍔🍔🍔

Rest. Phone #: (402) 461-3351
Location: Very short walk
Open: Daily: Breakfast, Lunch, and Dinner

PIREP   The Village Inn Restaurant is only 100 yards away from Hastings (HSI) Municipal Airport. In addition to the food provided, there is a shopping center in the same vicinity. Both are easy walking distance.

For a full afternoon or early evening of entertainment after your meal, attend the Leid I-max Theater. Courtesy car is available at the airport, and a visit to the I-max will add a special meaning to the day's flight. This is an enjoyable community in the central part of Nebraska and the excellent airport has two runways.

Enjoy!

Minden, NE (Pioneer Village Field—0V3)
Minden Country Club 🍔🍔🍔

Rest. Phone #: (308) 832-1965
Location: 1/2 mile
Open: Daily: Breakfast, Lunch, and Dinner

PIREP   A group of us were traveling to Oshkosh and ran across this airport by accident. We were impressed enough that we have decided to make it a destination. The Pioneer Village Museum claims to have a representation of every major technological advance in the past 100 years. I could not argue with them. The museum has about a hundred automobiles, lots of tractors, trucks, some aircraft, a steam train, merry-go-round, and about everything else under the sun you can think of. Plan on staying a day at least to see the museum. Minden is a nice, friendly Midwest small town.

Three courtesy cars are provided at the airport by the museum.

Pioneer Village Museum, one mile (308) 832-2750.

North Platte, NE (North Platte Regional Airport Lee Bird Field—LBF)
Airport Inn 🍔🍔🍔

Rest. Phone #: (308) 534-4340
Location: On the field
Open: Daily: Breakfast, Lunch, and Dinner

PIREP   Best dessert in our opinion goes to North Platte, NE (LBF). Good food in the main terminal, but definitely try the BLUEBERRY PIE!!!!!!!
     Nothing less than outstanding.

PIREP   Lee Bird Field, North Platte, NE has a very fine restaurant on the field. Always good food and courteous service. Sunday can be a crowd at noon.

Omaha, NE (Eppley Airfield—OMA)
Bohemian Cafe 😋😋😋

Rest. Phone #: (402) 342-9838
Location: Crew car
Open: Daily: Lunch and Dinner

PIREP   FBO Comments: Two late-model Dodge Neons are the courtesy cars at Elliot FBO.

PIREP   A good trip to Omaha, landing at Eppley with excellent facilities at Elliot FBO. The visit was to go to Henry Doorly Zoo (highly recommended: one of the best zoos in this country that I've seen), which is located off 13th Street (runs north-south) south of I-80. Also on 13th Street, a little farther north, is the Bohemian Cafe, a Czech restaurant with great food. I can't praise the duck enough—it's great, and try their daily special stews, which they do really well. Meals come with liver dumpling soup and German-style breads and sides (get extra gravy). Good (dark) beer on tap, well-priced wines for an overnighter. The market area (previous pirep) has all sorts of other choices (German, Indian, Japanese, Mexican, Persian, and so on) and is also off 13th Street a little farther north. And, for a quick bite as you turn off 13th at the entrance sign for the zoo, a little Italian sausage and ice cream stand is worth checking out—Tony's, I think is the name. I recommend the sausage and peppers sandwich and strawberry sundae there.
     Omaha is a four-burger city: good food here. Talking of burgers, on previous visit went to Louie M's Burger Lust, located in the same vicinity (1718 Vinton Street)—top-notch burgers here.

Oshkosh, NE (Garden County—OKS)
Blue Star Cafe and Bistro 😋😋😋

Rest. Phone #: (308) 772-0108
Location: Across the street
Open: Daily: Lunch and Dinner

PIREP   Located just across the street from the Oshkosh Airport. The Blue Star Cafe and Bistro offers a variety of sandwiches, grinders, and many other menu items. Good food, fast service, stellar atmosphere, and a collection of celebrity-autographed photographs and memorabilia.

Scottsbluff, NE (Western Nebraska Regional/William B. Heilig Field—BFF)
Skyport Cafe & Lounge 😋😋😋

Rest. Phone #: (308) 632-3673
Location: On the field
Open: Daily: Lunch and Dinner

PIREP   On January 7, 2006 we stopped at the new terminal restaurant. The facility is modern and first class. Seating is available for over 130, and there was a community lunch meeting (Red Hat Ladies) while we were eating without really stressing the service. It appears the terminal restaurant is not only a favorite for the aircraft folks, but for the local community.

The food was good, the prices reasonable, and the menu has a variety of country favorites.

The FBO was, as always, friendly and efficient.

PIREP   The Skyport restaurant has the best turkey/Swiss croissant that will ever cross your palate. The choices are quite varied, but each time I return I don't stray from my favorite. I challenge anyone to find a better croissant!!!!!

PIREP   Stopped in at the terminal restaurant on my way home from Oshkosh and was pleasantly surprised by the quality of food and interesting choices offered. Service was good and very prompt. Would highly recommend it even if I wasn't suffering from airport-induced starvation.

PIREP   I think I had the BLT (could have been a burger), but I was impressed with the little restaurant in the Scottsbluff, NE terminal. Good sandwich, good fries, good pie, good service.

PIREP   Flew into Scottsbluff on an IFR training flight. Had lunch at the little restaurant in the airport called the Skyport. They served the juiciest, tastiest hamburger I've ever eaten in a restaurant. Service was friendly. Nothing fancy but really good food. Worth making the trip.

Sidney, NE (Sidney Municipal/Lloyd W Carr Field—SNY)

Cabela's - High Plains Cache 🍔🍔🍔🍔

Rest. Phone #: (308) 254-5500
Location: Call on unicom; Cabela's sends a van
Open: Daily: Lunch

PIREP   Visit the Cabela's retail store. Request on unicom (122.8) prior to landing that Cabela's be given a heads up to your arrival, and they will meet you at the airport with a van. They have a tiny sandwich shop in a corner and the largest static animal displays I've ever seen. I suggest the hot smoked buffalo sandwich.

If you are an outdoorsman, expect your lunch to include some equipment purchases.

PIREP   Okay, so this pirep is not about food, but it is important…IF you are a pilot. I recently had the misfortune of being stranded in a blizzard for five nights in Scottsbluff and Sidney, NE while trying to get a Howard back home. The GOOD fortune, however, was being in Scottsbluff and Sidney. The people at Valley Airways in Scottsbluff were great. And there is a place there by the name of Candle Light Inn that not only has the nicest rooms you'll most likely ever stay in for $50 a night, but they also furnish you with an airport car.

Then, after making it about 70 miles further to the southeast to Sidney, my good fortune continued. A gentleman by the name of Ed Kelly and his family have the FBO there. And I must tell you, in spades, that I've never, ever, been treated better, anywhere, at anytime, than I was by these people. Sidney Aviation is tops! They put the Howard in their hangar, furnished me with a car, preheated the aircraft, and so on and so forth. Total charge for two nights and three days? $10 dollars plus the gas I purchased!! What more can be said?

I can also echo the report on going to Cabela's. That place should be the tenth wonder of the world. So, the bottom line is: if you are ever traveling up around or through the panhandle of Nebraska, stop in at Sidney. Even if you don't need to. And tell Ed Kelly he is now on the Internet and doesn't even know it.

PIREP   Sidney, as stated previously, does have Cabela's for the sportsman, and Cabela's will provide a shuttle bus from and to the airport. Just let them know when you call the unicom that you would like to go to Cabela's and they will call for the bus. By the time that you land and tie down, the bus will be there waiting for you.

The Cabela's store contains the High Plains Cache restaurant where you can get elk, wild turkey, and buffalo sandwiches and buffalo brats. The food is good and has a wild twist that makes it quite interesting. Prices are fairly low.

Hotels and motels within walking distance of Cabela's include AmericInn, Days Inn, Comfort Inn, and Hotel Six. The Cabela's shuttle can be persuaded to ferry shoppers to any of the other hotels in town including Holiday Inn, Super 8, and Fort Sidney Inn.

Wallace, NE (Wallace Muni—64V)
Wallace Cafe 🍔🍔🍔

Rest. Phone #: (308) 387-4392
Location: 2 blocks
Open: Daily: Breakfast, Lunch, and Dinner Mon.–Fri.; Lunch and Dinner Sat. and Sun.

PIREP   Wallace, NE has an excellent hometown restaurant located two blocks north of the airport (64V). Food, service, and atmosphere are enjoyable. The menu features breakfast, dinner, sandwiches, salads, baskets, and desserts. During the week they offer a noon special.

Hours are 7 AM until 8 PM Monday through Thursday. On Friday, they close at 6:30 PM, on Saturday at 4 PM, and all day on Sunday.

York, NE (York Municipal—JYR)
Chances "R" 🍔🍔🍔🍔🍔

Rest. Phone #: (402) 472-6435
Location: Crew car
Open: Daily: Lunch and Dinner

PIREP   The York Municipal Airport (JYR) has an excellent place to eat. The restaurant is called Chances "R" and is located three miles from the airport. The restaurant was rated the #1 restaurant in Nebraska. If you have been there, you can understand why. You can order anything from a greasy hamburger to prime rib. On Saturday nights, they have a prime rib buffet.

The York Airport has two courtesy cars for transportation. The two cars are usually gone every Friday and Saturday to pilots who fly in to eat. If both cars are gone, Chances "R" will pick up anyone who wishes to fly in at no charge.

Call (402) 363-2660 for reserving the courtesy cars.

PIREP   A friend and I tried the York, NE as our first adventure using *The $100 Hamburger*. The only problem is anything else in the book will be downhill.

The whole experience was great! Wonderful 6000' strip, good crew car, great attitude of all, great food.

We will be going back to try their Sunday brunch or Saturday night for their prime rib buffet.

Cal Nev Ari, NV (Kidwell—1L4)

Cal-Nev-Ari Casino 🍔🍔🍔🍔

Rest. Phone #: (702) 297-1118
Location: On the field
Open: Daily: Breakfast, Lunch, and Dinner

PIREP Nancy Kidwell owns the restaurant and casino. Make sure you stop in the casino and look at the old airplane pictures. If you see Nancy, say hi. It is a short flight from the Las Vegas area and is on the California-Nevada-Arizona border; hence the name Cal-Nev-Ari. It is pronounced "Calnevair." If you say "Calnevaree," they will correct you.

The landing strip is dirt but is graded. You can taxi your plane right up to the restaurant. They have about anything you want on the menu and it is reasonably priced.

100LL is available.

Bon appetite!

Carson City, NV (Carson—CXP)

Sonic Burger 🍔🍔

Rest. Phone #: (775) 841-9000
Location: Short walk
Open: Daily: Lunch and Dinner

PIREP Walked about half a mile up the road to Sonic Burger. It was okay.

Elko, NV (Elko Regional—EKO)

Silver Dollar Casino 🍔🍔🍔🍔

Rest. Phone #: (775) 738-8485
Location: 1/2 mile or call for pickup
Open: Daily: Breakfast, Lunch, and Dinner

PIREP   The best restaurants in Elko are: The Silver Dollar Casino and Bar for casual sit-down, Misty's at the Red Lion for a more formal sit-down dinner. For the best prime rib it's The Old Red Lion across from the New Red Lion for cheese sticks with ranch dressing. Zappata's, one block west on Idaho Street from Commercial, for Mexican.

All of the restaurants listed are on Idaho Street. Zappata's is approximately a half mile from airport; next is Commercial with late night, large breakfast meals. Directly behind the Commercial is the Stockman's Casino, with acceptable breakfast food. Beside the Commercial, both east and west are Basque restaurants if you are up to six-course meals (baked bread, beans, string beans, pasta, soup, salad, and your ordered meat): don't eat here unless you are very hungry.

If you have a rented car, about 20 miles east over the mountain and below the Ruby Mountains is Michael's, an old ranch turned into a bed and breakfast. It is the base for heli-skiing during the winter. Often frequented by Clint Eastwood when visiting the Ruby Ranch a couple of miles from Michael's.

Best Gaming

Slots for payouts: Commercial, Stockman's

Friendliness: Commercial

Tightest: Red Lion

Only Poker (Hold 'Em, Stud): Red Lion, Hold 'Em 2/5, locals play 5/10 for rake

Sport's Book: Red Lion, Commercial

Big Screen Sports: Red Lion

Largest gold mines in North America. Barrick Goldstrike and Newmont Mining Company and many more small ones. Elko has doubled in size over the past five years.

Crime is low and town was voted best town in U.S.

Hope you find this interesting. My wife and I ate out likely four times a week. I was a supervisor for a local small gold mine, and my wife was head secretary for the new underground mine at Barrick. So we didn't do a lot of cooking.

Gary Burghoff (Radar from *MASH*) was a regular at Red Lion. Demi Moore and Bruce used to hang out at Cactus Pete's in Jackpot as they owned property in Sun Valley, ID and Bruce loved to gamble.

In eight years I've seen in Elko the following: Clint Eastwood, George Jones, George Strait, Barbra Mandel, Pointer Sisters, Dr. Hook, Williams and Reece, Bruce Willis, and many others.

Ely, NV (Ely Airport/Yelland Field/—ELY)

The Jailhouse 🍔🍔

Rest. Phone #: (775) 289-3033
Location: In town, use crew car
Open: Daily: Breakfast, Lunch, and Dinner

PIREP   I recently had the opportunity to fly into Ely, NV (elev. 6255'). Watch the density altitude, though a 6000-foot runway is helpful. Ely is a quaint little town with warm hospitality. El Aero services provided a well-polished courtesy car for the short drive into town.

We breakfasted at the Jailhouse, a portion of which, I believe, was just that many years ago, though there are several other places to eat in town. I'd give it an easy three burgers, perhaps three and a half!

Hawthorne, NV (Hawthorne Industrial—HTH)

The El Capitan Casino 🍔🍔🍔🍔

Rest. Phone #: (775) 945-3321
Location: Call for pickup
Open: Daily: Breakfast, Lunch and Dinner

PIREP   Hawthorne, NV is right at the south end of Walker Lake. Walker Lake is noted for being the home of giant cutthroat trout.

The El Capitan Casino has good food and gambling. They will send a courtesy van to the airport to pick you up. And, after lightening your wallet, they will even take you back to the airport.

This is the home of a large ammunition depot where they store bombs and ammunition for the U.S. military—just hope that they don't have an earthquake while you are there!

Jackpot, NV (Jackpot/Hayden Field—06U)
Cactus Pete's Resort Casino 🍔🍔🍔🍔

Rest. Phone #: (775) 755-2321
Location: Short walk
Open: Daily: Lunch and Dinner

PIREP   I highly recommend the buffet at Cactus Pete's Resort Casino, a short walk from the tie-down area at Jackpot, NV (06U). The price is reasonable, about $10 per person, and they have a fabulous selection. The buffet is open for lunch, closes for the afternoon, and then opens for dinner about 5. There is a 24-hour restaurant on the premises which is okay. If you stay the night (very reasonable prices Sun.–Thurs.), there is live entertainment in the Gala Showroom with dinner shows at 8. The dinner/show can be a bit pricey, but the meal is great.

Jean, NV (Jean—0L7)
Gold Strike Casino 🍔🍔🍔

Rest. Phone #: (800) 634-1359
Location: Short walk
Open: Daily: Breakfast, Lunch, and Dinner

PIREP   FBO Comments: Self-serve fuel. Terminal building. Clean restrooms and a skydive center. Outdoor picnic area with gas-fired barbecues if you bring your own steaks to cook while watching the airport activity.

PIREP   Stopped into Jean, NV for lunch on the way home from KHND. Really nice little airport with plenty of parking and good runway. The terminal is staffed by people from the skydive center and are all very friendly. Short walk to the casino for food or to lighten your wallet or both. I've been here several times and won't recommend the casino buffet as it's pretty bad. Even for buffet food. The restaurant right next to it, though, has pretty decent food and is open from 0600 until 2200. Menu has everything from burgers to steaks. Service is just okay though. Pretty typical casino restaurant. I'll go back to drop a few quarters, but when it's time to eat, I'll skip the buffet for the full-service restaurant.

PIREP   We flew into Jean, NV a couple of week ago. It is a short walk to the Gold Strike Casino and restaurant. The food was reasonably priced and good. They have a prime rib dinner that is very good.

Jean, NV (0L7) is between the California state line and Las Vegas. They have a 4600' runway for airplanes and a parallel 3700' runway for gliders.

The Nevada Landing hotel and casino is an easy walk on a lighted sidewalk from the transient parking area (75 yards). We didn't stay in the hotel, but we ate in the cafe and had a good prime rib dinner.

This is an easy flight from Southern California as you can just follow Interstate 15 to Nevada if you choose. Jean is just outside of the Las Vegas Class B airspace.

Las Vegas, NV (McCarran Intl—LAS)
Airport Deli 🍔🍔🍔

Rest. Phone #: (702) 736-1830
Location: Inside the executive air terminal
Open: Daily: Breakfast, Lunch, and Dinner

PIREP   McCarran Airport has a good little grill located inside the Las Vegas executive air terminal. It is a well-priced, clean establishment with courteous staff.

The restaurant is called the Airport Deli. They serve BLTs, pastrami sandwiches, and much more. The sandwiches are so big you need a to-go box for them. For a BLT, a can of Dr. Pepper, and bottle of Pepsi it cost $8 with tax.

Mesquite, NV (Mesquite—67L)

Wolf Creek Club House 🍔🍔🍔🍔

Rest. Phone #: (702) 345-6701
Location: Short walk
Open: Daily: Breakfast, Lunch, and Dinner

PIREP   FBO Comments: Very small operation. Park anywhere at a tie-down. Helicopter landing pad in front of the FBO building. Runway and ramp in very good shape. Decent downslope on the 5100' runway 19. Watch the wind since the field sits on top of a mesa.

PIREP   Walked from the parking ramp to the Wolf Creek Country Club where the restaurant is. It's about a half-mile stroll down the hill south of the FBO and around the driving range. The restaurant is nice, comfortable, and quiet. Food is reasonably priced and good quality. Burgers are half-pounders, and there is a full bar (for the nonpilot types). They also offer a nice selection of other items on their lunch menu. Breakfast and dinners are available too. See website for more info.

PIREP   Wolf Creek is located next to the airport and is in walking distance—on approach you can't miss it! Wolf Creek has received awards for their golf course and ranked number three in the "Best New Upscale Courses" in *Golf Digest Magazine*. I have seen it myself and it is a great course.

Additionally, their restaurant has great food. I recommend the peach melba for dessert. They have a fine dining, dinner-only restaurant that is really a treat.

I would definitely recommend Mesquite as a destination for a $100 hamburger, and it will give you the cross-country time every pilot needs too.

Casablanca Resort Hotel Casino Golf and Spa 🍔🍔🍔

Rest. Phone #: (800) 459-7529
Location: Call for pickup
Open: Daily: Breakfast, Lunch, and Dinner

PIREP   Mesquite, NV (67L) is located on the NV/AZ/UT borders. Inside the airport building are telephones that are connected to three hotel/casinos. One of our favorites is Casablanca Resort Hotel Casino Golf and Spa. They will send a courtesy vehicle to pick you up at the airport, usually a Lincoln or a limo. They have a breakfast, lunch, and dinner buffet that is very good.

Minden, NV (Minden-Tahoe—MEV)

The Taildragger Cafe 🍔🍔🍔

Rest. Phone #: (775) 782-9500
Location: On the field
Open: Daily: Breakfast, Lunch, and Dinner

PIREP FBO Comments: Soar Minden is the FBO on the field. They allowed us to park the Bonanza in front of their trailer. Soar Minden is over a dollar cheaper than Reno/Tahoe on 100LL. They also have awesome glider flights!

PIREP The Taildragger Cafe has some of the best food on the planet. They have a unbelievable Western omelet. They also have great lunches and dinner with beer on tap. I recommend a draft of Fat Tire. Their hours are from 0600 till late at night. Anyone that eats at the Taildragger will not be disappointed. The atmosphere is old school.

PIREP We flew into Minden to attend the Reno Air Races. On the morning we left the weather was marginal VFR so we decided to have breakfast and wait for thing to improve. The food was very good. The sausage and bacon were different—tasted like maybe it was a local brand and not the average. I was looking forward to maybe getting a hamburger later but the weather was good enough for us to depart by around 9:30.

Reno, NV (Reno/Tahoe International—RNO)
Amelia's 🥪🥪🥪

Rest. Phone #: (775) 858-7316
Location: Inside the jet center
Open: Daily: Breakfast and Lunch

PIREP FBO Comments: The wonderful service I have come to expect from Mercury air centers is alive and well at KRNO.

PIREP Amelia's restaurant again has new ownership; they now have a very extensive Indian menu in addition to the old favorites. If you like Indian food, you'll love the new Amelia's. If you've never had it, fly in and give it a try. Get the dinner package that includes appetizer, bread, rice, your choice of entree, and dessert. You won't go away hungry!
Location: Right inside the Mercury building
Food: Perfect
Service: Very attentive
Ambiance: Nice
Price: Fair

PIREP Amelia's is now under new management and even better than ever! Although it had been closed on weekends most often under the previous management, it's open every Saturday and Sunday now and serves an outstanding breakfast menu, including a bay shrimp omelet that is delicious!!! The bar area offers a big-screen TV, expansive view of the entire Reno International Airport, and board games for waiting out the weather or passengers who venture into the casino district for a few hours. An outstanding place for breakfast, lunch, or dinner!

Yerington, NV (Yerington Muni—O43)
The Hangar Cafe 🥪🥪🥪

Rest. Phone #: (775) 463-0088
Location: On the field
Open: Daily: Breakfast, Lunch, and Dinner

PIREP FBO Comments: Typically the best fuel prices within 100-mile radius (self serve) are at Yerington Aero, Inc.

PIREP The Hangar Cafe is open Tuesday through Sunday 6 AM to 2 PM. The food is very good, and the staff is very friendly.

PIREP The Hangar Cafe at Yerington Airport (O43) looks like your basic truck stop/greasy spoon, but the breakfasts are very good and the servings VERY generous.

Joe Dini, the Speaker of the Nevada State Assembly, operates a casino and restaurant in downtown Yerington, just a three-quarter-mile walk from the airport. Joe makes sure the food is fresh and the prices are right. It's usual American fare, but the freshness of the ingredients makes all the difference. Walk another quarter mile south to visit the Lyon County Museum on the same side of Main Street before you leave. On the way home, fly over the huge Arimetco copper pit and leach pads, with the former Anaconda company town, Weed Heights, on terraces overlooking the mine works. If you are coming from or heading north, fly true north 10 nmi to the junction of US 95A and the Southern Pacific railroad, and then follow the latter to the northwest up to the Carson River Canyon, the ruins at Fort Churchill, and over to Lake Lahontan before you leave the area. A scenic flight over historic country well worth the trip, and only about 80 nm from South Lake Tahoe or Reno.

New Hampshire

Alton Bay, NH (Alton Bay—B18)

Shibley's at the Pier 🍔🍔🍔🍔🍔

Rest. Phone #: (603) 875-3636
Location: On the airport
Open: This is an ICE runway, only open in the winter; check NOTAMs

PIREP FBO Comments: Every year (?–16 years now) they plow a runway into Alton Bay of Lake Winnapesaukee, NH. Only FAA-certified ice runway in the country.

PIREP Shibley's at the Pier offers good food at reasonable prices, plus a prime view of the ice runway and other family activities at Alton Bay (snowmobiling, ice fishing, and so on). Soup and sandwich tasted as if homemade, including the bread.

Colebrook, NH (Gifford Field—4C4)

Li Wah Restaurant 🍔🍔🍔

Rest. Phone #: (603) 237-8220
Location: Short walk
Open: Daily: Breakfast, Lunch, and Dinner

PIREP For a great Chinese meal, try Li Wah Restaurant in Colebrook, NH. Colebrook is a 2400' grass field, surrounded by mountains. A good chance to practice short field approaches and departures. Li Wah is approximately three blocks from the field. Leave the airport, turn right on dirt road, go to stop, turn right. Walk about three blocks; Li Wah is on your left.

Hampton, NH (Hampton Airfield—7B3)

Airfield Cafe 🍔🍔🍔

Rest. Phone #: (603) 964-1654
Location: On the field
Open: Daily: Breakfast, Lunch, and Dinner

PIREP   Great food, great location on a fine turf strip.
   Lots of aviation stuff, outside eating in good weather. As you can see, I like this place to fly to and to drive to.
Closes early about 12 or 1 PM, otherwise perfect.

PIREP   FBO Comments: Much less expensive than Maine.

PIREP   Breakfast and lunch served all day. We thought the breakfast items were much more tempting by smell
and taste than the lunch items. Wait times vary between 30–45 minutes for a table.
   The airport was having a "flour bomb" event on Labor Day. Obviously, the larger number of restaurant patrons
were nonpilots from the surrounding area.
   Nuttin' wrong wit dat, just plan accordingly! VERY short 2200' grass strip (accounting for density altitude)
but we hain't crashed yet!

Keene, NH (Dillant-Hopkins—EEN)

Campy's Country Kettle 🍔🍔🍔🍔

Rest. Phone #: (603) 357-3339
Location: Short walk
Open: Daily: Breakfast, Lunch, and Dinner

PIREP   Five of us had lunch today at Campy's Country Kettle. Everyone really enjoyed the meal. We've been
going there for about four years now.

PIREP   The approach to this well-maintained airport is highlighted with fantastic New England mountain views.
It's nice any time of year but in the fall it is particularly inviting when the foliage is changing. The lunch menu is
very simple and includes an all-you-can-eat buffet. I'm guessing that they change the menu for dinner and that they
attract neighborhood locals, not just fly-in customers. They also have live entertainment on the weekends.

India Pavilion Restaurant 🍔🍔

Rest. Phone #: (603) 352-6483
Location: On the field
Open: Daily: Lunch and Dinner

PIREP   This restaurant is very substandard! I have eaten Indian food in cities throughout the U.S. and in half a
dozen countries overseas. It is one of my favorite ethnic foods.
   The only thing this place has going for it is that you can taxi close to it.

PIREP   Excellent Indian restaurant, reasonably priced, and right in the terminal. Park your plane 50 feet from
your table.

Laconia, NH (Laconia Muni—LCI)

Patrick's Eatery and Pub 🍔🍔

Rest. Phone #: (603) 293-0841
Location: 2 miles
Open: Daily: Lunch and Dinner

PIREP FBO Comments: Very friendly, no fees requested.

PIREP Patrick's still provides good eats! Long walk though—must be approaching four miles round trip—not a problem if you've got your walking shoes on. The long walk is the only reason it didn't get a top rating from us, but really, we enjoyed the walk. Their pub "freedom fries" (I still can't say "french fries" with a straight face) were nice and crispy; the cheeseburger club sub cooked exactly to order, and Brenda's veggie nachos were way too much to eat at one sitting.

PIREP Patrick's is a good place to eat, but we really enjoyed Sawyer's Ice Cream across from the restaurant. It's REALLY good ice cream. Plus they have great fried seafood.

When we landed, we went in the main building and went to Skybright. They gave us free transport to and from and would not take a tip. Might want to put a few bucks in gas to say thanks. They were extremely friendly.

Nashua, NH (Boire Field—ASH)
The Midfield Cafe 🍔🍔🍔

Rest. Phone #: (603) 594-0930
Location: On the field
Open: Daily: Breakfast, Lunch, and Dinner

PIREP I stopped in to the Midfield Cafe at Nashua (KASH) the other day and thought it was time for a quick update. The restaurant is located (you guessed it) midfield on the second story of a blue building; just ask ground for taxi directions. It is still owned and operated by Sandra Adams, although the hours have changed slightly.

PIREP We now have a great airport restaurant! The Midfield Cafe underwent new management earlier this year. The new owner, Sandra Adams, has made great improvements into the faculties as well as exponential improvements into food quality and preparation. I recently had a scallop dinner that was top quality. Others who have had burgers, pasta, and gourmet salads have made similar comments as to the great food and fantastic improvement. The patio overlooking the runway provides a nice spot to keep the kids occupied while you enjoy some of the best airport cafe food!

Sky's up, tree's down.

Newport, NH (Parlin Field—2B3)
Lil' Red Baron 🍔🍔🍔

Rest. Phone #: (603) 863-1302
Location: On the field
Open: Daily: Lunch and Dinner

PIREP Parlin Field is a nice small country airport with an excellent Mexican food restaurant right next door to the terminal building. They also serve ice cream during the summer.

There is a small river with a covered bridge at the end of the grass strip.

Portsmouth, NH (Pease International Tradeport—PSM)
Paddy's Restaurant 🍔🍔🍔🍔

Rest. Phone #: (603) 430-9450
Location: Close by
Open: Daily: Lunch and Dinner

PIREP   Land at KPSM on an 11,400 ft runway; taxi SE to either of two FBO's; PANAM Services or Port City Air.

Get a lift to either Paddy's or the Red Hook Brewery. Both are on the former SAC base confines. In fact Paddy's is the former officer's club. Both restaurants offer good cuisine.

Paddy's has a good selection of TVs and pool tables accompanying their dining room.

Red Hook Brewery has a pub atmosphere while offering what they do best: beer!

The Hundred Dollar Hamburger

New Jersey

Belmar/Farmingdale, NJ (Monmouth Executive—BLM)

Runway 34

Rest. Phone #: (732) 919-2828
Location: On the field
Open: Daily: Lunch and Dinner

PIREP FBO Comments: Run by Colombia Air. They do a great job.

PIREP Restaurant is located on the field near the fuel ramp. They are open for lunch and dinner only. They open at 11 AM and stop seating at 9 PM. Gourmet food is served, and there is also a bar. Service is good, and food is moderately priced. Large variety from steaks and pasta to seafood. Some desserts are homemade. Good presentation.

There is also live entertainment on Thurs., Fri. and Sat.

PIREP Runway 34 is a short walk from the FBO. The airport authority established mandatory landing fees regardless of fuel purchase. I wonder if they'll continue to collect them?

Blairstown, NJ (Blairstown—1N7)

Runway Cafe

Rest. Phone #: (908) 362-9170
Location: On the field
Open: Daily: Breakfast and Lunch

PIREP FBO Comments: Excellent parking spaces by the office and cafe—they are of the asphalt "pull-through" variety, with chocks just sitting there waiting for you.

PIREP Arrived for lunch on a Tuesday at 12:15 PM. Was seated immediately. The waitress was very nice though very rushed. The food was excellent and extremely affordable. The only tricky part was finding the airport in the valley when approaching from the east!

PIREP FBO Comments: Self-service fuel available.

PIREP Nice restaurant!!! Small but good food. We went for lunch; we had a burger and a turkey sandwich— good choices!!! We rated the burger at 3.5. The airport has a lot of gliding activity but it is manageable. The field is nice to walk around, interact with other pilots, and see those gliders land and take off.

Caldwell, NJ (Essex County—CDW)
The Tuscany Grill 🍔🍔🍔🍔

Rest. Phone #: (973) 808-7100
Location: Short walk
Open: Daily: Lunch and Dinner

PIREP FBO Comments: CAS is a very general-aviation friendly FBO.

PIREP Tuscany Grill is an excellent place to stop by and enjoy a great dinner. Have gone back quite a few times and have yet to have a bad meal. The owner walks around and greets you with a friendly "hello." I have dressed in just shorts and I have been there in shirt and tie. No matter how I go though, I always feel comfortable. It is a short walk from the tower and across the street. This is definitely worth trying.

PIREP The Chinese restaurant across the street from the airport has become the Tuscany Grill. The food is good Italian and reasonably priced, much better fare than at the 94th Bomber Group.

Cross Keys, NJ (Cross Keys—17N)
Long Delay Cafe 🍔🍔🍔

Rest. Phone #: (856) 629-3033
Location: On the field
Open: Daily: Breakfast and Lunch

PIREP The restaurant is located at the approach end of 27. The food is good, service is friendly, and prices are reasonable.
　　If you can get a window seat you can rate everyone else's landings. Watch for parachuting! The landing zone is just to the north of the runway at midfield.

PIREP The best part about the cafe is that it is now located at the back of the airport, where the skydiving school is located, so from the large deck you can watch the skydivers land.
　　The restaurant is now open 6 days a week Tues., Wed., and Thurs. from 8 AM to 3 PM; Fri., Sat., and Sun. from 8 AM to 5 PM.

Lincoln Park, NJ (Lincoln Park—N07)
The Clubhouse 🍔🍔🍔

Rest. Phone #: (973) 686-0700
Location: On the field
Open: Daily: Breakfast and Lunch

PIREP Date 3.19.06. Small diner-type restaurant on east side of runway at north end. Omelets, salads, burgers, and so on—cheap and good. Deck with awning and umbrellas facing the runway—nice during summer; deck closed during colder months. Two dining areas, one with great view of north end of runway.

PIREP This restaurant is awesome. I understand it has recently changed hands. The food is excellent. The atmosphere is casual, and most of the people who eat there are pilots, so you'll feel at home no matter where you're from. The restaurant is on the taxiway and the view of the tie-downs, taxiway, and runway is beautiful. The airport is nice to fly into; very much alive and well maintained. The overall package is absolutely a five-star rating.

Lumberton, NJ (Flying W—N14)

Flying W 🍔

Rest. Phone #: (609) 267-8787
Location: On the field
Open: Daily: Lunch and Dinner

PIREP   After a beautiful afternoon on the beach, my family was looking forward to a great dinner at Flying W on the way to our home airport. We took a chance by arriving without a reservation. The hostess said that they were really busy. Surprisingly, approximately only half of the tables were full.

We waited over 55 minutes for our meals. They still did not arrive after telling our waiter that we really needed to be served as we wanted to get home to take Fido out. We overheard that due to a large party, perhaps 16, that service was delayed. Two couples and their children who were seated next to us got up and left after waiting and still not receiving their meals. When I spoke with the owner, he said that two servers had not reported to work.

That was the only explaination or compensation that I was offered. I ended up paying $23 for beverages, my son's piece of pizza, and my appetizer. We flew home hungry, yet with a bad taste in our mouths. We will not return.

Manville, NJ (Central Jersey Regional—47N)

Pizza & Pasta 🍔🍔🍔

Rest. Phone #: Unknown
Location: Short walk
Open: Daily: Lunch and Dinner

PIREP   Had a great gourmet pizza at Pizza & Pasta, and a waitress nicer than Mom!

PIREP   Just at the entrance to the airport, about a quarter-mile walk from the FBO, is Pizza & Pasta, which features pizza cooked in wood burning ovens as well as a lot of other great food. Informal, and a favorite hangout for the locals. Pizza & Pasta is a good place to eat.

Millville, NJ (Millville Muni—MIV)

Antino's Cornerstone Grill 🍔🍔🍔🍔🍔

Rest. Phone #: (856) 293-7771
Location: On the field
Open: Daily: Lunch and Dinner

PIREP   We flew up from Norfolk (ORF) to Millville (MIV) to try the five-hamburger Antino's Cornerstone Grill. Millville is a really great airport (WWII era). We were surprised it was not controlled. We parked at Big Sky Aviation. The gent that came out directed us to what appeared to be a large outhouse. It was bright yellow and right on the tarmac. We got there just a little after 12 and it was filling up. We counted the chairs and it appears able to seat 30 at one time.

The workers were very nice and friendly. The food was very good and cheap (for New Jersey). We recommend it for a fun fly for lunch. Great food.

Just all-around fun and filling. Then back to Norfolk. Tomorrow, we go the Charly's in Williamsburg. Eaten there several times.

Always great.

PIREP   FBO Comments: bought gas at Big Sky Aviation—friendly, fast, right next to the P47 T-bolt museum and Antino's Grill.

PIREP   Food at Antino's Cornerstone Grill was GREAT for what looks like a greasy spoon grill resembling an old Woolworths counter! They specialize in panini, and they were uniformly wonderful. I asked them to make me a turkey, bacon, avocado, and tomato, and they are gonna add it to the menu—they liked it so much!

Wonderful stop for fuel/food, and so on on the NE/SE corridor—outside the manic NY and PHL areas.

PIREP   Millville Radio was excellent and professional, announcing other aircraft in the area, landings, takeoffs, and inbound. A pleasure when arriving and departing. The restaurant is located near the end of runway 14 so that you can see the view of aircraft arriving and departing from it. I took my parents and four-year-old son for lunch there and I was very pleased with the courtesy the FBO gave us when we landed and parked, as well as the fantastic lunch we had at Antino's.

We arrived shortly before closing time and sincerely enjoyed our lunch there. The waitress and the service we received was excellent. The food came out shortly after we ordered, and it was a cozy and relaxing environment. Nice menu selection to choose from. Try out their panini sandwiches—excellent! I would recommend for all to go and enjoy but carefully read their hours of operation so that you won't be disappointed when you get there. It was my parents' first time in a small aircraft (Piper Archer 2) and going there made it all the worthwhile. They were happy and so was I.

PIREP   Great place to grab breakfast or lunch—inexpensive, clean, quick. Good view of runway 14.

Mount Holly, NJ (South Jersey Regional—VAY)

Runway Cafe 🥪🥪🥪

Rest. Phone #: (609) 518-0400
Location: On the field
Open: Daily: Breakfast, Lunch

PIREP   The menu is mostly burgers and sandwiches, and the food is good. It's located right off the ramp in the FBO, so it's very convenient. The staff was very helpful, even giving us some ice for our cooler before we took off. When we stopped there, they were open for breakfast and lunch, closing at 4 PM.

PIREP   Excellent food and service. Staff is always friendly and willing to go the extra mile. Food comes out hot and is prepared fresh, but you don't have to wait long. Lunchtime has a good crowd of locals and pilots. Weekends are primarily pilots doing the $100 thing.

Ocean City, NJ (Ocean City Muni—26N)

Airport Restaurant 🥪🥪🥪

Rest. Phone #: (609) 399-3663
Location: On the field
Open: Daily: Breakfast, Lunch, and Dinner

PIREP   My fiancé and I spent yesterday at Ocean City, NJ. The airport staff were friendly and helpful, the airport was easy to find, and the tie-down ropes were as reported, stout and in good condition. The aircraft parking fee is now $10.

Breakfast at the on-field diner was excellent with quick service and a friendly staff.

PIREP Ocean City, NJ has a small cafe on the field. Serves good breakfasts. Typical of most airport cafes. Busy in the summer, especially on weekends. The airport is close to the beach, and a short cab ride gets you to a nice boardwalk with plenty to do.

Pittstown, NJ (Sky Manor—N40)
Sky Manor 🍔🍔🍔🍔

Rest. Phone #: (908) 996-3442
Location: On the field
Open: Daily: Breakfast and Lunch

PIREP I stopped for lunch en route KJYO from KMVY. Outstanding food and very reasonable prices. Restaurant closed Tues. and Wed.; dinner on Friday night only during the winter. Great view of the approach end of 25 while you eat.

Sky Manor (N40) is under new management and new ownership. The facility and business upgrades are impressive, and more are on the way. During my stop, seven airplanes came in, the CAP used the site for training, and instructors taught students, all while great food was served by an outstanding staff. Meanwhile, the largest collection of Beriev 103 seaplanes sits on the ramp for inspection, and numerous home-builts based there provide owners to talk to.

Sky Manor is routinely discussed as one of the better fly-in lunch destinations in the DC area.

Robbinsville, NJ (Trenton-Robbinsville—N87)
Divots Bar and Restaurant 🍔🍔🍔🍔

Rest. Phone #: (609) 259-1010
Location: On the field
Open: Daily: Breakfast, Lunch, and Dinner

PIREP Divots Bar and Restaurant is located on Miry Run Golf Course, which surrounds Robbinsville runways 29 and 11 (overlooking the approach end of 11). Enjoy great home-style dinner specials and sandwiches while overlooking the airport. This is a good place to enjoy a discount round of golf for parties flying in and a meal afterwards. Call from the Kenmarson AeroClub, and if available, they will even send a golf cart down to the airport to pick you up, although it is within an easy walk.

Sussex, NJ (Sussex—FWN)
Airport Diner 🍔🍔🍔

Rest. Phone #: (973) 702-7324
Location: On the field
Open: Daily: Breakfast and Lunch

PIREP I am 12 years old and this is the first restaurant I flew to with my dad. I liked the food and really liked the flight. I hope to get my pilot's license so I can fly to these places.

PIREP We arrived around 11:30 AM and both breakfast and lunch were still available. The waitress was courteous and relatively friendly. I had the short stack of pancakes and a side of sausage links, although I could have had patties. The pancakes were two of the largest I have seen in a long while. The two links of sausage were cooked to perfection and delicious.

My wife had two scrambled eggs with home fries and whole-wheat toast. She wanted the eggs very well done, but they were just a bit too moist for her; there seemed to be more than two eggs. Most people would have liked them. The home fries were good.

The atmosphere was typical of a small diner.

Teterboro, NJ (Teterboro—TEB)

Segovia's 🥟🥟🥟

Rest. Phone #: (201) 641-4266
Location: 5-minute drive
Open: Daily: Lunch and Dinner

PIREP   If you get into Teterboro (TEB), try Segovia's. It is about five minutes from Jet Aviation. Excellent shrimp. Prices moderate.

Keep the blue side up.

The Hundred Dollar Hamburger

New Mexico

Alamogordo, NM (Alamogordo-White Sands Regional—ALM)
Airport Grille 🍔🍔🍔🍔

Rest. Phone #: (505) 439-1093
Location: On the airport
Open: Mon.–Fri. 7:30 AM to 3 PM, Sat. 8 AM–2 PM

PIREP FBO Comments: Ed's Flying Service, not much for friendliness or service.

PIREP Linda Madron runs the Airport Grille and Linda's Catering. Try the Southwest chicken breast with green chiles for $6.75 or the Southwest bacon melt hamburger with green chiles for $5.95. Delicious desserts. Daily lunch specials.

Albuquerque, NM (Albuquerque Intl Sunport—ABQ)
The Prop Wash Cafe 🍔🍔🍔

Rest. Phone #: (928) 757-4420
Location: On the field
Open: Daily: Breakfast and Lunch

PIREP FBO Comments: very friendly line staff, very quick service. We will come back!

PIREP After a long four hours in the Cessna 182 from Las Vegas, NV, we appreciated the friendly welcome at the restaurant and the good food, promptly served and for a very good price. Free refills on the drinks! Good news for pilots who dehydrate fast. The building is nicely decorated; somebody has definitely done his homework here. I will come back with my family. My two little sons will love eating hamburgers and watching the busy traffic pattern at the some time.

Carrizozo, NM (Carrizozo Muni—F37)
Outpost Bar & Grill 🍔🍔

Rest. Phone #: (505) 648-9994
Location: About a mile walk
Open: Daily: Lunch and Dinner

PIREP Carrizozo, NM (F37) has a great restaurant in town—about a mile walk from the airport—that makes great green chile cheeseburgers! I don't remember the name, but there is really only one place in town to eat. The FBO will direct you there. The FBO is great at Carrizozo—great fuel prices.

Deming, NM (Deming Muni—DMN)
Si Senor Restaurant 🍔🍔🍔

Rest. Phone #: (505) 546-3938
Location: 1 mile
Open: Daily: Lunch and Dinner

PIREP The Si Senor Restaurant is the place recommended by the FBO, and it's a great family Mexican restaurant, worth stopping by. At the FBO, kindly ask for their courtesy van and directions to the restaurant, which is about a mile away in downtown Deming.

The address is 200 E. Pine, Deming, NM 88030.

Grants, NM (Grants-Milan Muni—GNT)
Iron Skillet 🍔🍔🍔

Rest. Phone #: (505) 285-6621
Location: 1 mile
Open: Daily: Breakfast, Lunch, and Dinner

PIREP I had lunch today in Grants, NM at a truck stop!

I called ten miles out and was surprised to get an answer. The man who runs the airport greeted me and asked if I would be needing gas or the use of the courtesy car. I borrowed the car and went about a mile down the road to the Petro Stop Truck stop. For people traveling through you can even buy a shower at the truck stop. The food is wholesome and rib-sticking.

The airport manager lives on the airport in a Fifth Wheel behind the terminal. He said he is there from 8 AM to 5 PM every day including weekends. Stop by and say hi.

Las Cruces, NM (Las Cruces International—LRU)
Crosswinds Grill 🍔🍔🍔🍔

Rest. Phone #: (505) 525-0500
Location: On the field
Open: Daily: Breakfast, Lunch, and Dinner

PIREP Crosswinds Grill is inside the Adventure Aviation FBO. Catering available with advance order 24/7. Open until 1500. After 1500, you can order from the "After Hours" menu. The food was yummy; great New Mexico fare.

PIREP Crosswinds Grill is still going strong! I flew a T-41 out there from Holloman AFB last week and enjoyed some wonderful New Mexican food. My wife had the grilled chicken quesadilla, while I had the bacon cheeseburger. Huge portions (watch the weight-n-balance). They even rolled out a red carpet and golf cart to haul us in! I got what my $100 paid for.

Lordsburg, NM (Lordsburg Muni—LSB)

El Charro 😋😋

Rest. Phone #: (505) 542-3400
Location: 3 miles
Open: Daily: Lunch and Dinner

PIREP We stopped in Lordsburg for fuel on a trip. The lady at the FBO recommended a small Mexican food place in town. It was a family owned, very small place, but the food was great. I don't remember the name of the restaurant, but the FBO could tell you. It is only about three miles from the airport and they have a courtesy car.

Raton, NM (Raton Municipal/Crews Field—RTN)

Pegasus Aviation 😋😋😋

Rest. Phone #: (505) 445-3076
Location: On the airport
Open: Daily: Breakfast and Lunch

PIREP Keith and Fern Manglesdorf are the owner/operators of Pegasus and are very gracious hosts. Though there isn't a set menu, Fern provides a good meal for a reasonable price. I had the $402.50 pancakes and sausage ($400 for the aircraft, $2.50 for the meal), but she offered several other choices for breakfast. For lunch there is always the quintessential hamburger, or a green chile cheeseburger if you want a true New Mexican treat, but call ahead and you might be surprised at what they can offer.

They also have a car available, should you want to drive the 10 miles in to town for fast food.

If you are a flatlander and haven't spent time flying in the Rockies, then perhaps Keith and Fern's son Martin can give you a short course in mountain flying. He is a CFII (with all the good manners his folks taught him) and well versed in both high altitude operations and mountain flying. You will soon appreciate why there are such long runways and high MEAs.

The bonus here is getting your picture taken by Fern. She is keeping a photo album, and you might find yourself posted beside Ted Turner, Tom Selleck, or any number of other celebrities that visit Vermejo Park Ranch, NRA's Whittington Center, or other nearby attractions.

Santa Fe, NM (Santa Fe Muni—SAF)

The Grill 😋😋😋

Rest. Phone #: (505) 471-7412
Location: On the field
Open: Daily: Breakfast and Lunch

PIREP Santa Fe, NM (SAF), is a great place to stop along the way to just about anywhere in New Mexico. The Grill in the terminal has a great breakfast burrito and several other items that I have not yet tried. From the Grill or the patio you can watch the commuter jets taking off and landing, catch the chatter of the commuter passengers as they pass through, or just enjoy the warm weather most of the year. I usually park the plane at the Santa Fe Jet Center and walk to the terminal building (about 50 yards).

Try it, you'll like it!

Pasqual's 😋😋😋😋

Rest. Phone #: (505) 983-9340
Location: In town—5 miles
Open: Daily: Lunch and Dinner

PIREP   FBO Comments: Santa Fe Jet center has excellent service, a wood burning fireplace, and some massaging chairs to ease the fatigue from those long flights. There are a few courtesy cars available for two hours at a time.

PIREP   Pasqual's is located in town (approximately five miles) and has some of the best New Mexican food around. Food, service, ambiance, and price are all excellent. The four-star rating reflects its location in relation to the airport.

Santa Rosa, NM (Santa Rosa Route 66—I58)

Truck Stop 🍔🍔🍔

Rest. Phone #: (505) 472-5627
Location: 1/4 mile and on historic Route 66
Open: Daily: Breakfast, Lunch, and Dinner

PIREP   There is a truck stop within walking distance of the Santa Rosa airport. If you cut across the vacant lot, it is about a quarter of a mile. The food is basic truck stop fare.

Socorro, NM (Socorro Muni—ONM)

Owl Bar and Grill 🍔🍔🍔

Rest. Phone #: (505) 835-9946
Location: In town
Open: Daily: Lunch and Dinner

PIREP   The best green chile cheeseburgers around.
    Call Socorro taxi for a ride.

Taos, NM (Taos Regional—SKX)

Ogilvie's 🍔🍔

Rest. Phone #: (505) 758-8866
Location: 6 miles, free crew car
Open: Daily: Lunch and Dinner

PIREP   Ogilvie's is actually on the plaza. This is a more up-market restaurant, generally an excellent dinner choice. The filet mignon is excellent.
    The plaza is about six miles SE of the airport and the airport has a courtesy car.
    From my own experience it's about 1.5 hours from Colorado Springs.

The Hundred Dollar Hamburger

New York

Buffalo, NY (Buffalo Niagara Intl—BUF)

Charlie the Butcher's

Rest. Phone #: (716) 855-8646
Location: Long walk or short drive
Open: Daily: Lunch and Dinner

PIREP I highly recommend Charlie the Butcher's. Their specialty is a Beef on Wek, which is a local specialty. The beef is pink, juicy, and very prime rib–like, while a "wek" is a Kummelwek, a local salted-roll.

Try it, you'll like it. I plan my trips with stops here around lunch time specifically so I can head over to Charlie's.

Park at Prior Aviation and the courtesy car over/back makes for a quick stop. It's not fancy, but it's not like any roast beef sandwich you'll get anywhere else.

PIREP Charlie the Butcher's is a short ride and a long walk from the Prior FBO at BUF. Friendly service and reasonable prices. Good place to stop after touring the falls. By the name, you can expect to see a menu with meat-based dishes. It's not fancy though. Just down-home cooking and decor.

Cicero, NY (Michael Airfield—1G6)

Plainville Turkey Farm Restaurant

Rest. Phone #: (315) 699-3852
Location: Just across the road
Open: Daily: Lunch and Dinner

PIREP The Plainville Turkey Farm Restaurant is just across Rte. 11 from Michael Field in Cicero, NY, just four miles north of Syracuse International Airport. The portions are huge, and the turkey is the best this side of Grandma's. (Ask any upstate mother about Plainville turkeys.)

Syracuse Approach gave vectors to find the field, as it is beneath/inside Syracuse Class C airspace.

Dansville, NY (Dansville Muni—DSV)

Buckhorn Restaurant

Rest. Phone #: (716) 335-6023
Location: Short walk
Open: Daily: Breakfast, Lunch and Dinner

PIREP   It's a truck stop!

Just a short walk west from the tie-down ramp at Dansville and under the expressway underpass, you will find the Buckhorn Restaurant. Not only is this a good food place to eat, it is a part of the Truck Stops of America chain of truck stops and therefore not only has large portions of great food, but also restrooms/showers, sleeping area, store, and telephones at many booths (you can update your weather briefing over the salad course). After trying to finish that bottomless glass of ice tea, the walk back to the airport is very good for digestion.

PIREP   Dansville, NY (DSV) has an abundance of eating places within walking distance of the airport. McDonald's, Arby's, Burger King, Subway, and a truck stop with a 24-hour restaurant and motel.

East Hampton, NY (East Hampton—HTO)
John Papa's Cafe 🍔🍔

Rest. Phone #: (516) 324-5400
Location: 10-minute cab ride
Open: Daily: Breakfast and Lunch

PIREP   This was our first trip to East Hampton. The staff at the airport was very friendly and called a taxi for us. The taxi arrived within five minutes (a Coupe DeVille taxi). The airport lies between the towns of Sag Harbor and East Hampton. We went into East Hampton, a short 10-minute ride in Cadillac style. We ate at John Papa's, a cafe located in the town square. Breakfast was very good and the service was great. After breakfast we walked around town, which was decorated for Christmas. There are several small shops within the square.

We met the taxi at the time we told him we'd like him to take us back to the airport and we were on our way home to finish wrapping presents (some of which we purchased in East Hampton). Round trip, airport to town taxi ran $22 (a bit high, but that's the Hamptons), but the quick, courteous service made up for it. I understand HTO is very busy in summer, with all the N.Y. City executives coming in for the weekends. This day was not busy, but there were some beautiful sea planes on site, and we took several pictures.

No sight of Jimmy Buffett's albatross; I guess he flew south for the winter. Hope to see it in the summer! I give it a two-burger rating, due to the location factor of the restaurant.

Elmira/Corning, NY (Elmira/Corning Regional—ELM)
National Warplane Museum 🍔🍔🍔

Rest. Phone #: (607) 739-8200
Location: On the field
Open: Daily: Lunch

PIREP   Within the National Warplane Museum, there is a great little snack bar that has wonderful burgers, fries, o-rings, and so on! On top of the great food, you will find a wonderful display of WWII-to-modern military aircraft. It is definitely worth the trip!

Fishers Island, NY (Elizabeth Field—0B8)
The Pequod Inn 🍔🍔🍔🍔

Rest. Phone #: (631) 788-7246
Location: Short walk
Open: Daily: Lunch and Dinner

PIREP A must-see for a little getaway. Fisher's Island, also known as Elizabeth Field, is the perfect place to land for a day or for the weekend. An uncontrolled field, with PCL, is nestled right on the edge of the island. Occupied by 700 people in the winter and 2100 in the summer, you would never know you were in New York. Only one real restaurant/bar/inn on the island, The Pequod Inn. Great food, great people, and great atmosphere.

Freehold, NY (Freehold—1I5)
Freehold Inn 😋😋😋😋

Rest. Phone #: (518) 634-2705
Location: Short walk
Open: Daily: Lunch and Dinner

PIREP You can picnic and swim in Catskill Creek without leaving the airport. Wonderful!

An uncontrolled country airport located at the base of the northeastern Catskill Mountains. A half mile in one direction is the Freehold Inn, a great new restaurant, and a mile in the other direction is the town of East Durham, which is a summer Irish enclave. Hotels, motels, and B&Bs abound.

Transportation can be a problem unless prior arrangements are made, but walking is not that bad.

Fulton, NY (Oswego County—FZY)
Caroline's Cafe 😋😋😋

Rest. Phone #: (315) 592-5277
Location: On the field
Open: Daily: Breakfast and Lunch; they stop serving at 2 PM

PIREP This is a classic. Taxi right past the restaurant, park, and have a great meal. We had a meal for two for less than $12. Food was homemade and wonderful. Full of friendly airport bums.

The ONLY downside is they stop serving at about 2, but they seem to tolerate late arrivals okay.

PIREP Recently stopped at Oswego-Fulton County airport (KFZY) in upstate New York. Cafe there recommended by a pilot at Syracuse (KSYR). Typical, old style, GA airport restaurant, nice, homemade food, good menu, daily specials, serving breakfast and lunch (I don't think they were open for dinner). Service marginal, but the airport waives the landing fee (twins and larger) if you eat at the restaurant! Good food, inexpensive, three-burgers/food; two-burgers/service.

Glens Falls, NY (Floyd Bennett Memorial—GFL)
Briefing Room Cafe 😋😋😋

Rest. Phone #: (518) 798-8190
Location: On the field
Open: 8 AM–2 PM Tues.–Sun.; closed Mon.

PIREP Stopped into Glens Falls yesterday and had myself a great burger at the restaurant in the terminal, The Briefing Room, I believe. I was met by someone from the FBO and given a ride to the restaurant (it was less than a quarter mile). They had seating inside and out, and the service was quick and friendly. It is kind of small though inside—on a cold day it might be tough to find a seat.

The field itself is in fine shape, a 5000' strip. It looks like it may have been an old military training field.

PIREP Rebecca and John Mandell are the owners of the Briefing Room Cafe. We roast our own meats and serve fresh deli breads.

Please come give us a try!

Great Valley, NY (Great Valley—N56)

Eddy's 😋😋😋

Rest. Phone #: (716) 945-5106
Location: Across the road
Open: Daily: Breakfast, Lunch, and Dinner

PIREP   Great Valley N56, near Jamestown, NY, has a good restaurant, EDDY'S, near the field. Good food at reasonable prices, friendly people.

Park airplane near end of 24 and walk across the street (500'). Restaurant and airport owned by Mr. Eddy.

Hamilton, NY (Hamilton Municipal—H30)

The Colgate Inn 😋😋😋

Rest. Phone #: (315) 824-2300
Location: Cab ride
Open: Daily: Lunch and Dinner

PIREP   FBO Comments: Small, very friendly, very comfortable basic FBO.

PIREP   The Colgate Inn is a wonderful place tucked away in the small village of Hamilton. They have a very interesting menu, with lots of interesting dishes including good vegetarian dishes. And the prices are very reasonable for such outstanding food. The main restaurant is really upscale but closed on Mondays. The Tap Room has the same good food but a more informal atmosphere. Located closer to the airport (quarter mile) is a diner that is highly recommended but also closed Monday.

You have to get a cab into town ((315) 824-2227), but it's only $10 round trip. I am betting this is a lovely town on a fall day. There are other small restaurants in town, so you have a lot of choices.

The airport is a nice one-hour or less flight from about anywhere in upstate NY, so it makes a perfect place to take someone special.

Hudson, NY (Columbia County—1B1)

Meadowgreens 😋😋😋

Rest. Phone #: (518) 828-0663
Location: Next door to the airport
Open: Daily: Breakfast, Lunch, and Dinner

PIREP   They have a great brunch buffet from about noon to 2 PM. They even serve free champagne (though I wouldn't recommend this for the flying pilots!). Park your plane on the east (?) taxiway, push it off to the side so other planes can pass, and walk across the grass to the restaurant abutting the airport.

Old Rhinebeck is only about 10 miles to the south.

PIREP   Had a very good brunch. Worth the trip for me. Fuel was reasonable. Great for anyone looking for a quick nine holes of golf also. Golf course next to airport. Only bad thing is paying extra for the soda. We had four people and drank eight sodas. It cost us $12 just for drinks.

Ithaca, NY (Ithaca Tompkins Regional—ITH)

Wanderlust Cafe 😋😋😋

Rest. Phone #: (607) 266-0007
Location: On the field
Open: Daily: Breakfast and Lunch

PIREP   The Landing has reopened under the name of Wanderlust Cafe; however, they close at 6 PM and stop serving prior to that, so come early.

Jamestown, NY (Chautauqua County/Jamestown—JHW)
The Airport Restaurant 🍔🍔

Rest. Phone #: (716) 664-4836
Location: On the field
Open: Daily: Breakfast and Lunch

PIREP   FBO Comments: No response from the ground on the CTAF frequency. The AWOS was wacky as it reported winds of 110. An airplane on the ground confirmed that the wind was favoring R13, but I landed on 13 with a tailwind. R31 or R25 would have been preferable. A very sleepy little airport with little activity.

PIREP   I was last here many years ago and the restaurant was named Tailwinds. It apparently has changed hands as the name is now The Airport Restaurant. Very creative. The inside has been changed to accommodate a larger secured waiting area for airline pax and now there are only a couple of booths on the window wall that overlook the ramp. This area of the restaurant was closed today, so you had to sit with a nice view of the pax lounge. The food was good and the prices very good ($21 for lunch for three including dessert for two of us). The service was lousy, though. I think the waitress was new—at least I hope so. We had to ask for everything: spoons to stir our coffee, silverware when our food came, ketchup, refills on the coffee, and my son had to chase her down to get dessert. I'd go back, but it won't be at the top of my list. Then again, it seems that airport restaurants are rapidly becoming extinct in my area (NY/PA).
    The restaurant is a short walk from the parking in front of the FBO.

Johnstown, NY (Fulton County—NY0)
Airport Diner 🍔🍔🍔

Rest. Phone #: (518) 762-4748
Location: On the field
Open: Daily: Breakfast, Lunch, and Dinner

PIREP   FBO Comments: Valley View Aviation a/p maintenance and fuel sales.

PIREP   The food is inexpensive and good. Breakfast served anytime. Located just off ramp and open daily 8 AM to 8 PM. Nice people.

Middlesex, NY (Middlesex Valley—4N2)
Airport Cafe 🍔

Rest. Phone #: Unknown
Location: On the field
Open: Saturday and Sunday breakfast May–Sept.

PIREP   Great old-fashioned grass strip. Very clean and great food. You'll think you're back in the 1940s.

Millbrook, NY (Sky Acres—44N)

The Perfect Landing 🍔🍔

Rest. Phone #: (845) 677-5010
Location: On the field
Open: Thurs.–Fri. 7:30 AM–3 PM, Sat.–Sun. 7:30 AM to 4 PM

PIREP I visited Sky Acres (44N) and found that it had new management and a new name. It is now called The Perfect Landing and is open Thursday and Friday from 7:30 to 3 and Saturday and Sunday from 7:30 to 4. Food was very good and so were the prices. Nice airport and friendly people. Give it a try.

Montauk, NY (Montauk—MTP)

Rick's Crabby Cowboy 🍔🍔🍔🍔

Rest. Phone #: (631) 668-3200
Location: At the end of the runway and on the water
Open: Daily: Breakfast, Lunch, and Dinner

PIREP FBO Comments: Didn't get fuel but don't forget the $15 landing fee at MTP. Nice little airport and worth the fee!

PIREP Great food! Nice location, at the end of runway 24, across the road and on Montauk Sound. Outside seating is available also. Nice airy atmosphere, great food, great service. Nice family-type place. Great view of the water.

   It was very quiet when we were there but it was after lunch and I suspect this place gets real busy for dinner. Rick came over and thanked us for coming and gave me a poster to take home and post at my home FBO.

PIREP The Crabby Cowboy is across the street, about a one-minute walk. Nothing fancy, but they have indoor and outdoor seating on the docks, on the water. The beach is a short walking distance, over the dunes at the end of the runway. The operator at the FBO will give you a ride to the beach in a courtesy van. Landing fee $15. The restaurant is open for breakfast lunch and dinner seven days. Very limited menu, five items for breakfast. Burgers and such for lunch and dinner. Walk to the counter to order, paper plates, plastic forks, and so on. All in all a good place to get something to eat and spend the afternoon at the beach.

Montgomery, NY (Orange County—MGJ)

Rick's Runway Restaurant 🍔🍔🍔

Rest. Phone #: (845) 457-6323
Location: On the field
Open: Tue.–Fri. 11:30 AM–8 PM, Mon. 11:30 AM–4 PM, Sat.–Sun. 7:30 AM–8:00 PM

PIREP Try the specials, written on a blackboard as you enter. The mushroom and gouda cheese panini, or the special grilled chicken with basil leaves and tomato sandwich, are well worth a trip! My son loves the burgers—and the fries are some of the best anywhere!.

PIREP Call ahead–they will make and hold your order if you're in a rush!

PIREP My wife and I flew into MGJ this morning to visit Rick's for breakfast. The cafe is nicely appointed, especially compared to the previous "decor", but the prices are very steep for just an ordinary breakfast. My wife had a cup of coffee, one egg, home fries and toast, I had French toast, side of bacon and small OJ, price was $15.14 plus tip and the service was just ok. I would say we were very disappointed, would rather go to other airports enroute for a better breakfast at a lower price. Compare to 1N7, Blairstown, NJ, yesterday with my buddies, three of us had pancakes (full stack), they each had fresh strawberries, and we all had a side of bacon (gotta' cut down on the bacon), two of us had a small OJ, and two coffees for a total of $16.23…and I spent less time on the taxiway burning fuel and did not have to walk around the building to enter.

PIREP   Rick's has opened—and are we ever glad to have our cafe back! I ate there and found it to be clean with good food. Even a picky semivegetarian can find fare other than the typical burger—though my instructor polished off his burger with flair—almost made me renounce my meat-free decision! The Portobello sandwich was wonderful, the staff friendly and smiling, and the beverages include some local specialty teas which are excellent! Rick's is open for extended lunch hours, and probably breakfast soon as well.

And MGJ is a friendly uncrowded airport with nice long runways—helpful after a filling lunch!

PIREP   Just returned to Farmingdale from MGJ and the newly opened Rick's. Well, we waited two years for someone to take over from Sue's and it was really worth it!

The new place is immaculate and the food excellent. I had a house salad and burger, my partner had the fish & chips, and a friend had apple pie a la mode.

The food was on the table in less than 15 minutes from when we sat down (and it was 3/4 full) and everything tasted as good as it looked (and that was real good). The menu is pretty simple but varied enough to suit everyone and MGJ.

This always has great sport, vintage, and experimental planes around. Today there was an autocross race on the seldom-used runway.

Newburgh, NY (Stewart Intl—SWF)

The Bella Cafe 🍔🍔🍔

Rest. Phone #: (845) 223-1606
Location: On the field
Open: Daily: Breakfast and Lunch

PIREP   The Bella Cafe inside CAS is a great place to impress new passengers. Not your run of the mill burger by any means. Beautiful facility with super accommodating staff from the ramp boys to the desk staff to the restaurant crew. They placed a red carpet at the door of my C172 before we deplaned! The $8 parking fee is waived if you use the restaurant.

They have a crew lounge with recliners and movies and super-clean modern bathrooms. Like I said, this will impress you and your passengers, but kill everywhere else for them.

Johnny D's 🍔🍔

Rest. Phone #: (845) 567-1600
Location: 5-minute drive
Open: Daily: Breakfast, Lunch, and Dinner

PIREP   I flew from Morristown, NJ (MMU) to Stewart International (SWF). Pulled up to the FBO (Ripton Aviation)—very nice FBO with weight room, sauna, pilots lounge—went to front desk and got a courtesy car. Left the airport and went to a rest called Johnny D's. It's a local diner that is out of this world! Prices are very reasonable and they are open 24 hours a day. The chicken Caesar salad is great, and their brown gravy is awesome on fries (gravy is ordered as a side dish).

SWF is also an ANG base so be prepared to see C-130 and C-5's on the tarmac and practicing t/o and landing. Would give a four-burger rating.

Niagara Falls, NY (Niagara Falls Intl—IAG)

Goose's Roost 🍔🍔🍔

Rest. Phone #: (716) 297-7497
Location: 1 block
Open: Daily: Breakfast, Lunch, and Dinner

PIREP　I stopped by IAG with a friend on our way to do an air tour of Niagara Falls. We pulled into the FBO and the guys there directed us to Goose's Roost restaurant, which is just about a block from the FBO.

It was a nice, neighborhood type place, with breakfast any time. It had both counter and table service. The food was good, cheap, and quick. Great place to stop in if you're in the area for a falls tour.

P.S. You can also get a fairly cheap cab, I'm told, to the Falls themselves from the FBO.

Penn Yan, NY (Penn Yan—PEO)

Sarrasin's 😋😋😋😋😋

Rest. Phone #: (315) 536-9494
Location: Less than half a mile
Open: Daily: Lunch and Dinner

PIREP　Great restaurant. Right on the lake. We traveled there on a Sunday. They had a little jazz band on the deck overlooking the lake. Beautiful dining experience. The food was reasonably priced.

Highly recommended.

PIREP　A friend and I went to Sarrasin's. It has a beautiful view and good food. Very nice, classy atmosphere and not too crowded. Highly recommended, within walking distance, only about a quarter mile.

PIREP　Wonderful place! I took my 7-year-old grandson and we had a nice meal on the deck. The view is great, as were the people there. I was sorry I'd ignored the advice about the suit as there's a little park next door with a small beach and even a lifeguard.

Highly recommended; this definitely goes on my favorites list!

PIREP　You must stop and eat at Sarrasin's. It's just 3/10 mi. from your aircraft parking space at Penn Yan (PEO) airport. Great food, reasonably priced, and excellent atmosphere. The restaurant is right on the shore of Keuka Lake. Beautiful, peaceful scenery. And you can even dine out on the back veranda.

Bring your suit because they rent jet-skis too.

The airport courtesy car is usually available.

Potsdam, NY (Potsdam Muni/Damon Field—PTD)

The Airport Diner 😋😋

Rest. Phone #: (315) 268-1000
Location: On the field
Open: Daily: Breakfast and Lunch

PIREP　Airport Diner is a small diner on the 06 end of the field. No ambiance except for the friendliness of the locals. Large portions. Park on the 06 end of the airport. The buildings on that end are hangers and there are tie-downs at the end of what looks to be a driveway but is an unmarked taxiway.

Poughkeepsie, NY (Dutchess County—POU)

Lilliana's 😋😋😋😋

Rest. Phone #: (845) 462-6600
Location: On the field
Open: 11 AM–10 PM, 7 days a week

PIREP Lilliana's has delicious Italian food served by a pleasant staff. Use General Aviation parking on the SE side of the field, just outside the gate and across the street, five minutes walk tops.

It is a terrific place to hit for lunch or dinner. I have been twice and will go again.

Lilliana's Deli 🍔🍔🍔

Rest. Phone #: (845) 462-6600
Location: On the field
Open: 6 AM–9 PM, 7 days a week

PIREP Located right next door to their really good restaurant, Lilliana's also has a deli. It looks like they offer pretty good sandwiches, but I can't say as I was there for breakfast, which was really quite good.

Rochester, NY (Greater Rochester International—ROC)

Crystal Place 🍔

Rest. Phone #: (585) 436-8877
Location: Very short walk
Open: Daily: Breakfast, Lunch, and Dinner

PIREP Try the Crystal Place at Greater Rochester International (ROC). Taxi to parking on the east tie-down. FBO is Corporate Wings East. Walk out the FBO to Scottsville Rd., make a left and walk—yes, walk—less than an eighth mile and cross the street to the Crystal Place. Good Greek diner. Breakfast through dinner, 7 days. Your passengers can even get a beer.

FBO can provide shuttle if asked.

PIREP The service was good, the place was clean, and the food was awful. The name of the place is Crystal Palace. It is *very close* to the FBO. While I am a big fan of inexpensive diners, I would give it only one hamburger. The directions are accurate.

Saranac Lake, NY (Adirondack Regional—SLK)

Paula's Cafe 🍔🍔🍔

Rest. Phone #: (518) 891-4600
Location: On the field
Open: Daily: Breakfast and Lunch

PIREP FBO Comments: I don't have the exact prices, but the fuel has always been reasonable and the service excellent. These guys appreciate the business and our C175 was as welcome and well cared for as the big guys.

PIREP Paula's Cafe has been open now for about a year. They serve daily from 7–2, have very reasonable prices, and we have enjoyed every meal we have had there. We hope that they can get enough business to be successful. A great stop in a beautiful area!

Saratoga Springs, NY (Saratoga County—5B2)

Gideon Putnam Hotel 🍔🍔🍔🍔

Rest. Phone #: (518) 584-3000
Location: Short drive, crew car available
Open: Daily: Breakfast, Lunch, and Dinner

PIREP FBO Comments: The North American Flight Service is an exceptional FBO. I had called in first to check about landing and ramp fees and so on. There are no fees, and they reserved me a courtesy car for three hours.

I called about 10 miles out and a 98 Sable was parked outside with the motor running. They had a map of the town ready for me with directions clearly marked to all the restaurants and the race track I had asked about. There are no restaurants on the field, but with this service who cares.

PIREP   I had breakfast at the Gideon Putnam Hotel. The sausage links were the best I ever had. In fact I even ordered a second side! The rest of the breakfast was typical. My wife did not like her omelet. I tried a bit of it and was not impressed either. I will give the place two points for the location, even though it's not ramp-side, only because the courtesy car situation from North American makes it better than ramp-side.

Shirley, NY (Brookhaven—HWV)
Blue Skies Diner 🍔🍔

Rest. Phone #: (631) 281-9857
Location: On the field
Open: Daily: Breakfast and Lunch

PIREP   Brookhaven Airport (HWV) has always had an airport cafe, but the service and quality were marginal. Recently it changed hands and has become a very good destination. The food is good, and the variety is appetizing. I would give it two burgers.

Sidney, NY (Sidney Muni—N23)
Toddy's Restaurant 🍔🍔

Rest. Phone #: (607) 563-8465
Location: Two miles
Open: Daily: Lunch and Dinner

PIREP   Toddy's Restaurant in the village of Sidney has fantastic food for breakfast and lunch. Small-town ambiance and small-town prices. Short drive to town (less than two miles) can be arranged with any of the friendly airport dwellers.
Located on a constantly updated, well-kept airport.

Wallkill, NY (Kobelt—N45)
CAVU 🍔🍔🍔

Rest. Phone #: (845) 895-9208
Location: On the field
Open: Daily: Breakfast and Lunch

PIREP   Fantastic lunch at Cavu today. I parked on the grass to the side of the restaurant. Best catfish sandwich I've ever had. Good service. The restaurant was clean and nicely decorated. Decent prices.
Highly recommended.

PIREP   FBO Comments: Flew here for first time. Very nice atmosphere.

PIREP   Flew to Kobelt and ate at the Cavu. Very nice place to fly to for a quick bite or a nice dinner. Service was good and the seating outside by the runway was cool. Will be flying there again soon and will bring some friends.

Westhampton Beach, NY (Francis S Gabreski—FOK)
Belle's Cafe 🍔🍔🍔🍔

Rest. Phone #: (631) 288-3927
Location: On the field
Open: Breakfast and Lunch 7 days 8 AM–3 PM, Dinner Thurs.–Sat. 6 PM–10 PM

PIREP   We flew into Westhampton for what we hoped would be a nice dinner based on reviews we read here. I have to tell you that overall, we couldn't have been happier with what we found. It's a small place, but it has atmosphere. The food is a mix of Creole and Jamaican with no typical "American" options, so be prepared. It is a bit pricey, but for the quality and quantity, it's worth it if you like the style of food.

Since you're reading this report here, then it probably won't matter that it's BYOB since we can't drink and fly, but if you're staying the night remember that they do not serve alcohol but they do encourage you to bring it with you (they will keep it cold for you).

They have live music on Friday and Saturday nights and I'm told on Sunday mornings as well during brunch. While we were eating, there was an excellent blues band playing that really added to the atmosphere (there is a $5 surcharge if music is playing). The menu changes slightly every night, so call ahead and find out what's cooking!

Parking is simple: right at the base of the tower and walk right into the small main terminal building for access to Belle's. Also, you may want to call ahead to find out about wait times. We were seated right away, but when we left there were quite a few people waiting in the lounge area.

PIREP   This is a super place that is a huge favorite with pilots, local and nonaviation types alike. The food is delicious and generously portioned. Try the coconut cream pie, it's killer. If you get to Southampton (no airport), about halfway between KFOK and KHTO, there is a Belle's East. A little more upscale but excellent food and atmosphere.

Post Stop Cafe 🍔🍔🍔

Rest. Phone #: (631) 288-9777
Location: Short drive
Open: Daily: Breakfast and Lunch

PIREP   FBO Comments: Fuel prices are quite high (but they are throughout the area). Friendly and courteous service.

PIREP   Went for lunch on the weekend to Westhampton Beach and had an absolutely fantastic time! The airport staff was very friendly and helpful. Rental cars available (though I would call ahead to reserve). It's a beautiful drive throughout the area. Went to the Post Stop Cafe for lunch. Food and service was excellent.

Highly recommended!

White Plains, NY (Westchester County—HPN)

Wings of Westchester 🍔🍔🍔🍔

Rest. Phone #: (914) 995-4850
Location: On the field
Open: Daily: Breakfast and Lunch

PIREP   HPN is schizophrenic—if you fly in your L3 with a handheld radio and ask to park at general aviation, they'll direct you to a ramp 100 yards from the 737s, MD80s, and ATRs.

The restaurant, on the second floor of the new terminal between the coffee shop and the bar (both usually busy), looks right out over the ramp. No jetways yet to block the view. I'd give the whole show four burgers.

Arriving at HPN is de facto Class C. You need to coordinate with the approach facility, NY Approach on 126.40, over the city of Stamford, CT or over the Hudson River (typically they want you to fly to the field from the Tappan Zee Bridge). Departure is standard Class D procedure.

North Carolina

Andrews, NC (Andrews-Murphy—RHP)

Cherokee Cafe 🍔

Rest. Phone #: (704) 321-4566
Location: 2 miles to town
Open: Daily: Breakfast, Lunch, and Dinner

PIREP One of my favorite places is the Cherokee Cafe in Andrews, NC. The cafe is located in downtown Andrews, approximately two miles east of Andrews-Murphy Airport (RHP). Transportation is usually available.

Plain and friendly, come as you are, good country cookin'. Popular with the locals. Airport is located in a very mountainous area of western NC (VFR ONLY).

Beaufort, NC (Michael J Smith Field—MRH)

Take Your Pick 🍔🍔🍔

Location: In town, call a cab
Open: Daily: Breakfast, Lunch, and Dinner

PIREP Getting into Beaufort, NC is pretty easy and well worth the trip. Cherry Point airbase is between MRH and EWN (New Bern), and they are friendly and helpful—don't let the alert airspace scare you away.

We arrived in MRH (Michael Smith Field) at noon with the help of flight following from Cherry Point approach. MRH has three runways, so winds are not much of a problem, but today we had almost none. They helped us park and filled our tanks and even called us a cab for the short ride to the waterfront and about two dozen restaurants. The folks at the FBO will give you some suggestions, as will the cab driver. The ride was only five minutes and cost $4. There are shops, restaurants, and boats all along the waterfront. You could easily spend the day there. Finding a good restaurant is easy.

Charlotte, NC (Charlotte/Douglas Intl—CLT)

Mister G's 🍔🍔🍔

Rest. Phone #: (704) 399-2542
Location: On the field
Open: Daily: Breakfast and Lunch

PIREP   Chicken sandwiches and burgers are great! Don't expect much from the decor—this is a true greasy spoon restaurant. Ceiling is yellow with grease, and Jerry Springer is on the TV in the dining room. But you get a ton of good comfort food for your money. If you're at the main terminal you'll probably need a ride; it's a bit of a walk. It's over by the GA hangars, near the museum, just outside the airport fence next to the National Guard facility.

Edenton, NC (Northeastern Regional—EDE)
The Waterman's Grille 🍔🍔

Rest. Phone #: (252) 482-7733
Location: In town, 4 miles away
Open: Daily: Lunch and Dinner

FBO Comments: Run by a genial crew of retirees, renegade Yankees, who cheerfully admit that while "they're not from North Carolina, they got here as fast as they could." There's a crew car you're welcome to borrow—if it's running and if nobody beat you to it.

PIREP   Edenton's a wonderful little town and makes a great, if serene, weekend destination. We stopped on a road trip whim to have lunch one day a couple of years back, and like many others, fell in love with the place, its gorgeous cypressy waterfront, its lovely historic district, the neat little downtown with shops and galleries, and the friendly locals. I'm pleased to say we're here to stay!

The Waterman's Grille has good, moderately priced seafood and so on, good sandwiches, raw bar, and a full bar and extensive wine selection. The town has half a dozen excellent bed and breakfasts, one of which, The Lord Proprietor's Inn, offers a breathtakingly expensive and exquisitely prepared multicourse dinner by reservation. Very special indeed!

Most of the B&Bs are within walking distance of downtown and will gladly provide transportation from the airport, four miles away.

Elizabeth City, NC (Elizabeth City Cg Air Station/Regional—ECG)
The Circle 🍔🍔🍔

Rest. Phone #: (252) 338-3060
Location: Hughes Blvd.
Open: Daily: Breakfast, Lunch, and Dinner

PIREP   We flew into Elizabeth City and found a full-fledged field. Fuel onsite, as is a crew car. Wonderfully managed and friendly. We recommend checking out the historic downtown area. Lots of cafes and mom-and-pop restaurants have sprung up. Entire downtown has been renovated and is really nice.

We ate at The Circle just on the outskirts of the old town area, and it was very good and very reasonable priced.

Elizabethtown, NC (Curtis L Brown Jr Field—EYF)
Melvin's 🍔🍔

Rest. Phone #: (910) 862-2763
Location: In town, crew car available
Open: Daily: Lunch and Dinner

PIREP   Melvin's is approximately three miles from the field. A courtesy car is available. Been serving hot dogs and hamburgers in the same location since 1938. They are so good, people eat them for breakfast.

PIREP   Heck with the restaurants…

Taylor Aviation was serving hot dogs and hamburgers off the gas grill, chips and slaw too! Free! (Tried to put some money in the kitty—wouldn't take it.)

Real nice folks, real nice facility—sit outside in the shade and enjoy the quiet…and the right kind of noise.

Fayetteville, NC (Fayetteville Regional/Grannis Field—FAY)

Airport Cafe 🍔🍔

Rest. Phone #: (910) 343-9881
Location: Short walk from the FBO
Open: Daily: Breakfast and Lunch

PIREP   I fly back and forth from South Florida to Leesburg, VA a lot. Seems that on the return trip Fayetteville is an easy second fuel stop in a 172 or similar airplane. This is definitely no more than a two-burger rating, standard airport terminal fare. But considering that there isn't much to choose from along the way, it's a very short walk from the FBO to the terminal, the fine approaches to the airport plus friendly controllers and cheap rental cars if you're stuck overnight, this isn't a bad place to stop or hang around until the weather improves.

Goldsboro, NC (Goldsboro-Wayne Muni—GWW)

Alton's BarB-Que 🍔🍔🍔

Rest. Phone #: (919) 734-4332
Location: 10-minute walk
Open: Daily: Lunch and Dinner

PIREP   Here is a great North Carolina barbecue joint off the departure end of rwy 23 of the Goldsboro/Wayne Airport (GWW). It is called Alton's BarB-Que. He is open Thursday–Saturday and you want to talk about good… The best collard greens and hush puppies in the area. Oh, and their hot sauce is called Mile High Hot Sauce.

PIREP   FBO Comments: Very nice guys at this FBO. They have a very nice courtesy van that has air conditioning!! I called ahead and they had it waiting for me.

PIREP   Alton's is great. The barbecue is very good, and I ended up buying an extra pound of it to take home. I purchased two bottles of the Mile High BBQ sauce at the FBO. That stuff is pretty darned good. I ate the remaining barbecue with the hot sauce the following morning.

Even if the courtesy car isn't available, you can walk to the restaurant in ten minutes. There are no sidewalks along the way, though.

Wilber's BarB-Que 🍔🍔🍔🍔

Rest. Phone #: (919) 735-7515
Location: In town, crew car available
Open: Daily: Lunch and Dinner

PIREP   FBO Comments: Doug is amazingly friendly. Wonderful service. There is a courtesy car, but call ahead. Doug said sometimes they get busy, and the car is in use when people want to use it. This was the first time I had an FBO offer to clean the windshields while we were off to lunch.

PIREP   Since Alton's was closed on Tuesday when we went, we took advantage of the courtesy car, and drove 10 minutes into town to Wilber's BarB-Que. The meal was fantastic and very inexpensive.

For $6 I received a plate full of barbecued pork, two vegetables, and a basket of hush puppies.

They had a full menu of barbecue and nonbarbecue items.

Southern Living magazine honored them as the best barbecue in the South.

Hickory, NC (Hickory Regional—HKY)
The Runway Cafe 🍔🍔🍔

Rest. Phone #: (828) 328-2872
Location: On the field
Open: Mon.–Fri., Sat. 10 AM–2 PM

PIREP   FBO Comments: Beautiful FBO, staff is absolutely marvelous!

PIREP   I flew into Hickory today on a training flight. My instructor and I ate at the Runway Cafe. It was magnificent. The owner treated us like we were family. The food tasted like it came from a very expensive restaurant. The price was worth it. I recommend everyone fly there to eat. Their hours are Mon.–Sat. 5 AM–6 PM.

PIREP   I flew to the Runway Cafe from Lexington, NC for lunch. The cafe personnel, three really nice ladies, were as friendly as could be. The food is all "home-cooked" vittles. I had a cheeseburger with corn nuggets, which I love and can't get anywhere. The burger was not from a patty, it was homemade, and everything was great. They have a weekday daily menu, and even if you go on Saturday, they were willing to prepare something from that.

Well worth the trip.

Jefferson, NC (Ashe County—GEV)
Shatley Springs Inn 🍔🍔🍔🍔🍔

Rest. Phone #: (336) 982-2236
Location: Call for pickup
Open: Daily, 7 AM–9 PM, mid-April–late Nov.

PIREP   Shatley Springs, 8 miles from Ashe County Airport, is the ultimate in Southern home-style cooking. Just call the restaurant and one of the staff will come to the airport and pick you up. This place is the ultimate Southern all-you-can-eat experience.

PIREP   Until just recently, I had last eaten at Shatley Springs over 20 years ago. It has not changed one bit! The folks there are as friendly as you'll find and the food is truly awesome.

This really is a unique Southern dining experience. If you can stay on a weekend during the summer, you can get some great rafting trips out of Jefferson on the New River.

PIREP   FBO Comments: 9–5 operation, generally. Marvin Stump runs the FBO.

PIREP   The people were as nice as advertised! The food was wonderful. There is also a store with all kinds of homemade jams and relishes. Live music in the evenings during the latter part of the week. We were delivered to the restaurant by a nice gentleman from the airport who we luckily ran into after hours and then returned to the airport cheerfully by an employee of Shatley Springs Inn.

Kenansville, NC (Duplin County—DPL)

The Country Squire 😋😋😋😋

Rest. Phone #: (910) 296-1727
Location: 3 miles via crew car
Open: Daily: Lunch and Dinner

PIREP   Land at Duplin Co., NC (DPL) for a really unusual restaurant. The FBO there is really helpful and maintains a courtesy car. Go about three miles to The Country Squire. It has a large and varied menu as well as a five-star atmosphere. The restaurant resembles a very old log cabin. The waitresses are wonderful; there is an open fire in the lounge area, and the food is very good indeed. I had pork tenderloin that was excellent, and my wife had blackened chicken, which was extremely spicy. They have all alcohol permits.

It is truly a unique place.

Kill Devil Hills, NC (First Flight—FFA)

Wright Brothers National Memorial 😋😋😋😋😋

Rest. Phone #: (252) 491-5165
Location: On the field
Open: Daily

PIREP   Take a walk over to the hallowed grounds where flight was first achieved, then call a cab for a short ride to the Black Pelican Cafe on the beach. The restaurant serves gourmet wood-fired pizzas, steamed seafood, a killer around-the-world–styled menu.

It is located in an old Coast Guard lifesaving station built in 1876. It was the communications point for all the telegraphing by the brothers while they were here achieving the first flight.

This is a pilgrimage every pilot would love to make.

Kinston, NC (Kinston Regional Jetport at Stallings Field—ISO)

Ham's Restaurant 😋😋😋

Rest. Phone #: (252) 939-9560
Location: Short drive via available crew car
Open: Daily: Lunch and Dinner

PIREP   We stopped at Ham's of Kinston on February 10 for a fuel stop on our way from Orlando to Pennridge Airport in Pennsylvania. A courtesy car was available for the short and easy drive to Ham's. This is a casual restaurant with an attractive and regional decor. We enjoyed their shrimp as well as macaroni and cheese. They make their own delicious potato chips.

Service was good.

Manteo, NC (Dare County Regional—MQI)

The Weeping Radish 😋😋😋😋

Rest. Phone #: (252) 473-1157
Location: Short drive, courtesy van
Open: Daily: Lunch and Dinner

PIREP WOW!!!!!! The Weeping Radish was the BEST place for food I have ever been to. I flew down to FFA Kitty Hawk to visit the Wright Monument and then got back in and hopped over the bay there to MQI, and the staff there was great, I let them know before I landed that I was going into town and they had a van ready for me and my friends. Great place to fly into, and the Weeping Radish has the best 100-dollar hamburger ever!!! The service was great and the food came out to us fast! I'm already planning another trip down there. If you're going to MQI, make the quick stop at FFA—you can see both fields from the air—just to pay your respects to the Wright Brothers.

PIREP We flew in to Dare County Regional (MQI) on Memorial Day weekend. While the ramp was full, all the tourists must have been on the beaches. The courtesy van was available (I haven't missed it yet) to take us on the short drive into town. The Weeping Radish is highly recommended. Excellent German-style fare at a fair price. The atmosphere is right out of the Bavarian Alps. Great food.

We also toured the Civil Air Patrol Museum and the North Carolina Marine Aquarium. Both of these attractions are located back at the field. The CAP Museum tells the story of the civilian pilots who patrolled the near offshore waters for U-boats during WWII. The Marine Museum has several exhibits that were just spectacular: otters, alligators, touch pools, many fish tanks, and a large offshore wreck aquarium. You don't want to miss either of these stops.

The terminal is also a great attraction. It has all the necessary services while providing a very open, airy beach house decor. I'm just scraping the surface.

Manteo is a great launching point for several other excursions such as premier offshore charter boat fishing, Wright Memorial National Park, Hatteras National Seashore, lighthouse adventures (Currituck, Bodie Island, Hatteras, Okracoke), Nags Head Beaches, Jockey's Ridge, and so on. Certainly, put MQI on your must-see list.

PIREP FBO Comments: 24-hour self-serve gas, courtesy car.

PIREP Manteo is a terrific little town on historic Roanoke Island. Very nice terminal with friendly staff. Crew car is still available, but we recommend you call ahead and arrange a rental car so you can take your time and see the many things to see, including the site of the Lost Colony and birthplace of the first English child in America, Virginia Dare. There's really nothing within walking distance of the field.

The Weeping Radish German restaurant is still going strong, food very good, beer excellent, brewed on-site. Darrell's Restaurant is a down-home style seafood restaurant. Everything is reasonably priced and the food is quite tasty. There are numerous other restaurants in city, so there's something for everyone!

Manteo is an easy flight out of Norfolk, VA, and it's on our list of places we plan to re-visit.

Mooresville, NC (Lake Norman Airpark—14A)
Midway Marina 🍔🍔

Rest. Phone #: (828) 478-2333
Location: Medium walk
Open: Daily: Lunch

PIREP Located on Lake Norman just west of Lake Norman Airport. Lake Norman is on the CLT. sec. on the 11deg. radl. off of CLT. vor 14A.

The Marina has an excellent sandwich shop. It is a good place to sit, watch the water, eat a light meal, and maybe "wet-a-hook."

Mount Airy, NC (Mount Airy/Surry County—MWK)
The Snappy Lunch 🍔🍔🍔🍔🍔

Rest. Phone #: (336) 786-4931
Location: In town, free car available
Open: Daily: Lunch

PIREP   FBO Comments: The famous Chico greeted us while parking, and was as friendly and hospitable as can be imagined. The airport and facilities are nice, with a great view of Pilot Mt. when we departed on 180.

PIREP   My wife and I flew to Mt. Airy on a Saturday. When we got to the airport, Chico greeted us while parking the plane and was more hospitable than we had imagined. He set us up with an old Nissan for transportation into town, free of charge. The car was reliable and probably looked better than the old 150 we were flying.

We went to Snappy Lunch first, and it was a good thing we did. We got there about 11:30 AM, just in time to beat a busload of tourists that were standing 20 deep in a line out the door when we left. The place closes at 1:30 PM on Saturdays, so don't lollygag about getting there. It was crowded when we got there, but we got seated shortly after. The marquis item on the menu is the pork chop sandwich, $3.50, and was fantastic. The owner is in the front frying the pork chops as you walk in, in plain view. The place is a hole in the wall, but the sandwich was great, and the service friendly and fast.

There are plenty of shops, a local winery, Emporium, and it is Andy Griffith heaven for fans of the show.

PIREP   I have eaten at the Snappy Diner and it is wonderful. Get there early for lunch because it fills up quickly and the seating is limited. The pork chop sandwich is delicious and is filling in itself.

For you Mayberry fans, go by the visitors' center. It has a lot of wonderful old Andy Griffith memorabilia and many wonderful behind-the-scenes photographs as well as a suit donated by Andy Griffith when he played Matlock and one by the man who played Otis, the lovable town drunk.

You can also spend the night at the house Andy was raised in until he got out of high school. For more information go to www.visitmayberry.com, or you can call them at (800) 948-0949.

PIREP   I'm a pilot and my family has a cabin in the Mt. Airy area. For you Andy Griffith/Mayberry buffs, Mt. Airy is the hometown of Andy Griffith and was the concept for the show.

It would be worthwhile to take a cab into town (three miles) and eat at the Snappy Lunch (mentioned on the show) where you dine on their delicious pork chop sandwich with fries and coleslaw. The sandwich has chili on top and is good and reasonable ($4 range I think). The Snappy Lunch was built in the '20s and has the wash sink right in the middle of the restaurant!

After lunch, if you need a haircut, the original Lloyd's Barber Shop is next door.

Oh yeah, stop at the Pine Drive-In on the main drag back to the airport and get a strawberry shake.

When flying out of Mt. Airy, look south approximately five miles and view Mt. Pilot (can't miss it).

This is no joke, it's all really there. They even have a big Mayberry festival in the fall season, complete with squad cars, show celebrities, and so on. Makes for an interesting place.

Oak Island, NC (Brunswick County—SUT)

Provision Company 😋😋😋

Rest. Phone #: (910) 842-7205
Location: 3 miles; use crew car or call a cab
Open: Daily: Lunch and Dinner

PIREP   FBO Comments: Nice friendly down-home NC folks. Big skydiving operation at airport.

PIREP   Grab one of the courtesy cars or call a cab and head into Southport. Go eat at the Provision Company. Right on the water.

A hole in the wall with a million dollar location. Great crab cake sandwiches with fabulous onion rings. Worth the trip from Oak Island airport to Southport (about five miles or less). Southport is a scenic little gem right on the Cape Fear/Atlantic Ocean intersection.

Come on down, y'all!

Ocean Isle Beach, NC (Ocean Isle—60J)
Victoria's Restaurant 🍔🍔🍔

Rest. Phone #: (910) 575-4746
Location: Across the street
Open: Daily: Lunch and Dinner

PIREP FBO Comments: No fuel, no service—nice airport, though—no cars, either.

PIREP Victoria's Restaurant is located across the street from airport. Easy walk. Good food, good service.

Ocean Isle Golf Course is on other side of airport. Within walking distance: cut across airport and cross drainage ditch (not too wide). You can also call ahead for the course to pick up at property edge. Good course—very neat—well maintained and reasonably priced; challenging but fair course.

Food 4.5/5
Golf 4.6/5

Ocracoke, NC (Ocracoke Island—W95)
Howard's Pub 🍔🍔🍔

Rest. Phone #: (252) 928-4441
Location: 1.2 mile
Open: Daily: Lunch and Dinner

PIREP There is no greater getaway than Ocracoke Island, NC. W95 is the identifier of this gem. The runway is paved and sports plenty of tie-down spots if you get there early on a nice spring, summer, or fall day. The runway is between the dunes and the main island road. The island is only accessible from plane or boat. You do not need a car to have a day on the beach and wonderful fresh seafood at Howard's Pub, which is only a half-mile walk into town. No fuel at the field but close by at Manteo, Washington, or New Bern.

Enjoy!

PIREP Airstrip right next to beach as everyone mentioned. First thing we did was make a quick walk to the beach—the dolphins were there to meet us. Our kids called it the "Ocracoke greeting party."

We made a short walk into town and had lunch at Howard's Pub. Great atmosphere, great beer selection, and food a notch above bar food. Good place for lunch—call it three burgers.

We tried Sargasso's for dinner. I would call it four–five burgers—my wife said they had a unique presentation. The crab cakes were the size of softballs and were real lump crab. They had excellent mussels, and the fried goat cheese was to die for. Good kids menu as well.

On the way out, we walked the beach again and, just as the kids hoped, the Ocracoke "going away party" was out, and we watched the dolphins again. We went in December when everything was shut down. I'm looking forward to another trip in the spring to see what it's really like.

Raeford, NC (P K Airpark—5W4)
Aviator's Pub and Grill 🍔🍔🍔

Rest. Phone #: (910) 904-6761
Location: On the field
Open: Daily: Lunch and Dinner

PIREP   Super hamburgers! Outstanding barbecue! They serve breakfast, lunch, and dinner and have a great bar, pool tables, and skydiving. The best part is, they are dirt cheap.
     Be careful of MOAs and restricted airspace around Ft. Bragg and Pope AFB on the way in. Aside from that, it's a great place to visit.

Roanoke Rapids, NC (Halifax County—RZZ)
Ralph's Barbeque 🍔

Rest. Phone #: (252) 536-2102
Location: $20 rental car
Open: Daily: Lunch and Dinner

PIREP   FBO Comments: Friendly FBO—no courtesy car, despite what the APOA directory says. Call ahead and the manager will arrange for $20 rental transportation for you. The airport was a little difficult to spot coming from the north.

PIREP   The $20 car rental is well worth the added cost to get to Ralph's, easily found and less than ten minutes from the airport. Ralph's has an absolutely SUPERB North Carolina barbeque all-you-can-eat buffet for less than $7 that will make you check your weight and balance before you leave.

Roxboro, NC (Person County—TDF)
The Homestead 🍔🍔🍔🍔

Rest. Phone #: (336) 364-8506
Location: Close by, call for pickup
Open: Daily: Lunch and Dinner

PIREP   Located only a couple of miles from Person County airport in North Carolina offers the most sumptuous buffet beginning each Sunday at 11 AM.
     The line is packed to overflowing with good country fare. In addition, the sideboard is loaded with cakes, pies, and puddings the likes of which are not found within a thousand miles.
     Just have the FBO call, and they pick you up in a van. For $6.50 a head, it's the best deal in the universe.
     Pack lightly, you may over-gross the airplane on the return trip.

PIREP   Update on the food at TDF. The Homestead is still in business and serving great food. The area is growing and many more places to eat have opened in Roxboro. Still no place in walking distance. Homestead Steakhouse is the closest at about a mile away. I recommend flying in for dinner. TDF is a nice airport with a good long runway, good lights, and ILS on 6. FBO offers free cookies and coffee most of the time. I agree with the four out of five burgers.

Rutherfordton, NC (Rutherford Co–Marchman Field—FQD)
57 Alpha Cafe 🍔🍔🍔🍔

Rest. Phone #: (828) 286-1677
Location: On the field
Open: 10 AM–3 PM, closed Mon., Tues.

PIREP   The 57A Cafe is OPEN on Sundays from 10 AM to 3 PM. It is an excellent little airport cafe, and its banana pudding seems to be legendary in the area. I was even asked by an inbound aircraft upon announcing my takeoff roll if I had left any banana pudding. As for the airport, it is an easy stopover. The staff and hangar flyers are VERY friendly and helpful. It has self-serve fuel for 24/7 service. I am very pleased to have discovered this great little GA airport.

PIREP   Ron and family are a hoot. We had a fine visit on a springtime day. I'm a longtime user of FQD and 57A Cafe. The food is fresh, and the surroundings are friendly and uplifting. Try this if you enjoy a day away from the rat race.

PIREP   FBO Comments: Nice runway, easy to see while approaching.

PIREP   We flew into Rutherfordton's airport to eat at the 57 Alpha Cafe on Sunday. The crowd was much larger than usual due to a "glider gathering," which caused them to run out of hamburgers. Bummer! But, since every thing else on the menu is as good as reported, we had no problem getting a good meal.

PIREP   While on a flight returning from Peach Tree Falcon Field near Atlanta, my wife and I checked the $100 Hamburger website, looking for a lunch stop on our way back to Virginia in our Cessna 195, and found the 57A Cafe at Rutherford County Airport (FQD). We decided to give it a try, since it has such a high rating.

I can tell you that we were not disappointed!

Ron's warm welcome and great food confirmed what had been previously said.

I can say without reservation that this spot will go on my list of "must stop for lunch!"

Siler City, NC (Siler City Municipal—5W8)

Hayley Bales Steak House 🍔🍔🍔🍔

Rest. Phone #: (919) 742-6033
Location: Short drive, crew car available
Open: Daily: Lunch and Dinner

PIREP   FBO Comments: Cardinal Air personnel were wonderful. We used their courtesy car to drive into town for dinner. Note on "Pay at the Pump" fuel service: Make sure the on/off switch above the credit card reader is in the "off" (down) position before swiping card. When card reader says "Lift Pump Handle," turn this switch on (up).

PIREP   Best steak dinner ever at Hayley Bales Steak House. Use the courtesy car from Cardinal Air (best to call ahead for availability): from the airport, turn right on Aviation Rd., right at dead end onto W. 3rd St., left on 2nd Ave. over to Hwy. 64. Turn right (east) on Hwy. 64 and look immediately on your right for Best Food Cafeteria. Don't be put off by the exterior. Hayley Bales is inside this building. Nice gift shop, reasonably priced menu, and friendly service. Excellent food.

PIREP   Siler City Airport (5W8) is 30nm west of Raleigh and south of Greensboro. No on-the-field restaurant yet, but you can always borrow a courtesy car for a quick trip to town for great barbecue or family-style seafood. You can also rent a car on the field on a weekend for $20 and drive 18 miles to the Asheboro Zoo! The field is lighted and 5000' long with nice, easy approaches.

Star, NC (Montgomery County—43A)

Martha's Grill 🍔🍔🍔

Rest. Phone #: (910) 428-1417
Location: 10-minute walk
Open: Daily: Breakfast, Lunch, and Dinner; closed Sundays

PIREP   Martha's is within just a short walking distance from the ramp; probably about ten minutes. You might take a glance at the transformed gas station to your left while you are on short final on 21.

Martha is the owner and the cook and she is running the show. The restaurant is open Mon.–Fri. 4 AM (yes) to 4 PM and on Saturday from 4 AM to 2 PM and yes, Martha is there all the time.

This is a typical family-run country restaurant. The food is good, the price unrealistic (I couldn't buy the ingredients for what she charges), and the people friendly.

Don't miss it. It's a gem.

Tarboro, NC (Tarboro-Edgecombe—ETC)
33 Grill and Oyster Bar 🍔🍔🍔🍔

Rest. Phone #:
Location: Tarboro, NC
Open: Tues.–Thurs. 6 AM–9 PM, Fri., Sat. 6 AM–10 PM, closed Sun., Mon.

PIREP   Nice little restaurant, about a quarter-mile walk from the Tarboro airport. Not much to look at; gravel parking lot, concrete floor on the inside. But, I like restaurants with a rustic charm. Good seafood menu with mahi mahi, crab meat, oysters, and hamburgers. An eight-ounce steak is $11.99. Very friendly folks there. There are five tables in the dining area and about 20 seats at the bar.

Wilmington, NC (Wilmington Intl—ILM)
Diamond Foods 🍔🍔🍔

Rest. Phone #: (910) 686-1817
Location: On the field
Open: Daily: Lunch and Dinner

PIREP   Diamond Foods in Wilmington, NC has the current gift and food service at ILM. We have a very broad menu compared to large airports. It's a little tough getting from fly-in parking to the main terminal, but, hey, I'll try anything once.

We are fortunate to have a very good Airport Director and his staff is first class.

Winston Salem, NC (Smith Reynolds—INT)
Smokey Joe's 🍔🍔🍔

Rest. Phone #: (336) 249-0315
Location: Crew car
Open: Daily: Lunch and Dinner

PIREP   Taxi to the Piedmont FBO, tell them you like barbecue, and they'll give you the keys to the courtesy car and printed directions to a great barbecue hotspot. A great restaurant for good old-fashioned barbecue, highly recommended.

North Dakota

Garrison, ND (Garrison Muni—D05)

Fezziwig's

Rest. Phone #: (701) 463-2980
Location: Limo provided
Open: 8:45 AM Sat. morning

PIREP Every Sunday morning at Garrison, ND there's an informal fly-in. Show up by 8:45 AM, get a free limo ride to the local cafe, enjoy a great breakfast, and meet new pilot friends.

Grand Forks, ND (Grand Forks Intl—GFK)

The Crosswinds Cafe

Rest. Phone #: (701) 746-7231
Location: In the terminal building
Open: Daily: Breakfast and Lunch

PIREP The Grand Forks International Airport has a closely guarded secret: the Crosswinds Cafe in the Airport Terminal building. The food is great, all homemade, and the staff is friendly and helpful. Feel free to stop by and don't forget to ask about the weekly special. Of course, no trip to Grand Forks International Airport is complete without stopping by the Airport Fire and Rescue Station to say hello to Little Ricky, the airport's resident cat. (He dines on the $100 mice!)

Minot, ND (Minot Intl—MOT)

Franchise Row

Location: Short walk
Open: Daily: Breakfast, Lunch, and Dinner

PIREP The Minot International Airport no longer has a restaurant on the field. We do have several great restaurants available within walking distance of our General Aviation Terminal, including Quizno's, Hardee's, Primo, Pizza Hut, and Kentucky Fried Chicken.

Akron, OH (Akron-Canton Regional—CAK)

356th Fighter Group Restaurant & Cabaret 🍔🍔🍔🍔🍔

Rest. Phone #: (303) 494-3500
Location: On the field
Open: Daily: Lunch and Dinner

PIREP   I recently visited the 356th Fighter Group Restaurant at CAK. While we were eating, the hostess sat a nice-looking gentleman at the table next to us. After ordering our lunch my daughter tugged on my sleeve and made me take another look at the fellow next to us; it was none other than HARRISON FORD!

The waitress said this is a favorite stop of his and he always takes time out to sign autographs and take pictures. Indeed there hangs an autographed photo on the wall by the lobby!

As we were paying our tab, Mr. Ford was readying to leave himself and we talked about his aircraft and his love of flying before he signed a napkin for us.

Our waitress said Mr. Ford keeps promising to bring his friend J.T. with him sometime. I said "JT?" and she said, "That's what he calls John Travolta!"

A great restaurant and—you never know—maybe a celebrity sighting!

PIREP   I spoke to Kim at Castle Aviation; this is a new and fantastic FBO open at Akron/Canton Airport. They are a nice new group that wants business and will take care of $100 hamburger pilots.

No tie-down fees, and they will bring the pilots over to The 356th Fighter Group. We also cater for Castle, so if any pilots need us to bring food over to Castle we can and will.

Bob Scofield
Owner/operator
The 356th Fighter Group Restaurant
and Commanders Catering.
PS. The phone number for Castle Aviation is (330) 498-9333, (800) 325-4703; fax (330) 498-9322.
Seth and Kim are great to work with.

PIREP   Growing up in the Akron/Canton area I've had the pleasure of eating at The 356th Fighter Group many times. But since moving away from the area a few years ago I've been denied the pleasure. Until recently when I flew into CAK to visit family and found myself sitting once again at a table along the glass windows that overlook runway 1. The food was fantastic and the service is always great. If you have a special request they will usually do their best to cater to you. They have great WWII aviation memorabilia and even the sounds of airplanes flying overhead in the restroom. I would highly recommend making the stop into CAK to dine here.

Batavia, OH (Clermont County—I69)

Sporty's 🍔🍔🍔

Rest. Phone #: (513) 735-9100
Location: On the field
Open: Store hours

PIREP Forgive me if this offends someone, but I think it needs to be said. I think it is a shame that Clermont County, home of the world's largest pilot shop (Sporty's), does not have a restaurant. True, they do have vending machines with sandwiches and a microwave. And true, they do serve free hot dogs, mets, and brats on Saturdays at lunchtime. But I would think that they get enough traffic to justify at least some sort of restaurant or diner.

I've been to Sporty's three times and was hungry every time I landed. But a microwaved, plastic-wrapped chicken sandwich wasn't exactly what I was looking for.

Sorry, Sporty's...I hope this doesn't take my name out of the hopper for the Skyhawk drawing!

PIREP Every Saturday at Clermont County Airport (Batavia I69) Sporty's serves up free hot dogs, mets, and brats. Dogs are served from about 11:30 until 2.

Carrollton, OH (Carroll County-Tolson—TSO)

The Blue Bird Restaurant 🍔

Rest. Phone #: (330) 627-7980
Location: Short walk through the woods
Open: Odd

PIREP First things first. This certainly appears to be a very nice place. The grounds are very impressive, and even the path through the woods that leads to and from the airport is extremely nice. It's a bit of an uphill grade to get back to the airport, but it's not too steep unless one of your passengers has difficulty with normal walking.

But the service was simply abysmal. We arrived about ten minutes before 3 PM on a Sunday, with their closing time still 70 minutes away. Maybe it was the fact that we had a small child with us, but the hostess—who had just admitted the party of two who were ahead of us in line—very bluntly told us that they were out of food and that we would have to come back some other time.

Telling a white lie is one thing, but this was a boldfaced falsehood spoken with no shame at all. When you walk in from the airport, you pass directly by the kitchen and can see in the window. There were entire cakes available in there with only one or two slices taken out of them. We weren't looking for a full meal anyway. Some iced tea and cake would have been just fine! Maybe they just didn't want a four-year-old loose in the precious house, or maybe they just wanted to close early that afternoon, but the fact of the matter is that they brazenly lied to us in an effort to make us go away. It worked.

What really struck me as strange is that only the weakest of apologies were extended, and there was not even an offer for us to sit down for a few minutes and have a cold drink or two. Heck, if it were my place and I had really run out of food, I would have been embarrassed and would have offered a table and a few Cokes or iced teas on the house, just to placate an otherwise disappointed potential customer.

If you are flying in on a weekend, I guess you will need a reservation to get a seat here. You may also want to leave any little ones at home, since ours certainly didn't seem all that welcome there.

PIREP We checked the Blue Bird Farm website to see whether the restaurant is open during the week. According to the web page, it's open Tuesday through Sunday 11 AM to 4 PM. We arrived about 3 and found them closed. We were told that they close at 3 during this time of year.

In response to my statement that the website says they're open until 4, I was told, "Yes, we know."

No expression of regret, no indication that they'll make sure the website gets corrected, evidently no interest in encouraging us to make a return visit.

Well, we won't disappoint them: we won't be back.

R. and F. Airport Restaurant 🍔🍔🍔

Rest. Phone #: (330) 627-5250
Location: On the field
Open: Daily: Breakfast and Lunch

PIREP Bottom line: This is just the type of small-town atmosphere you are looking for. Great home-cooked food, nice folks, and a good view of airplanes. You are glad to be there and glad to tell friends what a nice time you had. I'll go back very often.

Cincinnati, OH (Cincinnati-Blue Ash—ISZ)

Watson Brother's Brewhouse & Restaurant 🍔🍔🍔

Rest. Phone #: (513) 563-9797
Location: Short walk, near the approach end of 24
Open: Daily: Lunch and Dinner

PIREP Great food at an active GA airport. Co-op aviation FBO very friendly and will loan you their clean van for the three-minute drive to Watson Bros. We had a steak, ribs, chops. Bruschetta and ahi (tuna) as appetizers. All was excellent. Passenger had an ale and he said it was great. There is also a Buffalo Wings restaurant on the way if you do not want to walk as far. Watson Bros. is just off the approach end of runway 24 (right traffic for 24).

PIREP My wife and I flew into Ash Airport on a lovely Sunday winter afternoon. Ash is a nice (3500 × 75) airstrip on the north side of Cincinnati, OH. A friend of mine told me if I was ever close to there to stop and eat at Watson Brother's Brewhouse & Restaurant.

After parking the plane, I caught one of the ground crewman and asked him where the restaurant was located. He stated just a half mile down the road (just follow the sidewalk). Well, since it was a warm winter day (lower '50s), we decided a good walk back and forth would be a good thing, especially after sitting in a small plane for an hour.

After a brisk 15-minute walk, we walked right up do the door and stepped in. Immediately I was glad I did, because the interior is adorned with an aviation theme, and you can see the huge beer vats sitting in the back of the restaurant. The place was also very clean and roomy.

The waiter took us to our table and gave me the microbrew beer menu, to which of course I had to say, sorry I can't have any beer (I almost cried). They have a vast selection of microbrews that really look great (next time I come there I will make sure I'm a passenger).

He gave us the lunch menu, which had a good selection of appetizers, sandwiches, burgers, pizzas, and some salad and pasta dishes. All the dishes ran under $10 and averaged about $8.

I choose the Brewmaster Burger entree, a hefty half-pound burger (made to order) and side dish, and my wife picked a smoked barbecue chicken pizza.

Both our meals were delicious and well worth the $8 charge for each meal.

I also noticed they have an outside deck that lets you watch planes take off or land on runway 24, which would be great on warmer days or evenings.

So I must advise anyone that is near Ash Airport (ISZ), to stop in and take a short walk to Watson Brother's Brewhouse & Restaurant—you will be glad you did.

I give this place a four-plus hamburger rating!!

Cincinnati, OH (Cincinnati Muni Airport Lunken Field—LUK)

Sky Galley Inn 🍔🍔🍔

Rest. Phone #: (513) 871-7400
Location: On the field
Open: Daily: Lunch and Dinner

PIREP My wife and I recently visited Lunken Field. We had lunch at the Sky Harbor restaurant. This was our third trip there, as it is a nice destination and the food is reliably good if not outstanding. We parked at the FBO right next door, fueled, and then had the use of a courtesy car that took us to a local Banana Republic/Gap warehouse outlet. My wife "saved" so much money at the outlet (truth be told we bought $1400 worth of retail merchandise for $105), that it was almost like having a free hamburger at Sky Galley!

PIREP Had an early dinner there Saturday night. Food was very good (try the steamed shrimp—excellent). You can taxi right up to the restaurant.

ATC at the airport was very nice, and accommodating. They could tell it was my first time to the airport and provided very detailed instructions.

Sky Galley and Lunken airport will be a regular destination of mine.

PIREP Had a nice Sunday lunch at the Sky Galley, Lunken Field, Cincinnati, OH. Excellent food at a reasonable price. Restaurant is located in the old terminal building, which you can park right in front of. Easy in and out, ATC (both Lunken tower and Cincinnati approach) were very accommodating. Neat airport, neat atmosphere!

PIREP The airport restaurant called the Sky Galley has a great staff and gave me a lot of info over the phone for my fiancé's birthday. We had an amazing time and a great meal. I highly recommend it. If you go, try the seafood; it is better than a lot of fancy restaurants that we have been to and is very reasonably priced. All of the food is very fresh!

PIREP I drop in to Sky Galley frequently from Lexington, KY. They're located in the Old Terminal and all you have to do is pull your plane up to the gate and get out. No need of crew cars or long walks. During the summers you can eat outside on the patio.

Love their food and the atmosphere. During the evenings there is live music.

Cleveland, OH (Cuyahoga County—CGF)

J.B. Milano's 🍔🍔

Rest. Phone #: (216) 289-4000
Location: On the field
Open: Daily: Lunch and Dinner

PIREP Last night I had dinner at J.B. Milano's at KCGF (Cuyahoga County Airport in Cleveland, OH). It's an okay Italian restaurant, located right on the field; you can walk there from the FBO. I especially enjoyed the fresh-baked bread and the stuffed portabella mushroom appetizer. The prices are reasonable.

Cleveland, OH (Burke Lakefront—BKL)

Hornblower's Barge and Grill 🍔🍔🍔

Rest. Phone #: (216) 363-1151
Location: Short walk from the FBO
Open: Daily: Lunch and Dinner

PIREP   FBO Comments: Nice airport, excellent access to downtown, exciting, and scenic approach. Stopped at Business Aircraft Center—good and prompt service. Pilot planning area is skimpy, just a pair of stand-up counters. Pilot lounge was very nice, though. Crew car offered promptly by friendly receptionist.

PIREP   Good food, good service, easy walk from either FBO. Seating available inside or on outdoor upper level. Outdoor seating is nice, but mosquitoes were big and fierce! Inside seating would probably be better.

PIREP   I decided to take a short flight with my wife and two of my kids one beautiful Saturday a couple weeks ago, and decided to visit the Hornblower restaurant for dinner, based on pireps from this website. I was not disappointed. It was a wonderful evening. The restaurant's large picture windows provide a great view of the lake, downtown Cleveland, and a WWII submarine on display, the USS Cod. (My kids especially liked the submarine.) We also loved the nautical atmosphere, made all the more convincing by the fact that the place is really a large boat converted into a restaurant. The menu and food were good too.

There is so much to do within walking distance in the area, that I would like to return sometime. We could not see it all in the short time we were there. Also, now that Meigs is gone, Burke is one of those rare and rapidly disappearing downtown lakefront airports. Experience it. Who knows how much longer you may still have the chance.

Columbus, OH (Bolton Field—TZR)

JP's Ribs 🍔🍔🍔

Rest. Phone #: (614) 878-7422
Location: On the field
Open: Daily: Lunch and Dinner

PIREP   Couldn't agree more with Mr. Hogg's comments. In fact, it's a rather significant disappointment that the box is an infrequent guest at TZR because the practices were as fun to watch as the competitions.

One note on the ATC services (0730-1930 local) is that the tower controllers are very polite and will prove an asset to even the most inexperienced aviators or those who may be nervous about going into a controlled field.

PIREP   FBO Comments: Don't buy fuel at Bolton; go 15 miles west to Madison Co (KUYF) for the cheapest fuel around.

PIREP   Having been a hangar tenant at Bolton Field for the past 10 yrs I've had many experiences with JP's, all good. Consistently good, inexpensive food, friendly service and hard-working employees. JP's built an outdoor pavilion a few years ago that hosts private parties about every weekend during summer months, but there is still dining outside with runway views, or you can sit inside.

There used to be frequent aerobatic activity in the Bolton "box," but the addition of several hundred new homes next to airport eventually put a stop to our fun because of noise complaints. There's still some entertainment for your outdoor dining: students in the traffic pattern, banner planes picking up and dropping banners, and karaoke coming from JP's pavilion.

If you fly a plane without a radio or you're shy about talking to ATC, no problem—wait until after 7:30 PM when the tower closes and the field becomes uncontrolled. Be sure to remain clear of CMH Class C and Rickenbacker Class D just east of field. There's plenty of parking in front of the control tower.

PIREP   I found JP's Ribs at Bolton (KTZR) through *The $100 Hamburger*. Been there twice, most recently about a month ago. The food is still great and the prices are CHEAP. I order more than I can eat so I'll have leftovers to bring home.

Columbus, OH (Port Columbus Intl—CMH)
McDonald's 🍔

Rest. Phone #:
Location: Short walk
Open: Daily: Breakfast, Lunch, and Dinner

PIREP If you land at Port Columbus and park at Lane Aviation, it's just a short walk to the Mickey D's across from the ATC tower. Lane also has a courtesy car on a first-come first-served basis for brief jaunts to local off-airport restaurants.

Concourse Inn 🍔🍔🍔

Rest. Phone #: (800) 541-4574
Location: On the field
Open: Daily: Breakfast, Lunch, and Dinner

PIREP The airport hotel, the Concourse Inn, has a restaurant about two blocks walk from FBO, Lane Aviation. A fuel purchase at FBO will also provide a 25 percent discount coupon for the hotel restaurant. Good sandwiches and burgers.

Coshocton, OH (Richard Downing—I40)
Roscoe Village 🍔🍔

Rest. Phone #: (800) 237-7397
Location: Short distance, free transport
Open: Daily: Breakfast, Lunch, and Dinner

PIREP A good fly-in spot in Ohio is Richard Downing Airport near Coshocton, OH. The airport is located on top of a ridge above Roscoe Village. This is a historical canal town with small shops selling antiques, quilts, ice cream, and so on. You can even ride a canal boat. There is a motel and several restaurants. Call the shuttle van from a booth near the airport terminal building for the short ride down the hill. Good spot to get a bite and stroll around for awhile. The airport sells 80 and 100 octane avgas at a reasonable price. Lloyd Barnhart—the place is worth four hamburgers in my book!

Dayton, OH (Moraine Air Park—I73)
The Golden Nugget Pancake House 🍔🍔🍔

Rest. Phone #: (937) 298-0138
Location: 1.5 miles from airport
Open: Breakfast and Lunch, closes at 3 PM, closed Mon.

PIREP Moraine Airpark (I73) is bordered on the east, south, and west by the Miami River. It has all the usual small airport amenities and is located just south of Dayton, OH, and west of Interstate 75, Moraine Airpark (I73).

The real attraction here is The Golden Nugget Pancake House. Don't let the name fool you: pancakes are not the only thing they do, and they do it very well. Although they make about a half dozen kinds of pancakes, which are big and fluffy, they also make GREAT omelets (about a dozen selections), soups, sandwiches, and other American fair. It's Midwest cooking right down to the biscuits and gravy. I capitalize GREAT because this is real, delicious food, not that slapped-together stuff you get at the family restaurant chains. Their coffee is award winning (literally). The people and service are excellent. If you tell them you want your home fries slightly crispy, that's how you get them. If you want your eggs moist but not runny, that's how you get them. You can always tell a good restaurant by the length of the line of people waiting to get in, and depending on the time of day the line can be 15 to 45 minutes long, winter or summer. Their prices are the same as the family restaurants, which means you get a better value for your dollar. And to add to their credits, they DON'T cook with canola oil (that's what gives everything a slightly bitter taste at other places).

The restaurant is only open for breakfast and lunch and closes at 3 PM. They are closed on Mondays.

The restaurant is a short eight-minute drive from the airport, so you'll have to get a cab or try slipping ten bucks to someone at the FBO to borrow their car. If you like good food, it's worth it.

Here are the driving directions to the restaurant: leaving the airport, drive down Clearview Road for 0.4 miles. Turn left on Elter Drive for 0.2 miles. Turn right on Main, which is a divided four-lane road, and onto the overpass that goes over the river and Interstate 75 (0.7 miles). Turn right at the second red light, which is Springboro Pike. Almost immediately you'll come to a light where you will turn left on Dixie Drive. Follow Dixie for about two miles, and on the right, at the intersection of Dixie and W. Dorothy Lane, is the restaurant.

Findlay, OH (Findlay—FDY)

Rose Villa Cafe 🍔🍔🍔

Rest. Phone #: (419) 424-9284
Location: Just across the street
Open: Daily: Breakfast, Lunch, and Dinner

PIREP The Rose Villa Cafe is just across the street from the GA ramp at Ohio's Findlay Airport, (FDY). The food is excellent and economically priced. The fare includes a good variety of fare. Highly recommended. We hosted an International Comanche Society lunch hour fly-in and the hospitality was excellent!

As a bonus, the airport has a variety of approaches and a good crosswind runway.

Fremont, OH (Sandusky County Regional—S24)

Davenport House 🍔🍔

Rest. Phone #: (419) 547-4444
Location: 2 miles
Open: Daily: Breakfast and Dinner

PIREP My wife and I made an unplanned landing at S24, Sandusky Co. Regional, Fremont, OH due to weather last month. We discovered the Davenport House, 2 miles from the airport in Clyde, OH. Phone number: (419) 547-4444. The owner, Claudia Laurendeau, came out to the airport and picked us up.

To our surprise, the establishment was a beautiful 100-year-old, brick Victorian home. The house has an interesting history including a pilot that you will hear about.

We ordered filet mignon for dinner...excellently done. Breakfast and coffee were served on the front porch the next morning. The personal attention and pampering were second to none. We didn't want to leave, and intend to return for a visit again soon.

Kelleys Island, OH (Kelleys Island Land Field—89D)

The Village Pump 🍔

Rest. Phone #: (419) 746-2281
Location: 2.5 miles; walk!
Open: Daily: Breakfast and Dinner

PIREP FBO Comments: 2200' paved runway, no FBO, no fuel.

PIREP An interesting, somewhat out of the way place to spend an afternoon, but be prepared to walk, as the "downtown" area is about 2.5 miles from the airport. Several decent if slightly pricey restaurants downtown. Just north of the airport on the east shore is a large old home that is a B&B. Haven't stayed there, but talked with the owner. Looked like a pleasant place to spend a night or two. Sign at airport asks for a $5 per day parking fee, which you put in a locked box.

McArthur, OH (Vinton County—22I)
Ravenwood Castle 🍔🍔🍔

Rest. Phone #: (740) 596-2606
Location: 2 miles, call for courtesy van
Open: Lunch Thur.–Sun., 11 AM–3 PM, May–October

PIREP Ravenwood Castle is a must for all pilots. A bed and breakfast nestled in the foothills of the Appalachians with hospitality that outdoes all. We called a day in advance and the owner picked us up at the airport, fixed us brunch, and returned us to our plane. Vinton County Airport is a treat as well. Self-supporting, there is always an activity going on. These are the friendliest people I have ever met. They cook a mean hog at the airport and on departure the unicom broke the hum of the Archer, "Ya'll come back soon," the voice on the transmitting end exclaimed. We will!!

PIREP Due to requests from several of our fly-in guests, we now have a "courtesy car" for overnight fly-in guests.

We are near several state parks with caves, waterfalls, lakes, hiking trails, beautiful scenery, horseback riding, canoeing, fishing, antique, craft and artist malls and studios galore, a scenic railway, and much more to see and do. People could easily spend several days and nights here and not see it all.

There is no charge for borrowing our courtesy car as long as our fly-in guests stay at least one night. And they will still get their 10 percent room and food discount!

Percelli's 🍔🍔

Rest. Phone #: (740) 596-5281
Location: 5 miles, call for pickup
Open: 10 AM–11 PM except Sun.

PIREP After reading an article from *USA Today* on world champion pizza makers and learning the international classic gourmet pizza winner was Percelli's in McArthur, my wife and I flew from our home in Kentucky to sample this gourmet winner. Arranging beforehand with John Snider, upon our arrival at Vinton Co. Airport we called (740) 596-5281, and his lovely wife Marilyn promptly picked us up for the five-mile trip into town.

The Sniders are really fantastic people, and we had a wonderful visit and enjoyed a delicious Greek Supreme Pizza, the winner of the international contest. In addition, at John and Marilyn's insistence we had a sample of their spaghetti using the family's 100-year-old sauce recipe, another culinary delight! This sauce is also used for their lasagna and meatball dishes. Additionally, they have a "take and bake" pizza, and of course we took advantage of that and will again tonight enjoy a freshly baked tasty pizza. I would rate this experience as a five hamburger!

While Percelli's is not open until 4 on Sundays, they are readily available from 10 AM to 11 PM all other days. Upon arrival at Vinton Co. (22I), give them a call ((740) 596-5281) and one of the family will promptly pick you up for a pleasant visit and a tasty pizza or spaghetti meal.

Mansfield, OH (Mansfield Lahm Regional—MFD)
The Flying Turtle 🍔🍔🍔🍔

Rest. Phone #: (419) 524-2404
Location: In the terminal building
Open: Daily: Lunch and Dinner

PIREP   Flew to Mansfield with my three-year-old daughter. Had never been here before, but I was fairly impressed with the Flying Turtle. I had the Turtle Burger, a half-pound bacon cheeseburger with coleslaw on it and all the regular trimmings for under $6. It was good. They have fresh-cut fries (real good), and my daughter's chicken tenders looked like they may have been hand breaded (also very good). Prices were very reasonable.

I saw other people with some real good-looking salads (they have about five or six different kinds), and I also saw a decent seafood menu, including two lobster tails for $14.99, as well as Cajun whitefish, catfish, shrimp, and so on.

Service was extremely friendly, and the control tower was helpful, although one of the tower guys seemed a bit grumpy when I asked him to repeat a broken transmission. Maybe he had had a long day.

On the downside, the nonsmoking section still got quite a bit of smoke smell from the bar area. That's a BIG negative in my book (I can't stand the smell of cigarette smoke), but nevertheless I'm sure I'll be back.

Three burgers (would've been at least four if it wasn't for the smoke).

PIREP   We'd read about the Flying Turtle on 100dollarhamburger.com and decided to give it a try. On what must have been the first clear night in a month here in Northern Ohio, I took my wife and kids over. We were the third plane to taxi up to their parking area (heard them all announce their intention to the tower on our way in), and while we ate, three more planes stopped by for a nice dinner. Food quality is something like an Applebee's although with fewer selections.

The kids loved it, and when they started playing Jimmy Buffett, my wife decided she needed a glass of wine. They have a nice bar area and it seems to be a mix of locals and fly-in traffic, with the locals carrying the business.

As we ate, a C130 headed out for some local area air work. They shut down all airport landing and we got to watch a C130 shoot some short landing with night vision goggles. All in all, a neat little place that offers the perfect excuse for a nice evening flight.

Middletown, OH (Hook Field Muni—MWO)

Frisch's Big Boy 🍔🍔🍔

Rest. Phone #: (513) 423-6596
Location: Adjacent to the field
Open: Daily: Breakfast, Lunch, and Dinner

PIREP   Flew into Hook Field with a few buddies. Easy to navigate to from Columbus. Just fly to the Xenia NDB, then fly to the Hook Field NDB. When I got on the ground, I didn't know where the restaurant was so I got on the unicom and asked. Our answer came from Blue Ash (about 15 miles away). We just parked in front of some T-Hangars (no tie-downs, hope your parking brake works). Food and service were about average for Big Boy. Hook Field also has quite a few antiques on the field, including several DC3s and several Beech 18s.

Mount Victory, OH (Elliotts Landing—O74)

Plaza Restaurant 🍔🍔🍔

Rest. Phone #: (937) 354-2851
Location: Adjacent to the field
Open: Daily: Breakfast, Lunch, and Dinner

PIREP   FBO Comments: No FBO, hangars, and so on. Only a runway (grass strip) 2750 × 110.

PIREP   Went to Mount Victory today with my wife, partly on recommendation from *$100 Hamburger* and partly from recommendations from fellow pilot friends. It's a restaurant/truck stop adjacent to the grass strip at Elliot's Landing.

We had the buffet. Chicken was excellent! It was the only main course on the lunch buffet, but it was even better than the chicken at one of our local Amish restaurants, which is known for their chicken! Rest of the buffet (sides, salad, soup, homemade bread, and so on) was also very good. You can also order from the menu. Buffet was $6.95, not including drink. They also have a neat little country gift shop.

Runway was a little rough. I've been on much nicer grass strips. At least the grass was cut.

Four stars (would've been five if they had more than one main entree on lunch buffet, which they do at dinner). We'll be back!

PIREP   This is a very good restaurant, and convenient; you can even park next to the building. They offer a standard breakfast menu, or you can opt for their outstanding breakfast buffet. During the summer months the weekend evening buffet includes either prime rib or walleye/perch, and so on. Restaurant owner is usually there and goes out of his way to make sure you enjoy the experience. This one is definitely a favorite with local pilots.

New Philadelphia, OH (Harry Clever Field—PHD)

The Perfect Landing 😋😋😋

Rest. Phone #: (330) 308-9000
Location: On the field
Open: Daily: Breakfast and Lunch

PIREP   FBO Comments: Fuel is self service, but the friendly staff made sure I knew they were available for whatever I needed.

PIREP   Earlier reports are still accurate: low prices on the specials, good service. The restaurant door is about 100 feet from the FBO door. The guys at the FBO were friendly as well. A nice place to stop.

PIREP   Great food and large portions! French dip was fantastic. The Sunday morning breakfast platters I saw others getting looked even better…maybe next time. They close at 2 PM on Sunday, later on Saturdays. Very nice tie-downs with ropes just outside the restaurant. Airport was in fantastic shape. Restaurant wasn't very impressive looking, but the food and friendly service more than made up for it.

Ottawa, OH (Putnam County—OWX)

Red Pig Inn 😋😋😋😋

Rest. Phone #: (419) 523-6458
Location: Close by, call for crew car or pickup
Open: Daily: Lunch and Dinner

PIREP   I just flew into Ottawa and had dinner with a friend at the Red Pig Inn. This was my first venture following up on a tip from the *$100 Hamburger*, so was cautious. I called the airport in the afternoon to confirm that this restaurant was still open and that a courtesy car was available. The airport operator assured me that it was. When I told him my ETA, he offered to make arrangements to leave me the keys and a map as he wasn't going to be at the airport. Everything went off without a hitch, the ribs were great, and it was a very enjoyable trip.

PIREP   The Red Pig Inn has fantastic barbecue! It is a five-time NW Ohio Rib-off winner! The restaurant is off the field but offers free courtesy transportation; call them to be sure it is available at (419) 523-6458. The menu is varied with steaks, seafood, barbecue, pasta, salads, and much more. A casual dining atmosphere is punctuated with a great staff and friendly service. Their fabulous Sunday brunch is a must for weekend flyers!

Oxford, OH (Miami University—OXD)
Alexander House 🍔🍔🍔

Rest. Phone #: (513) 523-1200
Location: In town, ground transportation required
Open: Daily: Lunch and Dinner

PIREP   Still have kids in college so still flying into Oxford, one of the best college towns in America. They paved the ramp and the runway—nice, but the FBO is gone. Fuel boys are still there, though, and 100LL is cheap. Skyline Chili is always on the agenda for this, but for variety try Buffalo Wild Wings. Great wings and "college food" and a ton of TV monitors so the game is always on.

If you really want to go upscale, try Alexander House. This is gourmet level, pricey, but not out of line for the service and fare. Fly the spouse here for an anniversary or birthday dinner and you'll be a star. You can get a nice room there, too.

Also a beautiful new inn downtown owned by Holiday Inn, the Miami Inn, which is run by the university, that has a great little tavern with the best single malt scotch selection in Ohio.

Also the Sycamore Inn (Best Western) or Hampton Inn. Use the pay phone at airport to call them for pickup or get a taxi. There are two taxis—literally—in Oxford! If you are in a jam, just tell the ramp dudes you need a ride and for $5 one of their friends will take you anywhere!

Port Clinton, OH (Carl R Keller Field—PCW)
Mon Ami 🍔🍔🍔🍔

Rest. Phone #: (419) 797-4445
Location: They'll pick you up
Open: Daily: Lunch and Dinner

PIREP   FBO Comments: The guy on the phone was very helpful, but I never met anyone face-to-face. The FBO closed at 6 PM, and the place was deserted when I arrived at 8 PM. There are plenty of tie-downs (complete with chocks and ropes provided) free of charge for private aircraft.

PIREP   Mon Ami *rocks*! This dining experience is *way* better than the 3.5 burgers scored with the standard Burger Form. Mon Ami is about the food (and VERY GOOD food, at that). It's not on the airport, and it's not the cheapest meal you can find, but it's very convenient and it's a fair price for a very nice meal. And the restaurant staff fall all over themselves to take care of you.

Location: 0 (they pick you up and take you back to the airport)

Food: +2 (wow!)

Service: +1 (*way* helpful)

Ambiance: +0.5 (not pilot-related but nice atmosphere)

Price: 0 (everything costs about what it should; no surprises, good or bad)

Your dining experience starts with a phone call to Mon Ami ((419) 797-4445); they send out Brian (a guy with a sense of humor) to pick you up at the airport. My wife called while I was tying down the plane, and Brian arrived only a few minutes after I was done. Throttle response was good on the transportation.

The Lexus was unavailable the night I dined there, but Brian's crew-cab Dodge Ram pickup was still very spacious for the short drive between the airport and restaurant. Reservations are a good idea (the place was *packed* at 8 PM on Saturday night), but if you don't have a reservation, they put your name on the list when you call for a ride. We didn't have a reservation, but I think we got special treatment for transient pilots because we were seated about ten minutes after we reached the restaurant.

Rumor has it Mon Ami has some seating outside overlooking Lake Erie, but it was chilly the evening I was there, so we dined indoors. The dining rooms are tastefully decorated, and although the winery dates to 1872, the building appears to be well maintained. It has a historic feel but without the old rundown look that often accompanies old buildings. It's hard to put my finger on exactly why I liked the atmosphere (perhaps the stone architecture?), but I liked it.

The food is fabulous, and the prices are reasonable (not dirt cheap, but reasonable). I got stuffed silly on a delicious 10-ounce prime rib (with veggies and baked potato) for $14; a larger cut is available for a few dollars more. Most of the entrees are in the $15–20 range, and my wife loved the fresh walleye. On Saturday nights, you can "strap on the old feedbag" and feast on the seafood buffet for about $26 (it was only last evening, but I've already forgotten the exact price).

The seafood buffet was truly tempting, but I was afraid I would fall asleep on the return flight. The desserts and coffee were excellent, and our nonflying dinner guest said the pinot grigio was *very* good.

After the meal, Brian promptly returned us to the airport and we flew home (with a tailwind, no less). All things considered, I really enjoyed the dining experience at Mon Ami. More importantly, so did my wife. Mon Ami isn't the cheapest restaurant on the planet, and it's not right on the airport, but I highly recommend it if you're looking for fine dining that you can fly to.

We'll be back.

Portsmouth, OH (Greater Portsmouth Regional—PMH)
Skyline Restaurant 😋 😋 😋

Rest. Phone #: (740) 820-2203
Location: On the airport
Open: Daily: Breakfast, Lunch, and Dinner

PIREP Skyline Cafe has some of the best food I have ever had at a airport: five stars. I stop there all the time on my way up to Sporty's Pilot Shop (I69). Thanks, and I hope you add this to your list for other pilots to enjoy.

Put-in-Bay, OH (Put in Bay—3W2)
The Skyway 😋 😋 😋 😋

Rest. Phone #: (419) 285-4331
Location: Close
Open: Daily: Lunch and Dinner

PIREP Put-in-Bay is on Bass Island in Lake Erie, just off Port Clinton and Sandusky. A real nighttime hot spot and weekend party destination. Many restaurants in town, just a short golf cart ride or school bus taxi from the strip. At the airport, you can choose from a little burger joint with a swimming pool, just behind the main hangar. You eat outside—very inexpensive. A bit more upscale is the Skyway Restaurant and Bar, just off the northeast end of the runway. Great sandwiches. Their specialty is a crispy walleye that is a real treat—something you just can't get anywhere else. Fantastic. Nice waitresses, good service, and bartenders that evidently win all of the bartending Olympics on Bass Island, evidenced by the numerous plaques on the wall. (Probably a popular hang-out for America West pilots, though I cannot confirm this!) I give Skyway four burgers.

Put-in-Bay is a great stop. There is nothing much more exotic if you are taking a friend to breakfast, lunch, or dinner than landing on an island in a lake, versus, for example, Jackson, Michigan (which has a fine restaurant). Just eat at the strip or rent a cart and cruise the town and island. A lot of history and very nice people. Watch out for Perry's Monument on left base—most people will fly inside of it. And major restrictions about operations before dawn or after dusk. Don't even think about it! If you are looking for VFR flight following, recommended due to heavy traffic in the area, you will find that you are in a bit of a no-man's land between Toledo, Detroit, and Cleveland approach or centers. And that is the order you will find help, by the way. Toledo approach is fantastic, flexible, and does everything they can to help. Virtually always gives and receives handoffs. Detroit tries and generally does a good job. Cleveland Center…well, I guess they are a lot busier than Toledo or Detroit, and they will try to dump you if they can. Don't fantasize about a handoff from them. You will rarely get one.

Sebring, OH (Tri-City—3G6)
The Flight Deck Restaurant 🍔 🍔 🍔

Rest. Phone #: (330) 938-1982
Location: On the field
Open: Daily: Breakfast and Lunch

PIREP   FBO Comments: Unfortunately, not attended.

PIREP   New restaurant owners since late. I had a well-stuffed ham and cheese omelet. Normal. Nice *large* coffee cups eliminated the constant need for a waitress. Great. Wife had a double cheeseburger that didn't need to be!!! A single would have been plenty. It wasn't the typical processed, store-bought patty but rather a nice, thick handmade burger. The little taste she shared with me was great.
   Service was average, as was the crowd.
   Car and plane parking right at the door.

Urbana, OH (Grimes Field—I74)
Airport Cafe 🍔 🍔 🍔

Rest. Phone #: (937) 652-2300
Location: On the field
Open: Daily: Breakfast, Lunch

PIREP   I planned this small X-country two months ago, and I decided to take two of my friends. I heard great things about the Airport Cafe and their pies. I had to find out for myself. The restaurant had a country-style environment, which was pretty neat. The waitress was nice, and the food and prices were great. One of my friends and I had the peach pie with a scoop of ice cream, while my other friend had the butterscotch pie. Both pies were pretty good. I would recommend this restaurant for anyone who enjoys good food at reasonable prices.

Van Wert, OH (Van Wert County—VNW)
Willow Bend Country Club 🍔 🍔 🍔 🍔

Rest. Phone #: (419) 238-1041
Location: 2 miles call for pickup
Open: Daily: Breakfast and Lunch

PIREP   FBO Comments: Great airport, clean and friendly. Crew car available.

PIREP   I have dined at Willow Bend Country Club many times, and the food and service are excellent. Willow Bend is one of the best-kept secrets in NW Ohio. It has a beautifully manicured nine-hole golf course and a top notch pro shop. If you're looking for a very fun and easy place to fly to for dinner and or golf, check this place out. Give them a call for transportation, just a couple miles from the airport. You won't be disappointed!!

Wapakoneta, OH (Neil Armstrong—AXV)
Main Street Station 🍔

Rest. Phone #: (419) 753-2909
Location: In town, use crew car
Open:

PIREP   FBO Comments: Friendly service, good weather computers, nice lounge area. Functional crew car.

PIREP   Waiting out some thunderstorms, we took the crew car (a classic) a mile into New Knoxville (much closer than the town of Wapakoneta) to Main Street Station. This is your most basic of very basic small-town diners. After our first three attempts at ordering something, our very sincere and literal waitress informed us, "We're out of that." When I politely suggested perhaps she just tell me what they DO have, it clearly sent her brain into crisis mode, so I backed off.

This place was so bad, that it was good. Kind of like renting *Ishtar* at the video store. Worth the trip and the experience. So, if you want some really good food, drive on into Wapakoneta. If you want a strangely amusing, genuine small-town Americana experience, try the Main Street Station in New Knoxville. And ask Carla just one more time to explain that thing about the muffins and the cream cheese.

Waynesville, OH (Red Stewart Airfield—40I)
Holly Hills Golf Course 🥪🥪🥪

Rest. Phone #: (513) 897-4921
Location: Right across the street
Open: Daily: Breakfast and Lunch

PIREP   The Sky Cream Deli may be closed at the Waynesville Airport, but there are plenty of places just a short drive away, in the town of Waynesville. (The airport has a courtesy car you can borrow.) Der Dutchman (Amish cooking), The Village (where the locals go), and the Hammel House (restricted hours but very good food). As an added bonus, Holly Hills Golf Course is right across the street from the airport (not to mention all the antiques in Waynesville). Call ahead for the car and runway conditions (grass).

Ada, OK (Ada Muni—ADH)

Bob's BBQ

Rest. Phone #: (580) 332-6253
Location: Short walk
Open: Daily: Lunch and Dinner

PIREP   Well, we went back again! The FBO provided a crew car and we feasted for our second Friday night dinner flight at Bob's BBQ.

PIREP   The trip is a short 30 from Oklahoma City, just about the right distance for a dinner flight. The airport has two paved runways to choose from, with the longer well lit. The new owner picked us from the airport and drove us back after the meal, a friendly lady eager to offer transportation to visitors. We enjoyed some very good barbecue with lots of fixings. Servings were substantial and inexpensive. The establishment has some character, including a sink in the entryway for washing up before dining and washrooms outside and down a long covered corridor, like an afterthought but adding more flavor to the experience. This is a fun dinner flight.

PIREP   Bob's BBQ is out of this world!!
The meal was served in less than five minutes, and there was plenty to eat, even without the all-you-can eat sides. The airport staff is friendly and very helpful; they even have a courtesy car now!!
I highly recommend the stop.

PIREP   Ada offers a wide variety of dining for those who fly in. Our downtown area (two miles from the airport) offers fine Italian and Mexican food. Other restaurants are close by. The courtesy car is still available through the FBO.
Call (580) 310-6062 to reserve the car.
Terry Hall
City of Ada
Airport Manager

PIREP   The shortcut through the Ada High School baseball facility is no longer available. All the gates are locked (not sure if it's security for the airport or the baseball field).

Afton, OK (Grand Lake Regional—3O9)

Anna Banana's 🍔🍔🍔

Rest. Phone #: (918) 257-4618
Location: Short walk
Open: Daily: Breakfast, Lunch, and Dinner

PIREP   Anna Banana's is on Monkey Island, OK. It is walking distance from Grand Lake Regional Airport. They serve breakfast, lunch, and dinner and have a full bar (not for the pilot).

PIREP   This airport has a lot of potential, but it appears from a discussion with the "manager" that his real plan is to take the airport private and build a private community around it. As an attorney, I recalled a lot of legal problems with this airport, so when I got back, I checked on them and found that houses had been constructed against FAA regulations at the south end of the runway and posed an obstruction. The airport, according to the FAA, will probably never get an instrument approach again. Then I found that the airport was purchased with an FAA grant and was a public airport, not a private airport as it was represented. For my money, I will not use the facility again as it looks like there is some funny business going on.

PIREP   The airport has a nice runway, wide and plenty long enough. King Air's and small jets land there all the time. Anna Banana's has good barbecue that I recommend. As for the "funny dealings there," I have visited Monkey Island many times and have never heard or seen any dealings that would raise an eyebrow.

Alva, OK (Alva Regional—AVK)

Champs of Alva 🍔🍔

Rest. Phone #: (580) 327-2025
Location: 1 mile from the airport
Open: Daily: Breakfast and Lunch

PIREP   VIP's, which used to be on the field, has closed. It is about a mile down the road to Champs of Alva. The airport manager has a car and you can use it.

Ardmore, OK (Ardmore Muni—ADM)

Blue Pig BBQ 🍔🍔🍔🍔

Rest. Phone #: (580) 389-5555
Location: On the flight line, in the terminal building
Open: 10–5 Mon.–Sat.

PIREP   My first experience at Ardmore and The Blue Pig is one I will always remember. I was still a student pilot with my airplane based at Decatur, TX (KLUD). I needed one more solo CC and also three solo landings at a Control Tower AP. My instructor suggested Ardmore. I was nervous about talking to a tower alone for the first time. I need not have been. The guys in the tower were great, and when I told them I needed two touch-and-goes and one full stop, they were more than happy to help. And if I messed up on the radio, as I'm sure I did, they never let on. I finally taxied to the ramp and literally parked in front of The Blue Pig. It was 3:15 on a Saturday and when I pulled on the locked door of the restaurant I noticed they closed at 3. The owner himself came and insisted that I come on in. The barbecue was great, but I believe that the fries where the best I have ever eaten. Without a doubt, Ardmore and The Blue Pig will always be one of my favorite places to go.

PIREP   The cafe in the terminal is terrific. We had lunch and the food was great. They have wonderful barbecue and smoked meats, as well as fish, burgers, and more.

Bartlesville, OK (Bartlesville Muni—BVO)

Murphy's Steak House 😊😊😊

Rest. Phone #: (918) 336-4789
Location: 2 miles, crew car available
Open: Daily: Lunch and Dinner

PIREP Located about two miles from the airport (two courtesy cars available), this modest-looking place, on the low rent side of Bartlesville, nevertheless attracts hordes of hungry "Bartians" for lunch and dinner. As well as the steaks, you can sample a true Okie specialty: the Hot Hamburger. (Basically, it's grease-laden white bread topped with a lard-oozing beef patty, a monster heap of french fries, all smothered in rich brown gravy (roughly a gallon), with onions atop—an extravaganza; a veritable heart surgeon's dream come true). Get directions (it's easy to find) from the fine crew at the FBO.

BVO has a 17–35 6000' runway. CTAF Bartlesville Advisory 120.0 unicom (Phillips Aviation Services: fuel; taxi; courtesy car, and so on.) 123.0.

PIREP The taxiway paving and airport improvements at Bartlesville are all complete and they have an excellent facility with some very caring linespeople to take care of your airplane needs. They have two courtesy cars, and there are an unlimited number of places to eat in the town (all within five miles driving distance).

These guys will meet you on arrival and will come on the ramp and park and chock or tie your plane down. Bartlesville is now all *convenience*!! Can't wait for the next Biplane Fly-in.

Burneyville, OK (Falconhead—37K)

Falconhead Resort & Country Club 😊😊😊😊

Rest. Phone #: (580) 276-9284
Location: Long walk, but a walk
Open: Daily: Breakfast and Lunch, Sunday Brunch

PIREP FBO Comments: No services.

PIREP The Country Club Restaurant is open for breakfast or lunch every day and for dinner on Friday. Sunday brunch is also served. They also have a good 18-hole golf course! Be warned: it is a pretty good hike around the lake from the airport to the Club House.

Cookson, OK (Tenkiller Lake Airpark—44M)

SmokeHouse Restaurant 😊😊😊😊

Rest. Phone #: (918) 457-4134
Location: 1 mile
Open: Daily: Lunch and Dinner

PIREP We flew in to Tenkiller Airpark and took the courtesy to the SmokeHouse Restaurant. The catfish dinner was great, as well as their curlicue french fries. Great food at a reasonable price!

Folks at the airport are very friendly and helpful.

Try it!

PIREP A must to try!

Located in Cookson, OK, on the east side of Lake Tenkiller is the SmokeHouse Restaurant. This log cabin restaurant has excellent barbecue but also specializes in homemade biscuits and gravy, steaks, fish, and homemade apple dumplings. It is open seven days a week for breakfast, lunch, and dinner. It's a great place to eat, but you better bring BIG appetites with you when you come.

You can experience the cozy atmosphere of the SmokeHouse by flying to Tenkiller Airpark (one mile SW of Cookson), which has a turf runway (2600 × 75) with VASI lights. Runway 5 is generally the preferred runway for landing, with 23 the preferred take-off runway. 100LL is available.

Cushing, OK (Cushing Muni—CUH)
Rodolfo's Mexican Restaurant ☕☕☕

Rest. Phone #: (918) 225-4204
Location: 2300 E. Main St.
Open: Open daily, call for hours

PIREP Excellent Mexican restaurant on the east side of town. Cooks are Hispanic, and the food is first-class—and abundant! Airport's open seven days a week with a good courtesy car available. Well worth the short drive! If you want a hamburger, there's also the Steer Inn and Naifeh's Deli, both on Main Street.

Duncan, OK (Halliburton Field—DUC)
Phipps Bar-B-Que ☕☕☕

Rest. Phone #: (580) 470-5531
Location : 1/4 mile
Open: Daily: Lunch and Dinner

PIREP Phipps Bar-B-Que is a short walk out the gate and to the left, approximately a quarter mile from the FBO. The Smoked Steak is the best in Oklahoma.

You can ask for the courtesy car for the short drive, but if it's not raining you won't need it.

Well worth the trip.

Goodner's Steakhouse ☕

Rest. Phone #: (580) 255-6181
Location: Crew car
Open: Daily: Dinner

PIREP FBO Comments: Very friendly. Crew car was clean and ran well.

PIREP Goodner's is a short trip from the airport. We got there at 5 PM on a Sunday, and the place was empty. I had the brisket dinner and my date had the chopped beef sandwich. Both were good, but they were nothing special. It's not a bad spot to hit, but there are so many choices for barbecue within an hour flight from Dallas, I doubt I would return to this one.

Durant, OK (Eaker Field—DUA)
Chuck's Bar-B-Q ☕☕☕

Rest. Phone #: (580) 434-5698
Location: 1.5 miles
Open: Daily: Lunch and Dinner

PIREP We tried Chuck's Bar-B-Q just down the road from Eaker Airfield in Durant on the advice of the owners of the FBO. In addition, we were able to use the courtesy car. This particular restaurant was a delight. The brisket and smoked sausage were delicious, but the real treat was the *hot* barbecue sauce. If you like your barbecue spicy, then you have to go try this place out. If you don't like that kind of heat, make sure you order the *mild* sauce. Very friendly service. Not super cheap, but more than reasonable for barbecue. I had sliced brisket, sausage, beans, coleslaw and bread for about $7.99 (and this was the large order!). They also have an all-you-can-eat for about 12–13 dollars.

We will definitely return!

PIREP   FBO Name: DW Aviation.

FBO Comments: Nolan Avionics located on the field. Chuck's Bar-B-Q 1.5 miles away. Three Arrows Buffet 1.5 miles away. Courtesy car available.

Enid, OK (Enid Woodring Regional—WDG)

The Barnstormer 🍔🍔🍔

Rest. Phone #: (580) 234-9913
Location: On the ramp
Open: Mon.–Fri., Breakfast and Lunch

PIREP   FBO Comments: The airport has some of the lowest fuel prices in the nation and there are no tie-down fees.

PIREP   The restaurant is open Monday–Friday 8–2.

There is also a fly-in buffet breakfast the fourth Saturday of each month from 8–10:30.

PIREP   I was at Barnstormers last month on the fourth Saturday—there is a fly-in breakfast the fourth Saturday of each month. Barnstormers is open 8 AM–2 PM Monday through Friday, closed weekends, *except* for the fourth Saturday.

Hobart, OK (Hobart Muni—HBR)

Roy's Backyard Bar-B-Cue & Grill 🍔🍔🍔

Rest. Phone #: (580) 726-3277
Location: 2 miles
Open: Closed Tuesdays

PIREP   I had heard rumors, in Albuquerque of all places, about a small barbecue place in Hobart, OK. It's a small town in southwestern OK about 30 miles north of Altus. The airport was an auxiliary base during the war so it will handle anything except a 747, and it has instrument approach facilities handled out of Altus AFB. There is a small FBO on the field, but he told me they don't get a lot of "stop-in traffic" because most people tend to use Altus as a place to refuel on cross countries. Well, those who do need to do some different planning, because the real gem of this place is two miles north of the airport, right on the highway.

Take note: Roy's Backyard Bar-B-Cue & Grill is closed on Tuesdays. It's a typical small-town rustic place that will seat about 50 people, and the most expensive thing on the menu is a steak for $4.95. I had a large chopped beef sandwich, which includes a side order and drink, for $3.75, and it was absolutely wonderful. The FBO at the airport loaned me his pickup to make the short drive and said, "It's the best place to eat within miles." Yep, these small-town folks have an uncanny ability for the understatement. It's a four-burger place for certain and, by some standards, much higher than that. The restaurant owner is considering buying a car to leave at the airport just for his patrons. Typical Okie—just too darn trusting and friendly.

Want to call him? Ask for Roy at (580) 726-3277.

PIREP   One note, take cash—they don't take any plastic.

I read about Roy's Backyard Bar-B-Cue on the $100 Burger and had to give it a try. Definitely worth the trip!! We went on a Sunday, and on arrival at the airport we were met by Bryan (the airport manager), who provided us with keys to a car, directions to Roy's, and noted that the Sunday buffet was the best. 1-3/4 miles north, on the right. Roy's was full; however, we were seated promptly, and the food was great!! My wife and I definitely overindulged, and at least three people took time to see if we had enough of everything.

The bill (including T-shirt): $26.

This is a great place to eat—try it!

Kingston, OK (Lake Texoma State Park—F31)

Lake Texoma Lodge 🍔🍔🍔🍔

Rest. Phone #: (580) 564-2311
Location: 1/3 mile
Open: Daily: Breakfast, Lunch, and Dinner

PIREP We fly to Texoma each Fourth of July to watch the fireworks display. This year will be our last.

Someone running things at the lodge has no respect for aircraft or their owners. They allowed cars to park all over the ramp, taxiway, and runway. Kids were allowed to shoot off their fireworks around the aircraft on the ramp. The runway and taxiway lights were turned off and the field closed without any NOTAM issued.

The total disregard for aircraft really shocked me. There were county sheriffs and highway patrols everywhere, but no one said a word. I will never go to a big event there again and have to worry about leaving my aircraft on the ramp to even go eat.

This place is no longer airplane friendly, in my opinion.

PIREP The runway is in good condition. It is surrounded on all sides by tall trees, so it's a bit difficult to find from below about 2000' AGL or when approaching from the east or west. There are tie-downs available but no fuel or other services.

From the tie-down area, walk to the main east-west road and turn right (east). It's about a 1/3 mile walk to the Lake Texoma State Park Lodge, which is on the north side of the road. It looks like a small apartment building from a distance. If you don't want to walk, call the lodge front desk at (580) 564-2311 when you land, and they'll come down to the strip and pick you up.

The restaurant in the lodge has big picture windows overlooking the lake. It has a typical Southern/country menu. I had chicken-fried steak (which they call a "steak fritter") that included a salad and fries. My friend had fried shrimp. The steak was tender and flavorful. We went on a Monday night, so we were the only ones in the restaurant, but that meant we got great service. If the weather is pleasant, ask to sit outdoors on the patio, where you can commune with nature while enjoying your meal.

I asked about breakfast also—they do a breakfast buffet on Saturday and Sunday mornings with a couple different kinds of eggs, sausage, bacon, biscuits and gravy, and so on.

Overall, a beautiful fly-in destination with decent food and great ambiance. One note: if you smell smoke during your takeoff, it's most likely wood from a campfire in the park and not your airplane!

McAlester, OK (McAlester Regional—MLC)

Pete's Place 🍔🍔🍔🍔

Rest. Phone #: (918) 423-2042
Location: Call for pickup
Open: Daily: Breakfast, Lunch, and Dinner

PIREP Pete's serves some of the best Italian food in the state. This is attested to by all the autographed photos of famous people adorning the walls. When our group (C.A.P.) spends the weekend at McAlester, the unanimous choice for dinner is Pete's.

The airport is about eight miles from the restaurant, but Pete will gladly pick you up at the airport after you call. I recommend that you make reservations ahead of time. They will have your table ready when you arrive.

I won't waste time on the menu. If you like Italian, you will be in heaven when you get there. Hint: Come hungry; Pete doesn't skimp on the portions.

One other note: there is an FSS at McAlester Regional. They are real nice folks and will be happy to give you the dime tour.

PIREP   Fuel is fairly expensive, but the food at Pete's isn't. Outstanding food and service. Call the restaurant and they will pick you up in about ten minutes for some of the best Italian food I have ever eaten (I have eaten a lot of it, too!). Pete's brews their own beer, so plan on a designated copilot or stay at one of the hotels if you plan on sampling the brew. Food is served family style, and it takes a large family to eat it all. Don't plan on jumping into your plane when you are done. (Don't jump at all, come to think of it.) Plan for a nice leisurely stroll after eating. It is only seven miles back to the airport and the time it would take to walk back should be about right—that, or plan on a long nap.

Marietta, OK (Love County—4O2)
Marietta 😊😊

Rest. Phone #:
Location: Short walk
Open: 8-5 Daily

PIREP   You can't get a burger there. In fact, it's not even a restaurant. It's really more like a cookie factory!

Land on the grass at Love County in Marietta, OK and leave your plane next to the T-hangers. Go through the gate behind the hangers and cut through a narrow field to get to a farm road. (Watch your step. There's a herd of cattle that frequents that field.) Follow the farm road south until you get to a stop sign at a major road. Turn left. This road will take you under I-35. Keep walking until you see the sign ("COOKIES"). Total trip is about 1.5 miles. Do it on a nice day and you won't even notice. I believe the factory is called President's Bakery. There's a store at the factory that sells 4—5-pound bags of "broken" cookies for $1 each. The store is open every day until at least 5 PM (maybe later?).

Take enough to get the plane back to max takeoff weight, and you'll never need (or want) cookies again!

Marietta, OK (McGehee Catfish Restaurant—T40)
McGehee's Catfish Restaurant 😊😊😊😊😊

Rest. Phone #: (580) 276-2751
Location: On the field
Open: Daily: Lunch and Dinner

PIREP   I just went to McGehee's yesterday and wanted to advise readers on their current hours:

Mon.–Fri. 5–9:30

Sat.–Sun. 1–9

The runway is in ideal condition; they have paved the road that crosses it, so it is not a factor. The food was great and the landing was tight. The length is 2350', which, in my Bonanza F33a, doesn't give you much margin for error. The south runway (17) grade is definitely uphill.

Yesterday was the fifth time that I have landed at McGehee (three in a 172 and two in a Bonanza). This time I left DFW with half tanks and one passenger. Flight time was 23 minutes and we were light. The landing and roll out were perfect (we landed and took off on 17) and the takeoff roll was only 750 feet or so.

My reasoning for the play by play is that this airstrip has had *many* accidents. It is dangerous, very narrow with obstacles. Land and take off LIGHT, real light. The last time I landed McGehee, my butt cheeks were sore for about a week, due to the overwhelming desire to clench them on takeoff and landing. This time I was very light and no cheek clenching was observed. I would say that Bonanzas, Saratogas, and so on are pushing the envelope on this field. I would not land a twin on this field—just my opinion.

All of this aside, I still think this place is one of the best fly-in restaurants in the country because:

1. Their food is wonderful.
2. They have their own runway just for the restaurant.

Have fun and be safe.

PIREP   We flew in in a 2004 172 160 hp at max gross. We landed to the south and stopped just past the road that cuts through the middle with moderate braking.

There were three of us and our bill was $40 with drinks. The food was great and the view was nice at sunset.

We took off to the south and lifted off just before the road and held it in ground effect for a bit. No problems. I think the runway isn't that bad. Pretty straightforward.

There was a crash on Friday 6/10/05. They both lived. The plane is on the west side of the runway about 600' from the south end on the west side down in a ditch. From what I gather they were taking off to the north with a tailwind and slid off the edge and clipped a tree at dark. The plane was totaled.

Norman, OK (University of Oklahoma Westheimer—OUN)
Ozzie's 🍔🍔

Rest. Phone #: (405) 341-8145
Location: On the ramp
Open: Daily: Breakfast, Lunch, and Dinner

PIREP   Ozzie's all-you-can-eat breakfast is now $4.95, but it's still a fun stop. Served Monday–Friday 6–10:30, Saturday 6 AM–1 PM, and Sunday 6 AM–3 PM. Their other fare is priced inexpensively.

PIREP   FBO Comments: Punctual and friendly line personnel.

PIREP   Ozzie's still has the very good all-you-can-eat breakfast made to order, but it is now around $4—still not bad for really good breakfast food.

PIREP   It's a great place to eat breakfast. Still all you can eat, and you can taxi right to door. Crews will gas you up while you eat.

Oklahoma City, OK (Wiley Post—PWA)
Annie Okie's Runway Cafe 🍔🍔🍔🍔

Rest. Phone #: (405) 737-7732
Location: On the field
Open: Daily: Breakfast and Lunch

PIREP   Annie Okie's Runway Cafe is a favorite weekend breakfast and lunch stop. They're open through the week, but the big crowds tend to convene during the weekends. Their omelets rule, and the burgers will most definitely make you want to come back. You'll see many familiar faces if you come more than once!

The Runway Cafe is located in the terminal building at Wiley Post Airport (PWA), Oklahoma City. You can keep an eye on your airplane (and brag on it) from the restaurant atrium.

PIREP   Located in the tower building at PWA with an excellent view of the runway. Good food and good company from 7 to 2 Monday through Saturday, and 9 to 3 on Sunday.

Overbrook, OK (Lake Murray State Park—1F1)
Fireside 🍔🍔🍔🍔🍔

Rest. Phone #: (405) 226-4070
Location: Very short and pleasant walk
Open: Dinner only, Tues.–Sat.

PIREP For those unfortunate and discerning palates that have not discovered the real place to dine at Lake Murray, let me describe. How about succulent, mouth-watering prime rib, or Cajun-style blackened fish? Want to be healthy? Any number of broiled fish, chicken, or shrimp can fill the bill. A special occasion spent here is a must, whether it's a birthday, anniversary, or just a really romantic date.

Enough suspense already—I'm talking about Fireside Dining, an actual sit-down-and-enjoy-it restaurant. One hint: bring time and your heavy wallet. The absence of green will offset the weight added after a truly awesome fly-in culinary experience. I usually allow an hour of dining time for this multicourse extravaganza. And budget about 20 to 25 dollars per person without drinks—and if you're flying, please don't imbibe, for your safety and your loved ones.

Gary (the owner) and his staff always make every dining experience truly top notch. Upon arrival at the golf course clubhouse (it's the only building by the parking facility), you can either use the pay phone (the numbers for all restaurants and other services are listed conveniently near the phone) or you may request the clubhouse attendant to call Fireside Dining for you.

The restaurant is close enough to walk, but who can resist a ride in Gary's new Suburban in which he will arrive in mere minutes from your call to personally greet you and your party?

Runway 14-32 is 2500' feet and asphalt. The runway has a crude but effective reflector type VASI for the 14 approach—be sure to use it, especially at night. The trees at this end accompanied by the hill that the runway is draped over combine to help suck you down on short-short final. Be careful of the visual effect of the dramatic crown on this runway; it appears that you're headed straight down on short final.

Be sure to be adept at your short field procedures. While the runway is plenty long, after this big dinner and with starting out uphill and all…well, you get the idea.

The Fireside Dining Restaurant is open for seating at 5 PM Tuesday through Saturday, first come first served.

Those of us who fly in are usually specially pampered and seated upon arrival, or maybe it's just me! Try it and see.

And believe it or not, this isn't even a commercial.

PIREP You land on the park's own airstrip just inland from the lake shore and adjacent to the 18-hole golf course. Walk up the steps to the clubhouse and call the lodge—they'll send a van or car to pick you up. Or you can walk the mile down the road through the Okie woods. Beautiful lodge with a nice restaurant overlooking the lake. There are paddle boats and canoes nearby, or you might be able to convince them to run you down to the big marina where you can rent a boat or jet ski or just go swimming.

If you've got an amphibian, the lake and marina welcome float and seaplanes.

Ponca City, OK (Ponca City Regional—PNC)

Enrique's Mexican Restaurant 🍔🍔🍔🍔

Rest. Phone #: (580) 762-5507
Location: On the field
Open: Daily: Lunch and Dinner

PIREP Ponca City, EAA Fly-in. The local chapter of the EAA hosts a fly-in breakfast on the first Saturday of each month. The food is great ($5 donation) but it's the wide array of home-builts and vintage aircraft that make this a popular destination. The fly-in makes for a busy place and the field is not towered, so be watchful and communicate your intentions. Be sure to bring a camera! This isn't quite Oshkosh but you'll have a grand time! Note that breakfast is served to 10 AM.

PIREP   Flew to Enrique's again last night. Food was great, as usual. Friendly staff and good service. Only us and a 414 on the ramp, but the place was full of customers. If you like great authentic Mexican food. Now we just need to make the Saturday morning breakfast fly-in.

PIREP   We flew up to PNC for lunch this past Saturday. There were a lot of airplanes coming and going, which was surprising to me since I had not flown into that airport in about 20 years. Gotta tell you, the Mexican food is outstanding! And not Tex-Mex. These folks cook like the ones in Sante Fe and Albuquerque! But you better get there before noon or you'll find yourself walking some distance from your airplane. Large ramp, lots of planes, and no tie-downs.

PIREP   I, too, am an avid fan of Enrique's at the Ponca City terminal. I have flown there many times on any excuse to enjoy the food. The chips are puffed when they are fried and are different from any others I have seen. Plus, it's nice to be able to admire your plane parked a few feet away while you dine.

There is a very nice aviation memorabilia display in terminal lobby, including a WWII Norden bombsight!

The hours of operation (I believe, call ahead for safety: (405) 762-5507) are Monday–Friday 11 AM–2 PM, Monday–Thursday 4:30 PM–8 PM, Friday 4:30 PM–9 PM, Saturday 11 AM–9 PM.

Enrique's is top rate, and Ponca City is a first choice for a midday fuel stop when we are on C.A.P. missions.

Pryor, OK (Mid-America Industrial—H71)
Dutch Pantry 🍔🍔🍔🍔

Rest. Phone #: (918) 476-6441
Location: Crew car
Open: Daily: Breakfast and Lunch

PIREP   The name of the restaurant is the Dutch Pantry. It is located in Chouteau, OK and is about three miles south of the Pryor, OK airport. I think that the Pryor airport still has a courtesy car so you won't have to walk. It is an Amish restaurant and is served cafeteria style with two types of meat at every meal. The cost is $6, and you can have all of the soda refills and desserts and main courses that you want. They serve homemade breads with real homemade butter, and I could make a meal of the bread. I always gorge myself and turn into a real pig, but the food is so wonderful I can't resist. There is no tipping and seriously, I do not know how they make a profit, but apparently it is done on a volume basis.

I recommend a six-burger rating even though you have apparently only given that once.

J L's BBQ 🍔🍔🍔

Rest. Phone #: (918) 825-1829
Location: 2 miles
Open: Daily: Lunch and Dinner

PIREP   Let me suggest to you J.L.'s BBQ one mile north of the Mid-America Industrial Airport at Pryor, OK. They offer some of the best barbecue in the country. Their specialty is the rib plate, and they give you a giant portion on your plate.

The airport has a courtesy car available. I highly recommend J.L.'s restaurant for those persons traveling northeast Oklahoma.

Stillwater, OK (Stillwater Regional—SWO)
Eskimo Joe's 🍔🍔🍔🍔🍔

Rest. Phone #: (405) 372-8896
Location: Short drive
Open: Daily: Lunch and Dinner

PIREP   This restaurant is a short drive from the airport by means of a courtesy car, in Stillwater, OK. It is world renowned for its clothing store, which is next door. They have some of the best hamburgers you'll find, and the cheese fries are out of this world. According to their menu, the cheese fries are endorsed by ex-President Bush. I've been there many times, and it's one of my favorite places to go.

PIREP   We just barely got back from there and loved every minute of it. The folks at Stillwater Flight Service are wonderful, and there seemed to be plenty of parking for everyone. We thought a few numbers for this area might help others. Stillwater Flight Service is (405) 624-5463 or 747-8648. They let you reserve the courtesy car for whenever you need it. The phone number to Eskimo Joe's is (405) 372-8896, in case you would like to call and check business hours. The cheese fries are outstanding and we will definitely go back soon.

Freddie Paul's 🍔🍔🍔🍔

Rest. Phone #: (405) 377-8777
Location: 1707 E. 6th Ave.
Open: Daily: 11 AM–10 PM (closed on Sunday)

PIREP   You can fly to SWO, take the visitors car, and drive into town to Freddie Paul's for lunch or dinner. They have about anything you could possibly want on the menu (check out their website in advance: www.freddiepauls .com). The owners are Brian and Jodie Saliba. Brian is about the third generation of the Saliba family to operate a restaurant and is also a great chef. I have been there several times, and the food and service are great. They are open 11 AM till 10 PM, Monday through Saturday (closed Sunday).

Tahlequah, OK (Tahlequah Muni—TQH)

K.C. Harris Burgers 🍔🍔🍔

Rest. Phone #: (918) 456-7111
Location: In town, use crew car
Open: Daily: Lunch and Dinner

PIREP   If you can remember how good hamburgers tasted when you were younger, this is for you. Kriss Harris uses secret old family recipes to create the very best hamburgers, brisket sandwiches, old-fashioned ham and bean soup, outstanding cheese broccoli soup, out of this world genuine Northeastern Oklahoma ranch-style chili, and a dozen other indescribably delicious dishes. She uses only the freshest and finest ingredients available, and everything is prepared on the spot before your very eyes. Everything is very reasonably priced. She's there cooking Monday through Friday 1100–1900 hours, Sat 1100–1500 hours local. Try it once and you'll be back for more.

200 E. Downing St., (918) 456-7111.

At the airport, Dutch Wilhelm will provide you with the airport courtesy car and any other service you might need, including 100LL.

Airport phone (918) 456-8731.

Tulsa, OK (Tulsa Intl—TUL)

Sparks FBO Restaurant 🍔🍔🍔

Rest. Phone #: (918) 838-5000
Location: On the field
Open: Daily: Lunch

PIREP   Recently we visited TUL and had lunch at the restaurant located in the Sparks FBO. They have a daily special, salad bar, wide sandwich menu, a few entrees, and great burgers. The service was friendly and quick.

Vinita, OK (Vinita Muni—H04)

McDonald's 🍔🍔🍔

Rest. Phone #:
Location: 200 feet
Open: Daily: Breakfast, Lunch, and Dinner

PIREP This is the largest McDonald's in the world (29,135 sq. ft.) and is on the second floor of an arch (what else?!) that straddles the freeway south of Vinita.

At the airport, you park northeast of the runway and walk a couple of hundred feet to McDonald's. There is a large souvenir shop and tourist information available.

The menu...you've got to be kidding!!

Waynoka, OK (Waynoka Muni—1K5)

Cafe Bahnhof 🍔🍔🍔

Rest. Phone #: (580) 824-0063
Location: Call ahead for pickup, short drive
Open: Daily: Lunch and Dinner

PIREP Flew in at night. Runways lights are functioning. Great ride from Rebecca to restaurant. Food, drink, and host great! Wall in back courtyard available for artistic efforts. Open Thursday through Saturday, 5:30 PM. Next door is a liquor store with interesting selection. Well worth a visit.

PIREP Cafe Bahnhof is a wonderful little German restaurant in downtown Waynoka in northwestern Oklahoma. They serve authentic German food, beer, wine, and mixed drinks. If you call ahead (phone number is (580) 824-0063), they will pick you up at the Waynoka airport and take you back when you are finished. The restaurant has great food, beer, and atmosphere.

Weatherford, OK (Thomas P Stafford—OJA)

Alfredo's 🍔🍔

Rest. Phone #: (580) 772-3696
Location: 2 miles
Open: Daily: Breakfast, Lunch, and Dinner

PIREP Weatherford offers several great restaurants. There's Alfredo's, Casa Soto, Willie's (the best Southern cookin' around) and The Grill (for a great greasy burger for lunch). Plenty of free transportation. The Stafford Air and Space museum is really something to see if you've never been. They have artifacts in there you wouldn't believe a small town would have, including a real moon rock!

PIREP FBO Comments: York Ford and Precision Design donated a Lincoln Crown Victoria and a Ford minivan to the airport recently.

PIREP Alfredo's is a nice Mexican restaurant that is about two miles from the airport. The Thomas P. Stafford Museum (Gemini and Apollo astronaut) is located on the field and is a great bargain.

The Hundred Dollar Hamburger

Oregon

Albany, OR (Albany Muni—S12)

Lum Yuen's 🍔🍔

Rest. Phone #: (541) 928-8866
Location: Just off the airport
Open: Daily: Lunch and Dinner

PIREP Lum Yuen is a Chinese restaurant off the end of the runway at Albany Airport. You can taxi right up to the restaurant and park. Food and service are pretty good.

Astoria, OR (Astoria Regional—AST)

Runway Cafe 🍔🍔🍔🍔

Rest. Phone #: (503) 861-0599
Location: On the field
Open: Daily: Breakfast, Lunch, and Dinner

PIREP Tie down right in front of the cafe. Got there around noon, and the place was empty. Ordered a standard burger and cup of clam chowder. Very nice people. The guy cooked the burger on an outdoor grill and it was pretty good—give it four stars—it was a pretty big burger also. The clam chowder was decent too—now, I don't know for sure, but it sure didn't seem like out-of-can chowder, big pieces of clam. Owners seemed very nice. After I finished eating, planes started pulling up to the cafe—maybe the smoke from the grill is a call sign to pilots that the food is ready? Seemed like everyone knew each other—very pleasant atmosphere. Recommend if you want the pilot fellowship atmosphere, go a little late for lunch on Sundays.

PIREP Astoria has become one of my favorite destinations for a day trip from Seattle. It's just the right distance for a nice flight, and having an ILS approach is a great back-up when the fog or low clouds move in. The Runway Cafe does have excellent cheeseburgers, grilled outside on the deck, big and juicy and charred on the outside with a high-quality bun. The toppings are set out on a counter and you help yourself to how much you want of what, just like a backyard cookout. They have a fuel truck and will top off your tanks while you eat. They also have a beat-up old Plymouth Volare you can borrow for free to run into town (very cute town—check out the boardwalk along the riverfront with the old-fashioned trolley), or go to the beach.

Cave Junction, OR (Illinois Valley—3S4)

The Strip 😋

Rest. Phone #: (541) 592-2253
Location: On the field
Open: Daily: Breakfast and Lunch

PIREP   The Strip, a new restaurant, has opened. Joy Taylor, the new proprietor, has past food service experience and invested heavily to improve the building and facility. It looks great.

Christmas Valley, OR (Christmas Valley—62S)

The Lodge 😋 😋

Rest. Phone #: (541) 576-2216
Location: On the field
Open: Daily: Breakfast and Lunch

PIREP   Good food, great service; would rate it three burgers.

PIREP   The Lodge is one block north, then two blocks west. It offers a view of a duck pond. The food is pretty good and I like the scenery.

Chiloquin, OR (Chiloquin State—2S7)

Melita's 😋 😋 😋

Rest. Phone #: (541) 783-2401
Location: Across the highway
Open: Daily: Lunch and Dinner

PIREP   FBO Comments: No fuel available. Klamath Falls LMT is just a few miles south.

PIREP   We fly to Chiloquin quite often and have never had a bad meal. Melita's is just across Highway 97 for the airport parking area. Getting across the highway is the most dangerous part of the trip—watch out for fast moving cars and trucks.

PIREP   I've landed there and eaten many times over the years and without exception found the service to be friendly and food fairly tasty. Naturally the prices are good because it's in the middle of nowhere! You can't beat the convenience, as the restaurant is directly across the highway from the aircraft parking area. Be mindful of the density altitude in the summer and generally take off to the south, even with a light tailwind. There's usually a real sinker off the north end of the runway in hot wx.

Cottage Grove, OR (Cottage Grove State—61S)

The Village Green 😋 😋 😋

Rest. Phone #: (541) 942-2386
Location: Across the road
Open: Daily: Breakfast and Lunch

PIREP   The Village Green restaurant is well within walking distance. Lunches and dinners are served inside and out. They also serve drinks and great desserts. If you just want a quick burger, the bowling alley at the end of the taxiway serves 'em up hot.

PIREP   A taxiway leads from the airport proper to a parking area across the road from the Village Green restaurant. Popular Saturday morning breakfast stop. Runway 3200', lights, trees both ends.

Crescent Lake, OR (Crescent Lake State—5S2)

K.J.'s Cafe 🍔🍔🍔

Rest. Phone #: (541) 433-2005
Location: Short walk
Open: Daily: Breakfast and Lunch

PIREP   Crescent Lake, OR (5S2), 3800' × 35', paved.
   K.J.'s Cafe open 7 AM to 3 PM seven days a week. One block from tie-downs.

Enterprise, OR (Enterprise Muni—8S4)

Terminal Gravity Micro Brewery 🍔🍔

Rest. Phone #: (541) 426-0158
Location: Short walk
Open: Daily: Lunch and Dinner

PIREP   Terminal Gravity Micro Brewery and Public House, half-mile walk from 8S4 Enterprise Muni, great dinners, wonderful beer.

Florence, OR (Florence Muni—6S2)

Old Town 🍔🍔🍔

Rest. Phone #: (503) 997-8069
Location: Rent a bike from FBO or 20-minute walk
Open: Daily: Breakfast, Lunch, and Dinner

PIREP   An ideal way to spend a day or a week is to fly to Florence on the Oregon coast. Florence is located about midway between North Bend and Newport. The airport is about a mile from the "old town" waterfront district of town. It's an easy 20-minute walk on a nice day, or you can rent a bicycle from the FBO. A trip into Old Town makes for a nice day trip. There are a lot of shops and restaurants; as well as charter rides.

   Call the FBO ((503) 997-8069) to arrange for car rental from one of the downtown car dealers. With a car, you can make an extended trip of a weekend or as long as you like. There are sand dunes nearby where you can rent dune buggies. Or you may want to partake of the many other outdoor recreational activities in the area such as hiking and fishing.

Gleneden Beach, OR (Siletz Bay State—S45)

The Side Door Cafe 🍔🍔🍔🍔

Rest. Phone #: (541) 764-3825
Location: Very short walk
Open: Daily: Lunch and Dinner

PIREP   FBO Comments: Fuel was not available.

PIREP   Directly across Highway 101, adjacent to the airport, are stairs that lead up an embankment to the Side Door Cafe. It is a short walk and well worth it! The ambiance of the restaurant was cozy and the food was fabulous. This restaurant had excellent gourmet fare prepared with panache. The emphasis is on fresh seafood and decadent desserts. The service was excellent and there was good value for dollar. I highly recommend this restaurant and would even go back again for an overnight visit to further enjoy the wine menu and live entertainment.

PIREP   At Gleneden Beach, OR (Siletz Bay State: S45) there is a restaurant just across the highway. It's a fairly short walk and worth it. Walk down the road from the airport to the highway, then across Hwy 101 (*be careful!!*) and up a trail on the bank. At the top you'll come to the Side Door Cafe, a charming place with very good food. It's open for lunches and dinners, but last time I was there it was closed on Sundays and Tuesdays. Their phone is (541) 764-3825.

The beach is a short walk, as is a shopping mall.

Gold Beach, OR (Gold Beach Muni—4S1)

Nor'wester 🍔🍔🍔

Rest. Phone #: (541) 247-2333
Location: 1/2 mile
Open: Daily: Lunch and Dinner

PIREP   Gold Beach is a fine destination on the southern Oregon coast. The town is located less than a mile east of the airport with a lot of restaurant selections.

Closer to the airport, less than one half mile northeast, is the Nor'wester Restaurant with a very nice harbor view and top of the line food and service. Unfortunately, the Nor'wester is only open after 5 PM.

Right next to the Nor'wester is Jerry's Jet Boat Tours. You can take a half-day or full-day tour of the lower Rogue River. We took the full-day tour that goes 52 miles up the Rogue (as far as jet boats can go), which included several rapids.

Overnight stays at several lodges along the Rogue can be arranged. We stayed at Jot's Resort across the mouth of the Rogue from the airport. Pickup service can be provided. As with many small coastal airports, fog and weather can be a problem. So plan ahead if an extended stay is your intent.

PIREP   A little strip *right* on the beach, right at the mouth of the Rogue River. There's quite a selection of restaurants within a short walk of the airport—when we were there one of the local mechanics lent us his golf cart to ride into town. There's a local cafe that roasts its own coffee and has great pastries.

Also, in one of the hangars is a Ural importer (Russian motorcycles based on the BMW design)—it's one of those places that feels like a museum but is an operating, working shop!

Klamath Falls, OR (Klamath Falls—LMT)

The Satellite Restaurant 🍔🍔🍔

Rest. Phone #: (541) 884-7694
Location: On the top floor of the terminal building
Open: Daily: Breakfast, Lunch, and Dinner

PIREP   The Satellite Restaurant is on the top floor of the terminal building and serves a full menu. We found the food excellent. It has a great view of the airport.

Watch for a lot of Air Force types flying F-16s.

PIREP   Great steak and eggs on the weekend. Good homemade pies and soups. Wonderful grilled burgers.

Lakeside, OR (Lakeside State—9S3)

Mexican Cafe 🍔🍔

Rest. Phone #: (541) 271-0339
Location: Short walk
Open: Daily: Lunch and Dinner

PIREP A nice turf field within walking distance of town and five restaurants, one of which is Mexican. The lake also allows seaplane operations and has a dock available at the Lakeshore Lodge.

McMinnville, OR (McMinnville Muni—MMV)
The Spruce Goose 🍔🍔🍔🍔

Rest. Phone #: (503) 434-4180
Location: In the museum, short walk or free van
Open: Daily: Lunch

PIREP The Evergreen Museum is a terrific place to stop. You can eat in the museum restaurant in the shadow of the Spruce Goose. The prices are very reasonable and the food terrific.

The shuttle is incredibly convenient. The FBO, located at the northeast end of the field, will call the shuttle for you, and it arrived within five minutes. Most of the museum volunteers are ex-military pilots and really bring the planes to life.

PIREP The Spruce Goose is on display in the new museum building. The deli in the museum has pretty good sandwiches/salads. Short walk across the street from the north end of the airport.

Madras, OR (City-County—S33)
Hoffy's Restaurant 🍔🍔

Rest. Phone #: (541) 475-4610
Location: 2 miles, call for free pickup
Open: Daily: Breakfast, Lunch, and Dinner

PIREP There are no restaurants on the field, but two miles distant in the city is Hoffy's Restaurant, full service that is open from 6 AM to 10 PM daily. The food is very good. Adjacent is Hoffy's Motel.

They offer free pickup and delivery to and from the airport. Also the airport has a courtesy car available.

Manzanita, OR (Nehalem Bay State—3S7)
The Fireside 🍔🍔🍔

Rest. Phone #: (503) 368-1001
Location: 1 mile
Open: Daily: Breakfast, Lunch, and Dinner

PIREP From Nehalem Bay (3S7), walk west to the beach, about one-third mile on campground access roads, then north on the sand for about three-quarters of a mile. Turn right (or turn left if you want to go for a swim instead of eat). Straight east into Manzanita you will have about five restaurants to choose from. My favorite is The Fireside. The fish and chips are tops, but when crab is in season, go for the Crab Louis, which can't be beat anywhere.

PIREP Off US 101, three miles south of Manzanita Junction, on the north Oregon coast. On a sandspit between the bay and ocean, with six miles of ocean beach frontage. Sand-dune sheltered campground is open year-round; includes popular horse camp. Campsites: 291 electric (max. RV length 60'), 17 primitive in-horse camp,

6 primitive for fly-in campers using adjacent air strip, 9 yurts. Trailer dump station. Showers. Boat ramp. Bike trail. Meeting hall. Accessible campsites (3E), restrooms, picnic area, boat dock for people with disabilities.

Great place! I have been there several times. Beaches are great! Very few people.

10-minute walk to the beach or walk to the end of the runway for the Nehalem Bay beach.

Nehalem Bay State Park, 83000 3rd St. Necarney, Nehalem, OR 97131, (503) 368-5154/5943.

Medford, OR (Rogue Valley International–Medford—MFR)

The Red Baron 🍔🍔🍔

Rest. Phone #: (541) 772-7978
Location: On the field
Open: Daily: Lunch and Dinner

PIREP   A good restaurant! The view of the runways is great if you are seated at ground level. The prices are fair for the food quality.

Another two thumbs up.

PIREP   The Red Baron is alive and well. Not only that it was open on Easter when many other airport restaurants are closed, but great burgers and sandwiches and a very reasonable salad bar (which came with the burger), at least when I was there.

Newport, OR (Newport Muni—ONP)

The Canyon Way Bookstore 🍔🍔🍔

Rest. Phone #: (541) 265-8319
Location: Call a cab—it's 4 miles
Open: Daily: Lunch

PIREP   My wife and I make the short flight from Boeing Field (Seattle, WA) down to Newport Municipal as often as possible. One of our favorite restaurants in Newport is The Canyon Way Bookstore & Restaurant. It's a funky combination bookstore, used clothing store, delicatessen, and restaurant. The restaurant serves an outstanding lunch and a truly knock-out dinner. Their steaks and seafood are superb. The prices are midrange (not cheap but not totally extravagant either). I'd give them four stars (er, "flying hamburgers"). Reservations are an absolute necessity for dinner.

The restaurant is near the waterfront. You'll need to either call for a taxi cab or rent a car from the FBO. The address is Canyon Way Bookstore & Restaurant, 1216 SW Canyon Way, Newport, OR 97365-4636, (541) 265-8319.

Transportation to town is available from Yaquina Cab Company. Their phone is (541) 265-9552 for calling ahead and 265-9552 for calling from the airport. The phone is right next to the Pepsi machine. Can't miss it.

It's true the airport isn't very conveniently located to town, but it's also true that you can easily get a cab. You can get some of the "best clam chowder in Oregon," according to *Oregon Coast Magazine* (no relation to our newspaper) right here in Newport, or technically, Nye Beach, a historic neighborhood within the city limits of Newport. But here's the best news of all for pilots whose second love is a good burger. Four! Count 'em folks, *four* great burger joints can be found in Newport!

There's a place way out at the whole opposite end of Newport from the airport called Big Guy's. It's an old time 1950s drive-in type place with everything to match except carhops. You would expect to find a good burger just by how the place looks and feels. You will be perfectly satisfied to spend a dollar or two extra in cab fare for this one. The burgers are somewhat greasy, as in absolutely delicious and served hot!

The third closest burger worth flying into Newport for is at Newport's Chowder Bowl Restaurant located at what is known as the turnaround in the aforementioned Nye Beach. This is also the place *Oregon Coast Magazine* pinpointed as serving up the best clam chowder. Chowder Bowl's burgers are absolutely top-notch. Broiled. Buns toasted on a grill.

Fresh and generous makin's and fixin's. That happens to be true of all four of the Newport burgers mentioned herein. Chowder Bowl also serves up noteworthy fried potato rounds rather than French style or shoestrings.

The second closest $100 hamburger is on Newport's working bay front at a place called Whale's Tale, or maybe it's Whale's Tail. Anyway, they serve up their burger with a house dressing or sauce that is out of this world. They bake their buns on the premises. The ground beef, high quality stuff, and the only meat served at this otherwise seafood-only restaurant.

Finally, closest to the airport but all the way up the highway and over the bridge into Newport, is a Denny's kind of restaurant called Apple Peddler that serves up a Denny's kind of burger except as mentioned, the fixing's and makin's are superior, fresh, and generous. All four of these burgers are of the 3–5 napkin variety.

And clam chowder again? Even though Chowder Bowl got the award and deservedly so—their clams are like butter—you can indeed get an excellent and memorable bowl of chowder at Whale's Tail too. Big Guy's sometimes has a home cookin' type chowder, very good. The Apple Peddler also serves up an admirable bowl of chowder, but unlike the Chowder Bowl and Whale's Tale, which are absolutely and consistently held to the recipe, Apple Peddler's chowder depends on who is cooking that day.

But this report to you is about burgers, and Newport makes up for the distance away from the airport by offering four great cheeseburger choices. Did anyone mention cheese? Whale's Tail serves choices of combinations of cheese.

The cabbies themselves know all four and would be hard pressed to say which one is the best burger in Newport. All four and the restaurants they're served in make for a wonderful dining experience. And finally, all four of these restaurants are within 5 or 10 minutes of the airport, depending on who is driving, or flying low as they say.

Pacific City, OR (Pacific City State—PFC)

The Delicate Palate 🍔🍔🍔🍔🍔

Rest. Phone #: (503) 965-6464
Location: Next to the airport
Open: Daily: Lunch and Dinner

PIREP If you are looking to have a fantastic lunch or a dinner that rivals all, then you have to look no further! The cuisine is simply world class. The chef here is very personable, and he and his staff treat you like royalty. Fly in for a weekend get away, eat here, stay here too. Good accommodations.

The restaurant is about midfield to the runway on the east side. Walk to the north end of the runway, turn right. Walk about another block to the south, and you are there. The Delicate Palate is located within the Pacific City Inn.

You won't be sorry!

Village Restaurant and Bakery 🍔🍔🍔🍔🍔

Rest. Phone #: (503) 965-7635
Location: A few blocks
Open: Daily: Lunch and Dinner

PIREP The *best* breakfast on the Oregon Coast.

You really need to drop into Pacific City and look up Lori and her mother at Village Restaurant and Bakery. It is located only a few blocks off the north end of the runway.

Directions: walk off the north end of the runway and go right to the light. Then turn left, walk about two blocks north. The restaurant will be on your left.

A *great* place to eat and the best chicken fried steak on the coast. Careful, though! The "large" order is just that: *large*.

Paisley, OR (Paisley—22S)

The Homestead 🍔🍔

Rest. Phone #: (541) 943-3187
Location: About a 1-mile walk
Open: Daily: Breakfast, Lunch, and Dinner

PIREP   I have flown into Paisley a couple of times for breakfast and found the food and service excellent at the Homestead Cafe. The owners offered to pick me up at the airport and deliver me back to save me the time of walking back and forth. The runway is in good condition with a nice tie-down area.

PIREP   The Homestead is located 1.2 miles south of the Paisley (22S) Airport. It serves home-style meals, both inexpensive and good. My favorite is breakfast. Order *one* pancake—unless you are *really* hungry.

Pendleton, OR (Eastern Oregon Regional at Pendleton—PDT)

Elvis' at the Airport 🍔🍔🍔

Rest. Phone #: (541) 276-3104
Location: In the terminal
Open: Daily: Lunch and Dinner

PIREP   FBO Comments: Handy fuel by self-serve or truck. Oregon Army National Guard Aviation Division on field; watch for big helicopters.

PIREP   Elvis' at the Airport has all the great food you've been looking for, right in the terminal. Try the grilled chicken salad! Of course, the hamburgers are the best! Open lunch and dinner. Pendleton is home to the world-famous Pendleton Roundup—lots to see and do.
    Served by Horizon Air, rental cars at the terminal. Umatilla Indian Reservation and Museum 10 nm east.
    Jimmy Doolittle Museum, B-25 on the airport!

Portland-Mulino, OR (Portland-Mulino—4S9)

Airport Cafe 🍔🍔🍔

Rest. Phone #: (503) 829-7555
Location: On the field
Open: Daily: Breakfast and Lunch

PIREP   Small mom-and-pop cafe, popular with locals. Good ordinary fare. Nice local ambiance. In dry weather, one can taxi on the grass from the midfield parking lot to within 75 feet of the cafe.

PIREP   As you taxi onto the ramp, continue east. You will see a grass taxiway that leads to the Airport Cafe. If in doubt, ask the FBO for directions.
    It's a good burger, and the fries are terrific!

Portland, OR (Portland-Hillsboro—HIO)

Goose Hollow Inn 🍔🍔🍔

Rest. Phone #: (503) 228-7010
Location: Short walk
Open: Daily: Lunch and Dinner

PIREP   There is a good restaurant across the street at the hotel. It is just west of the approach end of runway 30 and across from Hillsboro Helicopters so you get an okay view. The food is not bad nor expensive, the atmosphere is late '80s but, hey, it's better than sittin' on a stool outside with a PBJ sandwich.

The last time I was there, a Jet Ranger and a film crew were dropping a bag of golf clubs from about a 100-foot hover, over and over again. Don't ask—probably a publicity stunt for one of the local golf clubs.

PIREP   An interesting alternative to the hotel cafe is a to take a short ride on the MAX light rail line. It is a short walk from HIO. A half-hour semiscenic ride brings you to the Goose Hollow station in Portland. Across the street is Bud Clark's Goose Hollow Inn. A nice place for lunch.

Portland, OR (Portland-Troutdale—TTD)
McMenniman's Edgefield 🍔🍔

Rest. Phone #: (503) 669-8610
Location: Call in advance to arrange pickup
Open: Daily: Lunch and Dinner

PIREP   There is a fantastic lodge (bed and breakfast) not far from Portland-Troutdale (TTD). It is very close to the field.

It is a revamped grounds that used to be the county poor farm. It includes a microbrewery that produces fantastic ales, a movie theater, lodging, and of course, dining. There are vineyards on the grounds, and many events occur throughout the year.

It is beautifully restored and offers a unique atmosphere.

McMenniman's Edgefield, 2126 SW Halsey, Troutdale, OR, (503) 669-8610 to arrange for airport pickup.

Redmond, OR (Roberts Field—RDM)
The Hangar 🍔🍔🍔

Rest. Phone #: (541) 504-4400
Location: At the commercial terminal
Open: Daily: Lunch and Dinner

PIREP   A fairly new full-service restaurant has been built at the commercial terminal building. It's called The Hangar. It has good food at reasonable prices. Out of its windows are some of the most spectacular mountain views Central Oregon has to offer.

Only one drawback: the terminal ramp/building is off limits to GA parking. One of Redmond's two FBOs (Redmond Air) is within walking distance of the terminal so the GA pilot can get to it. Luckily, Central Oregon weather is such that this walk is more frequently than not a pleasurable stretch of the legs.

Salem, OR (Mcnary Field—SLE)
Roscoe's Landing 🍔🍔🍔

Rest. Phone #: (503) 581-5721
Location: On the field
Open: Daily: Breakfast and Lunch

PIREP   FBO Comments: The self-serve pump is being updated, so only truck fuel available for the near future, and they close at 7 PM.

PIREP   Roscoe's is still good food at a good price. I live in Salem and eat here even when I'm not seeking the $100 hamburger. The smoke *still* wafts in from the bar side, which annoys my wife more than me, but we still go back.

PIREP   Breakfast and lunch are very good with reasonable pricing. I've never had dinner there so cannot comment, but the menu looks good and again, reasonable.

Generally there are more drive-in customers than fly-ins, but with CAVU skies it's a lot different. The entire dining area overlooks transient parking and the field with a view of both runways (16-34 (VFR) and 13-31(ILS)) and the main taxiway.

Located on the ramp (street side of transient parking) between VAL Electronics (Comm Radio line) and the fuel island. It's about 100 yards from a hangar housing a beautiful P-51D. II Morrow is located on the opposite (east) side of the field with taxi-up parking.

Scappoose, OR (Scappoose Industrial Airpark—SPB)
Josephine's at the Barnstormer Inn 🍔🍔🍔

Rest. Phone #: (503) 543-2740
Location: Next to the airport
Open: Daily: Lunch and Dinner, Sunday Brunch

PIREP My wife and I were very satisfied with the food and the service.

The lunch menu includes sandwiches, burgers, pasta dishes, soups, and salads. The dinner menu includes steaks, seafood, chicken, pasta dishes, sandwiches, and salads. Sunday brunch is not a buffet but is ordered off menu. The brunch menu includes typical breakfast items like pancakes, french toast, and omelets, as well as pasta dishes and salads.

Lunch is served 11 AM to 3 PM Tuesday through Saturday. Dinner is served 5 PM to 9 PM Friday and Saturday. Sunday Brunch is served 10 AM to 2 PM.

By the way, the Barnstormer Inn is a bed and breakfast, so if you're in need of an overnight en route stay, you might want to consider staying here.

PIREP They are open for lunch and dinner. We stopped there and had an excellent lunch while we waited for the ceiling to lift.

Seaside, OR (Seaside Municipal—56S)
Creekside Pizza and Chicken 🍔🍔

Rest. Phone #: (503) 738-7763
Location: Short walk
Open: Daily: Lunch and Dinner

PIREP FBO Comments: Lots of tie-down space.

PIREP Great "broasted" chicken. Short walk south of the airport on Highway 101 South. The pizza looks good, but I have been there twice and had the chicken both times.

Silver Lake, OR (Silver Lake F S Strip—45S)
Cowboy Dinner Tree 🍔🍔🍔

Rest. Phone #: (541) 576-2426
Location: Free pickup—bring a cell phone
Open: Daily: Dinner

PIREP Reservations required.

No hamburgers but the best steaks ever. The menu is cook's choice. Free pickup so bring a cell phone and be ready to eat large portions of outstanding food.

This is the place where Sunriver pilots go when we want a *meal*.

Airport phone: (541) 943-3105.

Sunriver, OR (Sunriver—S21)

Sunriver Lodge 🍔🍔🍔🍔🍔

Rest. Phone #: (541) 593-1000
Location: Free van service from the airport
Open: Daily: Breakfast, Lunch, and Dinner

PIREP One of my favorite eating destinations is Sunriver, OR. Popular upscale resort community. Good food, reasonable prices, and you will be treated like royalty. Free van service from the airport to the lodge/restaurant.
Check noise abatement procedures.

PIREP Sunriver is a resort community that offers something for everyone. The lodge-operated vans will take you to/from the onsite airport to anywhere in Sunriver. A tip is encouraged. Call (541) 593-1000 to be picked up or to talk with the lodge about accommodations.
The best restaurants, in my opinion, are
Trout House: $$$$—Northwest cuisine right on the river. Very popular, reservations required.
Sunriver Lodge: $$$$—Northwest cuisine in a gorgeous lodge setting. Reservations recommended.
Merchant Trader: $$—Also at the lodge; sandwiches, hamburgers, and the like with great views.
Michael's Place: $$$—In Sunriver Village. Very good Northwest cuisine with flair.
Chowder House: $$—In Sunriver Village. Seafood, sandwiches, hamburgers. Not bad.
There are all kinds of places to stay here. The lodge operates many rooms/condos and houses. Also, many owners rent out their houses (for less than what the lodge charges).
No matter what you like to do—golf, swimming, rafting, horseback riding, skiing, bike riding, hiking—it's all here.

The Dalles, OR (Columbia Gorge Regional/The Dalles Muni—DLS)

Country Kitchen 🍔🍔🍔

Rest. Phone #: (509) 767-4700
Location: On the field
Open: Daily: Breakfast and Lunch

PIREP The city of The Dalles is in Oregon but the airport is across the Columbia River in Washington. There is a nice little cafe on the field called the Country Kitchen. Be advised, they are open seven days a week but close at 2 PM.

PIREP The Country Kitchen features homemade soup, pies, cakes, and specials. Their flame-broiled burger is a definite winner. Their location is in the administration building next to the fuel pumps. They are open for breakfast and lunch daily.

Tillamook, OR (Tillamook—S47)

Air Base Cafe 🍔🍔🍔

Rest. Phone #: (503) 842-1130
Location: On the field
Open: Daily: Breakfast, Lunch, and Dinner

PIREP Tillamook S47 in Oregon now has a restaurant in the giant blimp hangar and museum called the Air Base Cafe. They offer the usual hot sandwiches and burgers served with chips. For dessert they offer Tillamook Ice Cream. The restaurant is open from 9 to 4 PM every day.

A very interesting place to visit, The Tillamook Naval Air Station Museum offers specially reduced prices for pilots who would like to visit the museum and also a discount on aviation fuel.

Wasco, OR (Wasco State—35S)

Goose Pit Saloon 🍽 🍽

Rest. Phone #: (541) 442-5709
Location: 1/2 mile
Open: Daily: Breakfast, Lunch and Dinner

PIREP   FBO Comments: runway 25 is steep downhill. Park at the bottom (west) end of the old runway (now taxiway/parking ramp) and it's a couple blocks downhill to the center of town. Taking off from 25, you might consider turning early to avoid over-flying the town.

PIREP   Goose Pit is a couple blocks (west, downhill) down the road and two blocks south on the main drag. Rough looking outside and in but good food and friendly people. We arrived after 10 on a Sunday, were the only operation at the airport, and didn't meet any of the vets.

PIREP   Goose Pit Saloon, opens 8 AM weekends, 10 weekdays. At 10 AM all of the retired farmers come in for coffee and tell stories of WWII and Korea.

Come about 10:30 when the help has time to catch a breath.

Don't let the outside scare you.

The prices are a bit lower than most.

Pennsylvania

Allentown, PA (Lehigh Valley International—ABE)
Gregory's Steakhouse 🍔🍔🍔🍔🍔

Rest. Phone #: (610) 264-9301
Location: Just off the airport, call for pickup
Open: Daily: Lunch and Dinner

PIREP Went to Gregory's for the famous 120-ounce steak.

WOW!

Four of us actually finished it off. Atmosphere is very casual and service is good. We stayed the night at the local Hilton Garden Inn; everything is within the airport grounds.

Highly recommended for those meat-and-potato people. At least four hamburgers.

PIREP Flew to Allentown and went to Hanger 7. They recommended Gregory's Steakhouse and they provided free transportation. Gregory's is just off the airport property and is definitely a five-star restaurant to visit. We have a group of us that fly from N07 Lincoln Park, NJ every other Saturday night for dinner. Also a great place to take your spouse for a nice flight and dinner. Our group has grown from 3 people in one plane to 10–12 people in four planes.

It is worth the trip. Service is excellent and food is great!!!!

Allentown, PA (Allentown Queen City Muni—1N9)
Queen City Diner 🍔🍔🍔

Rest. Phone #: (610) 791-0240
Location: 3/4-mile walk
Open: Daily: Breakfast, Lunch, and Dinner

PIREP Queen City Diner is within about 3/4-mile walk. IHOP is about 1-1/4 miles. No restaurants are closer. I am an ASEL pilot and am based at 1N9. It is a 10–15 minute walk to the nearest restaurant, which is average at best.

Bally, PA (Butter Valley Golf Port—7N8)
Runway Grill 🍔🍔🍔🍔🍔

Rest. Phone #: (610) 845-2491
Location: On the field
Open: 7 AM–1:45 PM weekdays, 6:30 AM–1:45 PM weekends

PIREP   FBO Comments: Fun rolling narrow runway.

PIREP   Nice people. Good breakfast. Reasonable prices.

PIREP   Butter Valley Golfport (7N8) has a runway right in the middle of the golf course. What a beautiful spot to fly to. After hours you can still get sandwiches, burgers, and ice cream.

   I landed the Lancair on runway 34 and was stopped before the end of the pavement with only moderate braking. There is plenty more runway (manicured turf) beyond the pavement. Highly recommended for breakfast, lunch, or just to stop in for a Coke.

Beaver Falls, PA (Beaver County—BVI)

Sal's 🍔🍔🍔

Rest. Phone #: (724) 843-4020
Location: Just off the field
Open: Daily: Lunch and Dinner

PIREP   This is a very busy training facility, with a flight school and an air traffic school both in operation.

   You will find the lowest price on 100LL and Jet A. Go to the self-serve pumps in front of the tower. A lot of the service is provided by college kids learning to fly, but they haven't learned about customer service, so if you have a problem with the pump or need any other help, be friendly and ask, and they will do anything for you.

   They can't seem to keep the restaurant open on the field, but there is a great facility just past the entrance to the airport named Sal's. First class Italian. Prices are reasonable.

   There is also a military air museum on the field.

Bellefonte, PA (Bellefonte—N96)

Airport Restaurant 🍔🍔

Rest. Phone #: (814) 355-7407
Location: On the field
Open: Daily: Breakfast and Lunch

PIREP   John and Marina run a nice restaurant on field. Open on weekends from 6 or 7 till noon. I love their breakfast almost as much as the burgers!

Butler, PA (Butler County/K W Scholter Field—BTP)

The Runway 🍔🍔🍔

Rest. Phone #: (724) 586-6599
Location: On the field
Open: Daily: Breakfast and Lunch

PIREP   I visit this restaurant quite often. Some of the greatest burgers you will ever have, service is *outstanding*, and the airport is a great facility as well. They have an ILS on field so you can work up your appetite before you stop in to eat.

   If departing NW and VFR, watch out for Beaver Airport (BVI). The Class D at Beaver can be tricky to get around. Best bet is to contact Pitt App and go through the Class B.

PIREP   Definitely better than most airport restaurants and conveniently located at the north end of the airport right by the ramp, I'd give The Runway Restaurant four stars. During a Sunday brunch the food was better than average, although there wasn't the variety that you'll find down at LBE's restaurant. Reasonably priced for the Sunday buffet.

The only downside to The Runway is that it's popular enough with the locals that you might end up waiting 20 minutes before you can be seated. If you're willing to sit at the bar though, odds are you'll be served immediately.

Centre Hall, PA (Penns Cave—N74)
Pack a Lunch 🍔🍔🍔🍔

Rest. Phone #: (814) 364-1664
Location: 1/4-mile walk
Open: Daily: Breakfast, Lunch, and Dinner

PIREP   Penn's Cave Airport (N74) is located in central Pennsylvania just east of State College. The strip is 2500' × 40' and slopes down at both ends. It sits between two parallel ridges, which make a long straight-in approach more comfortable than skimming the trees (towers?) on the ridgeline during your downwind leg. Unattended; call for fuel, lights, or service: (814) 364-1664. No landing fee; self-park on the grass and then walk the 1/4 mile to the cave.

Penn's Cave is an old-fashioned tourist stop, off the beaten path, which probably hasn't changed much since the '60s. The one-hour cave tour is entirely by boat, and its 52-degree temperature is a welcome relief on a hot summer's day.

Visitor's center has a full snack bar; burgers, dogs, shakes, ice cream, and more souvenirs than you can shake an authentic Indian tomahawk at. Picnics welcomed. There is also a nature preserve with elk, bear, mountain lion, and so on viewable via bus tours for an additional ticket. It's a nice place for kids or for a change of pace and a malt.

Coatesville, PA (Chester County G O Carlson—40N)
The Flying Machine 🍔🍔🍔

Rest. Phone #: (610) 380-7977
Location: On the field
Open: Daily: Breakfast and Lunch

PIREP   Flew in with a couple of friends based on the other pireps. The Flying Machine still lives up to its reputation. Highly recommend the Cajun burger. Can't beat the prices for the rest of the menu either.

PIREP   Lunch on a Sunday. Excellent burgers, fries. This restaurant is still right on the ramp. Lots of training flights and the great occasional jet as well. A nice and active airport. TFM (The Flying Machine) has a reputation among the local airports as a good place to eat lunch, with particularly good burgers. I fly out of Smoketown (Q08) and have never heard a bad word about this place. Service was prompt and pleasant and the food was great. I do not remember the exact hours but it was Sunday and they were open fairly early in the AM and ran till 7 PM or so at night.

Corry, PA (Corry-Lawrence—8G2)
Peek 'N Peak Resort 🍔

Rest. Phone #: (716) 355-4141
Location: Rent a car; it's 15 miles away
Open: Daily: Breakfast, Lunch, and Dinner

PIREP Just 15 miles from Corry Airport is Peek 'N Peak Resort and Conference Center in Findley Lake, NY. A four-season resort offering two 18-hole golf courses—an upper course and a lower course. The upper course is rated four-star by *Golf Digest* and was host to the Buy.Com tour in June 2002 and will host the Nationwide Tourney again this year.

Indoor tennis and swimming as well as many other activities are available year round. In winter, skiing is offered on 25 slopes for varied ski levels; snowboarding and tubing areas are also available. A variety of menu choices including fine dining are available at various restaurants throughout the resort complex.

Marlene's Restaurant 🍔🍔🍔

Rest. Phone #: (814) 664-3604
Location: 1-mile walk
Open: Daily: Breakfast and Lunch

PIREP I recommend a visit to Marlene's Restaurant just 1 mile down the hill in the center of Corry. The meals are delicious, the food is reasonably priced, and portion size is large. Smaller portions are available. The airport manager offers transportation if possible and there is cab service for a few dollars. The walk into town is pleasant, but climbing the hill back to the airport is a real workout.

Danville, PA (Danville—8N8)

Muffin Man 🍔🍔

Rest. Phone #: (570) 275-8811
Location: 1/2 mile
Open: Daily: Breakfast and Lunch

PIREP Danville is a 2200' grass strip with lots of experimental activity. It is a good idea to go around the pattern to check for no radio aircraft. The Muffin Man is a great place for lunch, about half-mile east of the airport.

Du Bois, PA (Du Bois-Jefferson County—DUJ)

The Flight Deck Restaurant & Lounge 🍔🍔🍔

Rest. Phone #: (814) 328-5281
Location: In the main terminal building
Open: Daily: Breakfast and Lunch

PIREP I have been there a few times recently, and it is really a great little stop. The buffalo chicken sandwich is *hot* if you like 'em that way. Prices are reasonable. Be sure to watch for a lot of commuter traffic in the area.

PIREP FBO Comments: Airport has separate door for GA pilots—nice, no walking around building like most airports.

PIREP My wife and I have been dining at the Flight Deck restaurant for 20 years and have always had above average food and below average prices. The current owner, now the Flight Deck, is better than ever. Friday night fish has even the locals waiting in line; Saturday after 4 PM is wing night and has the place packed also; and Sunday prime rib is as good as you can get anywhere. They also have holiday specials for two that can't be beat.

We fly to dinner quite often, and our hometown Flight Deck is hard to beat !!!

Easton, PA (Braden Airpark—N43)

Gennaro's 🍔🍔🍔

Rest. Phone #: (610) 253-2380
Location: Very short walk
Open: Daily: Lunch and Dinner

PIREP   Food is great—all kinds of food that you'd typically find in an authentic Italian restaurant (pizza, spaghetti, calzones, great steak sandwiches, chicken parm, and so on) in a nice casual setting. Hours are 11–10 during the week and 11–11 on the weekends. Restaurant also has a full service bar (you know, for your passengers).

It's a very short walk from the airport (maybe a quarter of a mile just south of the airport).

PIREP   Gennaro's is a brick oven pizza place. It is your basic pizza/pasta/red sauce joint. A visitor to Easton would swear that there is a city ordinance requiring a pizza place on every block!

Erie, PA (Erie Intl/Tom Ridge Field—ERI)
Serafini's Italian Restaurant 🍔🍔🍔🍔

Rest. Phone #: (814) 838-8111
Location: In the terminal building
Open: Daily: Lunch and Dinner, closed Sunday

PIREP   FBO Comments: Exceptionally friendly, responsive, helpful, and knowledgeable staffing at this FBO. Melanie and Thad provided excellent service and made this a very pleasant stop. Although the gas is a little expensive, the service and crew car made it worth the difference.

PIREP   Serafini's has not reduced their standards. We confirmed the accuracy of the previous pireps. Although they were having a busy day, the service was rapid and friendly. The server seemed to be enjoying the challenge of keeping everybody happy.

PIREP   Food is fabulous—all pasta is made on-site daily. Open 11 AM–10 PM Monday through Saturday. Five-burger stop!! If you like Italian food, this is a *must*! Excellent manicotti, pastas, meatballs; subs, pizza, and takeout are also available.

PIREP   There is a decent restaurant in the terminal building, an easy walk from the FBO. It's probably not as nice as the recommended Italian restaurants, but if you are just looking for a quick bite, this is a good bet.

Usually I follow the crowd to Serafini's Italian Restaurant, 2642 W. 12th Street, Erie, PA 16505.

Erwinna, PA (Vansant—9N1)
Airport Grille 🍔🍔

Rest. Phone #: (610) 847-8320
Location: On the field
Open: Weekends only: Breakfast and Lunch

PIREP   Vansant is a fabulous old-time grass airfield with a half-dozen Stearmans based there, as well as an assortment of other antiques and tail draggers.

On weekends during the flying season Linda is there with her grill serving up hot dogs, hamburgers, and veggie burgers. She usually has chips, cookies, fruit, and an assortment of drinks available.

Franklin, PA (Venango Regional—FKL)
Primo Barone's Restaurant 🍔🍔🍔🍔

Rest. Phone #: (814) 432-2588
Location: On the field
Open: Daily: Lunch and Dinner

PIREP   FBO Comments: We fly in here frequently for dinner. FBO people are always pleasant and welcoming.

PIREP   Flew over one Friday evening with stepdaughter and five-year-old granddaughter for dinner. Combo playing in restaurant (their normal Friday evening entertainment). Just as we started gathering ourselves to leave, the combo played "Itsy Bitsy Spider," to the great pleasure of our beautiful, charming five-year-old blonde. They really let us know they were glad we came!

PIREP   After reading all the high reviews on this eatery I decided to visit and see for myself. All the reviews are correct. The food is outstanding and the service matches the food. This airport is typical of other small airports in PA with limited airline service. It is in very good condition with good approaches and facilities.

PIREP   My wife and I flew into Franklin on a Sunday afternoon. It was our first time at the restaurant. It was outstanding! The view of the tarmac was wonderful, along with the food, the service, and the cleanliness of the restaurant. I would highly recommend it for lunch or even dinner.

Gettysburg, PA (Gettysburg Airport and Travel Center—W05)

Herr Tavern 😋😋😋😋😋
Rest. Phone #: (717) 334-4332
Location: Reasonable walk, 3/4 to 1 mile
Open: Daily: Lunch and Dinner

PIREP   I flew into the newly repaved Gettysburg airport today for lunch. Probably the smoothest runway I've ever landed on. Long and wide too (3100 × 60).
   The walk to the Herr Tavern is more than the quarter mile advertised, about three-quarters to one mile. The tavern itself, however, is extremely nice. The service was quite good, as was the food, not to mention being very inexpensive (without drinks, the check was only $21 for the two of us). The building itself was the coolest part, as it dates back to 1815.
   In addition to the dining rooms and bar, the tavern also offers an inn/bed and breakfast accommodations. Each room even has its own fireplace! Next time I come here I plan to stay for a weekend. Rooms range from $89 to $189/night, and they recommend reservations. The phone is (800) 362-9849.

PIREP   Historic (and delicious!) dining in Gettysburg. For the perfect weekday (and Saturday!) lunch stop, I recommend the Herr Tavern just outside Gettysburg, PA. Land at Doersom (W05) and enjoy the pleasant quarter-mile walk down Chambersburg Pike. Herr Tavern has wonderful sandwiches, as well as full-course meals and daily specials. The tavern also offers first class B&B accommodations. The Herr Tavern's phone is (717) 334-4332.
   By the way, there is a $5 landing fee at Mr. Doersom's airport, but it's well worth it for the convenience and hospitality of this little airport. Enjoy!

PIREP   Last evening we departed PNE (Northeast Philadelphia) to land at Doersom Airport in Gettysburg and take in dinner at Herr's Tavern. It was listed as five flying hamburgers.
   Doersom Airport, we report, has been sold, and there is now no landing fee whatsoever. The runway 6/24 is 3100 feet of skinny pavement, quite serviceable, with humps, but unlighted—an important consideration if you are planning on having dinner and then departing as we did.
   Herr's Tavern is far more than the quarter mile previously reported—I would peg it at about three-quarters of a mile—we took a cab, the dispatch headquarters of which was conveniently located right at the airport.
   Regrettably, Herr's Tavern was a dinner disappointment. We found the food quite average and needlessly pricey. Bus tour groups descended on us, which is usually an indication of trouble. My filet mignon entree $20

was as tough as kangaroo. Although the menu says Amex is taken, the waiter politely requested that we use a different card!!!

If you go to Gettysburg, I recommend you land at the renamed Gettysburg Airport as planned, avoid Herr's Tavern, get in one of the nice dozen taxis sitting near the hanger at the taxi HQ, and go further into town for a decent meal elsewhere.

Grove City, PA (Grove City—29D)
Elephant and Castle Pub 🍔🍔🍔

Rest. Phone #: (724) 748-1010
Location: 2 miles; use the free shuttle
Open: Daily: Lunch and Dinner

PIREP   This is almost the perfect afternoon trip for folks in western PA and eastern OH. Fly to Grove City (29D), go shopping at the Grove City Outlet Mall, and eat at the Elephant and Castle.

The folks at the FBO are more than willing to lend you their courtesy car, or call for the shuttle over to the outlet mall, maybe two miles away. After shopping, a meal at the Elephant and Castle Pub at the entrance to the mall is a must. The menu includes British-style food. It has always been good and the portions more than filling. Generally the service is good, but it can be a little slow when the Castle is busy.

There is an active skydiving club on the field and occasionally glider operations at this uncontrolled airport, so a little extra attention to unicom and sharp-eyed copilot are useful!

PIREP   I stopped in to Grove City airport this weekend on the way back from OSH and was amazed at the fantastic accommodations and restaurants. There is a Holiday Inn that faces the tarmac, and you can watch your airplane while you are in the pool.

For dinner we went to the Elephant and Castle Pub across the street. A great stop!

Harrisburg, PA (Capital City—CXY)
Pierre's Steak House 🍔🍔🍔

Rest. Phone #: (717) 774-0132
Location: About a block
Open: Daily: Dinner

PIREP   Looking for another new place to eat in Central PA? Try Pierre's Steak House in New Cumberland, PA by the Capital City Airport.

Pierre's is known locally for serving excellent steaks since 1929. It is located a block or so from the airport. If you taxi up and park at the Harrisburg Jet Center, the line boys are always glad to give you a ride to the restaurant.

Honesdale, PA (Cherry Ridge—N30)
Cherry Ridge Airport Cafe 🍔🍔🍔

Rest. Phone #: (570) 253-5517
Location: On the ramp
Open: Daily: Breakfast and Lunch

PIREP   FBO Comments: 24-hour self-serve fuel with credit card. 24-hour restrooms (below cafe). New airport beacon. New lighted windsock. Soon to be wind-tee.

PIREP   Airport Cafe now open every day of the week, 7 AM to 4 PM or until last person. Breakfast served all day. Open Friday nights for dinner.

PIREP   This place is really great! The location was always prime but the last time I tried to eat here, I wrote a flaming pirep because of nonexistent service. I'm happy to report that, under new management, the restaurant is now *excellent*. My burger was huge, juicy, and cooked just right. The service was prompt and friendly.

It's great to have this option back!

PIREP   The restaurant closed. Apparently, the manager was left alone one day when his staff didn't come in and he just closed up and walked out, never to return.

When I arrived, the doors were open and a very nice couple had put on some coffee and offered us donuts and told us the sad story. They are hoping that a local "turn-around specialist" will soon reopen the restaurant, but who knows?

From everything I heard, they used to serve great food, and it's right on the airport, overlooking the runway.

Johnstown, PA (John Murtha Johnstown-Cambria Co—JST)
Barnstormers 🍔🍔🍔

Rest. Phone #: (814) 539-5733
Location: On the field
Open: Daily: Lunch and Dinner

PIREP   FBO Comments: Very friendly, helpful FBO. Met us with baggage cart. Four-hour stay, no tie-down charge.

PIREP   Flew in yesterday. Stopped at Barnstormers for HDHBs and will definitely be back! Right now they offer lunch menu and buffet 11 AM–2 PM and 5 PM–7 PM happy hour with snacks. Restaurant is attractive and airy, service was delightful, hamburgers were worth the trip; cost was ten dollars and change for two HB, two sides, two nonbar drinks.

Rented from Hertz, pleasant agent, no problems. Spent day at Johnstown Flood Museum and Incline Plane. Nice low-key trip in the middle of a July week. Everyone we dealt with in town—parking, museum, incline plane ride, shop, ice cream stand—made us feel welcome. We plan to go back and spend time at the National Monument and have another great hamburger.

PIREP   The airport at Johnstown, PA has a great little restaurant. Their specialty is Italian.

Kutztown, PA (Kutztown—N31)
Airport Diner 🍔🍔🍔

Rest. Phone #: (610) 683-5450
Location: On the ramp
Open: Daily: Breakfast and Lunch

PIREP   FBO Comments: Very nice FBO people. Didn't have to pay a cent for a landing fee.

As a low time pilot, found it very stressful 1) finding the airport, 2) dodging the gliders all over the place, and 3) getting oriented to land when the active runway flip-flopped from 17 to 35 with each landing/takeoff preceding me.

PIREP   Diner location was kind of cool. Service was great, but the burgers were "eh"—typical of an old diner. Overall, a fun experience.

PIREP   I've flown in here at least two dozen times with dates or friends or alone and never once been charged a landing fee and if I were to be, I would gladly pay!

Excellent airport, have fun at night—during the school year, it is filled with college babes—you can always turn a head or two when you practically park your airplane in the parking lot of the diner. Food is cheap, excellent, typical classic diner.

Order the California burger!

PIREP Kutztown, PA is a great place for eats. 123.0 is the CTAF; it's printed in a current Airport Facility Directory. There's a *huge* water tower resembling an Eveready flashlight just north of the runway (on the turn from base to final for 17) and a small but deep quarry right up against the south end of the runway. Don't land short!! There's also a grass strip in the bowl, 10-28, I think.

Mind the gyrocopters and gliders (there are many on the weekends), and you'll do fine (that's the reason for the fee, it keeps the traffic low, but you've got $5, right? It's waived with purchase, that means fuel, a new AFD, a new VRF sectional, anything, even if it's under $5. You need a current sectional anyway, right?).

Land a bit long on 17-35, more toward the bowl in the middle. You'll be slow enough without braking even in a C-210. Mind the moisture content, but if it looks dry, taxi right toward the NW corner into the grass alongside the cars in the restaurant, turn back in toward the runway and shut her down. The FBO won't come and get you for the landing fee, but hey, $5? Courtesy dictates we can afford this fee, even if the "runway" looks like a piece of melted taffy. Go to the FBO and pay up first, maybe even get a T-shirt with the Waco "Mad Dog" on it and wear your $5 fee! Then go eat up!

The food is great, the coffee flows, and the service is like a Wright Whirlwind: it never misses a beat! If you go on the weekdays before (I think) 10 AM, the 222 Special is the best breakfast! Two eggs, two bacon (or sausage), and two toasts for $2.22! (that's the highway just above the airport that goes east to ABE and west to RDG).

Lancaster, PA (Lancaster—LNS)

Fiorentino's Bar and Grill 🍔🍔🍔🍔

Rest. Phone #: (717) 567-6732
Location: On the field
Open: Daily: Lunch and Dinner

PIREP Fiorentino's Bar & Grill had great food. Of course the service was unhurried. This is definitely not "fast" food. A great break, nice leisurely meal, and a balmy flight home. Airport people very friendly and helpful.

PIREP I went to Fiorentino's Bar & Grill on Sunday AM. I will say that the food was very good, and the prices were not that bad (for breakfast—I didn't look at the lunch/dinner side of the menu). Based on my experience, make sure that you are not in a hurry. The service was very kind and polite, but the food was very slow getting to us. Maybe it was just due to the Sunday church crowd, not sure.

However, I will go back there in the future.

Latrobe, PA (Arnold Palmer Regional—LBE)

DeNunzio's Italian Chophouse and Bar 🍔🍔🍔

Rest. Phone #: (724) 539-3980
Location: On the field
Open: Daily: Lunch and Dinner

PIREP Stopped there to grab a bite, was not aware that Blue Angels was gone. DeNunzio's has excellent food and service, will be back.

PIREP Just want to let everyone know that the restaurant at Arnold Palmer Regional Airport is open once again. It has been renamed DeNunzio's Italian Chophouse. DeNunzio's is a famous restaurant in the western

PA region. Some of the *best* Italian food you'll get and reasonably priced. Also, there is a luncheon buffet every day from 11–2 (not sure on the ending time for the luncheon buffet). My wife and I ate there the other day and thoroughly enjoyed it!

We will definitely be back.

Lock Haven, PA (William T. Piper Memorial—LHV)

The Texas Cafe 🍔🍔🍔🍔

Rest. Phone #: (570) 748-4418
Location: Close by
Open: Daily: Breakfast and Lunch

PIREP FBO Comments: On-field fuel service (100LL). An F.A.A. DER is on field (Mynah).

MIRL at night means you won't see the airport till you're almost on top of it; the mountains obscure it if you are coming from the other side of the hill, by the way, so stay >3500' till you clear them on the way in—they are very dark at night.

PIREP Nice tavern, decent food and service. Within walking distance behind Hangar 2 or the FBO next to Hangar 2. Phone numbers for (after hours) fuel on door of the FBO.

If you don't mind walking a bit west of the field into town, there is an *excellent* little diner called The Texas Cafe, 132 E Main St., also a great place for breakfast on weekend flights. If you're later than 9 AM, expect a wait—it's popular!! The tavern is a nice place to hang out while your lady comes to pick you up (assuming you are waiting for one to pick you up).

PIREP This is one of the reasons to go flying on the weekend!

Visit the Piper Museum on the SW corner of the field and then take a short walk into town for lunch. Fuel is available at the FBO on the NE side of the field. Parallel grass strip for tail draggers.

The museum offers a great glimpse into the history of Piper aircraft and general aviation. The town offers good distractions for those passengers not as interested in airplanes.

It is a top stop for anyone who is interested in GA.

Meadville, PA (Port Meadville—GKJ)

Eddie's Footlong Hot Dogs 🍔🍔🍔

Rest. Phone #: (814) 724-2057
Location: 2-1/2 blocks
Open: Daily: Lunch and Dinner

PIREP We have been flying in to GKJ airport at Port Meadville, PA and walking the two and a half blocks to eat outside at Eddie's Hot Dogs for many years. We love the place and have taken many first flight students there for a treat. Now there is another place across the street from Eddie's called Casey's. Casey's is the perfect place for dessert. Casey's serves hard ice cream, made onsite, in waffle cones that are also made right there. The smells when entering the shop are heavenly. Their ice creams are made from various recipes that you will not find anywhere else. My favorite is the Meadville Crunch. You can also purchase their fudge and divinity candy. They are open year round so that if you want to purchase gifts of candy you can do so. I promise that if you have a sweet tooth for ice cream and candy, you'll love walking over to this shop and sampling their fare.

Casey's is not fancy when it comes to eating accommodations. You will have to get your cone and eat it while sitting outside on the rocks or while walking back to the airport. Since it is only a 15-minute flight from my airport, I always plan on taking a cooler so that I can take one or two gallons home with me.

PIREP   The $100 hot dog!

Nice community airport with a newly paved 5000' runway, localizer and GPS approaches. Very reasonably priced gas. Open for gas 365 days, 8 to 5 only. Eddie's Footlong Hot Dogs, very popular with the locals, is about a 3/8-mile walk south of the terminal.

Great hot dogs can be ordered with a multitude of different fixings as well as good Sloppy Joes and chips. Very pleasant setting with outdoor picnic tables. To cool off your mouth after having a hot dog with Eddie's special sauce, there is an ice cream shop across the street. Eddie's is open only during the summer months.

Around the corner is a Red Lobster for those wanting a more complete meal and willing to spend a bit more.

Monongahela, PA (Rostraver—FWQ)

Eagle'S Landing 🍔🍔🍔

Rest. Phone #: (724) 379-8830
Location: On the field
Open: Daily: Breakfast and Lunch

PIREP   The Eagle's Landing is a popular fly-in spot for hungry pilots. The service is very good and so is the food. Fuel prices are usually below the average. Plenty of room to park on the ramp too.

Myerstown, PA (Deck—9D4)

Pennsylvania Dutch Country 🍔🍔🍔

Rest. Phone #: (717) 866-6071
Location: Off the field
Open: Daily: Breakfast, Lunch, and Dinner

PIREP   This is a nice public use, privately owned airport in the heart of Pennsylvania Dutch Country. There is a 3600' 01/10 runway with a VOR approach. 100LL and Jet A are available on the field. No landing fee. Small tie-down for an overnight stay.

A car is available for exploring the local farm country and going to any one of the many local PA Dutch-style restaurants (good food that locals judge by volume).

Good idea to call in advance to make sure the car is out and ready for you: (717) 866-6071. No credit cards for the fuel, but you can sign a slip along with your address and tail number and they will mail you a bill.

Pottstown, PA (Pottstown Muni—N47)

Pottstown Diner 🍔🍔

Rest. Phone #: (610) 327-1630
Location: Close by
Open: Daily: Breakfast, Lunch, and Dinner

PIREP   If you find yourself in the pattern at Pottstown Muni (N47) on downwind for 07, look about 1000' down out to your left just before you're abeam the numbers. That big road going by with the strip mall just there is Route 100, home of the Pottstown Diner!

If you can't hitch a ride from someone and you're on foot starting from the FBO, just head out the gate and straight ahead the one *long* block out to Route 100. It's at the southerly end of that strip mall right in front of you. I've always been in a car, which makes it a one- or two-minute drive, but I'd guess it would be a ten-minute walk or so.

This is a genuine diner in the traditions of the famous Pennsylvania diners. It's one of the later ones that's been expanded over the years rather than the trailer style. The atmosphere is comfortable and family oriented.

There's a very nice soup and salad bar there that shouldn't be missed and lots of home-style items to order from the menu. The prices are reasonable for just about any meal.

It's a nice, friendly place! I'd recommend it...say, I just did!

Clear skies!

Pottstown, PA (Pottstown Limerick—PTW)

Airport Hotel 🍔 🍔 🍔

Rest. Phone #: (610) 495-7626
Location: Across the street
Open: Daily: Breakfast, Lunch, and Dinner

PIREP There is a restaurant across the street in the Airport Hotel (imagine that.) They have great lunch specials during the week and a great dinner. Hours are 11 AM to 2 PM (lunch) and 5 PM to 9 PM (dinner). They are open weekends also but I would call ahead for seating. They have prime rib specials on Friday and Saturday. They have a seafood sampler to die for. Closed on Sunday. I would give this place five burgers. The line crew at PTW is very pleasant and will help you out, not to mention discounted fuel on the weekends. Can't miss airport. 6000 feet from Limerick Cooling Towers.

Selinsgrove, PA (Penn Valley—SEG)

Ulrich's Seafood 🍔 🍔

Rest. Phone #: (570) 374-5525
Location: 1/2 mile
Open: Daily: Lunch and Dinner

PIREP The restaurant is located maybe a half-mile walk south of the airport. Take the road south from the airport, cross the intersection with the traffic light, and it's ahead on the right. Good seafood at reasonable prices. Great small restaurant atmosphere.

Seven Springs Borough, PA (Seven Springs—7SP)

Seven Springs Resort 🍔 🍔 🍔 🍔

Rest. Phone #: (800) 452-2223
Location: 2 miles, free shuttle
Open: Daily: Breakfast, Lunch, and Dinner

PIREP FBO Comments: No fuel is available.

PIREP It seems there's always some kind of festival going on at Seven Springs. The airport is about two miles away, but a free phone call from the rustic airport lounge always makes a free van appear in less than ten minutes. Lunch and dinner are very average, and the prices are a bit on the high side, but the setting and the friendly experience is worth twice the price. We went last time for the fall festival in October and the color was overwhelming. I can't wait to go back. It's a shame the runway is closed during ski season. I'd fly there to ski. I'll just look forward to spring.

This has been my best-kept secret fly-in, but I guess it's time to share.

PIREP Seven Springs (7SP) is located southeast of Pittsburg, PA about halfway between Cumberland, MD and Pittsburg. It is an all-season resort with their own small airport. Unfortunately, the airport is closed in the winter, so a fly-and-ski trip is out. During the warmer months, the resort offers a beautiful golf course, hiking, horseback riding, and an alpine slide!

I flew the wife and kids up to 7SP for the Sunday brunch buffet at the main lodge. The selection was fabulous and the food was great! We stuffed ourselves silly. If we had more time, we would have stayed and ridden the alpine slide all day. We got a free shuttle to/from the lodge from the courtesy phone at the airport.

7SP is on top of a mountain ridge with a narrow runway and ponds on both sides—watch the crosswinds!

Smoketown, PA (Smoketown—S37)

T. Burkes 😋😋😋

Rest. Phone #: (717) 293-0976
Location: 2 blocks
Open: Daily: Breakfast, Lunch, and Dinner

PIREP FBO Comments: Self-serve 100LL at a good price. Friendly office staff.

PIREP Still great food and right by the airport property. T. Burkes accommodated 25+ for a fly-in along with their regular patrons with no fuss, speedy service, and tasty food. Try their Reuben sandwich.

PIREP Three hamburgers for Smoketown? More like five!

We've been going to T. Burkes for years and it's a favorite Saturday flight. If you're lucky, Tom might have crab cakes (broiled, fat lumps of crabmeat, mayonnaise filler) as good as Baltimore's best. Try the cordon bleu open-face sandwich: ham, chunky chicken salad, and broiled Swiss cheese on an English muffin, my mom's favorite.

It's a charming place, delicious food, and a short walk from Glick's Aviation.

PIREP I thought I'd send in a recommendation for one of the best restaurants I've been to. It's just a few hundred yards walk (mere minutes) from the Smoketown, PA airport in the heart of PA Dutch country.

It's called T. Burkes and is referred to locally as "the deli" at the Smoketown FBO. It's a little more upscale than the typical airport greasy spoon. The pleasant atmosphere combined with good food and reasonable prices make this a must stop if you want to impress your passengers. You're very likely to encounter Amish buggy traffic near this airport.

Tunkhannock, PA (Skyhaven—76N)

Minotti's Filling Station Restaurant 😋😋

Rest. Phone #: (570) 836-2762
Location: Short walk
Open: Daily: Lunch and Dinner

PIREP FBO Comments: No night operations. All grass areas are well drained; taxi anywhere that is mowed. Tail-wheel instruction in Piper J-3. Skydiving on weekends and Wednesday nights.

PIREP Minotti's Filling Station Restaurant is located about a half-mile walk from the airport. A pink Buick is always available for transportation. Some aviation items on display. Subway and Burger King are about 800 feet from the FBO. Town has old shops and things to see. Lots of home-builts, Shortwing Pipers and Stinson Gullwing on field. Rides in the Gullwing available. Grocery store, liquor store, Wal-Mart, pizza shop, and shoe stores located on field. Aircraft camping with showers available summer season.

Washington, PA (Washington County—AFJ)

Stone Crab Inn 😋😋😋

Rest. Phone #: (724) 228-5300
Location: 5 miles; use the crew car
Open: Daily: Lunch and Dinner

PIREP Got stuck for two days in Washington, PA en route to Washington DC due to weather. The folks at the FBO were great, willing to drive me anywhere local for meals or for a room. I even got a refresher course in weather forecasts from one of the resident instructors.

The two restaurants I visited while there were the Stone Crab Inn and Garfield's (in the mall). Both good food and reasonable prices. Both have bars.

Food was very good at both places. What really made the stay enjoyable, however, was the willingness of the FBO staff and management to do whatever they could to make my "visit" pleasant.

Waynesburg, PA (Greene County—WAY)
A.J.'s Landing 🍔🍔🍔

Rest. Phone #: (724) 627-8207
Location: On the field
Open: Daily: Breakfast and Lunch

PIREP The restaurant is still one of the best. It is always busy with local folks along with pilots. The food is great and the pies are awesome. And the prices are good. There are at least six different types of pies to choose from. If you are ever in southwest PA it would be a great place to stop. Great food, great fuel prices, and a friendly country atmosphere.

Wilkes-Barre/Scranton, PA (Wilkes-Barre/Scranton Intl—AVP)
Damon's 🍔🍔🍔

Rest. Phone #: (570) 654-3300
Location: 5-minute walk
Open: Daily: Lunch and Dinner

PIREP FBO Comments: Tech Aviation was very nice. Offered their courtesy car for the trip to Damon's.

PIREP Flew to AVP for lunch and was delighted to find a Damon's. They have very good ribs and the service was excellent. I will definitely be back. It is worth five stars. The food was great and the atmosphere was very good.

PIREP I frequently fly out to Scranton (AVP) from Sussex (FWN). It is just over 50 miles, making it a great cross country flight. ATC is very easy to work with. As for the food, inside the main terminal building, the airport cafe offers some pretty decent delights. Not a five-burger place, but it still has a lot to offer. It sits right on the tarmac. Sure beats walking. They run scheduled airliner service out of there. It's lots of fun to bring family and friends there and see them gasp at the size of the jets that arrive, knowing that minutes earlier our little beech had full run of the field. The main terminal is very impressive.

Wilkes-Barre, PA (Wilkes-Barre/Wyoming Valley—WBW)
Victory Pig Bar-B-Q 🍔🍔🍔

Rest. Phone #: (570) 288-3257
Location: Short walk
Open: Daily: Lunch and Dinner

PIREP Colonial Pancake House, Wilkes-Barre/Wyoming Valley Airport (WBW), PA. Restaurant is a short walk across the street from the ramp. Looking for a buffalo burger, try this place. Most pilots stop here for coffee.

No fees for short parking at airport.

PIREP Here's a place to get something a little unusual. If you fly into Wilkes-Barre/Wyoming Valley Airport (WBW), you can take a short walk (~.1 mile) north on Route 11 to the Victory Pig Bar-B-Q (next door to the place with the golden arches). Their pizza is famous throughout Wyoming Valley, and people drive for miles to get it. The only problem is that are only open Wednesday, Friday, and Saturday nights.

You can also get in a game of miniature golf or smack a few on Bob's driving range which is located next door.

Williamsport, PA (Williamsport Regional—IPT)
Skyview Restaurant and Lounge 😑😑😑

Rest. Phone #: (717) 368-2031
Location: On the field
Open: Daily: Lunch and Dinner

PIREP   My son and I flew from Pittsburgh on a beautiful Sunday afternoon for lunch. We enjoyed great hot sandwiches and prompt friendly service. View of the airfield is a plus. Based on the second floor of the commercial air terminal, this spot is as convenient as it gets. The only downside was that fuel service was not available at the ramp and we had to taxi to the FBO for fuel.

We highly recommend.

PIREP   The Skyview Restaurant and Lounge is on the second floor of the terminal building. The picture windows give a great view of the airport action.

I had dinner there last night, a delicious steak dinner, which was on their "specials" menu. The service was excellent. There were only a few local people there last night so it was hard to get a read on the atmosphere, but I would say that it is definitely a friendly place. You can taxi to the restaurant and get a table at the large windows the overlooks the field. I parked at the FBO. They gave me a courtesy car to drive around the perimeter of the airport to the restaurant. All in all, it was a nice night. I highly recommend the restaurant. They are open to 8 PM most nights of the week. They stop serving dinner at 7:45 PM.

York, PA (York—THV)
Orville's Restaurant 😑😑😑😑

Rest. Phone #: (717) 792-6233
Location: On the field
Open: Daily: Lunch and Dinner

PIREP   Park right outside the restaurant. The place was busy (Sunday morning), but we got right in. The server was friendly and the food was good. Best home fries I've ever had (not greasy). Prices were average for out in the country, downright inexpensive compared to a big city airport.

Will definitely be back again.

PIREP   This was my second trip to Orville's in less than a week. The first trip I had a bacon cheeseburger with fries. I asked for medium rare and was delivered a well done burger. The fries were less than tasty. Five days later I was back in York again. This time service was a little better as was the food. On this second trip I ordered the seafood bisque. It was really good with plenty of seafood. I tasted crab, oysters, and a few shrimp. When I go back I'll not have burgers but the bisque instead.

Zelienople, PA (Zelienople Muni—8G7)
El Toro Mexican Family Restaurant 😑😑😑😑😑

Rest. Phone #: (724) 452-5850
Location: Taxi up
Open: Daily: Lunch and Dinner

PIREP   A friend invited me and another friend out for my very first flight (in a noncommercial aircraft). We decided to fly from Erie to Zelienople to have lunch at El Toro Mexican Family Restaurant. It was a perfect day to fly because the leaves are changing and the day was just as beautiful as it could be! We even saw a couple of jumpers on our way there! We had a fantastic lunch at El Toro. The service was great and the food was excellent! I took home part of my huge lunch and ate it for dinner. It was just as good! I had a great time and would highly recommend it to anyone interested in attractive in-flight scenery and a delicious lunch!

PIREP   Good stop, great food, nice folks. Inexpensive!

Don't try to locate it during your approach because at first you'll think that the place must have shut down. As you're coming in on runway 35 you will suddenly spot it to the right, a yellowish building. The El Toro sign is outside the field, and when we flew in on a Sunday afternoon there were not too many patrons' cars parked, which added to the confusion.

Once inside, you will not regret the distance you flew. Pirep to those who arrive by helicopter: their ramp slopes more than it appears as you're setting down.

We will go there again.

PIREP   El Toro Mexican Family Restaurant is open for lunch and dinner seven days a week. El Toro offers great Mexican food, second to none. That's easy to understand; the restaurant is owned and operated by a Mexican-American family.

From the roasted pepper and cilantro-laced salsa (and free chips) to the deep-fried ice cream dessert, the food is authentic Mexican and reasonably priced. This isn't your Chi-Chi's Eastern version of bland Mexican food.

Aircraft parking is located at the restaurant, on the ramp, east of the runway. It's hard to find good Mexican food this side of Colorado and even harder to find it on an airport. Luckily, El Toro fills that bill on both accounts. And, judging by the crowd, the locals like the food also.

Block Island, RI (Block Island State—BID)

Bethany's Airport Diner 🍔🍔🍔🍔

Rest. Phone #: (401) 466-3100
Location: On the airport
Open: Daily: Breakfast and Lunch

PIREP FBO Comments: Very friendly FBO. If you eat within an hour, there are no landing fees!

PIREP Bethany's diner is a warm and friendly place with great food. My 15-year-old had a cheeseburger and of course I, the dad on Atkins, had the tuna salad. Both were very good.

PIREP Phil Schwartzman and I often fly from Farmingdale, NY Republic Airport to Block Island for breakfast or lunch at Bethany's Diner, right at the parking ramp. Great service and food, with many good Block Island folks to visit with.
No landing fee for two-hour diner stop.

PIREP They waive the state landing fee if you are only going to the restaurant.
There are many restaurants located in town, a 20-minute, 1-mile beautiful walk or an $8 taxi ride. They vary in price, but most feature good seafood and are located right on the ocean.
Block Island is well worth a day or weekend visit!

Newport, RI (Newport State—UUU)

Rhino's 🍔🍔🍔🍔🍔

Rest. Phone #: (401) 846-0707
Location: An easy walk
Open: Daily: Lunch and Dinner

PIREP FBO Comments: Parking a little tight. Call ahead.

PIREP   Landed at Newport. $15 cab ride to a great night. Started at Yesterday's, walked to the Rhino Bar, and ended at The Red Parrot before spending the night at a charming B&B (one of many) not too far from downtown.

$15 tie-down fee the next morning and we were off. Great time, each place unique and tasty. Also visited a great cigar shop called The Humidor.

Providence, RI (Theodore Francis Green State—PVD)

Hooter's 🍔🍔🍔🍔

Rest. Phone #: (401) 732-0088
Location: Very short walk
Open: Daily: Lunch and Dinner

PIREP   N2679H and I like to fly to Providence, RI. Hooters is a five-minute walk from the FBO.

South Carolina

Aiken, SC (Aiken Muni—AIK)

Baynham's Family Restaurant 🍔🍔

Rest. Phone #: (803) 649-7663
Location: 1 mile
Open: Daily: Breakfast, Lunch, and Dinner

PIREP Had the distinct pleasure of flying into Aiken SC on my dual cross-country. This was my first 100 dollar hamburger.

It was overall a wonderful experience. The line crew was top notch, the staff very friendly and helpful. We used the courtesy car to drive to a small family-owned restaurant about a mile away from the airport.

Baynham's Family Restaurant offers a vegetable/meat buffet, a good selection of sandwiches, and a very good burger. Staying with tradition, my CFI and I both had cheeseburgers. The restaurant offers breakfast, lunch, and dinner.

Overall a wonderful first cross-country flight.

Bamberg, SC (Bamberg County—99N)

Tommy Rose Barbeque 🍔🍔🍔

Rest. Phone #: (803) 245-6102
Location: Walking distance
Open: Daily: Lunch and Dinner

PIREP Tommy Rose Barbeque is open only on Thursday, Friday, and Saturday from 11–9 and Sunday for lunch.

PIREP Many of us have visited Tommy Rose Barbeque in Bamberg, SC (99N). The restaurant is within walking distance, but the owner is happy to come to the airport and pick you up, then deliver you back after your meal. The food is excellent, cheap, varied, and the service is great.

Barnwell, SC (Barnwell County—BNL)
Winton Inn Restaurant 🍔🍔🍔

Rest. Phone #: (803) 259-7181
Location: Call for pickup
Open: Daily: Lunch and Dinner

PIREP I have found a little-known place in SC, after flying into Barnwell Co. airport. The Winton Inn Restaurant will send a car to pick you up; a short drive later and you are enjoying an all-you-can-eat seafood meal including Alaskan King crab legs. The meal is $19 a plate. There's not a whole lot of "atmosphere" but the food is great.

Beaufort, SC (Beaufort County—73J)
Barbara Jean's 🍔🍔🍔

Rest. Phone #: (843) 524-2400
Location: Call for pickup
Open: Daily: Lunch and Dinner

PIREP FBO Comments: Yep, it's Frogmore International, with 3500' of very new blacktop (as of April '05), you'll likely not get any plane international! How many airports have *rocking chairs* on the ramp for folks to sit around and rate the landings?! We bought a house 5.6 miles away.

PIREP Barbara Jean's will *come get you*. They have *great* crab cakes—crab meat and seasonings and oil they cook them in, no bread, no fillers. *Cold* beer—very inexpensive, and they have some great black-eyed peas. Real good southern cookin' in a real nice little southern city.

Charleston, SC (Charleston Executive—JZI)
Cappy's Seafood 🍔🍔🍔

Rest. Phone #:
Location: Short drive
Open:

PIREP The airport is actually on John's Island, and there is no place to eat at the airport itself. But a short ride away is Cappy's Seafood. Great selection of fresh seafood, in a nice atmosphere. I was actually expecting it to be pricy, but it turned out to be quite cheap. Go on a nice VFR day and sit outside.

PIREP FBO Comments: Very nice FBO. Friendly and courteous.

PIREP The FBO has a nice deal with nearby Marriott Suites Hotel. The hotel picks you up and can take you to nearby restaurants. The hotel gives you a discount through the FBO, and it's a good deal. Plenty of three-star restaurants nearby.

Cheraw, SC (Cheraw Muni/Lynch Bellinger Field—47J)
Mom's Kountry Kettle 🍔🍔🍔

Rest. Phone #:
Location: On the field
Open:

PIREP   Very nice airport. Maintenance on field. At the end of 07 is a good country restaurant, Mom's Kountry Kettle. They are open Monday through Friday and Sunday 11–2.

Adjoining the airport is the Back Porch Steak House. It is open 5–9:30 Thursday through Saturday. I have eaten here many times. Sunday meal is great, but get there before church lets out or you'll have a wait.

PIREP   Stopped by Mom's Kountry Kettle this past Wednesday. The country buffet is $6, which includes tea or soft drink. Excellent country cooking, and the wait-staff is very friendly.

Bill's BBQ 😋😋

Rest. Phone #: (843) 921-9288
Location: 1/2 mile
Open: Daily: Lunch and Dinner

PIREP   Went looking for the Backyard Steak place, but the FBO was closed early, since it was only 5:15 PM on Saturday night.

Some very nice woman in an Expedition was driving around the facility and said we must be looking for Bill's BBQ, Steak & Seafood and offered us a lift the half mile or so down the road.

We ate there; my wife went for the steak (after all, I had promised her a steakhouse) and I had the country buffet. She wasn't super happy with her steak, but the buffet, complete with fried chicken and pulled BBQ, was fantastic. I would recommend it. Nice little local country restaurant with friendly folks.

Walk back to airport was under five minutes.

Clemson, SC (Oconee County Regional—CEU)
Glady's Kitchen 😋😋😋

Rest. Phone #: (864) 882-0058
Location: 3.5 miles crew car available
Open: Daily: Breakfast and Lunch

PIREP   Great food, friendly atmosphere. Courtesy car available at the airport, 3.5 miles from airport.

Columbia, SC (Columbia Metropolitan—CAE)
The Flightline Bar & Grill 😋😋😋

Rest. Phone #: (803) 926-3647
Location: 1/2 mile
Open: Daily: Breakfast and Lunch

PIREP   The Flightline Bar & Grill in Columbia, SC, located half mile up Hwy. 302 from Columbia Metro Airport, has killer food at decent prices. Try a hot sandwich with a good glass of wine when you get some time.

Columbia, SC (Columbia Owens Downtown—CUB)
Lizard's Thicket 😋😋

Rest. Phone #: (803) 779-6407
Location: Crew car
Open: Daily: Breakfast, Lunch, and Dinner

PIREP   Lizard Thicket, excellent home cooking at reasonable prices. Several in Columbia. Accessible from either downtown or metro airports. Courtesy cars usually available.

Greenville, SC (Greenville Downtown—GMU)

Jason's Deli 😋 😋 😋

Rest. Phone #: (864) 284-9870
Location: 3 minutes by crew car
Open: Daily: Breakfast, Lunch, and Dinner

PIREP FBO Comments: Greenville Jet Center is a great stop with friendly people. Nice crew car to grab a bite to eat.

PIREP Jason's Deli has all kinds of sandwiches, hot and cold; sides; fresh salad bar; and clean surroundings. It is a quick, short drive from the airport, about three minutes. They give you large portions to take back to the plane.

Greenville Jet Center has a very nice crew car they will lend you to go get something to eat. There are also many other restaurants around the airport area along with shopping.

Greer, SC (Greenville Spartanburg Intl—GSP)

Windows 😋 😋 😋

Rest. Phone #: (864) 877-7417
Location: In the main terminal
Open: Daily: Breakfast and Lunch

PIREP A great place to fly in for a meal. Windows is located in the main terminal building and has a beautiful view of a fountain garden overlooking the runway. They are open 11 AM–9 PM most days and always open till 7:30 PM. They serve a continental breakfast from 6 AM–11 AM. They not only have a five-star burger, but also have an ostrich burger on a multigrain bun for the health conscious. This is an upscale airport restaurant with a wide menu that would please most anyone.

PIREP The service was a little spotty when I had lunch at Windows, but the nice dining room and eclectic menu really surprised me. GSP International is not a large airport and Windows is the only place to eat other than a coffee wagon located in the concourse. However, I would highly recommend Windows while waiting on a flight or waiting to pick someone up.

Hartsville, SC (Hartsville Regional—HVS)

Shug's Smokehouse Grill & Tavern 😋 😋 😋

Rest. Phone #: (843) 383-3747
Location: Crew car, Caddy
Open: Daily: Lunch and Dinner

PIREP FBO Comments: Very plush Cadillac courtesy car. Johnny Payne of Hartsville Aviation was very helpful in reserving the courtesy car for us and recommending a restaurant.

PIREP We flew into Hartsville planning to eat dinner at The Beacon Restaurant, but it has closed down. Johnny Payne of Hartsville Aviation recommended that we try Shug's Smokehouse Grill & Tavern in Hartsville, which we did. We used Hartsville Aviation's oh-so-nice Cadillac courtesy car to drive to the restaurant for a delicious meal with a friendly wait-staff and great atmosphere.

Great airport, friendly and courteous service, very enjoyable restaurant.

Hilton Head Island, SC (Hilton Head—HXD)

The Crazy Crab 😋 😋 😋

Rest. Phone #: (843) 363-2722
Location: Crew car
Open: Daily: Lunch and Dinner

PIREP FBO Comments: The folks at Carolina Air Center were friendly and allowed us to use their crew van without any hassle. In addition, they provided a recommendation for a restaurant as well as giving a map with directions to it.

PIREP We drove a short distance, perhaps five miles, to The Crazy Crab restaurant. There are huge windows that afford most tables an outstanding view of a marsh area. The food was good, especially the broiled flounder. The hush puppies alone were worth the trip!

PIREP There are two FBOs to choose from on this one-runway paved field. Both are full-service FBOs.

A short walk down the road leads to a sort of shopping center. It has a Sam's Club along with several restaurants. I ate at the Pizza Hut; I believe there was also McDonald's and Taco Bell. Short on character, long on convenience.

Myrtle Beach, SC (Myrtle Beach Intl—MYR)

A 100 Beachy Restaurants 🍔🍔

Location: Crew car
Open: Daily: Breakfast, Lunch, and Dinner

PIREP I fly to Myrtle Beach Jetport (MYR), often, and if you like South Carolina beaches it can be a lot of fun. I stay at the Landmark Best Western ((803) 448-9441); they will pick you up at the airport, on request. The hotel is on the beach, with an indoor/outdoor heated pool. Also, the town has frequent bus shuttle service to other beachfront shopping areas and amusement parks. The public buses stop right in front of the hotel. I also understand that they have some great golf courses. Room prices run somewhere between $75 to $125 per night but since you don't need a rental car I feel it's a bargain.

The only snag is that the airport has just recently converted from a joint use military/commercial airport to civilian operation. I'm not sure if a fixed-based operator has started operating on the field as of yet. North Grand Strand airport is also open to private aircraft but I've not been there and don't know how far away from the beaches it is or if transportation is available.

Plenty of good eating close to airport as well as hotels and fun for the kids. Great place to go for lunch, the day, or even a weekend!

Newberry, SC (Newberry County—27J)

Summer's Restaurant 🍔🍔

Rest. Phone #: (803) 276-7658
Location: 1 mile from the airport
Open: Daily: Breakfast, Lunch, and Dinner

PIREP I recommend this place (City of Newberry) for the fun loving and young at heart. We have a great place for hamburgers only one mile from the field: Summer's Restaurant! They have other things on the menu, but *all* the locals go there for burgers!

The downtown has "rebuilt" itself and many attractions are open late when the Opera House has a show. With Hampton Inn right there downtown you can stay if you are having too much fun. The downtown drive is only two miles from the field. See you there.

North Myrtle Beach, SC (Grand Strand—CRE)
The Parson's Table 😋😋😋😋

Rest. Phone #: (803) 249-3702
Location: 20-minute drive
Open: Daily: Breakfast, Lunch, and Dinner

PIREP One of my all-time favorite restaurants is located relatively nearby: The Parson's Table in Little River, SC. It is about a 20-minute drive from the airport and quite easy to find (north on Route 17 right into Little River). This restaurant is quite unique in that it is located in what used to be a Presbyterian church. It has won numerous awards (for example, five diamond awards as one of the top 50 overall restaurants in the U.S.).

The menu is extensive and features broiled and sauteed seafood, prime rib, veal, steaks, duck, and chicken. They also have an extensive wine list that includes some reasonably priced wines.

There is a nonsmoking area, casual attire is the rule, and reservations are recommended, particularly during the summer. Their telephone number is (803) 249-3702.

Trust me on this: it is well worth the 20-minute drive from CRE to eat at this world-class restaurant. Best of all, it is surprisingly moderately priced!

For "the best burgers on the beach," try the River City Cafe located at 21st Ave. North in Myrtle Beach. This is about 15 minutes by car from CRE. Head south on Route 17 and stay left on Business 17 (*not* Bypass 17) until you get to 21st Street—then turn left and you're there.

They feature free salted-in-the-shell peanuts (be sure to toss the shells on the floor). They have numerous varieties of burgers and sandwiches. Their bloomin' onion appetizer is an absolute must, and their onion rings are superb.

PIREP Grand Strand Airport (CRE), North Myrtle Beach, SC is my favorite destination when I'm looking for someplace to go.

One paved runway (5-23), VOR on the field (CRE), tower (nonradar), two FBOs. I strongly recommend ramp 66 for services. Really sharp line crew will meet you as you taxi in and help you park. When it's busy (and it usually is in summer), they will park you at the door and then tow your plane to parking. They have a courtesy car available for free for one hour. If you will be longer, you can rent by the hour or longer. Car rental is $8/hour with the first hour free (keeps down abuse of courtesy car!). They had a pilot shop last summer. Closed now, and I'm not sure if they plan to reopen it for summer.

Many, *many* places to eat within a couple mile drive. A few are within walking distance. Especially a good ice cream place. The beach is about a three-quarter mile walk down a residential street. Watch out, a couple of people on the street have dogs that don't appear to be chained, but I've never had one come after me. They just bark a lot!

The main area of Myrtle Beach is about a five- to ten-minute drive away. One of the biggest tourist traps you'll ever see! They have *everything!*

Also have extensive banner tow operation. I can sit for hours and watch them drop and pick up the banners. Those guys are great. It gives you a whole new perspective on maximum performance operations watching a 150 pick up a 60-foot banner from the ground!

Spartanburg, SC (Spartanburg Downtown Memorial—SPA)
The Beacon 😋😋😋😋

Rest. Phone #: (864) 585-9387
Location: 2 miles north of the airport
Open: Daily: Lunch and Dinner

PIREP The Beacon Drive-in is located on John B. White, Sr. Boulevard, a street named for the founder of the Beacon. Food is great! I have been eating breakfast there for more than 30 years and don't suffer from any ill effects. Since his retirement several years ago, Mr. White and a group of followers meet there every morning at 6:30 for breakfast and to solve the world's problems.

I have been employed at the Spartanburg Downtown Airport since 1964 as an aircraft mechanic, commercial pilot, and now as airport director. My staff and I invite you to visit us and enjoy a unique culinary experience, The Beacon!

Frank G. Anderson

Airport Director SPA

PIREP   Here's fair warning: be kind to your gallbladder, and find some other place to eat in Spartanburg.

Here's the Beacon scene: Big mesh bags of yellow onions lie outside the door, stacked to the eaves. As you enter, a grizzled old fellow greets you with an unintelligible snarl. You finally realize he's asking you what you want. You follow the advice you read here, and order a Burger A' Plenty. The shout goes out to the rows of noisily genial fry cooks presiding over deep vats of bubbling fat. They manage to maintain their footing thanks to grease-trap mats liberally sprinkled with absorbent granules. Your humongous burger, totally obscured by a veritable mountain of glistening fries and onion rings, slides across the counter on a nearly translucent paper plate. Wisely, you help yourself to lots of napkins—and a side-order of Maalox. You finally find a cleanish table, unavoidably close to a TV set, and then a couple of rowdy soccer teams show up for a celebratory snack, slapping high fives and tossing unidentifiable stuff at each other. This is *not* the place to celebrate your anniversary.

The food? Give it a D for dreadful. The ambience? If possible, it's worse, but amusingly so. The prices? About what you'd expect—cheap.

Admittedly, though, if you clean your plate, you won't leave hungry. The Beacon claims to be world famous, and politicians of every persuasion have stopped here for decades to kick off their South Carolina campaigns.

What can you say good about the place? It's, umm…memorable.

Just hope there's not much turbulence for the trip home.

Billy D's West 🍔🍔🍔

Rest. Phone #: (864) 587-7766
Location: 1 mile from the airport
Open: Daily: Breakfast, Lunch, and Dinner

PIREP   Billy D's West is an excellent casual restaurant with a large selection of great food. From the terminal at SPA turn left on Kensington Road. Follow Kensington to the stop sign. Turn left on John B. White Blvd. (SC-296). Restaurant is one-eighth mile on the left, total distance one mile.

PIREP   Ate there this last weekend. Excellent food and service! We parked on the ramp and walked to the restaurant. There is a sidewalk the entire route and no problems with traffic. A very pleasant walk through a residential neighborhood to increase the appetite on the way to the restaurant and to walk off the excess food afterwards.

There are other restaurants along the way, including a Ryan's Buffet and Fuddruckers. Further down the street is a seafood place that appears to be open in the evenings and more of the ordinary fast food places.

The Hundred Dollar Hamburger

South Dakota

Aberdeen, SD (Aberdeen Regional—ABR)
Airport Restaurant 🍔🍔🍔

Rest. Phone #: (605) 225-7210
Location: In the terminal
Open: Daily: Breakfast and Lunch

PIREP Had the pleasure of stopping by the restaurant located in the main terminal building at the Aberdeen Regional Airport (ABR). After several rousing recommendations from local pilots, I decided to give it a try. The service was excellent; the food was very basic, tasty and delicious. They make excellent sandwiches and just about everything I had was great.

Aberdeen is a little remote but has excellent runways, nav aids, and facilities. This great restaurant makes it worth a little detour.

P.S. Some of the best pheasant hunting in the U.S. available there.

Huron, SD (Huron Regional—HON)
The Hangar 🍔🍔🍔

Rest. Phone #: (605) 352-1639
Location: In the terminal
Open: Daily: Breakfast and Lunch

PIREP The Hangar is located in the terminal at the Huron Regional Airport Huron, SD. This is a great restaurant! I highly recommend it.

Mitchell, SD (Mitchell Muni—MHE)
Cabela's Upland Cafe 🍔🍔🍔🍔

Rest. Phone #: (605) 996-0337
Location: Call for free shuttle
Open:

PIREP   The restaurant, Upland Cafe, is located on the premises of Cabela's giant sporting goods store. If you are an outdoorsman, expect your lunch to include some equipment purchases.

For a shuttle to and from Cabela's and the Upland Cafe, call Cabela's in advance of your arrival to make arrangements.

Rapid City, SD (Rapid City Regional—RAP)
Airport Cafe 😋😋😋

Rest. Phone #: (605) 225-7210
Location: In the terminal
Open: Daily: Breakfast and Lunch

PIREP   There is a decent restaurant in the terminal building at Rapid City, SD. The menu choices range from hamburgers to steak and seafood dinners. The prices are reasonable. You can walk the approx. one-third mile from Jet West FBO, or they will drive you to the terminal and pick you up when you call.

PIREP   Had occasion to stop in at Rapid City Airport this month on account of weather. The FBO was very helpful, insisting that we take their vehicle rather than walk to the restaurant in the terminal. Breakfast was good and reasonable, weather cleared and we came on to Des Moines.

Sioux Falls, SD (Joe Foss Field—FSD)
NAPs 😋😋

Rest. Phone #: (605) 332-NAPS
Location: In town; use the 2-hour free crew car
Open: Daily: Lunch and Dinner

PIREP   Business Aviation will loan you a crew car (two hours max); go to the corner of 41st St. and Western. They are a barbecue restaurant that has the *best* pulled pork sandwich I've ever had.

They also deliver.

Wall, SD (Wall Muni—6V4)
World Famous Wall Drug 😋😋😋😋

Rest. Phone #: (605) 279-2699
Location: 5 blocks
Open: Daily: Breakfast and Lunch

PIREP   A wonderful place to fly in for breakfast is Wall, SD (GV4). The airport is about five blocks from the center of town and the World Famous Wall Drug. For the uninitiated, Wall Drug is a unique combination of shops, restaurants, and tourist attraction that got its start by advertising "free ice water." Food is very reasonable in price and you can eat in one of several dining rooms, all decorated with wonderful collections of Western art. The town also has several other good eating places (Cactus Cafe, Elkton House) that can give you a change of pace from the good food at Wall Drug.

The airport is infrequently attended. There is a pay phone where you can call for gas. Often people see you land and insist on driving you the few blocks into town.

It's a fun change of pace.

The Hundred Dollar Hamburger

Tennessee

Bristol/Johnson/Kingsport, TN (Tri-Cities Rgnl TN/VA—TRI)
Rainbow Restaurant 🍔🍔🍔🍔

Rest. Phone #: (423) 325-6282
Location: In the terminal
Open: 6 AM–9 PM weekdays, 6 AM–7 PM Sunday

PIREP The restaurant is now on the main floor of the terminal and undergoing some remodeling. It was mid-day, so we had chicken salad sandwiches with homemade vegetable soup for about $5 and it was a good, quick lunch. But, *stay away* from Tri-City Aviation. They stuck us for a $10 ramp fee even though we topped off with their not-inexpensive fuel! That made it an unnecessarily expensive 40-minute fuel/lunch stop. AOPA does list other FBOs on the field.

PIREP Located on the second floor of the terminal building, a short walk from the FBO. Open 6 AM–9 PM weekdays, 6 AM–7 PM Sunday Great burgers and home cooking. We had the fried chicken dinner: half a chicken, vegetable, potato, salad, and bread—superb!

Chattanooga, TN (Lovell Field—CHA)
Rib and Loin 🍔🍔🍔

Rest. Phone #: (423) 499-6465
Location: 1 mile, crew car available
Open: Daily: Lunch and Dinner

PIREP The controllers at KCHA are very friendly and helpful. Free crew car available (and a new one at that). The food was excellent. If you're not up for the Rib and Loin, I can also say that there is a Lonestar steakhouse right up the road.
All-around great experience; highly recommended.

PIREP FBO Comments: Extremely friendly and helpful. Hospitality is very Southern. Gave us a crew car (brand new minivan) so we could get some lunch.

PIREP The ladies at Tac-Air recommended Rib and Loin BBQ and gave us a car to get there. Real live Southern barbecue. I recommend the ribs with banana pudding for dessert (Nilla wafers included). Definite stop-off if you are in the area!!

PIREP Crystal Aviation South has at least 4 crew cars. There's a Crystal North too, but we've never been there!

Lots of food is nearby. Good barbecue can be had at the Rib and Loin. There's a Krispy Kreme doughnut "factory" close to all the eating joints. Don't miss 'em when they bring 'em hot off the line.

Clarksville, TN (Outlaw Field—CKV)

Duke's Diner 🍔🍔🍔

Rest. Phone #: (931) 431-0299
Location: On the field
Open: Daily: Breakfast, Lunch, and Dinner

PIREP FBO Comments: FBO people are very friendly. Always call us on radio as we are leaving to thank us and invite us back. Tell Sam Robert sent ya! Crew car leaves something to be desired though.

PIREP Duke's Diner is great! On the field. We eat there three or four times a month. Duke is the star and she is very special. Thai food is wonderful but the hamburgers are the best this Southern boy has eaten. Huge and loaded! Stop in to say hi to Duke. They close at 3 PM but always have breakfast. I had breakfast and lunch there today as I usually do.

PIREP Once inside the diner you can tell who is in charge! I ordered a hamburger the first time I visited Duke's Diner and it was *great* (besides being too much to eat), with plenty of freedom fries and free refills too! I have since tried the Pad Thai, egg roll, and wonton, and her omelets (perfect every time) are to die for. All the food was beyond my expectations. A lot of food for the money. Duke's Diner is by far the best part of this airport!!!

Clifton, TN (Hassell Field—M29)

The Bear Trace at Ross Creek Landing 🍔🍔🍔🍔

Rest. Phone #: (931) 676-3174
Location: Next to the field
Open: Daily: Breakfast, Lunch, and Dinner

PIREP The Bear Trace at Ross Creek Landing is a relatively new (within the last ten years) course. The facilities and course are very good. This is a great golf course. One thing that makes it great is that you can fly to M29 (Clifton, TN), land, put your clubs directly on a golf cart and drive less than a half-mile to the clubhouse. The course and airstrip are separated by a fence and about 200 yards of grass along hole number 8. Call the clubhouse at (931) 676-3174, make your tee time, and ask them to have a golf cart (or two) at the airstrip, and they are very accommodating! Clifton is located 71 miles southwest of Nashville, TN directly on the Tennessee River.

Dickson, TN (Dickson Muni—M02)

Shell Gas Station 🍔🍔

Rest. Phone #: (615) 446-5421
Location: 1/4-mile walk
Open: Daily: Lunch and Dinner

PIREP FBO Comments: Very friendly folks who operate the FBO loaned us a car to run down to the local restaurant, which is disguised as a Shell gas station. We didn't even buy gas.

PIREP They are a "meat and 3" during the week, just hamburgers and sandwiches on the weekend. Good food, friendly but not too quick service, inexpensive. Colorful local hangout. About a quarter-mile from the FBO, walking distance if no car is available.

PIREP On the weekend when I visited, all they offered were those foil-wrapped, heat lamp–warmed burgers and biscuits that you'll find in just about any gas station or truck stop. There were several locals eating there, all of them smoking up a storm. I just bought a bag of chips and a soda, then walked right back to the airport.

Bottom line, this might do in a pinch, but if you're looking for a "destination" restaurant, head to nearby M93.

Dyersburg, TN (Dyersburg Regional—DYR)
Patsy's Sky Grille 🍔🍔🍔🍔

Rest. Phone #: (731) 287-3777
Location: On the field
Open: Daily: Breakfast, Lunch, and Dinner

PIREP FBO Comments: Need fuel after hours? No problem! Have the waitress call while you are dining!

PIREP We have an awesome restaurant here at the DYR airport. It is newly remodeled to see the entire field. It was named after Patsy Cline, for this was the last place she stopped before her plane crash near Camden, TN. The original phone booth where she made her last call is restored and plays her music when you open the door!

The food is beyond good! It is the nicest steakhouse in West Tennessee. Open 6 AM Monday through Friday, all day Saturday. Fly-ins Friday and Saturday nights for steak and seafood. The finest dining and well worth the flight in.

Clean fine dining. The new Patsy's has the best food in West TN!

Elizabethton, TN (Elizabethton Muni—0A9)
Betty's Burger Bar 🍔🍔🍔

Rest. Phone #: (423) 543-4561
Location: Right across the street
Open: Daily: Lunch and Dinner

PIREP Betty's Burger Bar is right across the street. Good food! Super milkshakes!

PIREP Betty's Burger Bar is still there serving excellent burgers and other home-cooked meals. It's only a five-minute walk from the FBO. Beautiful approach into this well-maintained airport. Beware, for a small-town uncontrolled field, it gets extremely busy. When I arrived there were five planes lined up at the runway ready to go. And that was a Thursday.

Gainesboro, TN (Jackson County—1A7)
Hubert's Eatery 🍔🍔

Rest. Phone #: (931) 268-6724
Location: Call for free pickup
Open: Daily: Breakfast, Lunch, and Dinner

PIREP The folks at Hubert's just couldn't be any friendlier. Call them from the airport and Hubert will send someone to pick you up and bring you to his place for lunch.

Maybe they don't expect it but leave 'em a tip and a little gas money—these are really nice folks.

Blue skies to all.

Knoxville, TN (McGhee Tyson—TYS)

P.J.'s Landing 🍔🍔🍔

Rest. Phone #: (865) 984-9001
Location: 5-minute drive, free crew car
Open: Daily: Lunch

PIREP I base my Warrior at Knoxville TYS. Cherokee Aviation has a courtesy car and can direct you to P.J.'s Landing. It is a marina not five miles from the airport, and they have a decent little deli at the dock.

Anyone at Cherokee should be able to give directions; however, it is on Louisville Road just past the Texaco station. Take Hunt Road (just past the baggage terminal) to Louisville Road. For a real adventure, rent a car and go through the town of Maryville to the mountain villages of Walland or Townsend. This will be about a 30–45 minute drive, but you will think you are in the middle of a mountainous nowhere.

Townsend has a couple of motels and some nice restaurants.

Lawrenceburg, TN (Lawrenceburg-Lawrence County—2M2)

Brass Lantern 🍔🍔

Rest. Phone #: (931) 766-0333
Location: 2 miles, 2 crew cars available
Open: Daily: Lunch and Dinner

PIREP FBO Comments: Nice FBO, reasonable fuel rates.

PIREP Two courtesy cars at airport, restaurant just a couple miles from airport. Excellent food, and I would strongly recommend lunch during the week or on Sundays.

Lebanon, TN (Lebanon Muni—M54)

Outback 🍔🍔🍔

Rest. Phone #: (615) 444-7193
Location: 1/2 mile
Open: Daily: Dinner

PIREP FBO Comments: Friendly airport with great service, willing to go the extra mile! Have a courtesy car, "Ole Blue." Best airport service around!

PIREP Outback is about half-mile, Chili's is a little farther.

McKinnon, TN (Houston County—M93)

Southernaire Hotel and Restaurant 🍔🍔🍔🍔

Rest. Phone #: (931) 721-3321
Location: 50 yards
Open: Daily: Breakfast, Lunch, and Dinner

PIREP   FBO Comments: No fuel, no FBO.

PIREP   We revisited the Southernaire Motel and Restaurant ((931) 721-3321), McKinnon, TN, last weekend and could have not have had a better time. Everyone has written about the food, which is absolutely outstanding, but nobody has written about the motel and environs.

My wife Mary and I saw a clear, crisp early fall weekend coming. I said, "Why don't we get away for the weekend, get outdoors, watch the football games and have a big steak? Let's try the Southernaire," I said.

We had been once before for lunch, but I had noticed while I was there that the place was built in the '50s as a fly-in resort on the river. Old posters feature water skiing, fishing, hunting, and the restaurant. Envision a seven-room motel, large picture windows, a large old-style kitchen designed for cooking from scratch, an indoor dining room with a fireplace, and a porch dining room for the summer, and hand-laid tile bathrooms in blue and white that would make the Jetsons proud. It makes me feel like years ago on a road trip with my parents in a car with fins on the back.

The rooms are clean and comfortable with large color TVs in each room and a big screen in the lobby, so the football is covered and appreciated by all in these parts. Ralph Powell, the proprietor, says, "This ain't no Holiday Inn." Well, it's sure not. The Southernaire is a most pleasant destination in itself, not a stop along the way. And at the price, $40 a night, you get much more than your money's worth.

The Southernaire is located at the back of a bay extending from the east side of Kentucky Lake. The airport (Houston County, M93) is immediately in front of the Southernaire. Just taxi down to the western end of the runway and park in the grass. You'll see a small walk gate in the airport fence. Walk through the gate and about 50 yards up the hill and you're there. There are no tie-downs here, so I brought a tie-down kit along to secure the plane.

When we checked in, the nice lady at the counter said, "You know there's nothing here," as if to say we really don't want to take your money unless you know what you are getting. I said, "Ma'am, that is exactly what we are looking for—that and a good steak." She smiled, and we checked in.

The Southernaire is in the middle of nowhere. It is located at the end of the only road in and out of the area that runs past the Southernaire, along the bay, and then dead ends at a boat launch ramp on the main lake. This area of Kentucky Lake has long been known for summertime water skiing, sailing, fishing, and good duck hunting. However, unlike the more accessible areas just to the north around Paris and Land-Between-the-Lakes, you won't find crowds here and no development except a few vacation homes. On the land side, this area has long been known for the best deer and turkey hunting in Tennessee.

After the Vandy-Ole Miss game (what an incredible fall it has been for us Vandy fans), we walked the road along the bay to the launch ramp at the main river.

This is as beautiful a section of Kentucky Lake as you will find. It is wide, dotted with islands. There is an abandoned railroad bridge that still spans about half the river, with the old draw section removed at the center. An old granary that stood at the riverside prior the enclosure of Kentucky Lake protrudes from the water in the middle of the lake.

Mary and I took a bottle of wine, sat on the old railroad bridge on the eastern shore, and watched a world-class sunset. And I've seen some good ones around the world in my life.

There at the at the road end, by the railroad bridge and launch ramp, you'll find the only other commercial establishment in the area: The Danville Grill and Bait Store ((931) 721-3865). I'm pretty sure that Danville was the community around the old granary, which was covered by the lake. There is certainly no Danville there now. They had fishing tackle and supplies, cold beer, an indoor stainless steel grill and outdoor charcoal smoker grill. I didn't try the hamburgers, but the fellows there looked as if they knew well how to operate a grill.

Mike Troupe is the proprietor of the Danville Grill and Bait Shop. He said the stripers and bass were biting well that day. I didn't ask about a duck guide, but it is certainly the area for it.

After the sunset, we walked back to get that steak and see Tennessee be embarrassed by Florida. The steaks are outstanding—big thick first-class steaks, not junk pieces soaked in soy sauce to mask bad quality. I had a 16-ounce New York strip, and Mary had a 16-ounce rib-eye, for $17 apiece. The top-of-the-line was a 20-ounce porterhouse for $25, which neither us felt ambitious enough to try. We asked about bringing in the wine bottle, and they said no problem as long as it was kept in a sack.

The next morning we had pancakes, eggs, hash browns, grits, and the best country sausage—hand patted, not from a tube, mild, rightly seasoned—I've had in years. And after that, we said our good-byes, walked across the street to our plane, and were off into another beautiful fall day.

The Southernaire is not for everyone. But if you want to get away from it all, eat fantastically, see some beautiful scenery, and spend next to nothing, this is your place.

We're going back when the leaves change. Ought to be great!

PIREP   I flew to M93 (about 50 miles west of Nashville, TN) with my grandson and daughter yesterday and had the biggest and best omelet on earth! Three hours later and I was back there having lunch with another grandson and son. After a great lunch, we flew about 15–20 nautical miles north-northwest and found two buffalo herds at the Land Between the Lakes State Park. They are located at GPS coordinates N 36.47.117 and W 88.03.927 as well as N 36.38.357 and W 87.58.756. This is located inside the Ft. Campbell restricted airspace, but you can talk to them on 118.1 and that is Ft. Campbell approach.

Memphis, TN (General Dewitt Spain—M01)
The Rendezvous 🍔🍔🍔🍔

Rest. Phone #: (901) 522-8840
Location: Downtown
Open: Daily: Lunch and Dinner

PIREP   Flying into Memphis, you can smell the ribs just as you have to call up for clearance into Bravo. Skip the Memphis International. Fly just north of downtown to a small 3800' strip locally know as Spain. General Dewitt Spain (M01) is primarily an agricultural airport. Talk nice and they may lend you a car. If not, take a cab.

The Rendezvous is situated in a basement accessed through an alley directly behind the Peabody Hotel. Ask a local for directions. They will know. Order a full rack of ribs and get ready. The waiters are rude, the place is hard to find, the beer is cold (may want to stay the evening at the Peabody or have a designated flyer).

The ribs are absolutely the best on earth.

Morristown, TN (Moore-Murrell—MOR)
Charlotte's 🍔🍔🍔

Rest. Phone #: (423) 581-6974
Location: 1/2-mile walk
Open: Daily: Lunch and Dinner

PIREP   I live about 30 minutes from the Morristown Airport and worked there over the summer on a DC-3 that is used for cargo flights. I have never flown in there since I have just started taking lessons, but I know of a nice little diner called Charlotte's and it is very good. It is a half-mile walk from the FBO. Be sure and ask for directions.

The airport is about 20 minutes northwest of Knoxville. If you do fly there or are ever in the area, go in the FBO and you are more than likely to meet one of the better-known living aviation legends, Ms. Evelyn Johnson (the oldest living active female pilot), who, I might add, still gives instruction. Then head over to the Maintenance Hangar, which is the old hangar beside the FBO that has a orange tin roof, and ask for Ron Tallent. He will almost always let you go and walk around in his DC-3. I know there are a lot of WWII vets that came out while I was working on it. If you are wondering what I was doing to it, I was removing the paint off it and buffing it up to a bright metallic finish (one of the most fun things I have ever done—I had never before been on top of an airplane and out on the wings, but it was a blast). The description of this airport may make it seem old, but it is very well maintained, and they even fly Lear jets out of here!

Mountain City, TN (Johnson County—6A4)

Suba's 🍔🍔

Rest. Phone #: (423) 727-5657
Location: 5-minute crew car drive
Open: Daily: Lunch and Dinner

PIREP A great place to fly in to eat is Johnson County Airport (6A4) in beautiful Mountain City, TN. The friendly FBO has a courtesy car. Suba's Restaurant is about five minutes by auto.

Suba's is owned/operated by a young couple who met while each was attending culinary school in Charleston, SC. He does the entrees while she specializes in desserts, and oh, what desserts! How about a Portabella croissant, Greek pasta, or juicy steak? Daily soup special served in a homemade bread bowl. Worth the trip.

Across the hwy is a barbecue joint called Mike's. Real hickory smoked-on-premises pork. Tomato-based home-made sauce (no, not that thin stuff they have in eastern NC—you can really taste this stuff and you will lick your fingers). Or, there is a country cafeteria near. The gang from Abingdon, VA flies in regularly for the roast beef.

Nashville, TN (Nashville Intl—BNA)

101 Airborne 🍔🍔🍔🍔

Rest. Phone #: (615) 361-4212
Location: On the field
Open: Daily: Lunch and Dinner

PIREP My wife and I went to the 101st Airborne at Nashville BNA for Thanksgiving dinner. We stopped at Signature and they were very friendly and accommodating. They lent us one of their three Ford Sprint crew cars for a couple of hours to drive the two miles to the restaurant. There was no charge for the car or a ramp fee as long as we purchased at least nine gallons of fuel. The fuel isn't cheap at $3.16 per gallon, but it beats a ramp fee and a taxi. Runway 2C or 20L are preferred to taxi to Signature.

The lady at the desk gave us excellent directions to the 101st. It is easy to find, and a sign sets on the main road directing you to a small windy road leading to the restaurant. The restaurant is adjacent to and overlooks the airport. The restaurant had great service, average prices, and good food. The buffet had many choices to match any taste. The hours of operation are Monday through Thursday 4–10 PM, Friday through Saturday 4–11 PM, and Sunday brunch 9 AM–3 PM. We made reservations as they were recommended for Thanksgiving Day. However, we arrived an hour early and they were able to accommodate us. The phone numbers for the 101st and Signature are located in AOPA's Airport Directory.

Getting into and out of BNA is relatively easy. We were asked to keep our speed up on final to mesh with the Southwest Airlines jet traffic. Print out a copy of the taxi diagram for BNA from the AOPA website. It is very helpful in navigating on the airport.

The reason we embarked on this adventure was the information in the *$100 Hamburger*.

Nashville, TN (John C Tune—JWN)

Bobbie's Dairy Dip 🍔🍔🍔🍔🍔

Rest. Phone #: (931) 726-3030
Location: 3 miles
Open: Daily

PIREP   Bobbie's Dairy Dip has been awarded best hamburger in Nashville!

They sell a fresh Angus burger with cheese or a double. Great milkshakes and sundaes. It is about three miles from the airport across from Wendall Smith's and Whitt's Barbeque, two other local favorites. It is not far from the airport, but certainly not walking distance. Bobbie's and Whitt's don't offer restaurant seating. Bobbie's does have tables outside.

Sewanee, TN (Franklin County—UOS)

Shenanigan's 🍔🍔🍔

Rest. Phone #: (931) 598-5774
Location: 3/4 mile
Open: Daily: Lunch and Dinner

PIREP   Great place in mid-/Eastern Tennessee to fly to. The FBO has a loaner car, but the cool thing to do is to walk on the old RR bed to Shenanigan's for lunch. Great soups and sandwiches, and beer for the nonpilot types. It is part of the rail to trails program of converting old railroad beds to trail. It's three-quarter mile of flat terrain to the restaurant, which is in an old general store.

Smyrna, TN (Smyrna—MQY)

Asia Chinese Buffet 🍔🍔

Rest. Phone #: (615) 223-0509
Location: 2–3 miles
Open: Daily: Lunch and Dinner

PIREP   The local grill next to the airport is now officially closed and the property leased to the Tennessee Rehab Center. However, there are several good restaurants within a couple of miles, and both FBOs will either provide transportation or a courtesy car. The jet jockeys from Columbus, MS AFB make this a regular stop for the food. The restaurant most frequented is the Asia Chinese Buffet in downtown Smyrna.

Tullahoma, TN (Tullahoma Regional Arpt/ William Northern Field—THA)

Piggy's Place 🍔🍔🍔

Rest. Phone #: (931) 455-5674
Location: Close by
Open: Daily: Lunch and Dinner

PIREP   Tullahoma, TN is a fine destination for good food, Staggerwing Museum, and Jack Daniels Distillery (Lynchburg, TN) tour. Airport ID THA has two runways with IFR approaches (SDF and so on). The FBO has two courtesy cars usually available for fuel buying customers.

Piggy's Place is great barbecue. They are close to the airport, have pork, beef, chicken, and turkey usually on Thursdays. Their desserts should not be missed—homemade and always fresh.

PIREP   City Diner. It's in "downtown" Tullahoma (about two miles from the airfield) and has some of the best Southern cooking I've had in a while. Very inexpensive for the quality (four burgers). For dinner we ate at the Catfish Restaurant (one mile from the field). Excellent catfish with generous portions (four burgers).

Thunderstorms were predicted during the evening, so the FBO offered to put the plane in a hangar overnight (no charge). Since we were leaving at 6 AM the next day (Sunday), I was hesitant because they wouldn't be in until after 9 AM to open the hangar. So they showed me how to open the doors without a key!

The food was great, but the FBO was the friendliest, most accommodating and helpful I've ever run across.

Abilene, TX (Abilene Regional—ABI)

Joe Allen's BBQ 🍔🍔🍔🍔🍔

Rest. Phone #: (915) 672-6082
Location: Short drive, get crew car
Open: Daily: Lunch and Dinner

PIREP   Went to Abilene for some Texas barbecue at Joe Allen's. The FBO, Abilene Aero, was great. Very friendly staff, new crew vans, and excellent facilities for checking the weather and filing the flight plan.

Unfortunately, Joe Allen's was a big disappointment. The service was slow and they had run out of brisket at 6 PM. How does a barbecue place run out of brisket at 6 PM? I would give it a two-hamburger rating, just for the Old West feel.

PIREP   Following the recommendations of the $100 Hamburger website, I headed off for Abilene to sample Joe Allen's barbecue. I almost made it. On the way to Joe Allen's BBQ is one of his other places, Lytle Land & Cattle Company. It sits on the right-hand side of the road from the airport into town. It advertises itself as a mesquite grill, not a barbecue place. Well, I figured why not. It was a great choice!! The rib-eye was, without a doubt, one of the best steaks I have ever had! I will freely admit to only managing the Sharon size (14 ounces) and not the Joe size (20 ounces). I really cannot say enough about it—it was delicious!!

Abilene Aero has a well-equipped FBO. Feels like a million dollars (probably where the profit from the rather high fuel prices goes!—and that is my only complaint!! But since the facility is good you can't really complain!). The crew cars are newish and have air conditioning!! There seems to be two of them (minivans) and since I arrived before the morning crush I think I was lucky. You may want to call ahead.

All in all, a five-hamburger recommendation.

PIREP   Best barbecue I've had in years. We knew we came to the right place when we saw the jammed-up parking lot and people carrying large coolers inside to load up with take-out.

Abilene Aero had plenty of maps and loaner cars—nice airport and nice facilities. We will go back often. There's also a catfish restaurant owned by the same people right around the corner—its parking lot was full, too!

PIREP   Get the loaner from Abilene Aero and go downtown to Joe Allen's BBQ. AA has maps available. Some of the best Texas barbecue anywhere, well worth the drive. Just be sure you land at the eastern airfield, not the western one (Dyess AFB).

Perini Ranch Steakhouse 🍔🍔🍔🍔

Rest. Phone #: (315) 572-3339
Location: 25-minute drive
Open: Daily: Dinner

PIREP FBO Comments: Exceptional FBO that helped considerably in planning our anniversary night out at the Perini Ranch.

PIREP For our anniversary this year, my wife and I decided to fire up N77VE, our Cherokee Six 300 (PA-32-300), for a trip to the Perini Ranch Steakhouse about 25 minutes away from Abilene Regional. We called ahead to Abilene Aero and received great help in setting up a car rental in lieu of their offered crew car. The FBO gets a two-hour rate from Hertz that you cannot get by calling directly, regardless of your status with Hertz. Great service on the ramp and behind the counter, which meant that getting to our evening out was perfect.

Absolutely superb dining at the Perini Ranch, which is rated as one of the top five rural restaurants in America. Definitely a meat-eaters' destination, with steak and catfish specialties. Great atmosphere, and you'll want to include dessert. This is now a great planned "Texas night out" for all our friends visiting from out of state.

Amarillo, TX (Rick Husband Amarillo Intl—AMA)

The Big Texan Steakhouse 🍔🍔🍔🍔

Rest. Phone #: (800) 657-7177
Location: It's a drive
Open: Daily: Lunch and Dinner

PIREP It has been a few months since I flew into KAMA, but I saw that there was not a pirep for what is by far the best restaurant in Amarillo; in fact, it might well be the best steakhouse in the world. It is home to the free 72-ounce steak (if eaten in one hour). It is The Big Texan. The restaurant is more than just a steakhouse, it is an experience. The interior is Western saloon–style, complete with authentic wagon wheel "chandeliers." The steak is, of course, wonderful; in fact, it is the best I have ever had. It can be cut with the side of a fork and is cooked perfectly every time. There is also an arcade, bar, gift shop, and motel on site, all in a Western motif as a good steakhouse in Texas should be. And of course, if you are really hungry and the fuel bill dipped into your wallet a little too far, you can eat a 72-ounce steak for free—just do it in one hour. Fair warning: over 30,000 people have tried it in 40 years and so far only 4800 succeeded, so bring your appetite.

English Field House Restaurant 🍔🍔🍔

Rest. Phone #: (806) 335-2996
Location: On the field
Open: 7 AM to 4 PM, seven days a week

PIREP I visited Amarillo, TX (KAMA) from Wichita (KICT). There is a good little diner right on the field, adjacent to TAC Air. It is called the English Field House Restaurant and is a 20-yard walk from TAC Air. The menu is Mexican influenced but has pretty much anything, and most dishes are less than $6. I had a good English breakfast served by friendly staff in quick time. The diner looks out over the main ramp area and runway 04/22.

Andrews, TX (Andrews County—E11)

Buddy's Drive Inn 🍔🍔🍔

Rest. Phone #: (432) 523-2840
Location: 7 or 8 blocks; use the crew car
Open: Daily: Lunch and Dinner

PIREP If you like *real* steak fingers, go to Buddy's Drive Inn. The steak fingers are sliced from round steak fillets and deep fried to a golden brown. The portions are huge—*you won't* go away hungry. The price is about $6 per person, and they have doggie bags on hand. The Andrews County Airport (E11) is about seven or eight blocks east of the cafe, but there is a real nice Ford Crown Victoria courtesy car waiting for your arrival. The airport has three runways, two lighted, and has recently undergone a complete overhaul. Buddy's rates at least a four-hamburger rating.

Angleton/Lake Jackson, TX (Brazoria County—LBX)
The Windsock Cafe 🍔🍔🍔

Rest. Phone #: (979) 849-1221
Location: On the field
Open: Mon.–Fri. 10 AM–2 PM, *no weekends*

PIREP Update to the opening hours of the Windsock Grill: it's now open Saturday 11–9. Also open Tuesday to Friday 10–2 for lunch and Wednesday to Friday 5–9 for dinner. Closed Sundays and Mondays. Had some very tasty friend jumbo shrimp (this Saturday's special), which came with a large salad, loaded baked potato, and hush puppies. Very reasonable prices. The only downside for me is that I'm based at LBX so it isn't actually a $100 hamburger destination for me!

Atlanta, TX (Hall-Miller Muni—ATA)
Grandy's 🍔🍔🍔

Rest. Phone #: (903) 796-5577
Location: Short walk
Open: Breakfast, Lunch, and Dinner

PIREP FBO Comments: No self-serve fuel for the public.

PIREP We flew to Hall-Miller and walked to Grandy's for breakfast. Grandy's is located at the entrance of the airport just a short distance from the terminal. Grandy's is next to a small truck stop. Both were very clean and in good shape. The food was excellent and they were excited that we had flown in to eat. Nice little airport.

Austin, TX (Lakeway Airpark—3R9)
Lakeway Resort 🍔🍔🍔🍔

Rest. Phone #: (512) 261-6600
Location: Call the resort—they'll pick you up
Open: Watch takeoff and landing times closely—heavy fine for a violation

PIREP As a frequent visitor to Lakeway from Sugarland Regional, I too have been concerned in the past about arriving safely well before sunset in order to avoid the fines and hassle that may ensue. However, it still is a nice little airpark and has much to offer. Its proximity to several attractions around Lake Travis makes it a fine place for pilots who wish to enjoy the Central Texas Highland Lakes and hill country.

It was unfortunate what happened to Mr. Johnston regarding his experience at Lakeway Airpark—it appears, given his testimony, that the local officials somewhat made an example of him. I am sympathetic to his plight. Moreover, I completely agree with your editorial comment—this problem is best avoided by landing there well before sunset.

However, in the interest of your readers and to provide some better understanding of why they do things the way they do at Lakeway, I thought it important to pass along that the entire Lakeway community is heavily populated by wild deer—they are simply everywhere and have been for quite some time. The deer population has been so high in the community that there have been several efforts over the years at controlling the population. This issue is simply a fact of life in Lakeway and the locals have come to terms with it.

Therefore, with regard to the airpark's policy on sunset to sunrise operations, I think any sensible pilot would agree that an ordinance (albeit fiscally harsh) to discourage such activity is probably warranted given the hazard that the high deer population poses to air safety after dark. In my opinion, it's tricky enough to keep watch out for deer on the runway at 3R9 during the daytime (and be ready to drop the hammer for that go-around), let alone attempt it at night!

Clear skies.

PIREP Landed at Lakeway in the middle of July. Sunset was 8:30 PM according to the newspaper and other official records. The Lakeway website said sunset was 2031. I touched down at about 8:20 PM and had the airplane tied down and was walking away when a policeman came up and issued me a ticket for landing after sunset. I told him I touched down at 8:20 PM and asked him for the exact time of sunset. He didn't know but issued me a $1200 ticket. We had talked for 10 minutes and the ticket was issued at 2045. There was no way I could possibly have landed after 2031 because I had the airplane chalked, tied down, windshield covers in place, and post-flight inspection done and was walking away by 2035.

I had three witnesses, but no lawyers would touch the case because of Lakeway's reputation (Judge Roy Bean atmosphere) of people losing cases in court. It is a court of record and they film all activity, so if you lose you not only have to pay the ticket but you also have to pay court costs that are extremely high.

They even give tickets to people walking in the park after sunset when signs are not in plain sight. It's $200, by the way, for that offense. They have changed the speed limit signs overnight and have ticketed the citizens without issuing notices.

If you land anywhere close to sunset, please have accurate touchdown times and a camera to prove your innocence. It takes more than witnesses to even get representation for a court hearing. I finally found an AOPA lawyer to represent me, but after talking to his lawyer friends, he convinced me to ask for deferred adjudication, which I did.

Lakeway should abide by FAA regulations of 30 minutes before and after sunset and should install an AWOS to announce the sunset times at Lakeway because they are obviously different than the real times.

Stay away from Lakeway!!!

James Johnston

Editorial Comment:

Folks, we put the preceding pirep up because it represents the feelings of many of our pilot readers. I certainly can't attest to its accuracy. Take it with a very large grain of salt. It may be 100 percent true, it may be 100 percent bull. In either case it certainly isn't the first time we've heard about this situation, nor is it a result of a recent rules change. 3R9 has had a strange view of sunset for many years. They post it on their website—on this website and others, I imagine. Not only do they post the rule, but they also post the *high* fines for breaking them and their absolute intention to enforce them. It is not a secret.

As we all know, key to any successful flight is flight planning, which includes becoming aware of the rules that affect operations at your destination airport. I am certain that this gentleman was well aware of the rule as he mentions clearly his absolute knowledge of what time sunset was on that day and at what time he landed, nine minutes before Lakeway's official sunset.

Perhaps the sheriff has an official clock that he monitors. I really don't know. What I do know is this. Lakeway is a terrific place to visit. It is one of the best resorts in that part of the world. It should not be missed. Flying into it for lunch, breakfast, or an overnight stay is wonderful. Go there and enjoy all that it has to offer, but be aware of the rules and be aware that the folks at the airport play hardball.

What's the solution? It is really kinda easy. Plan to land at 7:21, not 8:21. No problem.

Please don't misunderstand what I am saying. I do not agree with what Lakeway Airpark, Inc. is doing. At the very least we would all agree that its "sunset" regulation is capricious and seems crafted to cause conflict rather than avoid it. Personally I have wished for years that they would change it. It appears that they have no intention of doing so.

Mr. Johnston had a very bad and very costly experience. Don't let it happen to you. Either don't go to Lakeway (and that would be a pity), or time your trip to avoid conflict with Lakeway Airpark, Inc.. Land early and enjoy your stay.

That's my opinion.

What's yours?

John Purner

Editor

The $100 Hamburger

PIREP   Magnificent! If you are looking for a *great* place for a Sunday brunch, you definitely have to visit Lakeway in Austin, TX (3R9). Plenty of parking (park in any marked spot). I would suggest calling the hotel's courtesy car to pick you up unless you're really into long walks and hilly terrain. Brunch starts around 11:30 and the selections are endless. The food was wonderful. The desserts are four-star. The view of Lake Travis topped off a relaxing morning. As you take off and head out of Austin you'll be saying to yourself, "It just doesn't get any better than this."

Enjoy!

PIREP   Lakeway Airpark, Inc. is a nonprofit organization formed to own and manage the airport. We operate only sunrise to sunset (actual times—not 30 before or after). Tickets run $500 to $2000 for violating this rule, and you will be stopped from leaving!

Joe Bain

President, Lakeway Airpark

PIREP   Lakeway Resort on Lake Travis has one of the best Sunday buffets in Texas. Land at the airport and call the Inn. They will send a van at no charge. Dine while you look over Lake Travis and the hills of Texas. Four-star dining.

Bellville, TX (Grawunder Field—06R)

The Bellville Cafe 🍔🍔🍔

Rest. Phone #: (979) 865-8676
Location: Short walk
Open: Daily: Breakfast, Lunch, and Dinner

PIREP   Bellville, TX (06R) is a great place to fly to when you are just looking for an excuse to aviate. The airport is located right on the edge of town. A short walk gets you to the old Bellville town square, which is having a rebirth as an antique center. There are numerous good restaurants on or near the square. The Bellville Cafe has the definitive CFS (that's chicken-fried steak, the national food of Texas for you foreigners).

My wife and I spent the better part of a Saturday touring Bellville on foot and thoroughly enjoyed the place. Bellville is located in the heart of bonnet country, and is really scenic in the spring.

The airport has a single runway 17-35, about 2400' long, sloping uphill with the 17 approach end about 45 feet higher than the far end. Landings and takeoffs are not difficult, but they aren't routine either!

PIREP   Bellville is a beautiful fly-in from Houston. The airport is on the 900 block of Main Street and thus isn't far from the center of town. We visited on a Thursday, landing at about 1900 local, and were met by Henry

Grawunder—the field was named after his father! Unfortunately, the Bellville Cafe had closed at 3 PM so we fed on slow-food at the DQ; of three items ordered they forgot to cook two of them. So if you want more than just pink lemonade, leave plenty of time for the order slip to get to the kitchen!

The runway is lit at least until 11 PM and was fine for night departure—not sure if I'd want to tackle the notch in the trees that surrounds the southerly (downhill) end of the runway on a night approach!

Big Spring, TX (Big Spring McMahon-Wrinkle—BPG)
Al's & Son Bar-B-Q 😋😋😋

Rest. Phone #: (915) 267-8921
Location: Close by; call for pickup or use crew car
Open: Daily: Lunch and Dinner

PIREP Big Spring, TX has a barbecue restaurant that been serving the best barbecue in Texas for years. The place is Al's & Son Bar-B-Q and was started 35 years ago. It's now become a Big Spring tradition. Many that grew up in Big Spring and move away crave Al's Famous potato salad. Al's makes about the best hamburger in Texas with fresh-ground daily meat. The steak fingers top those sold in Andrews, and they serve a great sirloin steak with the just right seasoning.

The airport, which is operated by the city of Big Spring, is uncontrolled but there is an FBO. Lone Star Aviation has an operator on duty and you can radio them for information on the field. Also they have a loaner car for visiting pilots, but it is a first come, first serve, so call ahead: (915) 264-7124. They also have information on rental cars if needed. Mitchem Transportation offers taxi service for the area and you can ride to Al's for about $6.

If the rental car is not available, call Al's at (915) 267-8921. If possible, they will send someone to get you.

While you're there, be sure take a short ride up the Mountain in Big Spring State Park. For a flat land area, the view is quite special and even better at night. Big Spring has an excellent museum, the Heritage Museum, with history on local ranches and towns. Also you need to see the actual Big Spring, which served as the only water hole for Indians, cattle drives, and settlers for years.

Borger, TX (Hutchinson County—BGD)
Lorene's Kitchen 😋😋😋

Rest. Phone #: (806) 273-7106
Location: Short walk, about a block
Open: Daily: Lunch and Dinner

PIREP There is a Mexican food restaurant at Borger (BGD), Lorene's Kitchen. The best chile rellenos in the world. They have a courtesy car available, but it is only about a city block away, so not too far to walk.

Brenham, TX (Brenham Muni—11R)
Southern Flyer Diner 😋😋😋😋😋

Rest. Phone #: (979) 836-5462
Location: On the field
Open: Breakfast, Lunch, and Dinner

PIREP Excellent! The service, convenience, atmosphere, and prices at the Southern Flyer Diner make it a real treat! The food was great. You can get anything from a burger to a steak. One of the best on-field eateries I have visited.

PIREP FBO Comments: Southern Flyer has a comfortable pilot-friendly FBO with fuel, supplies (charts, AFDs, and so on), and a fine onsite restaurant in 1950s diner–style. The facility has a nice pilot lounge—TV, separate quiet room for sleeping, flight planning area, computerized weather access, fridge, Internet, telephones, and so on. Staff was friendly and helpful, right down to calling out, "Have a safe trip," on the CTAF as you take the active for departure. A first-class stop in every way.

PIREP On Fourth of July Sunday, I decided to grab a friend and make a $100 hamburger run from Addison (ADS) in the Dallas area to Brenham (11R). The FBO, Southern Flyer, is at the end of the parking apron fronted by a line of nice tie-downs. A youngster pulling a fuel trailer took on topping off the tanks as we got out of the plane, although self-serve fuel was also available (presumably cheaper, but I didn't check).

Walking into the FBO, I was greeted at the door by a cute young lady in a poodle skirt who seated us outside on the covered patio with ceiling fans overlooking the small lake behind the FBO. The diner has a 1950s theme, and the staff all dress the part. I thought it a good sign that the restaurant was chock full of local folks who had driven out to eat there after church. We ordered the fried pickles for an appetizer—yummy. I had chicken-fried steak and my copilot had a bacon cheeseburger. Both were homemade, fresh, hot, and as tasty as it gets. If you're into $100 hamburger trips, this one is certainly worth the price. You just gotta go to Southern Flyer!

Brownsville, TX (Brownsville/South Padre Island Intl—BRO)
Louie's Backyard 🍔 🍔 🍔 🍔

Rest. Phone #: (956) 761-6406
Location: In the terminal
Open: Daily: Lunch and Dinner

PIREP The BRO restaurant is located in the airline terminal building, with picture windows overlooking the field. It specializes in Mexican food, hamburgers, and french fries, but other items are also available on the menu. Prices are very reasonable for an airport restaurant. It is operated under contract by the well-known South Padre Island operator, Louie's Backyard. Louie's Lounge is adjacent to the restaurant.

As the field is air carrier, access is not available directly from the ramp but only through FBOs, which are located conveniently on each side of the terminal building.

Brownwood, TX (Brownwood Regional—BWD)
Underwood's Cafeteria 🍔 🍔

Rest. Phone #: (325) 646-1776
Location: 5 miles away—borrow the crew car
Open: Daily: Lunch and Dinner

PIREP Today this is a beautiful small airport with two very nice runways, but was once (during WWII) home to massive training facilities and German POWs…a lot of history here. You can view the past via many very descriptive photos inside the FBO—very interesting stuff. Currently there are two military aircraft, an F-111 and F-4, on display just outside the terminal.

Borrow the courtesy truck, a vintage 1985 Chevy diesel 4 × 4 complete with cattle guards, and at least look like a native West Texan for the five-mile drive into town for some of the best home cooking west of the Mississippi. Underwood's Cafeteria is a home-grown legend. With an atmosphere reminiscent of the 1950s along Route 66, this is a great place to eat. We had the fried chicken (about $10 per person and "all the side dishes you want") which was great—and don't forget the cobbler. It's 65 cents extra, but worth it. Get two.

And if the phone rings while your refueling, answer it—it just might be some cowboy who had his rodeo bull escape looking for a plane and pilot to help look for him. For those of you not familiar with Texas ways of doing things, we herd cattle with helicopters and Lear jets.

Keep your head on a swivel for birds and listen up for the three-scheduled airline arrivals—they use runway 17/35.

PIREP   Somewhat of a legend here in the central Texas area. Fly into Brownwood (BWD) 130 miles NW of Austin and get the courtesy car to go to town. Underwood's serves barbecue, chicken-fried steak, fried chicken, and a large variety of veggies all cafeteria style. The food is excellent. The hot cobbler (peach, cherry, and apple) and dinner rolls are the best you'll find anywhere. Don't come on Wednesday because they are closed, but they are open every other day. Lines are long on Sunday afternoon after church unless the Cowboy game starts early. Oh, and leave your American Express, MasterCard, and Visa at home. They don't take 'em. You can get an excellent meal and drink—more food than you can probably eat—for about eight bucks.

Bryan, TX (Coulter Field—CFD)
Sodolak's Beefmasters 😋😋

Rest. Phone #: (979) 778-4999
Location: A drive
Open: Daily: Lunch and Dinner, closed Sunday

PIREP   FBO Comments: A great airport with very friendly staff and patrons. They have a weekend discount. As mentioned in another post, there is a skydiving operation on the field, so keep your eyes and ears open. The jump pilot is in constant contact with Houston Center (120.4, I believe), so they can usually give you an update on the whereabouts of the jump plane.

PIREP   Sodolak's is about two miles from the airport, and it is worth the trip. Be careful what you order, though, or you may have to redo your weight and balance when you get out of there. Their *normal* rib-eye is 16 ounces, and their chicken-fried steak overflows the plate. We paid $21 for the 16-ounce rib-eye, four sides, and lemonade. The meal and service were excellent!

Burnet, TX (Burnet Muni Kate Craddock Field—BMQ)
The Hilltop Cafe 😋😋😋

Rest. Phone #: (830) 997-9242
Location: A couple of miles; use the cafe's crew car
Open: Daily: Lunch and Dinner

PIREP   The Hilltop Cafe provides a courtesy car at Burnet (BMQ). It has good food (lunch buffet or menu) at good prices, but they don't take credit cards. The Faulkners run the FBO and repair shop and will really take care of you. They are in the process of building a new pilots lounge for extra space.

Cameron, TX (Cameron Muni Airpark—T35)
Big Bob's Steakhouse 😋😋😋

Rest. Phone #: (817) 697-4669
Location: 3/4 of a mile off the field
Open: Tues.–Sat. 6 to 10

PIREP   Cameron Muni is a well-maintained hard surface strip, two miles north of Cameron, TX. The runway is 3200 × 50 and well lit. Big Bob's Steakhouse is a small, friendly place located about three-quarters of a mile from the field on Industrial Blvd. It is very clean with a truly country flair. The hospitality is gracious and the steak out of this world. The prices are reasonable and the portions generous. The cheesecake is homemade, so be sure not to miss it. Believe me, after such a big dinner the short walk back to the field is a welcome relief. But if you prefer not to walk, just call ahead and someone will shuttle you to and from.

Canadian, TX (Hemphill County—HHF)

Dairy Queen 😋😋

Rest. Phone #: (806) 323-5581
Location: 2 miles; crew car provided
Open: Daily: Breakfast, Lunch, and Dinner

PIREP Courtesy car provided by a local bank. Fuel available (cash only) by calling the sheriff's department. Nice terminal building. NDB approach. Airport is located right at the edge of town. No outstanding restaurants in town, but you won't starve either. Dairy Queen, Pizza Hut, and some local greasy spoons are available. One of the prettiest, friendliest, and richest towns anywhere.

Canton, TX (Canton-Hackney—7F5)

Ranchero Restaurant 😋😋😋

Rest. Phone #: (903) 567-5719
Location: About a mile walk
Open: Daily: Breakfast, Lunch, and Dinner

PIREP The airport at Canton, TX, famous for First Monday Trade Days, is a nice place to go for breakfast, lunch, or dinner. 8/10ths of a mile east of the field offers you a wonderful buffet at the Ranchero Restaurant at a very reasonable price ($6.95 for all you can eat). Plus, next door is a What-a-Burger!

The airport identifier is 7F5 and is designated Canton-Hackney Field. Comm frequency is 122.9. Check out further details in your airport directory. This is a sleeper. You will not be disappointed.

The Canton Trade Days is a huge flea market. It covers several acres and takes place the first weekend of every month. It has been going on for years. People especially in East Texas are very familiar with it.

Castroville, TX (Castroville Muni—T89)

Jimbo's Country Kitchen 😋😋😋

Rest. Phone #: (830) 538-2238
Location: Short ride into town
Open: Daily: Breakfast, Lunch, and Dinner

PIREP FBO Comments: 10500 Airport Road, Castroville, TX 78009, (830) 931-0234. They have a courtesy car. For longer stays, Enterprise will pick you up: (830) 426-3138.

PIREP We ate at the Cappuccino Haus, where they serve delicious, big sandwiches on homemade bread, at a reasonable price ($5.50 for sandwich, chips, and pickle). Their hours are 8 AM to 2 PM Monday to Friday.

Here is the contact info for the restaurants previously mentioned on the $100 Hamburger website:

Jimbo's Country Kitchen: 1314 Hwy 90 West, Castroville, TX 78009, (830) 538-2238.

La Normandie Restaurant: 1302 Fiorella St., Castroville, TX 78009, (830) 538-3070.

Sammy's: 202 Hwy 90 East, Castroville, TX 78009, (830) 538-2204.

You may also enjoy driving around this small town. Castroville is a historic little town, known as The Little Alsace of Texas.

Center, TX (Center Muni—F17)
Raceway Food Market 🍔🍔🍔

Rest. Phone #: (936) 598-2082
Location: Across the street
Open: Daily: Lunch and Dinner

PIREP There is a good barbecue restaurant, The Country Store, across the road from the airport in Center, TX. It is a combination small grocery store and barbecue place. It also has a sandwich counter where they slice ham, salami, and so on and will make a sandwich while you wait. The barbecue/hamburger counter is in the back of the store, along with several tables to sit and eat. My kids loved it because the owners made them feel right at home.

The airport has an inactive runway that extends to the south off the parking ramp. We just taxi down the closed runway and park off to the side on the grass.

The restaurant is a short walk. Prices are good and the food is great. It's run by some friendly people. We flew in right at closing one day and they fixed us something to eat and waited for us to finish.

Childress, TX (Childress Muni—CDS)
Ranch Hand Cafe 🍔🍔🍔

Rest. Phone #: (940) 937-8309
Location: Crew car required
Open: Daily: Breakfast and Lunch

PIREP We flew into Childress, TX (KCDS) and were directed to the Ranch Hand Cafe on Hwy 287. The food and service was great. Call ahead and tell them you're coming, and they will go out of their way to help you. I'm looking forward to going back.

Call ahead and they will warm the crew car up for you!

College Station, TX (Easterwood Field—CLL)
Koppe Bridge 🍔🍔🍔

Rest. Phone #: (979) 764-2933
Location: 2 miles, plenty of crew cars
Open: Daily: Lunch and Dinner

PIREP FBO Comments: Large, clean GA terminal with good weather and planning room. Plenty of crew cars available.

PIREP Koppe Bridge serves a burger and homemade fries that will make you have to redo your weight and balance. If a juicy burger with some of the thickest bacon you've ever seen doesn't suit your pallet, they have a mean chicken-fried steak or even a nice grilled chicken sandwich for the copilot. It's especially a good place if you have more than the required time before your flight, because they follow through on their promise to have the coldest beer around. They're located just a couple of miles from the terminal at Welbourne and FM 2818.

Corpus Christi, TX (Corpus Christi Intl—CRP)
Landry's 🍔🍔🍔

Rest. Phone #: (361) 882-6666
Location: A few miles
Open: Daily: Lunch and Dinner

PIREP I went down to Corpus Christi to go back to Landry's Seafood at the T-head piers downtown. I usually park at Signature at CRP. For a twin the size of the Baron, your minimum fuel purchase needed to waive the ramp fee is only 20 gallons. Since it was Friday night, they even reduced the fuel prices. And for this you get one of their no-excuses, well-kept crew cars for the trip downtown for free. (One time last year I told them I didn't need fuel and would go ahead and pay the $15 ramp fee to get a car. The woman at the counter (name withheld) just tossed me the keys to the car and said this one was on her.)

Corsicana, TX (C David Campbell Field-Corsicana Muni—CRS)
Fitzgerald Trading Post 🍔🍔🍔

Rest. Phone #: (903) 872-6222
Location: 1/4 mile from the end of runway 14
Open: 4 AM till midnight

PIREP I just returned from Corsicana after trying out the recommendations about Fitzgerald Trading Post. The burgers were extremely cheap and pretty good too. The proprietor is a nice guy who, when I mentioned I flew from Dallas for one of his burgers, proceeded to bring out pictures of people who had tried his two-and-a-half-pound Screaming Eagle burger and said someone came from Corpus Christi for one earlier in the week.

One caveat for going there is that there isn't much to the airport. The FBO closes at 6 PM, so if you want to use the crew car (a huge older Suburban), you had better get there early. I was lucky enough to run into someone using the phone inside who let me have the crew car even though they would be gone when I returned. Not much to the town, but the burgers made a good excuse for flying (as if any of us need one!!).

I'll give it three-and-a-half flying burgers.

PIREP Look out your left window as you turn final for runway 14 and you'll see it, Fitzgerald Trading Post, just one mile north of field. Good barbecue sandwiches and hamburgers. They'll even cook one for breakfast, or eggs and stuff like that if you prefer. Courtesy car available at airport. Also, four miles west in the big city of Angus, the Exxon filling station has the biggest, greasiest, most delicious hamburgers you ever saw. The smell of hamburgers cooking just hangs in the air and gets into your clothes. You can continue to enjoy it the rest of the day. Watch out for the duct tape on the seats if you have on nice clothes, but it adds to the ambiance.

Hours are from 4 AM till midnight (sometimes he doesn't get there till 5 if it's raining). You won't find me out flying in the rain at 5 AM.

Collins Street Bakery 🍔🍔🍔

Rest. Phone #: (254) 580-9346
Location: 4 miles; crew car available
Open: Daily

PIREP Fly into Corsicana some afternoon and check out the CAF museum adjacent to the FBO. Then grab the courtesy car and head into town for a snack at the world famous Collins Street Bakery. This is where the Queen of England gets her fruitcake from every year. They have lots of free samples and 10-cent coffee. Afterward, drive a mile down the street to the Cotton Patch Cafe. Although the atmosphere is nothing spectacular, the food and service are great. If you are really hungry, try the grande sized chicken-fried steak. I had the regular portion, and it was more than enough. The burgers are delicious also.

Cuero, TX (Cuero Muni—T71)

Wal-Mart 🍔🍔

Rest. Phone #: (361) 275-5796
Location: Across the road
Open: Daily: Breakfast, Lunch, and Dinner

PIREP Land at Cuero Muni (T71), tie down, and hop the fence. Right there is the big gray building, a Wal-Mart. You have probably seen one before. The good news is they have food. A small lunch counter makes burgers and hot dogs. The price is low. The quality is best judged by your hunger level: the hungrier you are, the better the food.

Now, don't forget that Ole' Sam Walton was a private pilot. He started with an Ercoupe (good choice) and finished with a Piper Malibu.

Dalhart, TX (Dalhart Muni—DHT)

Airport Cafe 🍔🍔🍔

Rest. Phone #: (806) 244-5521
Location: On the field
Open: Daily: Breakfast and Lunch

PIREP I stopped in Dalhart, TX (DHT) last week on my way back to Austin from Denver. This is a great place with friendly people and one of the best restaurants I've ever seen on an airport. I had a big chicken strip dinner for $4 and it was great. I heard that they had great homemade pies but after the $4 dinner, I couldn't handle anything else. They are open every day, during daylight. Ingram Flying Service owns the restaurant.

PIREP Boy! Talk about a great little airport for a stop! These have to be the friendliest people I've met in a while. We had to spend some time there last weekend for weather. The diner was great and very reasonably priced. I didn't try it, but others mentioned that they have the best pie in Texas. I can vouch for the sandwiches and fries.

Several motels are nearby (three to five miles), and some will pick you up from the airport. The FBO has a satellite radar/weather system.

The next time I'm anywhere nearby, you can bet I'll stop back in!

Dallas, TX (Air Park-Dallas—F69)

Christina's Fine Mexican 🍔🍔🍔

Rest. Phone #: (817) 430-3669
Location: On the field
Open: Daily: Lunch and Dinner

PIREP FBO Comments: no public fuel.

PIREP This is a great $100 hamburger stop just north of ADS—in fact, still in their Class D narrow runway but 3100'. The best part is that parking is at the north tie-down area (west side of the field close to the run up area and the abandoned building). Gives you a very short walk to a number of restaurants. An excellent Mexican restaurant, a Pei Wei Noodle house, Chuck's Hamburgers, even a Starbucks. You're also only about a quarter-mile walk from a large mall complete with all the shopping you'd want or need.

Contact approach like you would for Addison (ADS), and they'll hand you off to the CTAF when you have the field in sight.

And if you run Mogas, there is a quick trip across the street with good Mogas prices.

Dallas, TX (Addison—ADS)
The Outer Marker 😋😋😋

Rest. Phone #: (972) 380-0426
Location: On the field; ask tower for taxi instruct
Open: Daily: Breakfast and Lunch

PIREP   FBO Comments: Million Air, Mercury Air, or Addison Express, all the same price.

PIREP   The Outer Marker is the old Longshots but a lot better. The new owner has redone the place and with the great menu, it's the place to go. Large-screen TVs and three Golden Tees—you can't go wrong.

PIREP   Located on the airport property on the "other" side of the security fence. Sports bar atmosphere with plenty of television and computer gaming screens. Complimentary food during happy hour, weekdays, 1700 to 1900 (pizza, fajitas, tacos, and so on—buffet changes daily). Above average bar meals.; good menu. Cold beer, full liquor bar, really cute servers. Very popular with pilots and airport employees "after work."

Dallas, TX (Dallas Love Field—DAL)
Uncle Julio's Mexican Food 😋😋😋

Rest. Phone #: (214) 520-6620
Location: About 2 miles; use courtesy car
Open: Daily: Lunch and Dinner

PIREP   Uncle Julio's Mexican Food, southeast of Love Field on Lemmon Avenue, was wonderful. The grilled jalapeños were the best and their enchilada dinner was very good also. The salsa tasted quite unique, but I am only going to be able to guess how good their margaritas are since I was flying that day. We flew a Cessna Skylane into Dallas Airmotive and were transported via courtesy car to the restaurant, which was about two miles away.

Decatur, TX (Decatur Muni—LUD)
Mattie's 😋😋

Rest. Phone #: (940) 627-1522
Location: 3 miles, shuttle available
Open: Daily: Lunch and Dinner

PIREP   We went to Decatur Muni and by shuttle went to Mattie's on the square. It is approximately three miles from the airport and well worth it. The square is the center, like towns used to be, with the courthouse in the middle and all the shops around the square. And the food was good too. I had chicken-fried steak that I would give four stars. All meals are under $5, and kids meals are $1.75. Next time I'm going to try the ostrich burger ($3.95). They also have evening buffets for $7 to $11, depending on the theme (American, Mexican, seafood).
    We had a fun Saturday!

Denton, TX (Denton Muni—DTO)
Ranchman's Cafe 😋😋😋😋

Rest. Phone #: (940) 479-2221
Location: 10 miles and 100 years back in time; use crew
Open: Daily: Lunch and Dinner

PIREP   Both DTO FBOs offer crew cars, and rentals are also available for the short drive to a bevy of good eateries in town. For a real Texas treat you'll wanna head west from the airport the ten miles and 100 years back in time to Ponder, TX and Ranchman's Cafe for a true trail-drive satisfying meal. Life's slow and easy in the small town of Ponder. At least, that's what the menu at the Ranchman's Cafe claims. And any place with a creaky screened door out front can't be all wrong.

Folks have been traipsing through the door since 1948 when Grace "Pete" Jackson first opened the Ranchman's (dubbed locally the Ponder Steakhouse). Her down-home meals quickly won a regular following of appetites, and word spread to nearby Dallas and Fort Worth that her hand-cut sirloins, T-bones, and chicken-fried steaks were well worth the drive. After her retirement, Pete sold the cafe but visited regularly until her death at age 93 in 1998.

At Ranchman's, the atmosphere is pretty laid back. Oh, there's a few rules. Checks are *only* separated at the register, and if you want a baked potato with your steak, then you'll have to call ahead. Otherwise, pull up a chair to one of the vintage dinette tables. It's time to *eat*.

If a grilled steak doesn't suit you, Ranchman's also serves quail, grilled chicken breast, ham, chicken strips, and a fine selection of sandwiches and burgers. And pies? Of course, the Ranchman's got those. Cobblers, too.

On the front of the menu, a notice advises that "large print and Braille menus are available." Isn't that unusual for a restaurant? Owner Dave Ross explains, "My degree was in special education and speech therapy. So I'm sensitive to the varied needs of different people. I always look for disabled people to give them opportunities to work."

And how many restaurants do you know that have a huge hand-painted mural above the entrance showing the early days of its hometown? Ranchman's Cafe has one. You can't miss it.

Any place with a creaky screened door out front and an owner like Dave has to be special, right? Stop by the Ranchman's sometime and see.

Ranchman's Cafe (1948), 110 W. Bailey, Ponder, TX 76259, (940) 479-2221.

Dublin, TX (Dublin Muni—9F0)

Old Doc's Soda Shop 😋😋😋😋😋

Rest. Phone #: (254) 445-3466
Location: In town—call and they'll pick you up
Open: Mon.–Fri. 9–5, Sat. 10–5, Sun. 1–5

PIREP   01/02/2006: Flew in from Victoria, just as described—lonely airport but has self-service fuel and tie-downs. We called the number and in just a couple of minutes our ride was there. We had very good sandwiches and delicious Dr. Pepper floats in the little soda shop and toured the plant, strolled around the downtown. When we were ready to return, the Soda Shop provided a ride back to the airport. I would highly recommend this as a fun excursion.

PIREP   FBO Comments: Runway is in okay condition, but it is U-shaped with a definite sag in the middle (lengthwise). If you land on the first half, your landing will be slightly downhill. Land on the second half, and it'll be slightly uphill. Chains were in place to tie down. 100LL at self-service pumps.

PIREP   A very cool trip. It happened just the way others talked about. When we landed, we were the only plane (and people) there. We walked up, there was the phone, called the number listed there (toll free), and a girl was there picking us up in less than ten minutes. We ate lunch ($3/sandwich) at Old Doc's, toured the plant for $2, bought a few T-shirts, and a girl brought us back to our plane.

They also told us that the second weekend in June each year is their birthday celebration, where they shut down the town and have a parade and some other events on Friday and Saturday. We'll be there for it!

PIREP   The oldest Dr. Pepper Bottling Plant in the world is still going strong after nearly 113 years! I just went there and stocked up on this wonderful drink still made using the original recipe, which calls for imperial pure cane sugar. If you ever drank Dr. Pepper before the mid-70s, then this is what you grew up with.

When you land, tie your plane up in a tie-down and walk to the porch of the terminal building. There you will find a phone that Dr. Pepper and the Dublin pilots have provided free of charge to all transients. The numbers are posted on the cover, but if you want to call ahead here it is: (254) 445-3466. Tours are given daily about every 30–45 minutes, and you get a free Dublin Dr. Pepper with the tour. Also, the soda shop is a true soda fountain and the gift shop sells one-of-a-kind DP merchandise. Definitely make this an outing for the family or if you need to build time. Also, on June 12, the city changes its name to Dr. Pepper, TX. This is an awesome event to attend.

PIREP   Went to Dublin today and had a great experience. We got there and called the police with the above number with my cell phone and the officer was there in about two minutes, it seemed. He was very friendly and more than happy to give us a ride.

We ate at Old Doc's Soda Shop (I highly recommend the Frosty Pepper). The tour price is now $2 ($1.50 for seniors, $1 for ages 6–12 and under 6 is free). Tour times are every 45 minutes starting at 9:15 Monday through Friday (no tours between 11:30 and 1:15), 10:15 Saturday (no tour between 11:45 and 1:15), and 1:15 Sunday. Last tour each day is 4:15.

The tour is about 45 minutes and includes a tour of the plant and the museum and a free bottle of their cane sugar Dr. Pepper, but you can't take it into the museum, so drink up before you get to that part. Also, they only run the bottling on Tuesdays, so you might want to factor that into your plans if possible.

A case of the cans at the plant is about $10 ($13.50 for the bottles), but it's $8 at David's grocery store two blocks down the street (bottle prices seemed to be about the same at the grocery). Also, at the plant, you can only buy them by the case. The officer told us about the grocery store.

We called the police from Old Doc's and he was there in about a minute. On the way back to the airport, he showed us where the station was—it's one block away from the plant.

The pump on the field was locked and there was a number, but no one answered when we called it.

Great little trip! Thanks for the help and thanks to the previous commenters for Dublin!

PIREP   The oldest Dr. Pepper Bottling Company in the United States. They give tours for $1.50 for adults and 50 cents for children; this includes a free Dr. Pepper. They are the only plant to make Dr. Pepper with pure cane sugar. They have a fantastic soda fountain shop that makes sandwiches and wonderful shakes or malts. When you arrive at the airport call the Dublin Police Department at (817) 445-3455 and they will pick you up. The food and malt were superb; the ambiance is incredible. The museum houses the largest Dr. Pepper collection known.

Eastland, TX (Eastland Muni—ETN)

Louise's 🍔🍔🍔

Rest. Phone #: (254) 629-1588
Location: 1 mile; crew car available
Open: 8 AM–6 PM Mon.–Fri.

PIREP   We arrived to be greeted by Bode Zietz, the airport manager. He was eager to help out. He had the courtesy car ready for our fast trip to town, only about a mile. They have a cute historical town square with a wonderful little cafe called Louise's that serves great steak fingers and hamburgers. The airport has been recently renovated with a new runway. The strip is 4020' × 60' asphalt.

They have Air BP 100LL and Jet A.

El Paso, TX (El Paso Intl—ELP)

Marriott Hotel 🍔🍔🍔🍔

Rest. Phone #: (800) 228-9290
Location: 4 blocks from the FBO
Open: 6 AM–9 PM

PIREP   The Marriott Hotel on the airport property is a short four-block walk from FBO ramp. Attentive table service, excellent food, hotel prices.

Alternatively, there is a restaurant inside the airline terminal, about three blocks from FBO ramp. It has been "renovated" into a fast-food counter. Acceptable sandwiches and Mexican items.

El Taco Tote 🍔🍔🍔

Rest. Phone #: (915) 584-3244
Location: Next to the airport
Open: Daily: Lunch and Dinner

PIREP   At 9910 Montana Avenue, at the SE corner of the airport, is El Taco Tote. Although Taco Tote is a fast-food restaurant, it serves some of the best fajitas by far in the whole state of Texas. After you order beef, chicken, or both, there is a condiment island that will knock your socks off. Everything is made fresh on the premises, including the corn and flour tortillas.

Perhaps the reason Taco Tote is so good is because it is a fast-food restaurant chain from Juarez, just across the border. In 1995, Taco Tote was a small chain in Juarez only. Their buildings were colorfully painted and constructed to look futuristic. There was a big sign at the front showing a UFO out in space, and through the canopy were two aliens in the cockpit eating tacos. The message on the billboard read, "The best tacos in the universe. We have witnesses!" When I first saw this, I thought it was funny, but I didn't try them because I didn't want to go to a fast-food restaurant in Juarez. Today, they have the best fajitas in the state of Texas, and you're lucky because one of their three El Paso places is very close to the airport!

Fort Stockton, TX (Fort Stockton-Pecos County—FST)

K-Bob's Steakhouse 🍔🍔🍔

Rest. Phone #: (432) 336-6233
Location: 1-1/2 miles
Open: Daily: Lunch and Dinner

PIREP   Flying an R-22 helicopter from San Antonio, TX to Phoenix, AZ and back, we stopped for fuel at KFST on the way out. David Hardwick of Fort Stockton Aviation was very friendly and offered us a courtesy car for lunch, but we decided to push ahead. On the way back from Phoenix, we decided to stay overnight at KFST.

Greg/David gave us a courtesy car (we arrived at dusk); we stayed at the La Quinta and ate at K-Bob's Steakhouse a block from the hotel. Both were about a one-and-a-half miles from the airport. Fuel was very reasonable, too.

Call David at (432) 336-9900/9901, cell (432) 290-1536 or Greg at (432) 290-2343. Great guys, give them your business.

PIREP   My wife and I recently flew up to Llano, TX to try Cooper's BBQ, and it was as good as your book said. Definitely the best barbecue we have had in a while. However, a couple of days ago a friend and I were out flying around the Davis Mountains in West TX and decided to stop in Ft. Stockton, TX for lunch. We called up Dave's Taxi Service from the airport and asked him to take us to a barbecue place.

I can't remember the name of the place, but it is the newest if not the only barbecue place in town. It had the same setup as Cooper's in Llano but on a slightly smaller scale, and the brisket and green beans were to die for. I personally thought it was better than Cooper's. It is on the same road that the airport is on. Exit the airport and take a left. Cross IH10, and the restaurant is about a mile down on the left. Dave charged us $5 each way.

Fort Worth, TX (Bourland Field—50F)

Pate Museum of Transportation 🍔

Rest. Phone #: (817) 332-1161
Location: Adjacent to the field
Open: Daily: Lunch

PIREP There is a small airport SW of Fort Worth called Bourland (50F) that is adjacent to the Pate Museum of Transportation. There are many planes, cars, tanks, boats...just about every form of transportation. Some of the jets landed there for the museum. I've only been there by car (so far) but could be a fun afternoon for that something different to do. The museum is free. I don't know if there are any restaurants around. Cresson is the closest town (almost-town, anyway), but Granbury is not far.

Wooden Nickel 🍔🍔

Rest. Phone #: (409) 544-8011
Location: Close by, call for pickup
Open: 11 AM–9 PM Tues.–Sat.

PIREP Fly in to Crockett (Houston County) and check out the great food at the Wooden Nickel. If you call (409) 544-8011, they will pick you up for free and take you to the restaurant.

Fort Worth, TX (Hicks Airfield—T67)

Rio Concho 🍔🍔

Rest. Phone #: (817) 439-1041
Location: On the field
Open: Daily: Lunch

PIREP The restaurant is very popular with local pilots. The atmosphere is gregarious and food is great for lunch. It's more of a fine burger place in my view, but very enjoyable. Park your aircraft right in front and walk in!

Fort Worth, TX (Fort Worth Meacham Intl—FTW)

Cattleman's Steakhouse 🍔🍔🍔🍔

Rest. Phone #: (817) 624-3945
Location: Downtown; crew cars are easy
Open: Daily: Dinner

PIREP FBO Comments: Really good FBO. Very friendly people. Gas is expensive, but courtesy cars are easily available and they have no qualms about you keeping it overnight. My wife and I kept one for two days at their insistence! Makes the gas a lot cheaper. They also made reservations at a nice Marriott for us and gave my wife directions to three malls she hadn't spent money in before (a real treat for her, but I liked her shopping on the strip better).

PIREP Take your pick! Seafood, steaks, Mexican, barbecue, you name it. Risky's has a place for each of the above, the Cattleman's Steakhouse is known throughout the steak-eating world as one of the best anywhere, but my favorite is the H3. All are located on the "strip" in the stockyards. Texas Jet has courtesy cars, park for $5 a day on the strip, live entertainment with gun fights, cattle drives, and shopping all in about a five-block area. Stay at the Cattleman's Hotel if your wife shops. (They drink beer on the street and in the stores, which made shopping much more enjoyable for me, but ends the flying for the day.) Night life is abundant (based on rumor).

Joe T. Garcia's 🍔🍔🍔

Rest. Phone #: (800) 217-4356
Location: Downtown, crew car available
Open: Daily: Lunch and Dinner

PIREP If you're looking for some of the best Mexican, Tex-Mex, or good old Texas barbecue in the world, it's just minutes south of KFTW. Authentic Mexican joints like Chuy's or the unbelievable atmosphere of Joe T. Garcia's offer a superb, fun dining experience at a price that will also allow you to put gas back in the plane. The Fort Worth Historic Stock Yards are located about 2.5 miles south of Meacham and have a couple of the best steak houses you will ever eat at. Risky's is the place to go for barbecue ribs—bring a water hose! There are way too many more to go into, most as good as the ones mentioned! If you catch the right time you may even get treated to the Longhorn round-up that happens daily in the stock yards. To the north is a the little town of Saginaw that has some fast food joints as well. Texas Jet is a good FBO that will help you out and get you where you need to be. KFTW offers nearly any other service you may need for your airplane as well. Paint to avionics to you name it, they do it. There is a vintage airplane museum on the field with a B-17 and other great WWII warbirds. Air Source One is the local pilot shop and is priced fairly. Best of all, everything mentioned here is on the same road as Meacham. No getting lost! See y'all in Fort Worth!

Fort Worth, TX (Fort Worth Spinks—FWS)

Cracker Barrel 🍔🍔🍔

Rest. Phone #: (817) 249-3360
Location: Short walk
Open: Daily: Breakfast, Lunch, and Dinner

PIREP I live close to Spinks and drive by it several times a day. The area has built up considerably. The Cracker Barrel is within easy walking distance to the FBO, maybe 100 yards or so. Next to Cracker Barrel is a string of restaurants that have been built in the last few years. There's Chili's, On the Border Mexican Food, Outback Steakhouse, Spring Creek BBQ, Tia Pan Chinese, and a McDonald's (if you're into that sort of thing), all within a quarter mile or less of the FBO. And there's also a Lowe's right there for those hard-to-find Cessna parts. They are currently building a Mexican Inn next to Chili's—probably open spring of 2003. The locals here call this area "restaurant row," and it is just that. One next to the other. The FBO is nice enough and there is now a small pilot shop for those essentials.

There is also a Motel 8 there, so close it might need clearance lights.

Fort Worth, TX (Fort Worth Alliance—AFW)

Lee's Smokehouse 🍔🍔🍔

Rest. Phone #: (817) 439-5337
Location: Short drive, crew car available
Open: 10 AM—8 PM Mon.–Sat.

PIREP I've just looked up the restaurant ratings at AFW—Ft. Worth's Alliance Airport. Apparently, no one has found the best barbecue in Texas—or else they have but are keeping it to themselves. Well, here's the lowdown.

The place is called Lee's Smokehouse. To get there, borrow a crew car from the FBO. Head south on the main airport road (or get onto I-35W South) to Westport Parkway (it's one of the three AFW exits). Turn right (west) on Westport Pkwy for about two to three miles. At the second stop sign, there are railroad tracks in front of you. Don't cross the tracks but turn left and go 100 feet to Lee's. Open till 8 PM Monday through Saturday (closed Sundays). I believe he now closes the third Saturday each month as well.

This is a classic barbecue restaurant. The food is outstanding, and I've traveled to most of the best in Texas. The decor is very basic, and from the outside you might wonder if it's the right place. It is, and a great treat awaits you. This is not a chain like Sonny Bryan's, but a privately owned and operated restaurant. Tell Lee that "The $100 Hamburger sent you," because he'll want to know who the pilots are.

The ribs are the greatest.

Fredericksburg, TX (Gillespie County—T82)

The Airport Diner 😋😋😋😋😋

Rest. Phone #: (830) 997-4999
Location: On the field
Open: Wed.–Fri. 11 AM–2 PM, Sat.–Sun. 8 AM–4 PM

PIREP By far, this is the absolute best $100 hamburger joint I have ever been to! The Airport Diner at T82 and the Hangar Hotel next door are absolutely top shelf, first class joints with over-the-top friendly people, outstanding facilities, great service, and the food!

The food at the Airport Diner is *outstanding*! Prices are on average $2 to $3 more than your average burger joint, but we are talking the gourmet version of your airport burger, sandwich, salad, or whatever you order.

Speaking of ordering, the Airport Diner has an outrageously priced old-fashioned milkshake at $5, which is incredibly *good stuff*! If you want the real deal milkshake made from the most tasty real ice cream (no low fat artificial stuff here), then this milkshake has got to be on your must-do list!

If there is a list of must-do $100 hamburger joints on your "before I go to pilot heaven" list, then make the Airport Diner at T82 number one on your list.

PIREP Breakfast at the Airport Diner was very good. A great setting, good food, and fast, friendly service. We did not have lunch there but I can believe what others have said about it. Our ratings for breakfast:

Location: +2
Food: +1
Service: +1
Ambience: +1
Price: 0

Another great restaurant is AugustE, but that is out the other side of town so it does not really qualify for a place in the $100 Hamburger site. It's a long (20 minutes-plus) taxi ride but worth it.

If you need a taxi, phone Jim Cassello, who owns and operates the friendliest taxi service you could imagine. Jim's phone number is (830) 997-8044, and he'll be glad to pick you up any time, any day, anywhere, including Thanksgiving Day in our case. If you do get a ride with Jim ask him about his famous daughter—he has good reason to be proud.

The only letdown was the Hangar Hotel, which we would have to describe as uninviting and unfriendly. The establishment itself is all top drawer and there's the potential for great atmosphere, but it was a huge disappointment. I'm not going to bore you with the details, but we would not really want to stay there again.

PIREP My wife and I spent the Labor Day weekend at the Hangar Hotel in Fredericksburg, TX and dined several times during our stay at the Airport Diner while there. What a great facility for those of us who are addicted to flying. The hotel is absolutely beautiful, all done up in a World War II–era decor. Where else can you sit on a balcony overlooking the flight line sipping on a cool drink of your choice and grade other pilots on their landings as they come and go? When you get ready to check in or out, it's only a few steps to the front desk from your aircraft. And the hill country of Texas is beautiful to visit as well. Lots of history and architecture.

There is also a first class avionics shop on the field, Pippen-York Flying Machine Co., so you can spend a day or two at the Hangar Hotel while having avionics work done if you choose. Everyone we met there during our visit was as friendly as could be. Rented a car at the Motor Pool right next door in the terminal for use while there.

Lots to see and do and lots of good food all around the area. We'll go back again.

Fuel prices were okay, not as high or low as we saw on our trip. Two fuel-up sites on the field, both have self-serve, and one offers to refuel you from a tanker truck if you wish. We paid $3.41 on north end of field. Down by the terminal, I think the cost was $3.51 self-serve and maybe a little more from the truck. I didn't inquire.

The Airport Diner is a part of a conference center located on the other side of the hotel. It also overlooks the flight line. Even lots of local nonflyers have found it and come out to eat both for the good food and to watch the flight activities. It was good to see a lot of couples bring their kids out to see it. Maybe some future fliers in the families.

Keep up the good work.

Marty Shingler
Donalsonville, GA
Bonanza K35 N646Q

PIREP   The Hangar Hotel and Airport Diner destination is a great new place to visit for an overnight stay or simply a good meal while stepping back in time—to the 1940s, to be exact. Both the hotel (50 rooms) and diner were newly opened in the fall or thereabouts. My wife and I stayed at the hotel in November and were impressed by the efforts made to provide excellent service and accommodations at both the hotel and diner. Due to inclement weather that weekend, we drove from Dallas/Ft. Worth, but I have since flown down to T82 several times. Both the breakfast and lunch menus are good, but when busy, allow for extra time. On a flight to the airport diner in February, I was told that the day before (a beautiful Saturday) there had been 40 private aircraft at the airport. It would appear that the word is getting out! While I found that the prices for food and lodging are a little high, the obvious quality makes it worth it.

Also, I learned from the hotel staff while staying there that the hotel and diner are only the planned beginning. In addition to the golf course, a theater is being planned (negotiated) as are other attractions. The idea is to make this a one-stop entertainment destination that caters to pilots and their families.

PIREP   Airport Diner is a 1940s–style diner that serves breakfast, lunch, and dinner. The diner features a genuine soda fountain with drink and dessert treats.

Also on-airport, the Hangar Hotel. Their phone number is (830) 997-9990. Based on Fredericksburg's early aviation and rich military past, a recreation of an old WWII military hangar, familiar to many aviation and history buffs, has now opened. This one-of-a kind, 50-room hotel sets the standard for style and romance of the period. Unique to the Hangar Hotel is an Officers' Club where guests can relax and visit when not in their rooms. Outside, a searchlight, water tower, and palm trees add to the hotel's Pearl Harbor ambiance. Golf carts will soon be able to meet the pilot and passengers and take them to the adjacent 18-hole golf course. A staff car limo is being completed to transport guests to the Fredericksburg Brewing Company, the Nimitz Museum of the Pacific War, or for hire, to local vineyards, Enchanted Rock, the Fredericksburg Herb Farm, the Wild Seed Farm, or nearby downtown shopping. The WWII staff car is a 1940 black Cadillac limo with dual side mounts and is the official "Command Car of the Hangar Hotel." All guest rooms are king sized and have a nightly rate of $139 per room. This rate includes two five-dollar food ration stamps per night to be used for breakfast at the Airport Diner.

The Cotton Gin 🍔🍔🍔🍔

Rest. Phone #: (830) 990-5734
Location: 10–15-minute walk from the airport
Open: Daily: Lunch and Dinner

PIREP   We flew from Houston to Fredericksburg for Thanksgiving weekend and I would like to pass on my 100 dollar hamburger thoughts.

The best restaurant we ate at was the Cotton Gin (www.cottonginlodging.com), which is about a 10 to 15 minute walk from the airport, but the very friendly staff will gladly send someone over in a car and pick you up. The food was excellent as was the great service, and you really need to phone ahead and book a table. The Cotton Gin is also a bed and breakfast but as we did not stay there we cannot comment on that.

Gainesville, TX (Gainesville Muni—GLE)

Dieter Brothers BBQ 🍔🍔🍔

Rest. Phone #: (940) 665-5253
Location: Close by, car available
Open: Daily: Lunch and Dinner

PIREP FBO Comments: The lowest price 100LL I've seen in a long time! Nice airport terminal. They have two crew cars that are available even when the terminal is unattended.

PIREP Dieter Brothers Barbecue is west of the airport (take Weber out of the airport and turn west/right on Hwy 82—it's a mile or so west on 82, on the right). The food was great and very inexpensive. Everything was made from scratch—the corn was right off the cob, the potato salad tasted like they'd just made it a few minutes ago! Mmmmm!

The Smokehouse Pit BBQ 🍔🍔🍔

Rest. Phone #: (940) 665-9052
Location: Close by, car available
Open: Daily: Lunch and Dinner

PIREP FBO Comments: FBO owned by the city and extremely well kept. Got loan of Crown Vic with impressive City of Gainesville logo and Airport Courtesy Car emblazoned on it. (Also a "report if you see bad driving" hotline sticker, which was unfortunate as we managed to drive the wrong way around the zoo's one-way system! I hope no one picked up the phone.)

PIREP GLE is a nice airport and very popular with folks looking for fuel bargains. Lots of activity. The Smokehouse is next door to Dieter Bros. and, according to the gentleman in the FBO, they are both about on a par. We really enjoyed The Smokehouse. Excellent service, good prices, and tasty barbecue. Don't miss the onion rings. The sliced brisket and ribs were both particularly good, as was the German sausage. Their sauce was tasty. And pie for dessert was coconut creme. Give it four stars on combination of value for money with good food and service. It's just a short drive west on Hwy 82 out of the airport. If you go to downtown Gainesville (left in US 82), try to locate and then drive the historic tour for autos, which takes you on a sign-posted little scenic tour of the old mansions and interesting districts. The old train depot is also worth a peek. We did not spot many other dining places that looked interesting downtown, although the Ranch House was packing people in on the basis of all-you-can-eat ribs.

The Center Restaurant & Tavern 🍔🍔🍔

Rest. Phone #: (940) 759-2910
Location: Muenster
Open: Tues.–Thurs. 8 AM–9 PM, Fri. 8 AM–10 PM, Sat. 6 AM–10 PM, Sun. 6 AM–9 PM, closed Mon.

PIREP Muenster is ten miles west of Gainesville Municipal on Hwy 82. The FBO provides a crew car and is very friendly. The food is well worth the drive. Excellent German food, great service, and good prices. The restaurant has large capacity and the locals had it near capacity for our Sunday brunch. The town (pop 1556) features a popular Germanfest in April. The owner isn't set up to transport fly-in customers yet, but we should keep asking. This restaurant should be a popular fly-in dinner destination.

Galveston, TX (Scholes Intl at Galveston—GLS)
Terrace Restaurant 🍔🍔🍔🍔

Rest. Phone #: (800) 582-4673
Location: On the field
Open: Daily: Breakfast, Lunch, and Dinner

PIREP Landed at Galveston for lunch yesterday and everyone was super friendly. We were directed to parking by a young guy in a golf cart who tied the plane down while we sorted things in the cockpit. He then gave us a lift on the golf cart to the terminal building.

The lady in the FBO phoned Moody Gardens to get the courtesy bus to pick us up, even though that's not really necessary as the hotel is less than ten minutes walk away.

Terrace Gardens Restaurant has informal dining overlooking the hotel's swimming pool. The service was great and the food was excellent—and it was under $20 for the two of us. When you are there try the turkey burger with cranberry-chipotle sauce—wonderful!

Gaido's Seafood Restaurant 🍔🍔🍔🍔

Rest. Phone #: (409) 762-9625
Location: Close by, cabs available
Open: Daily: Lunch and Dinner

PIREP On the advice of persons contributing to the $100 Hamburger site, our group of nine flew from Baton Rouge, LA to Scholes, parked at Evergreen Air (really friendly folks), hopped in a rental van from Enterprise (they left it with the FBO so it was waiting for us when we arrived), and headed to Gaido's Seafood Restaurant. I can honestly say (and this from a person from south Louisiana!) I enjoyed one of the finest meals in my experience at this restaurant. We were there in the middle of the afternoon, between lunch and dinner, and the restaurant was full. I expect you would have to wait for a table during prime hours—they do not take reservations. We recommend the shrimp bisque to start with—from there, take your choice, it was all good! So much so that not a single one of us had room for dessert. After walking around the old town area shops, we spent the night at the Hilton (very nice and they will pick you up at the airport in their shuttle bus if you want them to), then an excellent breakfast at IHOP and a leisurely tour of the wonderful Lone Star Flight Museum. For the ladies, the San Luis Resort, which also owns the Hilton, has a fabulous spa—a great place to spend the day. A good tailwind had us back in Baton Rouge in less than two hours. We'll certainly go back to Galveston—what a great place for a weekend!

Giddings, TX (Giddings-Lee County—GYB)
Texas Cafe 🍔🍔🍔

Rest. Phone #: (979) 357-2535
Location: Crew car available
Open: Daily: Lunch and Dinner

PIREP Went to a place called the Texas Cafe in Giddings. Great American buffet that was only about $7, all you can eat, and most of the food appeared to be handmade, not frozen.

Airport in Giddings is *great*! Low fuel prices and one of the friendliest FBOs anywhere!! These people really take care of you, and a courtesy car to boot.

Jambalaya's 🍔🍔🍔

Rest. Phone #: (979) 542-0845
Location: Crew car available
Open: Daily: Lunch and Dinner

PIREP   This small airport has the friendliest people at their FBO and the best running courtesy car I have ever driven.

The airport is about two miles out of the town of Giddings. We drove into town and had a great meal at Jambalaya's, a Cajun restaurant serving the best gumbo I have ever tried in Texas.

Gordonville, TX (Cedar Mills—3T0)

Pelican's Landing Restaurant 🍔🍔🍔🍔🍔

Rest. Phone #: (903) 523-4500
Location: Next to the field—short walk!
Open: 8 AM till 8 PM, 7 days

PIREP   Flew to Cedar Mills from Addison (ADS) to take advantage of a beautiful morning and get a $100 hamburger. The 3000' × 75' grass strip is in excellent condition and posed no problem for this first time in on the turf Turbo Arrow pilot, although the east-west runway orientation could potentially make for a more interesting approach in a stiff crosswind. Winds were light today. Always approach westbound from the lake to land on runway 25 and depart eastbound toward the lake on runway 07, as there are no obstacles taller than a sailboat in that direction.

The Pelican's Landing Restaurant is a quarter-mile (ten-minute) stroll from the runway, through the trees and along the lake and marina. On our arrival, about 20 homebuilt experimental planes from the Vans Air Force were just finishing their breakfast fly-in. We had burgers out on the glassed-in porch. The food and service were both excellent, and the prices were reasonable. They serve breakfast, lunch, and dinner and are open from 8 till 8 on Sundays as well.

PIREP   The restaurant is a nine-minute stroll south of the aircraft parking area at Cedar Mills. It is part of the huge Cedar Mills marina. Opened in 1984, the eatery has been flooded several times and was last rebuilt in 1993. Fly-in visitors can nearly hit the restaurant with a thrown stone from parked aircraft; however, a small inlet lies between the parking area and this delightful little restaurant. The restaurant is open Wednesday through Sunday, starting at 11 AM weekdays and 8 AM weekends. An attached open-air dining room (zipped in plastic on chilly days) overlooks the attractive marina, bristling with tall masts. Inside and out, the restaurant is spotless. Diners are greeted by numerous smiling staff. A good view is to be had from the dining room, where nonsmoking customers are well away from their addicted brethren. We sat down at 10 AM in time for weekend breakfast. Our waitress greeted us and poured my coffee simultaneously. No bilge brew, either: the coffee was fresh, hot, and endlessly refilled. Service was exemplary. The day's chalkboard breakfast special was smoked pork chops, two eggs, and biscuits, all for $5.95. My wife, KayCee, ordered it and was pleased. I had a nicely done plate of breakfast tacos. KayCee offered me a bite of homemade biscuit, which prompted me to order my own, a la carte. I accused the waitress of having someone's sweet, old grandmother manacled to the stove. She laughed, but I spotted a strange, haunting glint in her eye...

With orange juice, tea, and coffee, our total bill came to $13.81.

Lunch and dinner is also moderately priced. Burgers are less than $4 and dinners begin at $9. Hangarmate Dan Meeks, a frequent visitor, claims the lunch and dinner is just as delicious as breakfast.

After eating, we walked across the street to the pristine Ship's Store and Boutique. A friendly clerk gave us brochures revealing that the resort has cabins, RV hookups, a sailing school. There are plans, she said, to build duplexes near the airstrip.

Then, en route back to our waiting 170, I let myself into the enclosure that houses the marina's pot-bellied pigs and ewe. The larger of the pigs is the size of a large oil-drum and will prompt you to rub her belly.

Taking off downhill, toward the lake, I was careful to ascertain there was no inbound traffic, landing opposite direction. A lazy turnout over the water afforded one last spectacular view of the entire resort.

(Special thanks to Chicken Hawk owner, buddy and hangarmate Dan Meeks of Fort Worth, TX, for first discovering Sherwood Shores. Ten of us busted our knuckles together in Hangar 29, KFWS, which you might remember from the movie about Pancho Barnes.)

Graford, TX (Possum Kingdom—F35)
Lumpy's Restaurant 🍔🍔🍔

Rest. Phone #: (940) 779-3535
Location: Call for pickup
Open: Daily: Breakfast, Lunch, and Dinner

PIREP Approximately 70 km west of Ft. Worth, there is a quaint little barbecue place on Possum Kingdom Lake (I agree, what a name!). Lumpy's Restaurant has fantastic barbecue, fried fish, and barbecue sandwiches. I have eaten here two times and have said I would go back each time. I can't remember the restaurant name but there isn't much else around there. The barbecue place is about a mile and half from the airport. There are flyers and a menu in the pilot's rest area. The first time we flew in the owner's wife came and picked us up just after finishing all her baking at home. We had to ride the mile and a half with eight home-baked pies! I tried the pecan pie after lunch, it was great. They wrapped up two barbecue chickens we carried home with us.

The second trip the owner came and picked us up at the airport. The owner is the fire chief and works with the county judge to help folks serve out their community service time. He had a couple of young men ticketed for excessive speeding washing and waxing fire trucks the day we were there. The food is great, reasonably priced, and the owner and his wife will tell you a story or two.

PIREP Possum Kingdom is a real treat. The airport is in excellent condition and the Possum Kingdom area is hill country–beautiful, located right on the edge of the lake. If you have never been to the area you should put it on your schedule.

Upon landing we called Cliffette Eads, the owner of Lumpy's Restaurant; she was more than happy to pick us up. We had a great meal, and afterwards Cliffette was insistent on giving us a personal tour of the area. We look forward to another trip to this jewel of an airport and taking another whack at Cliffette's homemade pies.

Graham, TX (Graham Muni—RPH)
Clayton's BBQ 🍔🍔🍔

Rest. Phone #: (940) 549-5090
Location: Short walk
Open: Daily: Lunch and Dinner

PIREP There is a new barbecue restaurant within walking distance of Graham Muni. Actually, it's just across the new, huge parking lot that the local Dairy Queen has just paved.

The food is great, the service is great, and the prices are hard to beat! A brisket plate with two sides and Texas toast is $5.95. Tea is another dollar (but it's *free* on Sunday). Desserts usually range from $1.09 for to-die-for peach cobbler to $1.29 for homemade banana pudding. I believe the rib plate at $8.95 was the most expensive entree I saw on the menu. The portions are very generous—you won't leave there hungry! If you're the only one in the plane that doesn't want barbecue, their burgers are *huge*! Give them a try!

PS: There is a Dairy Queen right across the street!

Grand Prairie, TX (Grand Prairie Muni—GPM)

The Oasis on Joe Pool Lake 🍔

Rest. Phone #: (817) 640-7676
Location: $15 cab ride
Open: Daily: Breakfast, Lunch, and Dinner

PIREP Arrived Grand Prairie on Saturday at 1745 local. There were no crew cars or rental cars available. Cab fare to the Oasis is $15 each way. The Oasis is located at a commercial marina. The ambiance is somewhat threadbare—tourist tacky.

The food? Ah yes, the food. The grilled fish, and the accompanying fries, were swimming in liquefied grease. The fries were soggy and rancid. My wife had the grilled shrimp with garlic potatoes. By the time we returned to the airport, she was physically ill.

'Nuff said.

PIREP For those making a hamburger run into DFW, GPM is south right between Dallas and Ft. Worth. If you don't feel like hitting the airport diner or are in the mood for something more exotic, try the Oasis.

Grab a crew car, make a right out of the airport, and go straight and keep going straight. The restaurant is part of the Joe Pool Lake marina, about four miles distant. In fact, if you are coming in from the south, you will fly right over it en route to the airport.

The menu is steaks, seafood, salad, and, of course fish. The theme is rustic mariner. Dinner runs $15–20/person. The environment is what makes the experience, though. On the walk over the piers you will be solicited by a few resident ducks and hundreds of large fish that swim up and beg for morsels. Kids love it, and people enjoying themselves add to the ambiance of any restaurant. My first visit was last week, and after about ten minutes my other half instructed me that we were coming back on her birthday.

If you feel like it, you can rent a boat or a jet ski for a while. I didn't catch all the prices, but a pontoon boat goes for about $100/four hours or $200/eight hours.

A local newspaper review posted on the wall says it all: "An excellent place to dine even if you don't have a boat."

Pepper's Grill and Bar 🍔🍔🍔

Rest. Phone #: (972)-602-PEPPER
Location: On the field
Open: Mon.–Thurs. 11 AM–11 PM, Fri.–Sat. 11 AM–2 AM, closed Sunday

PIREP I flew into Grand Prairie Municipal yesterday. Nice airport, very clean, friendly tower, and a good place to stop in for food. The cheeseburgers at Pepper's Bar and Grill were excellent, along with the iced tea; wait-staff is great. They have many other items on the menu that sounded really good also. There is a flight school on the field with a pilot shop inside. Plenty of aircraft parking. If you get a chance, stop in here, it's well worth it! If you're coming in from the IAH area, use the Leona3 departure with the Dodje3 STAR—pretty much straight shot in!

Greenville, TX (Majors—GVT)

Mary's of Puddin' Hill 🍔🍔

Rest. Phone #: (903) 455-6931
Location: A drive; crew car available
Open: Daily

PIREP Last Saturday we flew into GVT for lunch. Used their very nice courtesy car, a late model Caprice with a "certified speedometer." Had lunch at Mary of Puddin' Hill, about 7 miles from the airport. The deli was very good. Their desserts were heavenly!

Hillsboro, TX (Hillsboro Muni—INJ)

Lone Star Cafe 🍔🍔

Rest. Phone #: (254) 582-2030
Location: Short drive in the crew car
Open: Daily: Lunch and Dinner

PIREP FBO Comments: All fuel is self-service, 24-hour, well lighted, clean; accepts most credit cards. Airport is attended 8–5, Tuesday through Saturday, courtesy car available, terminal building is accessible odd hours by calling PD for door code. PD number listed on door. Fuel pricing very competitive!!

PIREP Lone Star Cafe is a short ride into town via crew car. Food and service very good. Place is typical Texas with chicken-fried steak, and so on. Pricing is very moderate—good value!

Hondo, TX (Hondo Muni—HDO)

The Flightline Cafe 🍔🍔🍔

Rest. Phone #: (830) 426-4020
Location: On the field
Open: Daily: Breakfast, Lunch

PIREP The Flightline Cafe (FC) did close on the field. However, they have reopened and claim to have the same good food as before. I don't doubt this since it's the cooks that determine the quality of food, not the location, and they kept the same cooks! Now, you arrive at the same old building, then call the FC at (830) 426-4020; they'll send someone to pick you up and take you to the chow!

PIREP Landing at Hondo Municipal Airport is like taking a trip back to 1942. The airport is dotted with wooden barracks and Quonset-style hangars. There are five extra-long 150' wide paved runways. In the center of 17R/35L stands a building that was once part of the navigator training complex. Today it is the Flightline Cafe, owned and operated by Betsy Hermann. It doesn't look much different from the Officer's Mess of 50 years ago.

Inside it is stacked with memorabilia. All the Cadet Class Books, beginning with August, 1942, are here. There is a photo collection that will really take you back.

The Flightline Cafe has a wide selection and is particularly famous for its baked goods: pies, cakes, pastries, and breads. I was immediately attracted to the *Texas-sized* chicken-fried steak. There is also a terrific buffet!

Houston, TX (Pearland Regional—LVJ)

La Casita's 🍔🍔

Rest. Phone #: (281) 482-2206
Location: Call for pickup
Open: Mon.–Sat. 11 AM–9 PM

PIREP When flying to Clover Field, take some time to visit La Casita Mexican restaurant. They have been there for many years, and their food is excellent. It is just a couple of miles from the field. Just park and call them from the terminal building; and they will come and pick you up.

Houston, TX (Ellington Field—EFD)

Randy's Smokehouse & B.B.Q. 😋😋😋

Rest. Phone #: (281) 486-8727
Location: 2 miles from the gate
Open: Daily: Lunch and Dinner

PIREP Good food, good prices. Best to come on a Saturday around lunchtime. That's when the mechanics from Commemorative Air Force and Collings Foundation West take a break from fixing airplanes at EFD and sit around and talk about airplanes.

It's about three miles from the ramp proper and two miles from the entrance to the airport (FM 1959 and Texas 3).

Houston, TX (Sugar Land Regional—SGR)

Pappasito's 😋😋😋

Rest. Phone #: (281) 565-9797
Location: Use the courtesy car
Open: Daily: Lunch and Dinner

PIREP On the Border is no more, but there are tons of good places to eat within ten minutes of the airport. Too many to list, but check with the line staff or airport desk for the latest—be it Thai, Chinese, Mexican, Italian, or plain old American.

PIREP Great Mexican place called Pappasito's. Very friendly folks at the FBO, and the linemen always remember me and ask about my Air Force adventures!! Good price for fuel. They have a flight school.

The FBO has a courtesy car and they'll give you directions to Highway 59. It's about four miles to On the Border and a whole ton of other restaurants. A perfect 10. Great Mexican food, awesome Margaritas!!

Houston, TX (Weiser Air Park—EYQ)

Carl's Bar-B-Que 😋😋😋😋

Rest. Phone #: (281) 890-2275
Location: Next to the field
Open: Daily: Lunch and Dinner

PIREP Being a local, I have several choices of eateries, what with Houston boasting over two thousand restaurants. Carl's Bar-B-Que, butting up against Weiser Airport (no kidding—about a 120-second walk from tie-down), ranks up there with classic Texas barbecues. My family enjoys the stuffed baked potato above all else. It comes in three varieties, the largest being heaped—I mean overloaded—with fresh chopped barbecue. I have a big appetite yet am frequently unable to finish it.

Enjoy!

PIREP About stopping at Weiser for lunch, at Carl's Bar-B-Que. As long as you're stopping to fill your stomach, Weiser is also a good place to top off your tank or to put in as much as you need. As someone flying out of Weiser, I know they tend to have the best prices on fuel in the Houston area. On the weekends, there is someone working the pumps. During the week, it's self-serve.

PIREP   When you are flying around a large city, such as Houston, TX, it is easy to pass by a small airport. Although in passing by you have missed out on a wonderful place to eat because you did not know it was there. Such is the case with Carl's Bar-B-Que.

Located along US 290, northwest of downtown Houston and Beltway 8 is Weiser Airpark. If you are flying in the Houston Class B airspace (with a clearance of course!), Weiser Airpark looks small from altitude. It is only after a closer look at the charts that you realize you can land there, no problem.

I had an occasion to visit Weiser the other day. My friend invited me along for a ride in his Ercoupe. Now this was a delight in itself, but along about lunch time, he suggested we stop at Weiser. I remarked, "Well, that's fine, but where do we eat?" It was then that I was educated that a great place resided right next to the airport property called Carl's Bar-B-Que.

After landing and walking around the airport for a short tour, we walked a very short distance to Carl's. Well, it was worth every step. This restaurant had barbecue from sandwiches to plates lunches, sure to please any Texas-sized appetite. If you are out flying, consider stopping at Weiser for a bite to eat at Carl's. You will not be disappointed!

Houston, TX (George Bush Intercontinental / Houston—IAH)
Airport Marriott 🍔🍔🍔

Rest. Phone #: (281) 443-2310
Location: On the field
Open: Daily: Breakfast, Lunch, and Dinner

PIREP   The restaurant is CK's in the Houston Airport Marriott, telephone (281) 443-5232.

When you're ready to put your Class B procedures to the acid test, fly on in to IAH. This is the big time! When you've finally worked up a Texas-sized appetite, I recommend that you head on over to the Marriott Hotel in the terminal complex, to CK's Revolving Rooftop Restaurant. Be advised that the Marriott does not provide transportation to or from the general aviation ramps, but both of the FBOs at the field will.

CK's is open for dinner from 5:30–10 Mon.–Sat. and for an all-you-can-eat lunch buffet from 10:30–1:30 Mon.–Fri. The food is superb Continental cuisine, easily the equal of the best restaurants in town. The ambiance is marvelous—a rooftop revolving restaurant overlooking the entire airport complex—and the service is impeccable. Be aware, though, that this is one $100 hamburger restaurant at which you could spend more for the food than for the flight—budget $50 per person for dinner. The lunch buffet, the last time I flew in, was $14.95.

If that's a little too rich for your blood, or if you're looking for a late lunch when CK's is closed, there is another restaurant at the Marriott, Allie's American Grill in the basement.

Houston, TX (William P Hobby—HOU)
Hilton Hobby Airport 🍔🍔🍔

Rest. Phone #: (713) 645-3000
Location: Take their courtesy van
Open: Daily: Breakfast, Lunch, and Dinner

PIREP   The restaurant is the Boardwalk restaurant in the Hilton Hobby Airport—telephone (713) 645-3000.

Just across the street from Houston's Hobby Airport is the Hilton Hobby Airport, home of the Boardwalk restaurant—a solid four-burger addition to the $100 hamburger club, open for breakfast, lunch, and dinner seven days a week.

If you park at Atlantic Aviation on the north side of the field, it's a short walk across the street; from any other of Hobby's FBOs the hotel will happily send their courtesy car to run you and your friends over for your meal. The food is top quality all the way around—the restaurant specializes in fresh Gulf seafood. Not to be missed are their all-you-can-eat buffets—prime rib on Tuesday nights; shrimp, shrimp, and more shrimp on Friday nights; and their theme lunch buffets served Monday–Friday from 11 to 2. Service is what you would expect from Hilton, first class at all times. Price is reasonable for what you get, comparable to other first-class restaurants in town. Hobby is a great place to practice your Class B procedures—come on by!

Houston, TX (David Wayne Hooks Memorial—DWH)
The Aviator's Grille 🍔🍔🍔🍔

Rest. Phone #: (281) 370-6279
Location: On the field
Open: Mon.–Thurs. 11 AM–2 PM, Fri., Sat. 11 AM to 9 PM

PIREP   The Aviator's Grille is a very friendly restaurant located directly on the airport grounds adjacent to the fuel services (north end of the airport). It has a great view of the runways. It's very clean and is decorated completely with the golden age of aviation murals and other decorations that make for a very pleasant airplane watching environment. The people that run The Aviator's Grille are extremely friendly, and the service is good. The food is inexpensive and well prepared. There isn't a huge menu, but most of the basics are there, ranging from the half-pound B-52 burger with fries to chicken-fried steak with mashed potatoes and white gravy. They have several different sandwiches as well as fried catfish and fried chicken. There is also a daily special. They are open for lunch during the week (11 AM to 2 PM Monday through Thursday) and lunch and dinner (11 AM to 9 PM).

In summary, it's a very friendly place with a good view that will serve you a good meal for not a lot of money. Recommended. And remember, David Wayne Hooks airport has a sea lane so you float flyers can feel welcome too!

Huntsville, TX (Huntsville Muni—UTS)
New Mount Zion Missionary Baptist Church 🍔🍔🍔🍔🍔

Rest. Phone #: (936) 295-7394
Location: Grab the crew or walk, but get there—4 miles
Open: Wed.–Sat.

PIREP   Friendly staff at Huntsville Aviation. They gave me their crew car to drive to Mt. Zion, about five miles away. Mt. Zion BBQ was good food, great atmosphere. All you can eat for $10! Only open Wednesdays through Saturdays.

PIREP   FBO Comments: Airport staff was very friendly and efficient. They had a reasonably nice courtesy car.

PIREP   The Mount Zion Holy BBQ was everything I was led to expect. The atmosphere is very cozy, with all seating family style. The brisket was very tender, and the potato salad was great. It is an easy flight from the Dallas/Fort Worth area.

PIREP   We flew to Huntsville for lunch at Mt. Zion church today. It was all that the review said it would be. Hours are Tuesday through Saturday 11 AM to 7 PM.

The line crew are very friendly. When you go, make sure to bring back a few bones for the airport dogs that are behind the counter.

Fly safe.

PIREP   This is the best barbecue feed in the world. Period! Just 75 miles north of Houston, it is worth a detour and a landing on any business day and a special trip on Saturdays. This is the "church of the Holy Barbecue." It has been smoked by an angel named Annie May Ward. She is the head pit diva at this decidedly down-home place.

What started 16 years ago as a one-time way to raise money for the congregation of this east Texas church has become a *for real* Q shrine. It is open from Tuesday through Saturday. For seven bucks you're served *all* you can eat.

Call Huntsville Aviation before you take off and reserve the airport car (713) 295-8136, or ask them to call a cab for you when you're ten miles out.

Iraan, TX (Iraan Muni—2F0)
Los Arcos Restaurant 😋😋😋

Rest. Phone #: (432) 639-2000
Location: 1/2 mile
Open: Daily: Lunch and Dinner

PIREP Got weathered in at Iraan and didn't expect much but went looking for breakfast. Had an easy walk north about a quarter/third of a mile (three or four long blocks) to the intersection of Hwys 349 and 190. On the NE side of the intersection, just off the corner, was Los Arcos.

Wonderful pancakes, *excellent* chorizo y papas burrito, and excellent sauce to go with the burrito. Friendly folks, excellent food. Sold us a jar of the sauce when we asked (and found a jar for us so it would be stable in the baggage area).

Jasper, TX (Jasper County-Bell Field—JAS)
Cedar Tree Restaurant 😋😋😋

Rest. Phone #: (409) 384-8832
Location: Crew car
Open: Daily: Breakfast, Lunch, and Dinner

PIREP Flew to JAS New Year's day, having called Steve Seale of LEMA Aviation in advance. Self-serve fuel and a very nice terminal building (lounge, weather, and so on.). Steve went out of his way and left us the keys to the courtesy car for our short ride to the Cedar Tree Restaurant on Hwy 96 in Jasper, TX. We pulled up just before 11 AM and there were already some folks in the parking lot waiting for the doors to be unlocked. Good buffet with pies and pudding for your choosing (even a soft-serve cone machine). The service was as pleasant and friendly as one could expect. Total for the four of us with drinks about $35–$36. We fueled the car before returning it to the field.

Thanks to Steve Seale!! Phone: (409) 381-8151 (from business card he left for me).

Kenedy, TX (Karnes County—2R9)
Barth's Restaurant 😋😋😋

Rest. Phone #: (830) 583-2468
Location: Next to the airport
Open: Daily: Breakfast, Lunch, and Dinner

PIREP Opened in 1936, it's one of "Texas' old-time restaurants." You cannot see Barth's Restaurant from the ramp, but it is on the other side of the trees just 150 yards east across a grass field from the FBO. An easy walk. Barth's is a good old highway cafe, famous for its chicken-fried steak.

PIREP The FBO can be counted on to fill or fix your airplane and/or give you a ride to Barth's. It's more fun just to park and walk through the weeds the quarter mile to this friendly place.

Killeen, TX (Skylark Field—ILE)

The Airport Coffee Shop 🍔🍔🍔

Rest. Phone #: (254) 699-4803
Location: In the terminal
Open: Closed Sat., Sun.

PIREP   I dropped in on Killeen on a short hop out of Houston. As we got to the Airport Coffee Shop there at high noon, they were putting up the "closed" sign on the door. It turns out that they close at lunch on Saturday (?) and are closed all day Sunday.

However, the fine people at Texas Aero gave us the keys to their courtesy car after we had already told them that we wouldn't need any line service. There are numerous places along the main highway that goes into town from along the south periphery of the field.

Kountze/Silsbee, TX (Hawthorne Field—45R)

Mama Jack's 🍔🍔

Rest. Phone #: (409) 246-3450
Location: Short drive to town
Open: 7 AM on, all week

PIREP   This small airport just about 15 miles north of Beaumont, TX has opened again after being closed for several years. The FBO is Big Thicket Aviation, which is managed by Kenny Seaman. It has rental planes, hangars, and flight training. There is always fresh coffee, cookies, and a big-screen TV. They have a courtesy car and nearby, in the town of Kountze, there are several restaurants. It is best to call ahead and let them know you are on your way: (409) 246-2330 (fax: (409) 246-8444).

I hate eating alone, so I took Kenny into town and we went to Mama Jack's. This restaurant has a breakfast buffet from 7 AM on and a lunch buffet as of 11 AM daily, all week. They also have a menu with great things such as a jalapeno burger, catfish po' boy, and barbecue potato. Friday and Saturday nights they have a seafood buffet, which is their best turnout. They make their own cinnamon rolls and always have cobbler. The restaurant is run by Barbara and Jerry Jackson, and their phone number is (409) 246-3450. I would rate the food at least two burgers.

La Grange, TX (Fayette Regional Air Center—3T5)

Tim's On The Square 🍔🍔

Rest. Phone #: (979) 968-9665
Location: In town, La Grange Taxi (979) 968-5309
Open: Daily: Breakfast, Lunch, and Dinner

PIREP   Tim's On the Square—my rating is 5 stars. La Grange is a happening place again for the fly-in gourmet crowd now that Tim's On The Square is open.

First, about the airport: GA terminal that is open from the hours of 7 AM to 7 PM 7 days. Fuel is available. Both VOR and GPS approaches are available off the industry VOR for the instrument rated. Runway is 16-34—long, wide, and beautifully maintained. Rental cars are available from EZ Rental: (888) 832-0333, or local: (979) 242-4034. On a budget for your $100 hamburger? Then call La Grange Taxi: (979) 968-5309. You'll need transportation since the airport is a few miles from town.

About the restaurant: located in Historic La Grange Town Square at 155 N. Main. We flew into a real bargain for Sunday brunch: a nice self-serve salad bar, a wonderful choice of five entrees that included beef tenderloin marsala, poached salmon, some sumptuous poultry dishes, a choice of great vegetables. Make sure you ask about the pablano and cheese potatoes. There is also an omelet bar with a chef to make one to your specs. A choice of several desserts is included and might be old-fashioned banana pudding, chocolate cheesecake, and so on. The food was fresh and delicious, the restaurant light, clean and quaint. This is a great fly-in spot.

Other local restaurants: for those of you that remember the heyday of the Cottonwood Inn, try Boss's Steak House. Boss Hrbacek, former Cottonwood owner and La Grange icon, runs the place, and they still specialize in those sizzlin' steaks you may remember.

Other attractions: Spoetzl Brewery, a brewery museum from the early days of Texas History, and Monument Hill, a scenic overlook on the bluff.

In close proximity (about 10–15 miles) down FM 159 toward Brenham is Round Top—a bed and breakfast and antiques mecca. Also, there is a summer concert series there at Festival Hill.

The Bon Ton 😋😋😋😋

Rest. Phone #: (409) 968-8875
Location: They'll pick you up
Open: Sun.–Thurs. 5:30 AM–9 PM, Fri. and Sat 5:30 AM–10 PM

PIREP We tried La Grange on 19 February and enjoyed parking on the ramp next to a DC-3 and Hun-on-a-stick.

At the south side of the ramp is a very nice little building with automatic locks so it doesn't have to be manned. Inside are nice restrooms, pilots' lounge, computer for weather checks, and telephone with free local calls. A sign on one closed office says Rental Cars. We were the only ones there, although an airplane was refueling with self-serve at the opposite end of the ramp.

We used the numbers Ed posted above, and a car from Weikels' Bon Ton Restaurant took only about five minutes to pick us up. We learned that the restaurant came back into the ownership of the original family last April. The buffet and their homemade bread was good. All-you-can-eat buffet with pie is great.

The Bon Ton is the closest restaurant (Taco Bell doesn't count) to the airport, and the only one offering free rides, but the old town square is a couple of miles farther, past the river. There's a neat old butcher shop/barbecue restaurant on the town square that really gives the flavor of the area. The Bon Ton is a pleasant new building, but the old butcher shop has the high metal-covered ceilings, a lunch counter, and a back room with long tables with interesting pepper sauce.

PIREP I am happy to report the rebirth of the Bon Ton Restaurant in La Grange. It is now open for business and looking better than ever!

The name is Weikels' Bon Ton Restaurant. The pies and homemade breads are there, as is the friendly staff. They will still pick up and deliver to the airport with a telephone call: (409) 968-8875 or (888) 330-8875.

Lampasas, TX (Lampasas—LZZ)

Storm's 😋😋😋

Rest. Phone #: (512) 556-6269
Location: 1-1/4 mile walk
Open: Daily: Breakfast, Lunch, and Dinner

PIREP Want a *real* hamburger? Try a storm burger at Storm's! It's more than you can usually handle, but try! I can't say there's a shuttle running, but if you ask Bob Brame the FBO, he'll see if he can get a ride for you. It is only about 1.25 miles south of Lampasas County Airport so you could walk it!

Lancaster, TX (Lancaster—LNC)

Happy Landings Cafe 😋😋😋

Rest. Phone #: (972) 227-5722
Location: On the field
Open: Tues.–Sat. 7 AM–3 PM

PIREP   Awesome restaurant! Great food, even better prices, and it is right on the field! Nice long runway, too.

PIREP   Very friendly place! We taxied right up to the building, and someone from the FBO came out to meet us. We had breakfast at Happy Landings and it was great! Try the omelets.

PIREP   Stopped by at Lancaster's Happy Landings Cafe and discovered they have added a few more items to their menu—salads and other good things, in addition to the outstanding hamburgers. Although this is a great breakfast and lunch cafe, normally closing at 2 PM, they have recently added Friday nights to their hours. Steve (owner/cook) and the gang are serving up a generous steak dinner with potato, salad, veggies, and so on. But if you like your steak medium rare, be sure to tell Steve to take it easy on your black Angus steak (he tends to cook more toward the medium-well side)!

La Porte, TX (La Porte Muni—T41)
The Runway Bar and Grill 🍔🍔🍔

Rest. Phone #: (281) 867-9600
Location: On the field
Open: Daily: Breakfast, Lunch, and Dinner

PIREP   New owners took over the old Rixters. They are on the field with ramp parking nearby. They have outside patio dining now near the approach end of runway 5. Excellent service every time we go. They are open for breakfast early and stay open until 10 PM. We try to go once a week to watch the traffic. I would like to see them install a speaker so that we could hear the unicom. Other than that, this is what a 100 dollar hamburger joint is all about. Great food and great selection. Oh yeah, and it is the only restaurant I can find that serves sweet tea. That in itself deserves ten stars. They will soon have beer to serve those flyers that have tucked their birds in for the day.

Livingston, TX (Livingston Muni—00R)
The Wet Deck 🍔

Rest. Phone #: (936) 967-2337
Location: It's a ways
Open: Daily: Breakfast, Lunch, and Dinner

PIREP   I flew into Livingston airport this morning to find that they are quite active. There is a mechanic on the field, 100LL gas, and nearby there is a bar called the Wet Deck that serves hamburgers, lunches, and so on.

The Wet Deck is a short drive from the airport down toward the lake and, although they do not have a courtesy car, I received several offers for a ride down to the bar. I was also told that the barbecue place will be reopening next month so there will be two options.

Lockhart, TX (Lockhart Muni—50R)
Chisholm Trail Bar-B-Que 🍔🍔🍔

Rest. Phone #: (512) 398-6027
Location: 30-minute walk
Open: Daily: Lunch and Dinner

PIREP FBO Comments: Martin & Martin Aviation: (512) 376-9608.

PIREP A TV show featuring the "great taste of Texas" caught our attention on Lockhart. For years we have been flying over Lockhart, and today my wife and I decided to give them a try—boy, had we been missing out!

Lockhart was named the barbecue capital of Texas by the state legislature and for good reason. There are several barbecue restaurants to choose from.

We tried Chisholm Trail Bar-B-Que by recommendation. It was great and had very reasonable pricing. The folks at Martin & Martin FBO were very friendly and helpful, and the loaner car made it easy to get to the restaurant and around town.

Longview, TX (East Texas Regional—GGG)
Papa Jack's Fishin' Hole 🍔🍔🍔

Rest. Phone #: (903) 663-9300
Location: 100 yards away
Open: Daily: Lunch and Dinner

PIREP Now open—great fried fish and other good stuff at fair prices. You can land and park at Stebben's Aviation and then walk 100 yards to the Fish Place.

Lubbock, TX (Lubbock Preston Smith Intl—LBB)
County Line 🍔🍔

Rest. Phone #: (806) 762-6201
Location: Use the crew car
Open: Daily: Breakfast and Lunch

PIREP This is a place for great barbecue. We got a sampler platter with a little bit of every meat. Fantastic! They also serve steaks, but I couldn't get past the barbecue. Great service—I wish every restaurant was this friendly. We had probably the nicest waitress we have ever had the pleasure of meeting.

It is about five minutes from the west side of the Lubbock airport. Lubbock Aero has excellent service, very friendly staff, and nice crew cars. They gave us a van to go to the restaurant. This is a must stop when you are in the area.

Lufkin, TX (Angelina County—LFK)
Airport Grill 🍔🍔🍔

Rest. Phone #: (936) 634-7511
Location: On the airport
Open: Daily: Lunch and Dinner

PIREP FBO Comments: Reasonable fuel prices, good service, attentive linemen—the kind of service you wish you got at Million Air or Signature.

PIREP Some of the nicest folks you will ever meet. Cafe is on the airport and hours on weekends are 11 AM to 3 PM. Once a month (fourth Saturday of every month, I think), the local EAA chapter has a Fajita Fly-In.

Food is great—had a double cheeseburger, but when I ordered it, I didn't know that it was so big that it would make me overgross when I took off out of there. Good food, good value, good fuel prices, great people. All in all, wonderful place to go for that $100 hamburger.

PIREP   I flew into Lufkin, Angelina County today, Saturday, to find that the restaurant in the terminal was closed. It seems like they close on the weekends. I borrowed the courtesy county pickup and abstained from plugging in the bubble gum machine on the roof. I was told about the nearby restaurants but as I was driving into town I noticed across the road a sign that said Don's BBQ. It is a small roadside place with tables outside—barbecue stand combined with antique store. Since I was there, I decided to try it. Don is a great old guy who is retired but could not keep still. He is a real people person and enjoys company anytime.

The barbecue was in the refrigerator and he warmed it up in the microwave. I thought to myself, "How bad can it be?" Well, I was surprised. It was the best barbecue ribs and brisket I have eaten in many years. He gave me plenty of napkins because he knew I would need them.

There is a sink with running water attached to the awning and it is for us eaters who wear our stains as medals.

I was given a complimentary bowl of apple cobbler that I enjoyed while browsing through the antiques that Don's wife has for sale. Sometimes the little hole-in-the-wall places have the best food. I found one this time.

McAllen, TX (McAllen Miller Intl—MFE)
Tony Roma Ribs

Rest. Phone #: (956) 631-2121
Location: Across the street
Open: Daily: Lunch and Dinner

PIREP   Flew down to McAllen Miller (MFE) today. Parked at McCreery Aviation which has been in business 50 years. They have a very interesting picture wall. Everybody from Bob Hope to George W. Bush. Tony Roma Ribs is just across the street from McCreery. There is a mall just down the street also if you want to shop.

Marfa, TX (Marfa Muni—MRF)
Mike's Place

Rest. Phone #: (432) 729-8146
Location: In town; use the crew car
Open: Daily: Lunch and Dinner

PIREP   If you ever make it over to Marfa, Mr. Moore has a loaner car that you can use to go into town. Had a hamburger at Mike's Place, and it was very good. For you James Dean fans, the movie *Giant* was filmed here, and they have all kinds of photos on the wall at the Paisano Hotel. I have not seen the Mystery Lights but maybe you will.

Mesquite, TX (Mesquite Metro—HQZ)
Mama's Daughters

Rest. Phone #: (972) 289-6262
Location: 5 minutes by crew car
Open: Daily: Breakfast, Lunch, and Dinner

PIREP   Mesquite is a handy stop for people transiting DFW or those making a hamburger run into the Metroplex. It sits on the eastern edge of DFW, easy to find because of its location (just south of Lake Ray Hubbard) and its runway (long and wide). The ILS makes it an all weather fly-in strip. The city has a couple of clean crew cars available for the short run into town. Within about five minutes are two excellent independent eateries.

Mama's Daughters is a local eatery in the home-cookin' genre. Meatloaf, beef tips, chicken-fried steak, okra, red potatoes, cornbread, and shamelessly sweet desserts are the standard fare. Their lunch specials with all the trimmings are under $6. Take the crew car west on US 80 to Galloway, then south 200 feet.

Catfish Cove is another favorite. They have the kind of mouth-watering catfish recipes that put the big chains to shame. Dishes are around $9, with relish bar and all-you-can-eat homemade hushpuppies included. I recommend the lemon-pepper catfish if you're trying to eat healthy, with a dash of Tabasco on the french fries if you're feeling lucky. Live bluegrass on Saturday nights, but be early 'cause the place gets packed. Crew car west on US 80 to Beltline, then north just past Town East.

By the way, husbands beware. Town East is named for the huge retail district and mall nearby, with virtually every big and small name in commercial retail. If your wife catches the aroma of shopping, you've had it. The whole area is the river Styx for husbands, strewn with the carnage of purse-holding men waiting, with glassy eyes and thousand-yard stares, for their women to finish perusing clothes, trinkets, and shoes.

But if your wife is one of those who doesn't care for your airplane, the lure of shopping may be a carrot you can dangle in front of her, leading her right into the cockpit. Just wait till she hits the checkout line of the first store to discover that you "forgot" your wallet in the airplane…

Mexia, TX (Mexia-Limestone County—LXY)

The Tamale Inn 🍔🍔🍔

Rest. Phone #: (254) 562-6022
Location: Adjacent to the airport
Open: Daily: Lunch and Dinner

PIREP   Flew with a friend to Mexia on the recommendation of the *$100 Hamburger* book I recently purchased. Bill at Trimaire was great, allowing us to use the airport minivan, although he recommends that you call ahead to reserve it. The food was very good, we had the chimichangas for lunch, and the wait-staff was great and very attentive. The Inn itself looks run down, but it is your typical hole-in-the-wall restaurant: poor looking on the outside, great food on the inside.

Midlothian / Waxahachie, TX (Mid-Way Regional—JWY)

Several 🍔🍔

Rest. Phone #:
Location: Off the field; crew cars available
Open: Daily: Breakfast, Lunch, and Dinner

PIREP   Mid-Way Regional Airport is just south of the DFW area. The airport staff will offer a courtesy car that you can take to many fine restaurants in historic Waxahachie or quaint Midlothian.

PIREP   Try the Burger Stand. Biggest hamburger is called the Monster Burger. You need to do a weight and balance check before departing. Hours of operation: 11 AM–6 PM. Several courtesy cars available. Of all the place I fly, this has the friendliest service and personnel, hands down.

Midland, TX (Midland International—MAF)

Fabela's Restaurant 🍔🍔🍔

Rest. Phone #: (432) 563-2882
Location: On the field
Open: Daily: Breakfast, Lunch, and Dinner

PIREP   FBO Comments: Avion Flight Centre (under Chevron sign), unicom 122.95, (915) 563-2033, (800) 75-WE FLY.

PIREP   Fabela's is in the Avion Flight Centre FBO and is open from 7 AM to 2 PM. They serve breakfast burritos, hamburgers, and Mexican plates. They will also provide catering.

There is also a food court in the airport terminal (an easy walk). They have pizza, hamburger, and bakery items. But you can't actually get into the food court without an airline ticket, so you have to use a phone from the nearby lounge to place an order.

You can request a shuttle ride from the FBO to the American Airpower Heritage Museum (formerly known as CAF Museum).

Mineral Wells, TX (Mineral Wells—MWL)
Woodie's 🍔

Rest. Phone #: (940) 325-9817
Location: 3 miles, crew car available
Open: Daily: Lunch and Dinner

PIREP   We waited out the weather at Mineral Wells and decided to use the crew car to have lunch at Woodies. Things have changed. The food is *bad*. I had to land at Coleman (COM) so my copilot could get rid of the so-called "Best Hamburger in Texas." Greasy Wendy's-style burger. Small, old dried-out onions and tasteless tomatoes. Don't waste a dollar there.

Mount Pleasant, TX (Mount Pleasant Regional—OSA)
Mt. Pleasant 🍔 🍔

Rest. Phone #: (903) 572-7860
Location: It's a low-fare cab ride
Open: Daily: Lunch and Dinner

PIREP   The place was typical of most small East Texas towns in that the folks were very warm and genuine. We were treated to a very clean and reliable old Crown Victoria courtesy car and went into town for a bite to eat. The gentleman at the FBO could not have been any more pleasant or inviting, and I am looking forward to my next trip to this quaint little town.

PIREP   Mt. Pleasant, Texas is a very nice, "pleasant" town, east/northeast of Dallas. About 45 minutes to an hour in a 172, 29 nm east of the Sulphur Springs VOR.

The airport has a nice, 3800' strip and a nice FBO that might remind you of the *Petticoat Junction* television series. It has 100LL self-serve.

The people, however, just did not seem quite as friendly in this town as most other Texas towns, but it seems to be a nice place nonetheless. For example, the people at Bodacious would not (or perhaps could not) come pick us up at the airport. But we called Eveready Cab at (903) 572-3623, and for $5, they were prompt and very friendly.

Bodacious BBQ is first-class barbecue. I had a beef/sausage combo plate that was mouthwatering. The home-made cornbread was out of this world. Don't bother with the fried pies for dessert, however. They were more like glazed donuts filled with jam rather than traditional fried pies.

If it weren't for the unfriendly people at Bodacious, I'd give it a four star. But instead, I give it a two.

Nacogdoches, TX (A L Mangham Jr. Regional—OCH)

The Country Kwik Stop

Rest. Phone #: (936) 569-1827
Location: Use the crew car—short drive
Open: Daily: Breakfast, Lunch, and Dinner

PIREP The Country Kwik Stop is located at 3322 Center Hwy (Loop 224 and Hwy 7 East). From the airport make a left, then a right on Loop 224, stay on it until you get to Hwy 7 East. The store will be on your left-hand side. The airport has a crew car, so getting there is no problem!

From the outside it just looks like a filling station, on the inside they have picnic tables for dining. Wonderful service, very friendly help, the best hamburgers around and *homemade* pies daily! Ask for pecan! Very reasonable prices. Cheeseburger (gigantic, homemade, fresh!), fries (huge helping), and drink $5.50! They also have other items on the menu. I forgot to mention they serve breakfast as well, if you want to do a morning fly-out!

New Braunfels, TX (New Braunfels Muni—BAZ)

Aviators Cafe

Rest. Phone #: (830) 606-6666
Location: On the field
Open: Daily: Breakfast and Lunch

PIREP FBO Comments: New Braunfels Aero Service is under the Texaco sign on the east side of the airport.

PIREP The grill at Aviators Cafe is open every day 7 AM to 2 PM. After those hours, they can make you "cold" items, such as sandwiches, desserts, and so on. They have lunch specials on Saturdays and Sundays. Enjoy free pancakes and coffee at the Aviators Cafe from 7 AM–12 PM on the second Saturday of each month. Don't forget to tank up your plane with their discount fuel offer during their Pancake Fly-in. On the fourth Saturday of every month, bring $100 cash and receive a burger, fries, and drink for all plane occupants, and top off your plane (up to 40 gallons). This FBO now has a courtesy car that you can borrow to drive to nearby Clear Springs Restaurant (one mile) or into New Braunfels.

PIREP I think you're getting some poor information about eating establishments that is letting you unfairly misrepresent some of the airports. I have no interest in the airports; I just use *The Hundred Dollar Hamburger* to help me plan outings for my wife and me, mostly here in Texas.

I base at Sport Flyers, where there is no food available, so I'm not promoting anything.

You've marked off New Braunfels as a place for fly-in meals because they've closed their onsite restaurant, but there's a really neat restaurant right down the road where the road to the airport meets the highway. Someone called it the Clear Creek, but I think it's actually Clear Springs, and it has a much nicer, more unique atmosphere and better food than most places on the list. A mile or two farther, on the edge of New Braunfels, are several other restaurants, at least two of which are really neat, one German food and one well-known smokehouse. Yet you've marked off New Braunfels, while you maintain other airports that require much more travel from the airport to get to the restaurant.

Don't take this as criticism, because I know you can only relate the information you get.

We're on the way to check out the restaurant at La Grange. They were happy to learn that they're in *The Hundred Dollar Hamburger*.

Orange, TX (Orange County—ORG)

JB's Pit BB-Q

Rest. Phone #: (409) 883-6964
Location: Short drive, crew car available
Open: Daily: Lunch and Dinner

PIREP   JB has a whole pile of trophies from the Houston Livestock Show and the Terlingua barbecue cook-off. J.B. Arrington makes some of the best barbecued beef in the South. Also, on Friday he has *awesome* rotisserie smoked chicken. Well worth the trip! ORG.13.8 DME 049 Radial BPT VOR.

Chuck Baldwin's FBO will gladly lend you a car, Suburban, station wagon, van, motorcycle, or skateboard. The crew cars range from an old Cutlass wagon with no A/C to a fire engine red Caddy to a BMW 735i to a supercharged Suburban. If they are out of cars, Chuck or Chipper Baldwin can get you a rental from the local dealership sometimes at no charge. Contact Baldwin Aviation: (409) 883-6964.

Palacios, TX (Palacios Muni—PSX)

Outrigger Restaurant 🍔 🍔 🍔

Rest. Phone #: (361) 972-1479
Location: Call for pickup
Open:

PIREP   Very friendly people. The food is great and the atmosphere really neat. I'll definitely be going back soon.

PIREP   FBO Comments: Self-serve with credit card. Pilot lounge with phone and restrooms.

PIREP   We landed and called the Outrigger. They were there in ten minutes. The food was great, the service was excellent. We had a relaxing meal and enjoyed the atmosphere. The girl that picked us up and took us back even gave us a quick tour of the fishing fleet there at Palacios. Very friendly people. We will go again and bring friends next time.

PIREP   I am the owner of Outrigger Restaurant in Palacios, TX. I just found out about your website and of course I want to tell you we are a great place to come. We hope to be known as well as Petersen's. Mr. and Mrs. Petersen are now regular customers of ours.

We specialize in fresh Palacios shrimp and other fresh Texas seafood. We also open for breakfast *and* we will pick you up from the airport. Please call ahead if possible.

It is at least two miles, maybe three. We can pick up any flyer if they call (361) 972-1479. Also, Palacios has a taxi: (361) 972-2197.

We have outdoor dining and live entertainment on Friday night. Y'all come.

Pampa, TX (Perry Lefors Field—PPA)

Coney Island 🍔 🍔

Rest. Phone #:
Location: Use the crew car
Open:

PIREP   Here's a pirep from someone who has eaten a lot at the Coney Island in downtown Pampa. There for more years than most can remember, the same family fixes great coneys, wonderful stew and chili, and fabulous homemade pies. We used to carry them back to the Metroplex by the bagful and freeze. There's a doctor in Amarillo that does the same and keeps them in his office freezer.

Definitely worth a trip to Pampa!

Paris, TX (Cox Field—PRX)

Applebee's 🍔🍔

Rest. Phone #: (903) 784-1005
Location: 4 to 5 miles—use the crew car
Open: Daily: Lunch and Dinner

PIREP Went to Paris today—Paris, TX, that is. Heard they have a crew car to go into town with, so I said that's good enough for me. I called ahead to confirm the crew car. The FBO said they normally close at 5 PM in the winter, but they would stay an extra half for me to get the car. Excellent!

Got the car and it was a four- to five-mile drive into town. We didn't go too far into town, but we found an Applebee's, and we were happy with it. Went back out to an empty terminal and came back home for a wonderful night flight back to the (bright) Dallas area.

Nice airport. Too bad they don't have a restaurant right on the field, but the crew car gets you to town.

Pecos, TX (Pecos Muni—PEQ)

Swiss Clock Inn 🍔🍔

Rest. Phone #: (432) 447-2215
Location: Short drive in the crew car
Open: Daily: Breakfast, Lunch, and Dinner

PIREP Another good place to go eat is in Pecos, TX. Fly in to the Pecos Municipal Airport, where the service is friendly and they have an airport car you can borrow. Just a short distance away is the Swiss Clock Inn, which has a pretty decent restaurant. A good place to stop in the middle of West Texas!

Port Aransas, TX (Mustang Beach—RAS)

Trout Street 🍔🍔🍔🍔

Rest. Phone #: (361) 749-7800
Location: Take the free trolley
Open: Daily: Lunch and Dinner

PIREP FBO Comments: The FBO is a building that can be accessed for restrooms with the CTAF code and *.

PIREP The fuel is outrageously expensive. T43—T.P. McCambell Airport is 8.7 nm to the west with 60 cents per gallon cheaper fuel.

There is a city trolley that attends the airport approximately every 45 minutes. A good place to get dropped off for some chow is Little Joe's BBQ. Just ask the trolley driver to take you there.

Rating: 2

P.S. I pilot a Lancair 320 and am the owner and operator of Coastal Bend Flying at Mustang Beach Airport (KRAS) in Port Aransas, TX. Come see us—will take you out for good seafood or barbecue!

PIREP All of the restaurants listed are still there, with the exception of QuarterDeck. This is now called Pelican Club, and it is not open for lunch on weekdays. We ate at Trout Street; the shrimp was delicious!

You can get the trolley route map from the Chamber of Commerce or from the Web. It has all of the restaurants annotated, so you will know where to get on/off the trolley. Download the trolley schedule/route map from: www.cityofportaransas.org/transportation.html.

PIREP   Senor Dudes has moved and been renamed Temerario's. They are located on Cotter Street just across from the ferry landing. Good Tex-Mex, or true Mexican food. Try the Shrimp Alhambe.

Trout Street Bar and Grill is located on the waterfront and is the newest and most upscale restaurant in town. If the weather is nice, ask for a table out on the deck. You will enjoy both the food and the sights.

Shells is a small restaurant hidden away on the corner of Avenue G and 11th Street (don't panic, the trolley goes right by). They have very limited seating, and the food is the major attraction. The quality is superb, the prices are really quite reasonable, and you will leave with that wonderful satisfied feeling. The portions are so generous that they offer one-half size helpings of most items. Check the chalkboard for daily specials. Do order the garlic bread as an appetizer.

The free trolley schedule has been revised. The trolley starts at the RV park one mile south of the airport on the highway at 10 AM. You can walk over to the highway (about a block) and catch the first trolley in at about 10:03. After that, the trolley will come by the airport terminal building at 45 minutes past the hour until 4:45. The remainder of the route is unchanged.

Porter, TX (Williams—9X1)
The San Martin 😋😋

Rest. Phone #: (936) 231-1533
Location: 4 miles away
Open: Daily: Breakfast and Lunch

PIREP   The Country Girl Cafe reported before has closed down. A new restaurant has opened up, The San Martin, owned and operated by Miguel Peres. 16945 FM 1314 at Allendale. This is four miles north of Williams on 1314 and about six miles south from Conroe CXO. Great Mexican food, plentiful, and great service. Prices are quite reasonable. Open Sundays also.

FBO Comments: No credit cards, but credit is available to visitors and you will be billed monthly.

Ranger, TX (Ranger Muni—F23)
El Rancho Cafe 😋😋😋

Rest. Phone #: (254) 647-1355
Location: 6 blocks
Open: Daily: Lunch and Dinner

PIREP   FBO Comments: No services. Well-maintained turf runways. Restroom, phone, and some snacks/drinks available.

PIREP   El Rancho Cafe is only five or six blocks from the aircraft parking area. No service or transportation available at the airport. Turf runways are well maintained. Office is unattended but gives you a welcome feeling. Food, service, and prices at the cafe were all good. I will return.

Roanoke, TX (Northwest Regional—52F)
Blue Hangar Cafe 😋😋😋

Rest. Phone #: (817) 490-0066
Location: On the field
Open: Daily: Breakfast and Lunch

PIREP The Blue Hangar is definitely a diamond in the rough. Small, clean, friendly family-run place with a full menu plus a tasty homemade daily lunch special. The prices are very reasonable and the service is down-home friendly. One of the few places where you can taxi to the fuel pump and then tie 'er down and enjoy a nice meal and some good hangar talk before you take off again. Tell Christy or Gene that Nick sent you. Maybe they'll give me a cup of coffee to go with their free cookies!

PIREP I think that the Blue Hangar Cafe is a really nice place. They have a pretty simple menu. The food is good, service friendly, and their prices won't break the bank. They have tables outside with a perfect view of the runway, too.

PIREP FBO Comments: Unfortunately, fuel costs have to go up to keep up with OPEC, but they do their best to keep the price as low as possible.

PIREP This is the best dang restaurant around, especially with Floyd and Norris at the helm—stop in and experience it for yourself!

Robert Lee, TX (Robert Lee—54F)

Country Cafe 🍔🍔

Rest. Phone #: (915) 453-2687
Location: Call for pickup
Open: Daily: Breakfast, Lunch, and Dinner

PIREP I knew of a great place to eat in Robert Lee (between Abilene and San Angelo, airport code 54F) but hadn't ever thought of flying there before. I called and talked to the owners, Lonnie and Martha, and they agreed to shuttle, between the airport and their place, anyone who flies in. Just give them a call ahead of time: (915) 453-2687. There is no phone at the airport so you have to have a cell phone or call ahead and give an ETA. They're trying to get the phone company to install a pay phone out there.

My family went there yesterday. It's two miles from the airport, so it didn't take long to get there and back (the town is only about three miles wide).

About the restaurant: I would rate the chicken-fried steak a four-and-a-half and the hamburgers a five. But you better come hungry—the helpings are huge. The hamburgers are on what must be a six-inch bun with a half pound of meat. They have charbroiled steaks too, but I haven't tried them yet. This is a real small town, with real friendly folk there. It was a neat experience! I would highly recommend this one!

Robstown, TX (Nueces County—RBO)

Joe Cotten's 🍔🍔🍔🍔

Rest. Phone #: (512) 767-9973
Location: Call for pickup
Open: Daily: Lunch and Dinner

PIREP Fly 15 nautical miles from Corpus Christi, TX on the 231 radial and you'll find the Nueces County Airport. It has a well-maintained, lighted, 3050', hard surfaced runway. There are a few hangars, an FBO and fuel.

I came to eat not visit! The best barbecue joint in the civilized world, Joe Cotten's, is here. Find the phone, drop a quarter and dial (512) 767-9973. Within minutes, the free transportation wagon will arrive.

Cotten's building is as rustic as his waiters are eager. You'll be shown to a table immediately! There's a large selection of meats that have been slow cooked over mesquite wood. If you expected to be served on a plate, you're out of luck. Here, they do it differently. White butcher paper will be spread before you. Meat and condiments will be slapped on it.

Q sauce is presented but not needed. The meat is choice and well seasoned. I had a chopped beef sandwich, beans, and salad. The best I have ever eaten for a reasonable $5.95, iced tea included. Full meals go for $7.95. Be warned: they are *huge*!

Joe Cotten's has been doing it right in South Texas for more than 40 years. Experience accounts for the quality of his product. Cecil Cotten is the host these days. Pilots have been flying here since his daddy started all those years ago. We wish Cecil a lot of luck keeping the legend alive.

Rockport, TX (Aransas County—RKP)
Sand Dollar Pavilion 🍔🍔🍔🍔

Rest. Phone #: (361) 729-8909
Location: Call for pickup
Open: Daily: Lunch and Dinner

PIREP   It's now called the Sand Dollar Pavilion but looks much the same. This time we thought the boiled shrimp was a bit pricey and not as good as before, but the po' boys and grilled shrimp kebabs were great. They'll still pick you up and run you back to the airport with a call to (361) 729-8909, although the walk, once you get down to the water, is quite pleasant. Avgas was $2.13 self-serve at the airport.

San Angelo, TX (San Angelo Regional/Mathis Field—SJT)
Mathis Field Cafe 🍔🍔🍔

Rest. Phone #: (915) 949-8529
Location: In the terminal
Open: Daily: Breakfast and Lunch

PIREP   FBO Comments: Ranger Aviation, Skyline Aviation, Precision Air.

PIREP   Update. The terminal building has been completely upgraded and the food is very good. The locals eat here and that says it all. Come by and check it out. There are also courtesy cars available at a couple of the FBOs.

San Antonio, TX (Boerne Stage Field—5C1)
The Scenic Loop Cafe 🍔🍔🍔

Rest. Phone #: (210) 687-1818
Location: Use the Lincoln
Open: Daily: Lunch and Dinner

PIREP   My girlfriend and I arrived at 5C1 late afternoon. We were provided with a courtesy car, which is a 1997 Lincoln. It had cold A/C and was a nice ride. We were given directions to the Scenic Loop Cafe by a friendly lady standing outside the FBO. We jumped in the car and headed out for dinner. The cafe is about three miles from the airport and is a top-notch place to eat. Very nice atmosphere. They even have live music on the weekends. The cafe had great food and they give you more than you can eat!!! I had the chicken-fried steak, baked potato, and salad. It is a bit pricey, but we both thought, given the portions on our plate we couldn't finish, that the price was well worth it.

5C1 has a 3900' runway. The markings are all faded and the taxiways could use some care but overall it is easy to get in and out of. I do recommend the cafe to anyone looking for a secluded dinner in a friendly place.

San Antonio, TX (Stinson Muni—SSF)

Tower Cafe 🍔🍔🍔

Rest. Phone #: (210) 826-7175
Location: On the field
Open: Daily: Breakfast and Lunch

PIREP FBO Comments: I had a very nice surprise upon landing at Stinson. The very friendly controller actually welcomed me to Stinson field! I had made a brief comment about having not been to the field before and they immediately made me feel at home. The people at San Antonio Aviation were also very friendly and helpful. They are the kind of people you want to do business with. They boasted of having the cheapest avgas in the San Antonio area. My top off was only $46 for 22.2 gallons. That makes them very reasonable compared to my home field.

PIREP The cafe under the control tower at Stinson is great. Just park your plane and walk in. The family that owns and runs the cafe is friendly and the food is great. I had the Mexican plate with lots of homemade salsa and flour tortillas. Yummy! Very reasonable prices too. They have daily specials that change every day so things don't get boring. I saw lots of folks driving in and eating lunch besides the fly-in customers. That always speaks well of the food. Try this place! You won't regret it.

San Antonio, TX (San Antonio Intl—SAT)

Chuy's 🍔🍔🍔

Rest. Phone #: (210) 545-0300
Location: Short drive
Open: Daily: Lunch and Dinner

PIREP I have traveled to SAT many times. Reason for this pirep is rental cars.

We park at Piedmont and the pickup/drop-off is usually quick. This time, I tried to save a few dollars by switching to Advantage. After waiting 30 minutes for the shuttle bus, our reserved full size became a two-door coup—I could upgrade for $5 more. I have two kids with car seats, so the doors are a big issue. There was a $2.50 charge that was unidentified on the bill. I also asked to add my wife as a driver. The representative did not mention it at the time, but that is a $5 charge. Somehow my $22.50-a-day car cost $40. Just wanted to warn others about this place. It sounds the cheapest on the Internet but what a rip-off.

P.S. Once we got the rental car and got to Chuy's, the enchiladas were outstanding. The margarita helped a lot.

Bill Tasso's Barndoor 🍔🍔🍔🍔

Rest. Phone #: (210) 824-0116
Location: Rental car required
Open: Daily: Lunch and Dinner

PIREP This site is not on the airport, but it's close enough, and the filet mignon is unquestionably worth the effort. I have traveled the world and so far, in 30 years of flying, have not found another filet mignon quite up to the size and quality *at any price* as you will find at Bill Tasso's Barn Door. I've made this recommendation to many corporate drivers and so far they have all come back to say thanks for the recommendation.

Exit San Antonio Intl via the main entrance (going south), cross over IH 410 and turn left (go east) onto the feeder, then on to IH 410 and continue east for approximately 1 mile. Exit to the right at Broadway. At the stoplight for the intersection of the feeder road and Broadway, you will see a street that makes an approximate 30-degree angle to the right (southeast). Take that street for about a third of a mile until you come to Nacogdoches Road. The Barn Door is on that corner on the right-hand side.

You simply can't beat the Ribbon Filet, but if you are *really* hungry, they have a 72-ounce steak, which, if you can eat it in an hour, is free!

Oh yeah—dress code can be anything from formal to boots and jeans (typical San Antonio). The prices are most reasonable.

San Marcos, TX (San Marcos Muni—HYI)
Phantom Cafe 😋😋😋

Rest. Phone #: (512) 396-2473
Location: Short walk
Open: Daily: Breakfast and Lunch

PIREP The cafe is a very short walk from the terminal building and a great place to have breakfast or lunch. Jeff, the new owner, and the rest of the staff are very friendly. We had the hamburgers, which were very good and reasonably priced. The cafe is in a small yellow building that we were told was once used for packing parachutes when the airport was the old Gary field. I believe the hours are 7:30 AM–2 PM Monday through Friday and 9 to 2 on Saturday.

Sherman, TX (Sherman Muni—SWI)
The Paper Plate 😋😋😋

Rest. Phone #: (903) 465-6855
Location: On the field
Open: Daily: Lunch and Dinner

PIREP The restaurant reopened at Grayson County Airport. It's called The Paper Plate and it's good. Must try the chicken salad sandwich.

Snyder, TX (Winston Field—SNK)
China Dragon 😋😋

Rest. Phone #: (325) 573-8999
Location: Short drive
Open: Daily: Lunch and Dinner

PIREP The Wagon Wheel is no longer in business. We were told that The Shack is the best place in town, but it is closed on Sunday. Went to the China Dragon for outstanding Chinese. Great FBO, and fuel price is low!

Sonora, TX (Sonora Muni—SOA)
Sutton County Steakhouse 😋😋😋😋

Rest. Phone #: (325) 387-3833
Location: Next to the airport
Open: 6 AM–~3 PM, re-opens ~5 PM–10 PM

PIREP FBO Comments: Full service from a friendly airport manager.

PIREP Flew into Sonora on Friday evening as part of a fly-out from Hicks Airfield in Ft. Worth to the Caverns of Sonora. The airport is well maintained, and the manager is a nice gentlemen that lives on the property.

The Devil's River Inn (a short walk across the runway) is not fancy but clean and comfortable. The food was great at the adjacent Sutton County Steakhouse. Nice place to stop.

PIREP   My wife and I flew a night flight to Sonora (E29) tonight and checked on the Sutton County Steakhouse.

The airport was completely deserted when we arrived, yet it was extremely well lit with PAPI, runway/taxiway lighting, and strobes at either end of the runway. Once we taxied to parking and shut down (we were the only people there), a very kind gentleman (who we later discovered is the airport manager) came out of his house (located on the field, adjacent to the FBO's single-room building) and struck up a conversation and checked to see if we needed anything. This was pleasantly surprising, considering it was approximately 8 PM (FBO closes at 7 PM). Gotta love this Texas hospitality.

After getting directions from the airport manager and instructions to "just walk across the runway," we walked to the south end of the field where the restaurant is located in a hotel, adjacent to I-10. The restaurant is definitely a four-burger, considering the quality and quantity of food versus the price. Worth the trip. My steak with onions, smothered in brown gravy (along with fries and salad), was almost more than I could eat.

Restaurant is open every day from 6 AM to approximately 3 PM, then reopens around 5 PM until 10 PM. Only the airport fence (which has several gates) separates the runway from the hotel and restaurant.

PIREP   When I purchased my airplane (1947 Cessna 140) in Bakersfield, CA, I had to fly it back to my home base (at the time) of Houston, TX. On my second night, I landed at Sonora, TX (E29). A friend of mine had recommended I check out the little steakhouse located practically on the field: Devil's River Inn (Sutton County Steakhouse). Sure enough, the steaks are terrific! What's more, I doubt I paid more than ten bucks for it. I made a point of keeping their business card: Devil's River Inn/Sutton County Steakhouse (darn close to the airport (E29)), (915) 387-3516; I-10 (Exit 400) at Golf Course Road, Sonora, TX 76950.

Spicewood, TX (Spicewood—88R)
Barton Creek Golf Resort 🍔🍔🍔🍔

Rest. Phone #: (830) 693-3528
Location: Close—you can walk, but call for van
Open: Daily: Breakfast, Lunch, and Dinner

PIREP   Good food, excellent view overlooking the upper end of Lake Travis and the golf course's #1 and #9 fairways. Call resort for a courtesy van: (830) 693-3528. It's about a two-minute drive.

Comments:

1. Runway is not as bad as the reports say.
2. There is no phone on ground; bring a cellular to call resort.
3. Have fuel in your tanks; there's none on the ground at this airport

Stephenville, TX (Clark Field Muni—SEP)
The Hard Eight 🍔🍔🍔🍔🍔

Rest. Phone #: (254) 968-5552
Location: On the field
Open: Daily: Lunch and Dinner

PIREP   FBO Comments: Airport is uncrowded, easy in and out, with friendly and helpful FBO staff, and a "bunch of good ole boys sittin' round havin' coffee" atmosphere. FBO closed on Sunday. The restaurant leaves a golf cart next to the FBO office for the easy quarter-mile trip to the restaurant. If the golf cart is in use, or stolen as it occasionally is, one of the good ole boys at the FBO will gladly drive you there, or it's an easy 15-minute walk up the road.

PIREP   The barbecue here is cooked on outdoor pits right outside the main entrance. Big fat pork chops, ribs, sausage, and beef brisket. You just walk up to the pit on your way in, select your meat, then carry the meat inside on a tray where you select your side orders (as if side orders are necessary with good barbecue) and pay for your meal. I had one of those big pork chops and was licking my fingers after the last morsel was gone. Ranch-style beans and jalapeno peppers are plentiful and free—this *is* Texas, after all! In the summer, the owner also has crawfish boils with outdoor seating under the big patio. This is a great place to stop for the best in traditional Texas barbecue.

PIREP   An airport, barbecue, and Dublin Dr. Pepper...what else could you ask for? Flying into KSEP is always easy, and the free parking has never been a problem. Once at the airport, you can give The Hard Eight a call ((254) 968-5552), and they'll come pick you up. Or, if it's available, the airport guys will lend you the golf cart for the short drive to some of the best barbecue around. In addition to thick pork chops, moist brisket, and great sweet tea, The Hard 8 also serves up real Dublin Dr. Pepper! Saturday nights, they often host live music and a crawfish boil. If you go in midspring, you'll be treated to a field of bluebonnets just off their back patio.

PIREP   Flew to Clark Field Today. It was Sunday and the FBO was closed, but there was a golf cart with instructions from the Hard Eight barbecue. We loaded up our Igloo cooler (to keep the barbecue warm) and headed out to the restaurant. We filled our Igloo with some of everything from the pit. They put in very generous portions of all the fixin's, and even threw in a whole loaf of bread. They ran around getting our potato salad and coleslaw and beans and treated us like we owned the place. When we loaded our cooler and went back to the airport, the owner and his lovely wife were taxiing up in their Bonanza. They got out and greeted us and were friendly as all git out. By the time we got back to Waxahachie, the meat was still hot and steamy. Sure was good. I think we'll be going back real soon.

PIREP   Just to piggyback Leo's note about fine barbecue in Stephenville: this is a wonderful place owned and operated by a fellow pilot. He has two new golf carts for the sole purpose of shuttling people to and from the airport. But plan on walking. It's a relaxing stroll past a piece of Americana in the form of the old Interstate motel, with its distinctive roofline and art deco styling.

The guys who run the airport are finalists for America's Friendliest—make sure you pop in and say hi.

There is only one way out of the airport, so don't fret about getting lost. At the highway, the barbecue place is several hundred yards to the left, and the Tex-Mex place (great breakfast tacos) is to the right. The latter is a filthy-looking building, but a steady stream of police cruisers coming and going reveal it's a local favorite.

On the other hand, the barbecue place is brand new. Be ready for the drill when you arrive: approaching the front door, you'll pause outside while an attendant opens a steel-doored pit from which you select your meat. That goes onto a piece of waxed paper on a tray, and then you continue indoors to retrieve silverware, drinks, chips, and so on. Near the drink machine is a pot of beans you dip out yourselves. You can dine indoors or outside on a large patio.

The owner is a big, energetic Texan who loves to rub elbows with his customers, particularly if they reek of 100LL. He commutes daily in his own A-36 Bonanza. Why on earth he picked Stephenville, TX to serve up world-class barbecue, I can't understand. But, it's there, and you're nuts if you don't give it a try.

PIREP   Stephenville has a new barbecue place, The Hard Eight. Brand new. This place is much like Cooper's in Llano except it is much larger. Great barbecue. A short walk from the airport. Highly recommended.

Taylor, TX (Taylor Muni—T74)
Louie Mueller's BBQ 🍔🍔🍔

Rest. Phone #: (512) 352-6206
Location: Call for crew car
Open: Daily: Lunch and Dinner

PIREP Texas *Monthly* was correct: Louie Mueller's barbecue in Taylor, TX is one of the three best in the state.

Awesome food. Get there early, like before 12 on Saturday, or the food may be sold out and the place will close. If you miss out, just walk a block to Mikeska's and pig out. Excellent as well, but not prone to sell out like Mueller's.

I borrowed a truck from a very nice fellow at Centex Flying Service for the two-mile drive into Taylor. Picked up a couple quarts of Aeroshell 100 when I got back. Nice folks and bulletproof barbecue.

PIREP The city of Taylor operates the airport. Self-serve fuel prices vary, according to the price they pay for each tanker. Latest price is $2.05, and the credit card terminal at the pumps is frequently out of service. Don't count on being able to get fuel. Although the city has used Centex's old office space as a pilot's lounge, they may not have things like oil for sale. They do have a DTN weather computer and a courtesy car to go to Mueller's barbecue. I've been to both Rudy Mikeska's and Mueller's barbecue locations. Mikeska's has been poor both times I went. On the other hand, Mueller's has been excellent. The price is about the same for either, so make your own choice.

Taylor is renovating the airport and the low bidder doing the work is far behind schedule. If and when they finish Phase One by completing the runway length taxiway, they'll start on the main runway. As I understand, flight ops can be done on the taxiway until the main runway is finished…and that could be a long while. There would be no night ops on the taxiway.

Teague, TX (Teague Muni—68F)
Hyden's Family Barbeque 😋😋😋

Rest. Phone #: (254) 739-3102
Location: Call for pickup
Open: Daily: Lunch and Dinner

PIREP Ate at Hyden's—their barbecue is awesome. It's a good idea to let them know about a day before you leave so they can be expecting to pick you up. If you have time, they have a real neat little Railroad Museum. The Chevy dealer in town used to rent cars and would deliver to the airport. The dealership is just a few miles from the airport so you can probably call when you land. The name of the dealership is Lonestar Chevrolet. Don't know the number, though.

PIREP FBO Comments: Our modernization and revitalization of the Teague Municipal Airport is complete. New 24-hour self-service fuel system, new asphalt over entire airport, new 80' × 140' turnarounds on each end of runway 15/33, new parking lot, new tie-downs, new entrance road, and new paving around hangars. Our new pilot lounge is also completed (look for key on porch ledge across from door).

PIREP Courtesy phone available to contact family members or Hyden's Family Barbeque ((254) 739-3102) for a ride to some of the best Texas brisket, ham, turkey, sausage, and ribs around. And best of all, the 100LL avgas is cheap!

Tyler, TX (Tyler Pounds Regional—TYR)
Edom Bakery & Grill 😋😋😋😋

Rest. Phone #: (903) 595-0504
Location: On the field
Open: Daily: Breakfast and Lunch

PIREP FBO Comments: Always excellent service unless the service man is at the main terminal servicing the American Eagle Service.

PIREP The Commercial Service Terminal Cafe is now closed and is not scheduled to reopen with a new vendor and the old (previous location/Tower terminal building) is also closed. So, the closest location is as described in the previous posting. There are more that require more driving, but access is easy via four-lane roads. Sweet Sue's has a great buffet-style breakfast from 6 to 11 AM every day followed by continued buffet-style/or menu service until 8 PM, except Sundays when they close at 3 PM. Take Highway 64 east to the loop and turn right (south) and proceed on the loop. Sue's is located on the left just after passing the Dodge dealership at the lighted intersection. Very homey and relaxed as well as are the owners and staff.

Victoria, TX (Victoria Regional—VCT)

Leo's Feed Lot 🍔🍔🍔

Rest. Phone #: (361) 575-3031
Location: Just across the road, easy walk
Open: Daily: Lunch and Dinner

PIREP The Feed Lot was about 200 yards across a grass field from the terminal, maybe 300 if you walk along the road. Service was quick and friendly. The establishment was clean. And the prices were low!

PIREP We are based in San Marcos, TX and were looking for a new fly-in restaurant and decided to try the Feed Lot. It is an easy walk from the airport and the attendant at the FBO let us borrow the courtesy car. He was afraid we might encounter a skunk while walking across the field to the restaurant.

Food was excellent. Jane got the petite cut of prime rib, which was large enough for both of us to eat. I got the oysters. Both were excellent choices and the food was superb. Service great. The restaurant is a gathering spot for locals, nothing fancy, but great food, service, and very reasonable prices. Most entrees are priced below $12 and that includes a choice of sides and soup and salad bar.

One side note, watch the weather carefully. Typically on the flight from San Marcos to Victoria the weather conditions worsen. I also know of a number of VFR pilots that have become stranded at VCT because of decaying weather conditions that reduced visibility to IFR.

Hours of operation: Monday–Wednesday 11 AM–9 PM, Thursday 11 AM–10 PM, Friday 11 AM–11 PM, Saturday 5–11 PM, closed Sunday.

236 Foster Field Rd., Victoria.

Waco, TX (Waco Regional—ACT)

Aerodrome Cafe 🍔🍔🍔

Rest. Phone #: (254) 754-4999
Location: In the terminal
Open: Daily: Breakfast and Lunch

PIREP I ate there while waiting out thunderstorms. It's more of a snack bar in the terminal than anything else—burgers and sandwiches. The BLT was pretty good, though.

There's also a NWS station and a control tower that is open to visits to pass the time if you're weathered in like we were.

PIREP If the terminal greaseburger just won't do (they're not that bad), borrow the car from Texas Aero (say hi to Jennifer for me) and drive into downtown Waco to Buzzard Billie's. Very good Cajun food. I particularly like the crawfish etouffee. The only problem there is that is seems criminal not to have a beer with that spicy food.

Weatherford, TX (Horseshoe Bend—F78)

Horseshoe Bend Restaurant 🍔🍔🍔

Rest. Phone #: (817) 594-6454
Location: On the field
Open: Daily: Breakfast, Lunch, and Dinner

PIREP Horseshoe Bend Country Club Estates is mostly a weekend retreat and retirement community of mobile homes located 12 nm SW of Weatherford on the Brazos River. The FAA designator is F78. Lat/Long: 32-22.48 N/ 097-52-21.122 W, elevation 715, CTAF 122.9.

I own a lot with a mobile home near the north end of the airstrip. The strip is mostly gravel with some grass. The club is about a quarter of a mile west of the airstrip. The airport is unattended; however, usually there is someone around the north end of the strip that will give visitors a ride to the club. The club also has a 9-hole golf course that is right off the north end of the airstrip.

There is a large pecan tree on final approach to runway 17 as well as a TV antenna. On the south end is a rather large hill, but the runway is approximately 3500' long, so no problem. The clubhouse restaurant opens at 7 AM and closes around dark. The menu includes the normal things that golfers like—breakfast food, burgers, fries, club sandwiches, and so on.

Horseshoe Bend is sometimes confused with Horseshoe Bay, which is located near Marble Falls, TX in the hill country. The name is the only thing they have in common.

Weatherford, TX (Parker County—WEA)

Driver's Diner 🍔🍔🍔

Rest. Phone #: (817) 594-8853
Location: Adjacent to the strip
Open: 24/7

PIREP We went to Driver's Diner on Saturday. The lunch buffet was all-you-can-eat chicken-fried steak, chicken, and barbeque (all homemade) for $8.95. This is not your ordinary truck stop. I was pleased (and stuffed). They had some really great-looking pie—too bad I didn't have any more room to put it. The people were friendly, too.

The runway is narrow and not very level, so those that fly out of the nicer airports will get to practice some landing skills.

PIREP One of my favorite fly-in restaurants is Driver's Diner, which is part of Driver's Truck Stop near Weatherford, TX. Driver's Diner is open 24 hours a day, 7 days a week and is located adjacent to the Parker County Airport (WEA), which is paved and lighted for around-the-clock operation. Driver's Diner offers the usual truck stop atmosphere and fare. Sunday mornings can get interesting with the number of different aircraft that show up for breakfast.

Wichita Falls, TX (Kickapoo Downtown Airpark—T47)

The Branding Iron 🍔🍔

Rest. Phone #: (940) 723-0338
Location: Crew car
Open: Daily: Lunch and Dinner

PIREP   FBO Comments: Very friendly FBO, world class service, very clean and comfortable facilities. Tossed us the keys to the courtesy car without even checking license or getting our names. After we saw it we understood why, but still it ran, and the service was great.

PIREP   If you are ever at Kickapoo Municipal Airport, Wichita Falls, a good barbecue restaurant option is The Branding Iron. It had been over 15 years since I had traveled on business through Wichita Falls, and I had eaten there once before. It hasn't changed. Don't let the looks fool you. It is good ole Texas style barbecue, pit cooked, served with all the classic sides. It is about five miles from the airport. The FBO staff will kindly loan the courtesy car and give directions. Afterward, when we arrived back at the airport, the wife napped in air-conditioned comfort in the private FBO lounge as I was having our C-150 topped off for our return to Denton.

Woodville, TX (Tyler County—09R)
The Pickett House 🍔🍔🍔🍔

Rest. Phone #: (409) 283-3371
Location: Call for pickup
Open: Daily: Lunch and Dinner

PIREP   The billing, "Country cooking served boarding house style," is earned by the Pickett House of Woodville, TX. To get there, fly to the well-maintained but unattended Tyler County Airport. The Pickett people will come and pick you up for the price of a phone call. Seems easy enough. It isn't! In Woodville you dial the number, wait for your party to answer, *then* drop the quarter!

The Pickett House is about a mile away and is more than one building. It is the centerpiece of the Heritage Village Museum. The restaurant itself is yellow clapboard trimmed with white and set off by a rust red, steep, tin roof and a standard 1920s "sit a while" porch. Inside it is vintage Grandma's house. The walls are covered by the Bubba Voss Circus Poster Collection. Some date to before the turn of the century.

Lunch is served family style even if you're traveling alone. Fried chicken, bowls of vegetables, and warm freshly baked bread. The table trophy goes to the chicken and dumplings and the blueberry cobbler. *First rate!* Either is worth the trip. The price is wallet pleasing.

Yoakum, TX (Yoakum Muni—T85)
Leather Capital Steakhouse 🍔🍔🍔

Rest. Phone #: (361) 293-9339
Location: Call for pickup
Open: Daily: Lunch and Dinner

PIREP   We flew down Sunday and had lunch. We called the phone number that was provided and were picked up in 5 minutes and given a ride back when we were done. The food was very good. I had a large chicken-fried sirloin steak. My wife had a honey mustard grilled chicken sandwich. They were both very good. The restaurant is in a very old building in downtown Yoakum, TX. We enjoyed going.

PIREP   Reasonably priced—fantastic food. Items range from soup and salad, charbroiled half pound (and full pound) burgers, steaks, chicken-fried sirloin, chicken (fried or charbroiled), fish (fried or charbroiled), fried hand-breaded shrimp, salads, and mouthwatering desserts. Beer and wine available. Everything is fresh, made to order in our kitchen.

Located in downtown Yoakum (the leather capital of the world, with 15 leather companies making everything from belts and wallets to saddles). We're across the street from Double D Ranchware Outlet (a women's clothing designer/manufacturer—clothes sold in Nordstrom across the U.S.) and down the street from Tandy Brands Leather Outlet. Yoakum is located 90 miles southwest of Houston, 90 miles southeast of San Antonio, and 40 miles from Victoria.

Transportation from airport is available—call (361) 293-9339 for a ride.

Bryce Canyon, UT (Bryce Canyon—BCE)
Ruby's Inn 😋😋😋

Rest. Phone #: (435) 834-5341
Location: 1 mile, walk or call for free pickup
Open: Daily: Breakfast, Lunch, and Dinner

PIREP We went to Bryce Canyon (BCE), UT for breakfast. The scenery was incredible and the food at Ruby's Inn was excellent and not too expensive. There is a free shuttle that picks you up at the airport and takes you down the road one mile to Ruby's Inn. The buffet, which included cooked to order omelets and smoked ham (and bacon, sausage, hash browns, waffles, biscuits and gravy, juice, and so on), was only $7 per person.

Just watch your density altitude, as the field elevation is 7586'.

Cedar City, UT (Cedar City Regional—CDC)
Shoney's 😋😋

Rest. Phone #: (435) 586-8012
Location: Crew car
Open: Daily: Breakfast, Lunch, and Dinner

PIREP In Cedar City, the FBO has a courtesy car available for the short drive into town. Numerous restaurants, with Shoney's serving up a great breakfast. Just east and southeast are Bryce Canyon and Zion National Park.

Gorgeous scenery!

Green River, UT (Green River Muni—U34)
Ray's Tavern 😋😋

Rest. Phone #: (435) 564-3511
Location: Car required
Open: Daily: Lunch and Dinner

PIREP I highly recommend flying to Green River, UT, getting a car from Redtail Aviation (the FBO), and going into town and having a cheeseburger with fantastic homemade fries at Ray's Tavern.

They make a great meal, from PBJs to steaks! The wonderful thing is they make their own fries, not some Simplot-manufactured potato product, but real homemade ones. With a burger they give you more fries than you really need. But that is how they serve a meal—lots of it and it has great flavor.

They now have a garden area to the side. So when the sun starts to drop, you can sit out and relax with a good meal.

I recommend you get a couple friends with planes and make a weekend of sightseeing in the Canyonlands.

Heber, UT (Heber City Muni—Russ McDonald Field—36U)
Kneader's Sandwich Shop 🥪🥪

Rest. Phone #: (801) 812-2200
Location: In town a few miles away, crew car
Open: Daily: Lunch

PIREP They usually have a courtesy car or a rental car, and the city itself practically borders on the airport. But it is just a bit too far to walk because the road does not go direct.

There are a number of restaurants in town, but my favorites are Kneader's Sandwich Shop or, if you want something more, there is Don Pedro's Mexican. They have excellent Mexican dishes, moderately priced and large portions. Very good.

Kanab, UT (Kanab Muni—KNB)
Houston's Trails End Cafe 🥪🥪🥪

Rest. Phone #: (435) 644-2488
Location: 1 mile to town
Open: Daily: Lunch and Dinner

PIREP The airport is only about a mile from town. The Houston's Trails End Cafe in downtown Kanab is super. At least four burgers. Kanab is a very nice municipal airport. Most of the old western movies were shot in this area. You are about in the middle of a triangle of the north rim of the Grand Canyon, Bryce Canyon, and Zion National Park. Grand Canyon, Bryce, and Zion all have airports close. Last time I stopped in Kanab they loaned me a car to go to town. Very friendly place and spectacularly beautiful.

Moab, UT (Canyonlands Field—CNY)
La Hacienda 🥪

Rest. Phone #: (435) 259-6319
Location: Rent a car—it's a ways!
Open: Daily: Lunch and Dinner

PIREP La Hacienda is a Mexican restaurant with mild chile seasoning for those with tender palates. It's well prepared and one of the better Mexican restaurants outside of New Mexico. Rental cars are available at the airport. While you are there, rent a four-wheel drive and tour Canyonlands! Great place for mountain biking and other outdoors activities also!

Monticello, UT (Monticello—U43)
Lamplighter's 🥪

Rest. Phone #:
Location: It's a drive—rent a four-wheeler
Open: Daily: Lunch and Dinner

PIREP   I think it's only open for dinner and may be open seasonally during the warmer months. Be prepared for your visit by starving yourself before you go. The food is all homemade, even the ice cream, and is probably the best food I have ever eaten. Unfortunately, we had to pass up the dessert that comes with the meal because we were already past full by that time.

Among other restaurants in Monticello is a steakhouse (located on Hwy 191). Sorry, I forgot its name, but the food there is also great!

I think you can rent a car or four-wheeler from Mike Young. The airport is currently without an onsite manager, but fuel is available. A phone number is posted in a building near the fuel.

Ogden, UT (Ogden-Hinckley—OGD)
The Auger Inn 🍔🍔🍔🍔

Rest. Phone #: (801) 334-9790
Location: In the terminal
Open: 7 AM to 3 PM, 7 days

PIREP   I fly often throughout the mountain west, and the Auger Inn is easily the best place to get a bite to eat near an airport. The service is great, the selection is great, and the locals will help you with anything.

PIREP   The food's good at the Auger Inn and the area is wonderful!

Three ski resorts within 30 minutes of airport: Dinosaur Park about eight miles; Hill AFB Aerospace Museum just three miles; Union Station Museum (trains, Browning Arms, automobiles) about four miles.

Price, UT (Carbon County—PUC)
Grogg's 🍔🍔

Rest. Phone #: (435) 613-1475
Location: 4 miles, crew car available
Open: Daily: Lunch and Dinner

PIREP   Fly in to Price, UT (Carbon County PUC). They usually have a vehicle you can use.

Go into town and head north taking the old highway between Price and Helper. About four miles from Price (or a little less) on your left will be Grogg's.

A very unique little restaurant specializing in sandwiches, but many other dishes are on the menu and each is very, very good. You will be glad you made the trip.

Wendover, UT (Wendover—ENV)
Peppermill Casino 🍔🍔🍔🍔

Rest. Phone #: (775) 664-2255
Location: Free shuttle
Open: 24/7

PIREP   Flew in early evening and had the FBO call the shuttle to the Peppermill. Looks like numbers for the Peppermill and others were also posted outside for when the FBO is closed. Food was buffet style all you can eat. Make sure you are proficient with your instrument skills at night, though—not many lights out there to create a horizon.

PIREP   This airport is right on the border of Utah and Nevada. Since Utah does not allow gambling and Nevada does, it attracts a lot of Utahans on the weekends. The casinos offer nice eating and transportation to and from the airport. When I think about the $100 hamburger, this is where we go the most.

Restaurants: Peppermill Casino, 3/4 miles (702) 664-2255; Rainbow Casino, 1 mile (702) 664-4000; Red Garter Casino (702) 664-2111; Silversmith Casino (702) 664-2221; State Line Hotel & Casino (702) 664-2221. Most, if not all, have some sort of restaurant and/or buffet and will pick you up at the airport. My wife's and my personal favorite is the State Line Casino's buffet. I have also heard that the Peppermill is good. The Silversmith is connected to the State Line, but as far as the others go, I do not have experience with them.

Transportation: Hotel shuttle buses. Courtesy car (I have never seen it). Taxis (702) 664-3333.

Local attractions: Airport was the training ground for the 509th Composite Group (the group that dropped the atomic bombs during WWII), and there is a memorial about one mile away. The Bonneville Salt Flats Speedway is about six miles away. Gambling at the casinos.

Note: All the area codes given are guessed. They would be either 702 or 801.

The Hundred Dollar Hamburger

Vermont

Burlington, VT (Burlington Intl—BTV)

The Flight Deck 🍔🍔🍔

Rest. Phone #: (802) 863-2874
Location: In the terminal
Open: Daily: Breakfast and Lunch

PIREP The FBO at the airport is now AVFBO, which closes at 10 PM. One cannot leave the ramp unless the FBO is open, so arriving late means a call-out at $90–$100.

The Flight Deck, in the main terminal, is still available. It's a quarter-mile walk from the FBO to the terminal.

Middlebury, VT (Middlebury State—6B0)

The Waybury Inn 🍔🍔🍔🍔

Rest. Phone #: (802) 388-4015
Location: 1.5 miles, cab available
Open: Daily: Breakfast, Lunch, and Dinner

PIREP Land at Middlebury, runway 1/19 (2500' runway in excellent condition at the foot of the Green Mountains, which was difficult to see when approaching from the west because of trees blocking the view). The Waybury Inn, a quintessential Vermont inn, is located 1.5 miles from the airport in East Middlebury.

We called Beaver's Cab at (802) 388-7320 or (800) 286-7320 before we discovered how close the inn was to the airport. Walking is probably the best bet. Make a left out of the airport access road, walk about one mile to four corners, make a left onto Rte. 125 and walk about one-half mile to Waybury Inn on your left.

There is an a la carte menu, and the Sunday buffet breakfast was served from 11 AM to 2 PM and cost $13.95. The menu changes from week to week, but be assured of a lavish feast. Some items on the buffet were sun-dried tomato and cheese omelet, French toast filled with blueberry-strawberry cream cheese filling, potatoes, asparagus, bourbon steak, chicken, shrimp, salmon, spinach salad, Caesar salad, roasted garlic, homemade soup, wild rice salad, various pasta salads, homemade rolls and breads, fruit salads; and for dessert, chocolate pecan pie, fruit with whipped cream, homemade cakes and cookies, creme brulee. The service was excellent and the management very happy to accommodate our group of 12.

We happened upon a very quiet weekend because of the timing of graduation from Middlebury College and so reservations were not necessary. However, we were told that reservations are recommended. Call the inn at (802) 388-4015 or (800) 348-1810 or fax at (802) 388-1248. The inn also has 15 rooms for overnight guests. Lunch and dinner are served as well.

Rutland, VT (Rutland State—RUT)

The Whistle Stop 🍔🍔

Rest. Phone #: (802) 747-3422
Location: 1-1/2 miles away
Open: Daily: Breakfast, Lunch

PIREP FBO Comments: Columbia Aviation always takes good care of us when we visit.

PIREP The Whistle Stop is the closest food now. Have eaten a number of times at the place. Maybe a mile to one-and-a-half from the airport. Take a right out of the airport, then a left onto Rte. 103 when the road ends, and the restaurant is an almost immediate right. Inexpensive and always good food. Place is an old train station. Ice cream window outside in the summer. Food, service, and atmosphere are excellent but you probably would not want to walk it, unless you really needed to stretch your legs. I give it three burgers in the ratings.

PIREP FBO Comments: Columbia Air Services. These guys were nice; they let us use the crew car to go to a small restaurant (I believe it was called The Whistle Stop) about a mile from the field and didn't ask for anything in return. I would have bought fuel, but the airport was only 35 minutes from home.

PIREP The restaurant on the field closed about two years ago. The nearest restaurant, The Whistle Stop, was about a mile away and offered lunch and dinner. We had a couple of burgers with chips and a soda. Basic stuff with good service. The burgers were pretty inexpensive.

West Dover, VT (Mount Snow—4V8)

West Dover Inn 🍔🍔🍔🍔

Rest. Phone #: (802) 464-5407
Location: Call for pickup
Open: Daily: Breakfast, Lunch, and Dinner

PIREP My wife and I chose to escape the little one for a couple of days and ended up flying in to Mt. Snow. What a beautiful location! Fall landings must be spectacular. Nominal fee for tie-down and parking but a pool table in the FBO to help pass the time. We stayed at the West Dover Inn and ate there as well as several other places. The inn is down on Rte. 100, maybe a mile or a mile and a half from the airport. Ed Riley, the owner of the place, came and hauled us down. This place is not your usual 100 dollar hamburger. This is a place to go and visit and be pampered. The attention was personal; the rooms, with gas fireplaces, were great, as was the living room with wood fire at night. The food was simply awesome. Filet mignon one night and an amazing seafood pasta the other.

A short but very top-notch and varied menu. They make their own breads. Top wine list, great microbrews, just perfect meals. Very detail oriented. If you want to fly somewhere to get away from it all, West Dover Inn is the place. This is a five-star joint where you will eat so much that I recommend you not go in with full fuel. Ed has a smallish dining room so, if you are not staying there, reservations are recommended.

Skiing is the main diversion if you stay in winter. Golf in the summer. Ed does some golf packages. If they are as top flight as the rest of the service we saw, they would no doubt be excellent.

Tell Ed Jim and Janet the pilots sent you.

As part of our stay, we sampled some of the local places.

Mt. Snow Brewery: Lunch in Wilmington at the Brewery, on one of the two main streets. Very decent pub food and excellent microbrewed beer. No, we did not fly that day, but it was worth being grounded for this fare. It is not walkable, but the food, service, and price would put it three to four burgers. As far as walking goes, if you are willing to walk the mile downhill to Rte. 100 from the airport, the MooVer, the local bus, is free and hourly. We used it and it was great. On time, even.

Dot's Diner: We ate lunch here the day we arrived. Good diner food, one location an arguably long walk from the airport (we walked it from the inn and it was reasonable for us). The MooVer was available as well. Free bus transport is not all bad. So I will give it the minus two for the walk, but it is still good for three burgers for good food, excellent and pleasant service, decent ambiance (locals and tourists), and decent prices.

Many good, from low end to very high end, places to eat in Mt. Snow, West Dover, Wilmington area. In two days we could not begin to tap them all. Skiing, golf, including a course that borders the airport. This place is a relatively undiscovered little secret that we pilots would do well to take advantage of.

The Hundred Dollar Hamburger

Virginia

Charlottesville, VA (Charlottesville-Albemarle—CHO)

Martin's Grill 🍔🍔

Rest. Phone #: (434) 974-9955
Location: Crew car
Open: Daily: Lunch and Dinner

PIREP For those of you flying into CHO, I wanted to let you know about a *great* new place close to the airport: Martin's Grill. I know it's not heart-healthy, but these homemade burgers and fries are worth the price of the gas! The crew at the FBO will lend you a rental car, and it's less than two miles away—definitely worth the trip!

Danville, VA (Danville Regional—DAN)

McDonald's 🍔🍔

Location: 1/4 mile
Open: Daily: Breakfast, Lunch, and Dinner

PIREP McDonald's and an Italian submarine sandwich shop a quarter-mile away (within walking), or take free courtesy car and leave a buck or so inside vehicle. The sub shop very reasonably priced and very nice.

Emporia, VA (Emporia-Greensville Regional—EMV)

Applebee's 🍔🍔

Rest. Phone #: (434) 336-9540
Location: Crew car
Open: Daily: Lunch and Dinner

PIREP FBO Comments: Excellent service by the airport manager, William, and the weekend person, Sam. The airport has two courtesy vehicles. A must-stop traveling southeast-northeast routes. Runway in great shape and 5100' long.

PIREP   Plenty of food places, from fast-food restaurants such as Hardee's to Applebee's. Not a lot of sight-seeing areas but a good place to relax and stretch out. Nice small facility at the airport, too.

Franklin, VA (Franklin Muni-John Beverly Rose—FKN)
Joe's Pizza & Pasta Palace 😋😋😋

Rest. Phone #: (757) 562-5020
Location: Short drive, crew car available
Open: Daily: Lunch and Dinner

PIREP   FBO Comments: Very friendly and accommodating manager at the terminal, courtesy car is still available. Fuel and service available. Nice terminal.

PIREP   Joe's is family owned by real Italians. They offer your typical Italian restaurant fare, but the pizza is highly recommended. It was delicious. It is walking distance from the terminal, if you like long walks!!! (It's about a mile walk, so I suggest taking the courtesy car.)

The town is very quaint, almost Mayberryish. We took the tour looking for the restaurants mentioned on this site, but they were not found. There were a couple of small restaurants, but they were not open, which was unusual for a Saturday afternoon. The town has a drugstore with a fountain. You'll feel like you're back in the '50s! Franklin is a very nice little berg, and you feel like you've gone back in time 50 years. We recommend that you check it out and eat at Joe's!

Leesburg, VA (Leesburg Executive—JYO)
Deli Express 😋😋😋

Rest. Phone #: (703) 669-6360
Location: 200 yards from the terminal
Open: Mon.–Fri. 7 AM to 4:30 PM, Sat. 9 AM–2 PM

PIREP   Deli Express is located 200 yards from the terminal building in the Miller Office Park. They offer excellent hamburgers and gyros, as well as full breakfasts. It's a short walk, nice people, good food, and convenient.

Luray, VA (Luray Caverns—W45)
Luray Caverns Restaurant 😋😋😋

Rest. Phone #: (540) 743-6551
Location: 1 mile, free shuttle
Open: Daily

PIREP   Try Luray Caverns Airport (W45), VA for a pleasant few hours. Land at the airport and take a free ride to Luray Caverns (about a mile). The Caverns are a beautiful one-hour (or so) tour. Also, there is a carriage house and old car museum included in the Caverns admission that takes another half-hour. There is a restaurant at the caverns that is supposed to be good, but we didn't eat there. Call the airport for a ride back, or the Caverns folks will give you a lift.

Martinsville, VA (Blue Ridge—MTV)
Runway Cafe 😋

Rest. Phone #: (276) 957-2233
Location: On the ramp
Open: Daily: Breakfast and Lunch

PIREP There is a $20 ramp fee. I understand that it is waived with a fuel purchase. Call first to check it out, as this is one of the highest ramp fees in the United States! Come to think of it, we don't know of a higher one.

PIREP I flew in with two friends to Martinsville on the recommendation of my flight instructor. I was pretty disappointed. We arrived for Sunday lunch right in the rush of things. This is apparently a place for churchgoers to come after service. The buffet table was missing chicken and meat loaf, there were no knives, and the food I had just wasn't that good. The tables are no longer adorned with tablecloths. On the positive side, I was not charged a ramp fee, and there was quite an audience there watching you land and take off (if you like that). I might try this again on a day when it is not buffet.

PIREP FBO Comments: Nice runway, FBO was clean and attractive. There was no ramp fee.

PIREP We ate at the Runway and found everything to be very good. The menu had items from just a well-priced sandwich to a more elaborate lunch. We had the blue cheese burgers and Cajun fries and smoked turkey sandwiches, all of which were liked by all. One of the turkey sandwiches was wrong but brought back correctly in just a couple of minutes. The prices were very reasonable, and the service was friendly.
    We will go back.

Moneta, VA (Smith Mountain Lake—W91)

The Virginia Dare 🍔

Rest. Phone #: (540) 297-7100
Location: 2 miles
Open: Daily: Dinner

PIREP Smith Mountain Lake is 30 miles SE of Roanoke, VA. The lake has numerous restaurants and water sport opportunities. There is a dinner cruise boat, the Virginia Dare ((540) 297-7100) that docks about two miles by road from the airport ((540) 297-4500). A two-hour cruise to the dam and delicious dinner make for a pleasant evening.

Petersburg, VA (Dinwiddie County—PTB)

Famous King's BBQ 🍔

Rest. Phone #: (804) 732-7333
Location: 1/4 mile
Open: Daily: Lunch and Dinner

PIREP Restaurant within a quarter-mile. Great subs, fries, and Italian menu. Pamplin Park Civil War Exhibit.

Richlands, VA (Tazewell County—6V3)

Cuz's BBQ 🍔🍔

Rest. Phone #: (276) 964-9014
Location: A drive; crew car available
Open: Daily: Lunch and Dinner

PIREP   FBO Comments: Very nice people. Loaned us a car to get to the restaurants. One of the great landing strips in the area, it's located on a flat-top ridge and has an incredible approach to their runway 25.

PIREP   Cuz's BBQ is *not* open for lunch and not open at all on Monday and Tuesday. Some good information before you venture off to this flying stop.

Roanoke, VA (Roanoke Regional/Woodrum Field—ROA)
Texas Tavern 🍔🍔

Rest. Phone #: (540) 342-4825
Location: 10-minute drive, 3 crew cars available
Open: Daily: Lunch and Dinner

PIREP   Depending on the availability of the courtesy car at Piedmont Hawthorne, a short 10-minute drive to downtown Roanoke where there are two excellent no-frills spots. Each offers hot dogs, hamburgers, and chili. Both have been around for over 60 years and are still going strong. Roanoke Wiener Stand (closed Sunday) has the best chili dogs on the East Coast, and the Texas Tavern has great hamburgers and chili and is open 24/7. The costs will certainly fit your pocketbook and leave money for fuel.

PIREP   FBO Comments: Nice FBO (Piedmont Hawthorne), three courtesy cars.

PIREP   The FBO will give you a map to all the local restaurants, all within five minutes. We chose Texas Tavern.

Saluda, VA (Hummel Field—W75)
Pilot House Inn 🍔🍔🍔🍔

Rest. Phone #: (804) 758-2262
Location: On the field
Open: Daily: Breakfast, Lunch, and Dinner

PIREP   FBO Comments: Friendly and helpful, el cheapo fuel!

PIREP   We landed and walked across the field to the Pilot House Inn. They had a really tasty breakfast buffet laid out so we thoroughly stuffed ourselves at a fair price. Eckhard's across the street does not open until 1600, btw.
  The highlight of the trip was a quick ride into historic and picturesque Urbanna, VA, roughly 15 miles away. It is an old tobacco port town on the Rappahannock dating back to the seventeenth century. Lots to see and do there; the 25-cent trolley car will take you all over the small town, and plenty of fine restaurants and neat places to shop. You can call ahead for car rental to Enterprise of Gloucester (you can get the number from the FBO) or call ahead to Saluda Auto Rental at (804) 758-4824. They can have a car at Hummel waiting for you.

PIREP   FBO Comments: Fuel pit located midfield. 24-hour automated credit card self-serve.

PIREP   Flew down to Hummel looking for a place to go. Restaurant and motel on the field. A leftover from the "good ol' days" of aviation when such configurations were common at little airports. Restaurant is a cozy little place. Food is above average, mostly hamburgers, sandwiches, seafood, and diner food. Breakfast buffet in the mornings and seafood buffet on Friday and Saturday nights. You can park your air machine near the restaurant in the grass on the approach end of 01 or in the midfield grass parking area (just don't use any labeled "reserved").
  I thought it was well worth the trip (don't know about Eckhard's as it was closed when we were there) and would definitely go back.
  You can stay at the motel for $45 a night. Restaurant is the meeting place for the local KofC and the Virginia Aeronautical Historical Society.

Eckhard's 🍔🍔🍔

Rest. Phone #: (804) 758-4060
Location: Across the street
Open: Daily: Lunch and Dinner

PIREP Pilot House Inn is still there and doing fine. The German restaurant was sold and is now reopened under the same name, Echart's. It is very nicely decorated inside and the food is excellent; so is the service.

PIREP FBO Comments: Didn't buy gas but got prompt attention after landing from FBO. Very friendly.

PIREP I stopped in today and Eckhard's is back open for business, with five German entrees and eight Italian entrees. The service was superb, and the appetizer I had (lump crab meat in filo dough, with a delicious sauce that I can't quite describe) was absolutely delicious. A little on the pricey side (entrees from $14 to $23), but compared to the cost of 100LL, not bad, especially for what you get.

Highly recommended.

Suffolk, VA (Suffolk Executive—SFQ)
Throttle Back Cafe 🍔🍔🍔

Rest. Phone #: (757) 934-3461
Location: On the field
Open: Daily: Breakfast and Lunch

PIREP Now called Throttle Back Cafe. Located within the executive terminal. Seems like good basic diner fare. We had breakfast—the food (omelet) was fine and the prices were very reasonable. Bottomless help-yourself coffee. We arrived a little after breakfast would have been over, but they fixed breakfast for us anyway. Weekday hours are 7–3.

Tangier, VA (Tangier Island—TGI)
The Channel Marker 🍔🍔🍔🍔

Rest. Phone #: (757) 891-2900
Location: Walkable
Open: Daily: Lunch and Dinner

PIREP FBO Comments: No fuel.

PIREP The newest restaurant in town, the Channel Marker, has the best fly-in crab cakes I've found. Shell-less back fin with virtually no filler. Crab salad is to die for. Hard crabs also available at a very good price. Open till 7 PM in season so a great dinner spot as well.

PIREP Tangier can only be described as one of the best places to visit in a plane. The island sits out in the middle of the Chesapeake Bay surrounded by Pax River restricted airspace. Check to see if it's hot; they bomb things there. The island is small, quaint, and easily walkable from end to end. No roads to speak of, just a paved pathway that goes around the island. Hilda Crockett's Chesapeake House is the family-style restaurant. A real treat. Costs about $12 or $13, and you best be hungry. Fisherman's Corner, a short walk (actually, they all are a short walk) from the runway, has wonderful food and is cheaper. After your meal, take the walk around the island and meet some of the 700 full-time residents who live their lives relatively secluded from the rest of us. Some still speak with the old English accent of their English settled ancestors. Overnighting is fun with a stay at the Sunset Inn on the south end of the island. Be sure to reserve in advance. The island is truly a destination in itself.

Tappahannock, VA (Tappahannock Muni—W79)
Lowry's 🍔🍔🍔

Rest. Phone #: (804) 443-2800
Location: 15-minute walk
Open: Daily: Lunch and Dinner

PIREP FBO Comments: I have stopped in here on occasion for gas and a couple of times have gone to the nearest restaurant. The people are nice, but the facility is very poor and has seemed to really have gone downhill, especially in the past couple of years. It's very dirty and messy. A few weeks back I stopped in and they had even run out of gas, putting me in somewhat of a bind. Grass is starting to grow through the runway. They say they are going to build a new airport but can't seem to say when it will be built. If the same folks run the new airport it probably won't be any better than this one, if they won't keep it up. Bathrooms were okay, but that's about all I can say nice about the facility.

PIREP Lowry's Restaurant is about a 15-minute walk from the airport and I'm told is the closest one to the field. Food is okay but nothing special. They have a pretty varied menu and serve lunch and dinner. I'm not sure if they serve breakfast or not. The airport has no loaner car.

Wakefield, VA (Wakefield Muni—AKQ)
The Virginia Diner 🍔🍔🍔🍔

Rest. Phone #: (888) VADINER
Location: 1-mile walk
Open: Daily: Lunch and Dinner

PIREP Slow service, but the home cooking makes it worth the wait. Great local foods with a huge variety of delicious homemade sides. Save room for dessert! Open for dinner as well.

PIREP The Virginia Diner in Wakefield is about 40 nm SE of Richmond. They are about one mile from the airport (AKQ) and will come and get you and take you back (but you may want to walk back to work off some of the extra load). They feature Virginia ham, barbecue, chicken, and other regional country dishes, and all the peanuts you can eat. The gift shop sells the hams and peanuts and other things. Good place for breakfast/brunch.

Williamsburg, VA (Williamsburg-Jamestown—JGG)
Charly's 🍔🍔🍔🍔🍔

Rest. Phone #: (757) 258-0034
Location: On the field
Open: 11 AM—3 PM

PIREP I just received my private pilot ticket and have been busy flying my long-suffering husband all over the state. He's getting better—he only kisses the ramp once when we land now.

Yesterday was a beautiful VFR day. We decided to combine one of my favorite cross-countries with a trip that my husband made one Sunday with my CFI while I was at work. We flew down the James to Norfolk, and then back up to Williamsburg/Jamestown. We called Charliy's at JGG from Norfolk to make sure they'd still be open when we arrived.

We did barely squeak in on time, and the staff welcomed us like we had all the time in the world. There was a crowd in the little FBO/restaurant, pilots and friends who had flown in with friends and family who had driven to meet them. There is a nice outside seating area in addition to the indoor dining area. The staff had been waiting for our arrival. There was another plane with pilot and passenger that arrived after us. They radioed in after the restaurant was officially closed, and the grill was cleaned, but the cook said to tell them he'd be happy to serve them something from the cold deli menu.

The restaurant area and the FBO are quaint and remind one of a general store or old drug store soda counter. Both are spotless. The sandwiches are great— fresh, delicious, and generous portions. The waitress offered to refill our soft drinks before we had a chance to ask. The couple who came in after us were given the same warm welcome.

I have not been to the restaurant on Sundays, but I understand the prime rib special is "to fly for."

Blue skies to all!

PIREP   FBO Comments: 100LL and JetA are available during FBO hours.

PIREP   Excellent restaurant! My girlfriend and I were welcomed by a very polite waitress, and we chose to sit outside on the patio. I went with the prime rib with bread, seasoned potatoes, and a salad, and my girlfriend went with the French dip sandwich. We both saved a little room for dessert. I had the coconut pie while my girlfriend went with the carrot cake—both were to fly for! The food was excellent! Great ambiance! Terrific service!

The morning was not favorable for flying so I drove there instead from Virginia Beach. By the time I got there the sky was clear and the sun was shining and it wasn't long before you could here four or five voices yelling, "Clear prop," followed by the sweet rumble only 100LL could produce!

I would definitely enforce Charly's five-burger rating! I will definitely return in the near future!

Winchester, VA (Winchester Regional—OKV)

Chason's 🍔

Rest. Phone #: (540) 667-2553
Location: 2 miles
Open: Daily: Lunch and Dinner

PIREP   Take the free crew car to any one of more than 50 places to eat within five miles. Within two miles there is Chason's, an all-you-can-eat buffet with 50 items, including a big dessert bar. Plain, good food, very clean, about $8 for *full* lunch. Also, try the Cork Street Tavern's ribs (no, you don't have to drink)—four miles.

Wise, VA (Lonesome Pine—LNP)

Nellie's Market and Deli 🍔 🍔 🍔

Rest. Phone #: (276) 328-9699
Location: 5-minute walk
Open: Daily: Breakfast, Lunch, and Dinner

PIREP   Good food, reasonable price. Clean little restaurant; also is a Quik-Mart–style store that sells some grocery store items.

Transportation from the FBO is walking, loaner bicycles that the restaurant keeps at the FBO, and courtesy car. Walking takes about five minutes. The special when we went was homemade lasagna and apple cobbler.

PIREP   FBO Comments: The airport has a long, wide runway and no obstructions except a well-marked water tower in the nearby industrial park.

PIREP Nellie's is a new restaurant established to serve businesses near the airport. It is open seven days a week, serving breakfast, lunch, and dinner. It also sells snacks and soft drinks to go. Their specialty is barbecue, and they serve one of the best tasting double cheeseburgers I have ever had. Nellie's is about a 50-yard walk from the terminal building, and the owner is the chief cook. It has a very friendly, down-home atmosphere that you will enjoy as much as the food.

PIREP Nellie's is an easy five-minute walk from the FBO. Great menu with full breakfast, lunch, and great burgers!

The Hundred Dollar Hamburger

Washington

Arlington, WA (Arlington Muni—AWO)

Taildraggers Restaurant and Lounge 😋😋😋😋

Rest. Phone #: (360) 403-8970
Location: On the field
Open: Daily: Breakfast and Lunch

PIREP FBO Comments: Great airport, a little congested sometimes when departing 34 and 29, which bifurcate. MOGAS is very reasonably priced.

PIREP Taildraggers is a great restaurant! Have taken two friends there three or four times and going back with my wife for the fettuccine (excellent!). Great service, informal environment but not old or cheesy. Try the French dip.

Also: nice and clean bathroom.

PIREP Food was awesome. I had the Monte Cristo with fries. Nice-sized helping. Great atmosphere. Customer service outstanding!!!

Auburn, WA (Auburn Muni—S50)

BB McGraw's 😋😋😋

Rest. Phone #: (253) 804-8709
Location: Very close, easy walk
Open: Daily: Lunch and Dinner

PIREP FBO Comments: Friendly folks monitor the transient parking. $5 overnight tie-downs.

PIREP BB McGraw's at the south end of the airport has very good food at fair prices. It's a sports bar with many big-screen TVs showing everything that's on at the same time. Several hotels in the near vicinity.

Bellingham, WA (Bellingham Intl—BLI)

Mykonos 😊😊😊

Rest. Phone #: (360) 715-3071
Location: 10-minute walk
Open: Daily: Lunch and Dinner

PIREP   We flew up there on Saturday to check procedures with U.S. Customs, now available 24/7 at BLI, making it the best entry point from western Canada. We were very impressed by the helpfulness of the fuelers working for the Port of Bellingham. Despite our not needing any fuel, they shuttled us out to a nearby restaurant and picked us up when we were done!

   The restaurant, Mykonos, would only be about a ten-minute walk away and serves Greek food as well as burgers and so on. It's a brand new place with friendly service, and we both enjoyed the lamb souvlaki.

Blaine, WA (Blaine Muni—4W6)

Burger King 😊😊

Rest. Phone #:
Location: Touches the field
Open: Daily: Breakfast, Lunch, and Dinner

PIREP   Blaine is the furthest northwest airport in the lower 48. It sits right at the U.S./Canadian border and, in fact, if you land on runway 14, you will turn base and final over the truck Customs building. They want you to follow that truck route on final right down to the airport, with a little jog to line up with the runway at the last minute, for noise abatement and to keep you away from the houses. On very short final you will notice a Burger King! That's it for eats at this airport, but it's an easy walk.

   This airport is not for the faint of heart—fairly short (2100') and narrow, and large trees to swoop over with a seemingly ever-present downdraft if you have to land to the north.

Bremerton, WA (Bremerton National—PWT)

Airport Diner 😊😊😊

Rest. Phone #: (360) 674-3720
Location: On the field
Open: Daily: Breakfast and Lunch

PIREP   Best fish and chips I have ever had. Hands down! I continuously return for the crispy golden prizes. Careful, Bremerton can get a little crowded on busy days, and the nontowered field can quickly turn into an "uncontrolled" field, if you know what I mean.

PIREP   Friendly, fast service and great fish and chips at the Airport Diner (the best we've had in Washington state) prepared with a deliciously crisp, light batter. Very reasonably priced too—a platter with 12 pieces, enough for three to four adults, was $15. We had two small rug rats along and the waitress was out with the high chairs and some little bags of Goldfish crackers within minutes.

   We enjoyed convenient parking for our plane right in front of the diner and window seats to watch the aerial comings-and-goings throughout lunch. The only downside I can think of is being aware of the TFR's and Class Bravo airspace in the vicinity. A definite four-hamburger destination, as the last few reviewers have already noted!

Burlington / Mount Vernon, WA (Skagit Regional—BVS)

Crosswinds 🍔🍔🍔🍔

Rest. Phone #: (360) 757-1693
Location: On the field
Open: 11 AM–3 PM, 7 days a week

PIREP This restaurant continues to be excellent in all areas since my last report a few years ago. Food, service, value, you can't beat it for an airport dining experience—hamburger, fish, or other popular items not normally found in airport facilities. I want to point out they are now open seven days a week for lunch, from 11 AM to 3 PM. Some days the hours are much longer, but for the flying public the lunch hour is when most of us do our 100 dollar hamburger thing.

PIREP We flew in and dined at Crosswinds, Skagit Regional (BVS) and were very impressed. Starched white linens and the food to match. What was amazing was the prices were very reasonable, given the quality of food, atmosphere, and service. We highly recommend Crosswinds. They are open until 9:30 PM every night.

Cashmere, WA (Cashmere-Dryden—8S2)

Walnut Cafe 🍔🍔🍔🍔

Rest. Phone #: (509) 782-2020
Location: 20-minute walk
Open: Daily: Breakfast, Lunch, and Dinner

PIREP FBO Comments: We did not purchase fuel at the airport, but it appeared to be available.

PIREP The Walnut Cafe is about a 20-minute walk from the airport, but the town is pleasant and easy to maneuver. Bicycles and a van are available at no charge if walking is a problem. Looking at the humble town, the restaurant is very unexpected. Gourmet selections, from toasted pecan goat cheese with balsamic reduction to a spinach oyster salad with grapefruit, are just some of the unexpected selections. Ambiance is nice, service is excellent, and food is incredible. Turn right upon exiting the airport to find downtown.

PIREP FBO Comments: No FBO that I could see. No fuel that I could see. No problem! 1800' runway 07/25, short, but doable, in the heat??? Runway 07 is slightly downhill.

PIREP On asking at the airport, I was told that I could park "just anywhere," then got help pushing my plane back into a parking spot. Downtown, where the restaurants are, is about three-quarters of a mile from the airport, a little longer than I had thought, but good exercise! Go out the east gate of the airport, head north on Sullivan St., east on Pioneer Ave., go around the left bend, by Vale School, onto Division St. Downtown is just across the tracks, mostly on Cottage Ave., where Walnut Cafe is. Don't forget to check out the Aplets & Cotlets outlet, directly behind Walnut Cafe on Mission Ave. The Walnut Cafe is in a 1918 building, has great ambiance. They had lots of variety on their menu. There are several other options, pizza places, and so on. Don't really understand your rating system, so won't attempt it.

PIREP Cashmere bills itself as the friendliest airport in the world and lives up to its billing. A courtesy van and courtesy bicycle are available. There is one transient tie-down behind the pilot's lounge. Cashmere town center is a healthy walk. Leavenworth, the Bavarian theme town, is a quick drive or bus ride away.

Concrete, WA (Concrete Muni—3W5)

Hal's Cafe 🍔🍔🍔

Rest. Phone #: (360) 853-1122
Location: 10-minute walk
Open: 10 AM–9 PM

PIREP If you are flying over the North Cascades Highway in the mountainous and scenic northern part of the state and want to drop in on a rural but friendly little town for a bite to eat, the historic town of Concrete would be a good choice. Less than a ten-minute walk north from your airplane is a small restaurant that will take you back to the Elvis and Route 66 days in its decor and food choice items. They even have all the bottled pop from the '30s to '60s on display! The place is super clean and friendly, and you get plenty of food for the money. As with any canyon airport, unless there is some runway chatter on the unicom, I recommend a fly-over to see what the windsock is showing before making a landing decision.

Hours: 10 to 9 seven days a week. Restaurant phone (360) 853-1122.

Darrington, WA (Darrington Muni—1S2)

Backwoods Cafe 🍔🍔🍔

Rest. Phone #: (360) 436-1845
Location: 5-minute walk
Open: Daily: Breakfast, Lunch, and Dinner

PIREP Not too far northeast of Seattle is the neat little tucked-away mountain logging town of Darrington. Their airport runway was recently paved and is now quite accessible if you aren't too squeamish about a 2500' × 40' runway with tall trees on one end and a mountain just off the other! There's a windsock midfield, and I recommend you fly over it before landing, as the surface winds can be a bit tricky.

Only a short five-minute walk across the highway and you are at the very friendly Backwoods Cafe, which makes the trip worthwhile. Their dining room is nonsmoking with the smokers relegated out to the front counter/booth area.

PIREP Have been in there twice recently and caution readers about the Backwoods Cafe. They don't take anything except cash, service is very slow, and food is not good. In fact, the first trip yielded a milkshake with sour milk, and the second trip yielded a cold hamburger after 45 minutes of waiting. Nice logging town ambiance, however.

PIREP Now Mike, you know I would never take issue with you, but my experiences at the Backwoods Cafe have been quite different. As you know, Evergreen Soaring did a glider operation there several months ago. We found the food to be slightly better than average, and the milkshakes were really good. Ron and I flew the Colt in there today and had a really good burger. But you are absolutely right, bring cash, and don't be in a hurry.

1S2 is a beautiful field to fly into. The scenery is simply breathtaking. Overall, I think Darrington is a very worthwhile destination.

Eastsound, WA (Orcas Island—ORS)

Bilbo's 🍔🍔🍔

Rest. Phone #: (360) 376-4728
Location: 10–15-minute walk
Open: Daily: Breakfast, Lunch, and Dinner

PIREP Orcas is one of the San Juan Islands northwest of Seattle. It is actually further north than Victoria B.C., but is still part of the U.S.

Bilbo's is not right at the airport. There is a 10- to 15-minute walk from the airport to town, and I would not have known which way to go without a local giving me a ride one way.

Bilbo's is a Mexican restaurant with good food and reasonable prices. The portions are not the overwhelming sizes typical of most Mexican restaurants. Be careful when you arrive, as they were open for lunch 11:30–2:30 the Saturday I was there. Call first just to make sure (360) 376-4728.

The biggest attraction in my opinion is the airport itself. It is on the north end of Orcas Island with water on approaches to either runway. The runway has a noticeable slope downhill to the north and you end up climbing out low over the water. The water at the south end of the airport is Puget Sound. The island is shaped like a big horseshoe, and it's fun to fly between the two fingers. I like the approach from the north over the Sound as well.

The airport has a variety of aircraft, from an ample supply of Bonanzas to a T-6 to biplanes. Biplane rides are offered. When I was there a couple was camping out of their Tri-Pacer on the grass.

There are other restaurants in town and shops to check out.

PIREP If you want an informal alternative to the tourist prices found at Bilbo's and others, try out the locals' secret: Portofino Pizza. This pizza restaurant is actually the closest one to the airport and has a killer view! As you walk into town, it's on the second floor of the local convenience store, reached via an outside staircase at the west end of the building. My teenage son and I discovered this place after being grossed out by sanitation at several other local restaurants (not Bilbo's).

Prices are comparable to pizzerias in any major metro area, with similar service as well. They sport an outside deck, as well as large windows that look over the top of other buildings, to the water beyond.

Eatonville, WA (Swanson—2W3)

Between the Bread 😋😋😋

Rest. Phone #: (360) 832-3777
Location: Very short walk
Open: Daily: Lunch and Dinner

PIREP Eatonville is a great place to fly in. The town is only two blocks from the airport. There are four restaurants to choose from. My favorite is Between the Bread, owned by a local pilot on the airport. Great food! Another is Tall Timbers, which is also in town and also offers great food.

PIREP We just flew into Eatonville today looking for Between the Bread. We asked directions to food; following directions, we then walked directly off the end of the airfield through and over an ankle-turning rock pile (hill) covered with lots of blackberry bushes.

Skip the trip up the hill to Mountain Take Out. The burgers were okay but pricey; however, they do have a wide variety of milkshakes. The only seating is outside. Avoid the restroom, as it is primitive and seriously lacks cleaning.

We got clarity too late about the location of Between the Bread—it has had an ownership and name change to Noodles on the Move and is down in town by the Chevron station. Supposedly it has the same menu as before, but we can't swear by it as we unfortunately ate at the take-out place.

Elma, WA (Elma Muni—4W8)

Elma Airport Diner 😋😋😋

Rest. Phone #: (360) 482-1431
Location: On the field
Open: Daily: Breakfast and Lunch

PIREP   FBO Comments: No fuel available.

PIREP   Elma's Airport Diner, with old-fashioned burgers, fries, shakes and pies, has opened as of late August 2004. The place is bright, and the food is really good at a great price. I've flown in there about five times now for meals with many different passengers, from friends to family. Everyone has enjoyed the food, including the grandkids. Their hours are 8 AM–8 PM, open everyday except Tuesday. They use really fresh produce, and the fruit in the pies has never seen the inside of a can! Not too sweet, either—just right!

Look out for the burger called "The Diner"—hubby had to remove the big slab of onions in order to get his mouth around it, and he enjoyed the onion rings, too! The owners say that's nothing, they offer a triple burger too that may be hopelessly too tall. The country potatoes on the breakfast menu are really good and worth going back for.

It's a fun strip to land on, and there are pilot-controlled runway lights for the dinner runs (or, since winter is coming, late lunch runs may need those lights, too). I park on the grass right in front of the window. Service is wonderful and always with a smile and a chat if you'd like and if they aren't too busy. Hey, folks, let's support this diner. Not many destinations with such long hours, and the town is small and doesn't seem to have discovered the place as of yet. Let's help these folks sling their burgers for a long time to come!

PIREP   Nice parking in the grass right in front of the restaurant. I only had the onion rings but they were great! Don't order the large unless you really mean it—they are big. Very friendly and clean place. Will stop in again.

Everett, WA (Snohomish County–Paine Field—PAE)

Jet Deck 😑😑😑

Rest. Phone #: (425) 353-0770
Location: On the field
Open: Daily: Breakfast and Lunch

PIREP   Wow! This place has made quite a turn-around. I only went there because I needed to kill a couple hours while getting a propeller balanced, and I was very impressed at what they've done. The cheeseburger was juicy and tasty, the french fries were outstanding, the waitresses were hot; they'll bring you real coffee from the espresso stand next door, and the place has a full bar and turns into a live music club at night. Even the water was good! I hadn't been there in years, and had only heard negative things since, but I'll definitely go back just for a $100 hamburger because it's a nice place now with terrific views.

PIREP   I have been going to this place for six or seven years. During that time it has changed hands at least twice. The current owner has let the place run down to the point where it is filthy. The place reeks of stale smoke, the bathrooms are not kept clean, and the rug is so dirty, it's gross! I can't imagine that in all that filth the food is clean.

Reduce your rating to no more than one star—better yet, advise people what to expect if they do go there.

Hoquiam, WA (Bowerman—HQM)

Lana's Hangar Cafe 😑😑😑

Rest. Phone #: (360) 533-8907
Location: On the field
Open: Daily: Breakfast and Lunch

PIREP   I have been here several times. They offer good food, good service, and surprisingly good prices.
My rating: 5

PIREP   We were taking a tour of the Washington coast via ground transportation when we came through the town of Hoquiam. While driving through we noticed a fairly obvious sign for Lana's Hangar Cafe. We decided to turn around and try it out.

It's a little ways off the beaten path, but once you get there you won't be disappointed. There was a nice crowd and some folks had to even wait to be seated. The food was good and service was jovial. Stop by, driving or flying—it's definitely worth it!

Lynden, WA (Lynden—38W)
Homestead Golf and Country Club 🥪🥪🥪

Rest. Phone #: (800) 354-1196
Location: 2 blocks
Open: Daily: Breakfast, Lunch, and Dinner

PIREP   Walk just five minutes north of transient parking and you are at the Duffer's Cafe, which is part of the Homestead Golf and Country Club (public welcome). They have a complete menu including burgers, fish, steaks, stir-fry, sandwiches, and a continuous $6 all-you-can-eat spaghetti feed. All prices are low to reasonable, and the food is of high quality. You can get 100LL at the airport, and they have a friendly pilot shop. Hours 7 to 7, seven days a week.

PIREP   The Homestead is a good place to eat, but the neat place to go is downtown Lynden. All of the downtown is geared toward Dutch decor and items. Windmills, wooden shoes, and so on. It is about a 20-minute walk south of the airport. Also there is a bus that runs by the airport about every 30 minutes. If I am reading the schedule correctly, it goes by the strip at 22 and 52 minutes past the hour (never on Sunday).

Once in town there are many restaurants. Mostly with a Dutch theme but saw Chinese, Greek, and coffee shops also. Lots of shops, antique stores, gift shops, and so on. At the west end of the main street is a large windmill with a hotel and restaurant, cafe, fish canal, and more shops in it. We ate at the Hollandia. Very Dutch-like recipes, we were told by the Tourist Bureau. The meatball soup was excellent. The sandwich croquets must be an acquired taste. Not bad, just different. We are sure to go back and will be getting one of the free bikes to tour the town.

Gas was reasonable also. The locals prefer you land on 07 and depart 25, wind permitting of course.

Mansfield, WA (Mansfield—8W3)
Sunflower Cafe 🥪🥪🥪

Rest. Phone #: (509) 683-1068
Location: About 3 blocks
Open: Daily: Breakfast, Lunch, and Dinner

PIREP   Want to get away from it all? This town is maybe five blocks wide, a charming farm community in central WA in the middle of nowhere. We found the one restaurant in town for lunch about three blocks from the airport (so, in the center of town). Don't order the Reuben, the cook will come out and ask you how to make it! After a good lunch, walk around town and soak up the wonderful sound of silence and then head back to the mayhem where you live.

Mattawa, WA (Desert Aire—M94)
Desert Aire Golf Club 🥪🥪

Rest. Phone #: (509) 575-3010
Location: 30-minute walk
Open: Daily: Breakfast, Lunch, and Dinner

PIREP   I stopped at this well-maintained airport last Saturday after turbulence at 4m feet from 100 degree temperatures woke me up.

Great golf course, 18 holes, near Columbia River with camping on river available. Restaurant is a half-hour walk for great food, motel, and a Texaco.

Ocean Shores, WA (Ocean Shores Muni—W04)

The Pirate's Cove 🍔🍔🍔

Rest. Phone #: (360) 289-4400
Location: 1 mile
Open: Daily: Breakfast, Lunch, and Dinner

PIREP   We have a newly opened restaurant and pub in Ocean Shores, WA (W04), The Pirate's Cove. It is located just over one mile from the airport, on the east side of Ocean Shores Blvd. just north of the Shiloh Inn. The phone number is (360) 289-4400.

The restaurant has a dinner special each evening priced under $8. Breakfast is available all day. The food is excellent and a great value. You will not leave hungry!

Packwood, WA (Packwood—55S)

Peter's 🍔🍔🍔

Rest. Phone #: (360) 494-4000
Location: 1/2-mile walk
Open: Daily: Lunch and Dinner

PIREP   The airport is about two blocks off State Hwy 12. Packwood has some restaurants along the highway. My favorite is Peter's, which is at the east end of town. Logger-type food is served here. That means okay! Some of these restaurants could be a half-mile walk, but some are closer.

Port Angeles, WA (William R Fairchild Intl—CLM)

Airport Cafeteria 🍔🍔🍔

Rest. Phone #: (360) 457-1138
Location: In the terminal
Open: Daily: Breakfast, Lunch, and Dinner

PIREP   Fairchild International Airport in Port Angeles, WA is home to a small GA fleet and some Horizon Air commuter traffic. The airport is well maintained and easy to find. Watch out for the Horizon guys doing five-mile straight-ins from the east, IFR students shooting approaches, and the 6000' cumulo-granite to the south.

The terminal has an okay cafeteria, a gift shop, and so on. Definitely a small-town feel: the folks are friendly, and the traffic is light. It is a nice stop before continuing west to the Pacific Coast or turning back to the more crowded skies over Puget Sound.

Port Townsend, WA (Jefferson County Intl—0S9)
Spruce Goose Cafe 😋😋😋😋

Rest. Phone #: (360) 385-3185
Location: On the ramp
Open: Daily: Breakfast and Lunch

PIREP FBO Comments: Fuel is available and a fair price.

PIREP The Spruce Goose Restaurant is rampside with a nice ambiance. Outdoor seating is available to enjoy the sunshine and ample airplane activity. The food is simple, tasty, and prepared quickly. The wait-staff is incredibly friendly. Save room for dessert—their specialty is marionberry cobbler a la mode, and it is yummy!

PIREP This is one of the best fly-in restaurants I have ever been to. It has a very nice atmosphere, nice people, great prices, and absolutely delicious hamburgers. It is only a 10-minute flight from my home airport, but I have been to many other airport restaurants in Washington and this is one of the best. I would recommend it to anyone.

Puyallup, WA (Pierce County–Thun Field—1S0)
Airport Cafe 😋😋😋

Rest. Phone #: (253) 848-7516
Location: On the field
Open: Daily: Breakfast and Lunch

PIREP Restaurant's been open about a year now, and it's pretty good. I like the Monte Cristo and the burgers. Patty melt is okay too but not as good as the one at the Jet Deck restaurant at Paine Field. Excellent view of the runway and outdoor seating during the summer.

    The Airport Cafe is also the regular meeting place for the Northwest Antique Aircraft Association. Second Friday evening, suspended for summer because of flying. This is one of the most active, friendly flying clubs around, whether or not you own an antique. Many members fly their beautiful antiques and classic aircraft to the meeting, too. The restaurant is very accommodating to groups and puts on a really nice meal at a reasonable price.

Renton, WA (Renton Muni—RNT)
Mimosa's 😋😋😋

Rest. Phone #: (425) 255-2221
Location: On the field
Open: Daily: Lunch and Dinner

PIREP Food is mostly the same, basic greasy-spoon diner kind of place with oversized portions and plenty of meat and potatoes. Great for breakfast. You can probably park at the south end of the runway at Proflight and just walk across Perimeter Road, up and over the berm, and then across Airport Way (which is kind of busy, but you could go east half a block and use the crosswalk/stoplight).

Richland, WA (Richland—RLD)
Almost Gourmet 😋😋😋😋

Rest. Phone #: (509) 943-7604
Location: On the field
Open: Weekdays only

PIREP   This restaurant is right by the gas pumps. It is excellent but is only open for lunch during the weekdays.
It has been there for several years and has a good local reputation. Well worth flying in for.

Roche Harbor, WA (Roche Harbor—W39)

Roche Harbor Resort 🍔🍔🍔🍔🍔

Rest. Phone #: (800) 451-8910
Location: On the field
Open: Call for permission before landing; they'll give you the hours, which change throughout the year

PIREP   Roche Harbor is a terrific flying destination. The runway is narrow and has a significant slope and some
humps, but it's plenty long at 4300'. You just call ahead to the resort and they write down your N number. I think
it would be okay if you didn't call ahead, but you better put some money in the box or you'll get a bill in the
mail. There was a person monitoring the comings and goings and scribbling on a pad. The landing fee is $10 for
singles, $15 for multi-engine airplanes. Same day parking is free, but the overnight fee is an additional $5 per day.
The parking area is bumpy turf, so take it easy if you don't have a lot of prop clearance. Definitely bring chocks,
as some of the spots have a considerable downhill slope toward the runway and there's a lot of competition for
big rocks and such.

They gave me a short cautionary briefing about the airport and how operating there is at my own risk, but
never mentioned any rules regarding runway choice as per the previous pirep. The wind was maybe two knots
from the west, and I landed and also departed on 24. I had previously thought that I would depart runway 6, to
avoid the long back-taxi and to reduce noise over the resort area—until I saw the place from the ground. I was
flying a Beechcraft Duchess, and the significant eastbound uphill slope and rising terrain to the east with a wall
of tall trees on top of that quickly changed my mind. The strip is basically surrounded by 50+-foot high trees all
around (except to the west) and there is at least a 70-foot high hill a quarter mile off the east end of the runway
covered with those same tall trees. If you lost an engine below 200 feet during an eastbound departure, you'd
probably be toast. On the other hand, during a westbound departure, the terrain is falling away from you and there
is a wide gap between the trees and then the ground drops all the way to open water. Much healthier scenario. I
don't think we disturbed anybody too much since we were climbing through 500 feet before we reached the end
of the runway and pulled the props back quite a bit as we flew over the harbor.

The village/resort is storybook quaint, with really old buildings, brick pavement and lots of gardening. The
view of the harbor is very relaxing. My Paradise cheeseburger was pretty good, but not sublime. The Bloody Mary
I didn't even get to sip was raved as the best damn Bloody my girlfriend ever had in her life. It was certainly a
salad-in-a-glass and very spicy looking.

Overall, I give the place five stars.

PIREP   Roche Harbor Resort is a five-minute walk west from the airport. This airport is owned and maintained
by the resort folks. After landing, taxi to the west end, and park in the grassy parking area. There is an honor-
system landing fee; $5 in the box as you walk off the airport boundary. I'm not really sure what's here off season.
The resort may still be open, but I'm pretty sure the restaurant is not. During the summer, this is a great place to
have lunch or dinner. You can sit in the dining room, or if it's nice, sit downstairs out on the deck. The restaurant
is closed November through February. Good food and pretty reasonably priced. Of all of the airports in the San
Juans, this one is the closest to amenities. Plus, there is a good kayak rental at the harbor and you can easily paddle
into the Haro Straight where the best orca watching in the islands happens. Pretty exciting stuff to be sitting in a
small sea kayak with killer whales cavorting about.

At the resort is also a small marina, a variety of shops in an old barrel warehouse (or something like that), a little chapel, and a very nice flower garden. Last time I was there, there was a kiosk that looked like they were selling tickets for some kind of bus tour or shuttle or something. If you're there in the evening, keep your eyes open for deer and other wildlife as you walk back to the airport. After sunset, as with all of the other San Juan Islands destinations, a clear night presents one with some awesome star-gazing. Services at airport: zippo. This is a strictly "for getting in and out" airport, operations at your own risk. Fortunately, it's actually maintained pretty well, except the windsock was a little chewed up last time I was there. Not the best place to be landing after dark. The approach to 24 is from over a hill with trees, so you're probably better off coming in from over the harbor/resort. Even that approach isn't easy, since the airport is poorly lit, and there's no VASI. Also, the runway has a significant slope uphill to the east. Unless the winds strongly oppose you, I'd recommend landing on 06 and departing on 24. Also, be polite. Remember that this is a resort. If your plane can safely climb out with reduced power after liftoff, that's a nice gesture. Otherwise, just remember not to turn before 500' AGL. Departures to the east from runway 06 aren't a problem noisewise, but there is that hill with the trees to worry about.

Seattle, WA (Boeing Field/King County Intl—BFI)
Museum of Flight Cafe 🍔🍔🍔🍔

Rest. Phone #: (206) 764-5720
Location: On the field
Open: Daily: Lunch

PIREP First, and foremost, the restaurant at MOF does not have burgers. What they do have are some nice salads and usually a tasty soup. I get the impression that it is under new management, as the help seems eager to please.

The restaurant sits on the north end of the museum, and in nice weather you can sit with a view of the B taxiway and the long runway.

Being based at Boeing Field, I tend to eat there (and Randy's) fairly often. While it is not a five-burger place, it should rate maybe three.

CAVU Cafe 🍔🍔🍔

Rest. Phone #: (206) 764-4929
Location: On the field in the main terminal
Open: Daily: Breakfast and Lunch

PIREP We were at the biz-jet preview and stopped in at the Cavu Cafe for lunch. Same great selection of Italian-style sandwiches at reasonable prices. The Cavu Cafe is in the main-terminal building and the Aviator's Store is a three-minute walk. I believe the hours were 7 AM–4 PM Monday through Friday, 7 AM–2 PM Saturday, closed Sunday. Limited menu but highly recommended!

Snohomish, WA (Harvey Field—S43)
The Buzz Inn 🍔🍔

Rest. Phone #: (360) 568-3970
Location: On the field
Open: Daily: Breakfast and Lunch

PIREP The Buzz Inn is good enough food, especially the patio seating in good weather.

But the real gems are a five-minute walk into town, where there are all sorts of good places to eat: Mexican, Greek, Italian, generic American, ice cream, coffee shops, and so on.

PIREP   The food is good at the Buzz Inn, but the service is fairly slow. It took almost 20 minutes just for a cheeseburger. The view from the outside deck is hard to beat, especially when the drop zone is active.

Spokane, WA (Felts Field—SFF)

Skyway Cafe 🍔🍔🍔🍔

Rest. Phone #: (509) 534-5986
Location: On the field
Open: 6 AM–2 PM

PIREP   If you're in Spokane, this one is a should-do! The Skyway Cafe is, in my opinion, a crown jewel for Spokane. Great food, great atmosphere, great folks working there. You can find members of EAA, the 99s, and all kinds of stray cats and dogs there on a regular basis.

PIREP   Very good food, very good service, a great meal!

Tacoma, WA (Tacoma Narrows—TIW)

Narrows Landing Restaurant & Bar 🍔🍔🍔🍔

Rest. Phone #: (253) 853-4113
Location: On the field
Open: Mon.–Sat. 6:30 AM–9 PM, Sun. 9 AM–3 PM

PIREP   The views are awesome. Plenty of corporate aircraft, as well as general aviation. Meet the local pilots while you enjoy some of the best-tasting coffee I have had. The food is excellent and the service is some of the best that I have seen. There are times when it gets busy and they seem understaffed, but the food and the service is well worth the wait, if you have to. Usually, they are take very good care of you from the time you walk in the door until you leave.

PIREP   Please make a date to eat at the Narrows Landing, the recently opened restaurant at TIW. The food is very good, reasonably priced, and the owners very friendly. The decor is worth a trip in itself. I took an "aviation-impaired" friend there for lunch, and he remarked on how impressive the decor was.

   Check it out.

Vancouver, WA (Evergreen Field—59S)

Chinese 🍔🍔

Rest. Phone #: (360) 892-1201
Location: Short walk
Open: Daily: Lunch and Dinner

PIREP   This place has the feel of an early '50s airfield. The asphalt runway is hidden behind a turf tie-down area. The owner of Evergreen Flying service believes every pilot should learn to fly a tail dragger. It's worth visiting.

   Oh yes. A Chinese restaurant adjacent, and a chain restaurant in a shopping center a quarter-mile west.

   There are lots of restaurants in walking distance to Evergreen Field. Applebee's and just about every kind of food from fast food (Wendy's to excellent Mexican food). Evergreen is a great airport and lodging is nearby at the Phoenix Inn. Lots of shopping around, too. This is my favorite place. A great place to take the spouse as she/he can shop or eat while you hang around the airport and talk to the owner, Wally Olson, who has taught flying for over 55 years. A great place.

Walla Walla, WA (Walla Walla Regional—ALW)

Florentyna's at the Airport 🍔🍔🍔

Rest. Phone #: (509) 525-2294
Location: On the field
Open: Daily: Breakfast and Lunch

PIREP FBO Comments: Walla Walla has the friendliest contract tower. Handy fueling north end of ramp. Not cheap, but handy bathroom. *Very* noisy pump, so be ready. Blue Ridge Aviation fixes anything, small or large. Plenty of ramp space.

PIREP Florentyna's is in the brand new terminal at the south end with a view of the ramp. General Aviation parking allowed for passenger pickup or dining. *Great* Greek salad. Try the prime rib dip with grilled Walla Walla sweet onions! Can't be beat!

Wenatchee, WA (Pangborn Memorial—EAT)

Bobbie Sue's 🍔🍔🍔

Rest. Phone #: (509) 886-7323
Location: On the field
Open: 8 AM–3 PM

PIREP I had a business trip into Pangborn and was thinking of a place to go for lunch. I was told to try this new restaurant in the Pangborn Terminal. I've had terminal food before and I didn't think much of the idea. Well, I did try it out and to my complete surprise, I found this place very, very good. The staff and owner are great people and they do prepare home-cooked meals and desserts. This place is worth going out of your way to try out.

Fuel Price: high.

Westport, WA (Westport—14S)

Kings Pancake and Steak House 🍔🍔🍔

Rest. Phone #: (360) 268-2556
Location: Short walk
Open: Daily: Breakfast, Lunch, and Dinner

PIREP Less than a half-mile north of the field is Kings Pancake and Steak House. A small, nice place with good food and friendly people. Warning: They have karaoke on Saturdays.

PIREP Westport is a great place to go if you're into deep-sea fishing, whale watching, or any other sea charter activities. There's a waterfront district about three-quarters of a mile northwest of the airport. Walk to the right on the main road that passes the airport parking lot. There's a good selection of gift shops, candy stores, restaurants, motels, and boat charter establishments. The waterfront area is compact and self-contained, so transportation is really not needed for a weekend of fishing. However, you can rent bicycles at a go-cart track that you'll pass going into town.

There's an entrance to a state park about halfway to the waterfront area. The distance to the state park is about the same as to the waterfront. The park is for day use only.

At the northernmost part of town there's an observation tower overlooking the entrance to Gray's Harbor. The view can be spectacular when the waves pound the breakwater.

Woodland, WA (Woodland State—W27)

The Oak Tree 🍔🍔🍔

Rest. Phone #: (360) 887-8661
Location: 1/4-mile walk
Open: Daily: Breakfast, Lunch, and Dinner

PIREP   About a quarter-mile walk from the tie-down at Woodland, WA are several good eating establishments, including the Oak Tree.

The Hundred Dollar Hamburger

West Virginia

Fairmont, WV (Fairmont Muni-Frankman Field—4G7)

DJ's Diner 🍔🍔🍔

Rest. Phone #: (304) 366-8110
Location: On the field
Open: Daily: Breakfast, Lunch, and Dinner

PIREP 50 or 70 yards from the FBO is DJ's Diner, a '50s-style stainless steel diner with great atmosphere, excellent food, and low prices. And right beside DJ's is a Comfort Inn. Also, the FBO has a new 24-hour self-serve 100LL pump with very reasonable prices.

Huntington, WV (Tri-State/Milton J Ferguson Field—HTS)

Tri-State Airport Restaurant 🍔🍔🍔

Rest. Phone #: (304) 453-4183
Location: On the field
Open: Daily: Breakfast, Lunch, and Dinner

PIREP Try it, you'll like it. This is a great place to grab a meal. They feature burgers, chicken, salads, and so on. Food is very good, tastefully prepared and presented. Service is very good. The restaurant is in the main terminal, an easy walk from the FBO.

To the best of my knowledge, the FBO has any type of fuel or lubricants you may need.

Lewisburg, WV (Greenbrier Valley—LWB)

The Greenbrier Hotel 🍔🍔🍔🍔🍔

Rest. Phone #: (304) 536-1110
Location: Call for pickup
Open: Daily: Breakfast, Lunch, and Dinner

PIREP The Greenbrier is a real neat place to fly in to. Fly in to LWB (Lewisburg) White Sulpher Springs in West Virginia. The hotel will pick you up. This is the "former" place for the Congress to go, in the event of a nuclear attack. It is a beautiful hotel and restaurant on top, and all the meeting rooms, dorms, and the rest are all hidden below. It is a fascinating place to go.

PIREP For those with Biz Jets and pocket books to match, the Greenbrier Hotel is located in White Sulfur Springs about 12 miles from the airport. They provide limousine service. Bring your golf clubs and your American Express Card—the platinum one!

For those who like caving, you can tour Organ Cave and Lost World Caverns. If you like wild cave tours, call Ed Swepson at (304) 645-6984. His outfit, Venture Underground, will take you on a three-hour tour. Bring coveralls, kneepads, and your love of mud.

General Lewis Inn 🍔🍔🍔

Rest. Phone #: (304) 645-2600
Location: Borrow the crew car
Open: Daily: Lunch and Dinner

PIREP FBO Comments: High fuel prices. The airport taxi charged us $12 per person each way for the five-mile trip to the Inn. This was the only taxi service running on Sunday. The FBO staff were a little slow.

The General Lewis Inn is very nice but the service was a disaster. We were seated right away but the waitress came by 37 minutes later only to discover we did not even have menus. It took another 15 minutes for her to come back to take our orders. After that things went well. The food was average to good but the prices were a bit high. We had four people so the roundtrip taxi fare was $96. I'd take this off the $100 Hamburger list and add it to the $250 Hamburger list.

PIREP The Greenbrier Valley is one of the most beautiful places to fly to on the East Coast. At LWB a hungry flight crew has several options. There is a small restaurant on the field, but five miles down the road is the General Lewis Inn, a quaint bed and breakfast inn. The food is very good, and the rooms rent for about $55 a night in the off season. The inn has an old-fashioned reception desk at the entrance. Civil War memorabilia line the wall on the first floor. The rooms have double-size four-poster beds with wonderful white canopies in some of the rooms. You can sit out back next to the ornamental pond sipping mint juleps prepared by the innkeeper—of course, not when you will be flying the next day. Spend a weekend! Dinner is good but pricey. While in Lewisburg, you can stroll the streets and take a look at the shops. It's a small town and a nice getaway.

The Fort Savannah Inn is on Rte. 219 just before you get to the General Lewis. It offers nice food and comfortable lodging.

Martinsburg, WV (Eastern WV Regional/Shepherd Field—MRB)
The Airport Cafe 🍔🍔🍔

Rest. Phone #: (304) 264-9585
Location: On the field
Open: Daily: Breakfast and Lunch

PIREP Cafe is open and filled with friendly service. Also, they are open again on Sundays. The restaurant is under new ownership and appears to be working hard to provide food and service their customers want at a very reasonable price!!!

PIREP Appears the cafe may be closed again. Call ahead to the FBO to check.

PIREP The Airport Cafe is alive and well at Martinsburg, WV. Located next to Aero-Smith FBO—tower controller can take you there. Good food, inexpensive, very friendly atmosphere, and a meeting place for aviators. I give it two burgers.

PIREP Airport Cafe status undetermined. Construction at airport has resulted in probable closing or changes.

PIREP   A great place for good food at inexpensive prices. Staff is friendly and you are sure to run into some great aviators there. Tell tower you want to taxi to restaurant.

PIREP   The Airport Cafe is located in the old Martinsburg FSS building located next to Aero-Smith FBO. They serve breakfast and lunch and are open 7 days a week at 7 AM for those early flyers, but I did not note what time they close.

Home cooking is their thing, such as daily soups and specials. It's a small restaurant that is cheap (helps pay for the avgas!).

With the building of the Cheetah aircraft and Swearinger Jet directly across the field, this place could take off. Three burgers' worth.

PIREP   We recently flew our Cherokee from California to Delaware so we got to make a lot of pit stops along the way. We refueled at MRB on a great fall afternoon in September. The controllers and FBO were great but the little cafe in the back was a real delight. Good food (we had burgers) and big portions. Coming from California we couldn't believe the prices—a PBJ for 75 cents? Highly recommended.

PIREP   The Airport Cafe on the southeast side of the Martinsburg, WV (MRB) airport has the finest (read: greasiest) hamburgers and freshest everything else around. And best of all, from a poor aviator's standpoint, the prices are low, since the owners grow many of the ingredients themselves. Typical lunch for two is about $8–$9.

Morgantown, WV (Morgantown Muni-Walter L Bill Hart Field—MGW)

Voyagers 🍔🍔🍔🍔

Rest. Phone #: (304) 291-9078
Location: In the terminal
Open: Daily: Lunch and Dinner

PIREP   Excellent food. Huge variety. Good prices. Only wish it was open on Sunday evening (after the buffet) for dinner. Everything from burgers to pecan-encrusted salmon. A true gem. I will go out of my way to stop there for lunch.

PIREP   FBO Comments: The FBO services and pilot facilities also were very good.

PIREP   The Sunday buffet at the Voyagers Restaurant, located in the terminal, is excellent and very reasonably priced. Service was excellent, and the desserts were to die for.

I highly recommend it.

PIREP   Located across the end of the road to the airport in Morgantown, WV. Morgantown is a college town, but this is on top of the hill, far enough away from the campus that there aren't that many college students as customers. Nautical theme. Two floors to the restaurant. Varied menus. Steak and seafood are excellent. Prices are fairly reasonable. Portions are generous. Would highly recommend to anyone in the vicinity. Airport usually has a shuttle service, but it's not a bad walk on a nice day.

PIREP   The new airport restaurant is called Voyagers. The food is predominately Mediterranean and is simply fantastic. They have a fairly large menu with a variety of American and Mediterranean selections. Our lunch service was fast and very courteous. We hope to return for dinner some time. On a five-star rating it should receive four and a half. If you're in the Morgantown area it's worth your time to stop.

PIREP   Voyagers is the new name of the on-field restaurant. We got there late afternoon in time for the end of the buffet. Very limited selection but quality food—real country-fried chicken and, of all things, lamb. Also, veggies and so on—if it wasn't on the buffet and you asked for it, you got it. Interesting ambiance, cozy. Great price!!

15 bucks for two to gorge on chicken and lamb. Service was real slow though—only one guy waiting tables. All in all, if I was in the area I would go back, but I'm not flying there just for Voyagers.

PIREP   The airport restaurant is closed (? formerly Patty's Landing). Friendly tower controller says may open next week under new management. However, Back Bay is still great and really a short walk. The line crew guys were glad to offer to take us because we didn't know the way, but it is just across the road opposite the numbers for runway 18. Worth the trip definitely.

PIREP   At Aunt Patty's Landing the lunches are great and the pies are wonderful.

PIREP   The Back Bay Restaurant is wonderful, but they are sometimes booked solid, especially on game days and graduation days. I suggest calling ahead. If you can get in, you'll enjoy the food and service. There is also a wonderful Middle Eastern restaurant called Ali Babba on High Street in downtown Morgantown, but it'll cost you a $10 cab ride (each way) to get there.

As for Aunt Patty's Landing at the airport, the service is sometimes a little funky, but the food is quite good, and the prices are very reasonable.

PIREP   We flew into MGW with the intent to eat at the Back Bay Restaurant, which is located across the street from the airport entrance. We planned on walking, but the FBO offered a ride that we fortunately accepted because the restaurant wasn't open for dinner yet. Too bad, because my wife ate there once and said it was really good. Sadly, we returned to the airport.

After suffering the letdown, we decided to chance Aunt Patty's Landing in the airport terminal. The food there was actually very good, and the desserts just short of excellent. I'd give the place a recommendation and three winged burgers.

If anyone makes it to the Back Bay, how about a pirep?

PIREP   Morgantown, WV has a great airport with the friendliest tower I've ever encountered. They also have a great place to eat on the field in the terminal building. You taxi right up to the front door and the guys from the FBO will even put out the red carpet for you as you exit the plane—even my '62 Cherokee 160. Prices are reasonable, the food is good, and the dessert menu is out of sight. If you go on a Saturday in the fall, you can even try to catch some great Mountaineer football over at West Virginia University.

Food gets a good four and the football gets a solid ten.

Parkersburg, WV (Mid-Ohio Valley Regional—PKB)

Helen's 🍔🍔🍔

Rest. Phone #: (304) 464-4413
Location: Terminal building
Open: Daily: Breakfast and Lunch

PIREP   Helen's in the terminal building is excellent. The burgers are unreal. Especially the Cargo Loader, which is a half-pounder topped with chili. If you dare a closed cockpit or are flying solo, it is worth it. The homemade pies a la mode are also wonderful. Can't miss for a quick turn, good food, friendly FBO.

PIREP   FBO Comments: Friendly, helpful, good pilot information services, enjoyable FBO.

PIREP   My long solo cross-country (a stressful time), I chose this airfield and was delighted to see friendly faces, helpful and courteous people, and a very good restaurant. I will be back many times.

PIREP   Great restaurant in the terminal building, adjacent to the GA ramp. Excellent food (great burger!), reasonable prices, good view of the field. Apparently the owner is into radio-controlled models as there are plenty of models suspended from the ceiling.

Appleton, WI (Outagamie County Regional—ATW)

Victoria's 🍔🍔🍔

Rest. Phone #: (920) 730-9595
Location: Use the shuttle
Open: Daily: Lunch and Dinner

PIREP While this particular restaurant isn't right on the field, you can access it via a hotel shuttle at the public terminal. Known as Victoria's, this Italian restaurant consistently rates "outstanding" by locals and visitors alike. It's quite easy to spend less than $10–$12 per person for a dinner that few visitors have been able to finish. And while quantity is often used to make up for lack of quality, this isn't the case here.

ATW is *the* alternate airport for visitors to the annual EAA convention, and is actually the best place to tie-down when visiting—less air traffic, friendly tower operators, and lots of room for tie-downs.

Ashland, WI (John F Kennedy Memorial—ASX)

Deep Water Grille 🍔🍔

Rest. Phone #: (715) 682-4200
Location: Take a cab
Open: Daily: Lunch and Dinner

PIREP Nice airport, with log cabin terminal. Took a taxi from the airport to the Deep Water Grille for great food (four burgers) downtown (808 W. Main St.). Stayed at the Hotel Chequamegon on the shores of Lake Superior. Had excellent fun!

Baraboo, WI (Baraboo Wisconsin Dells—DLL)

Ho-Chunk Casino 🍔🍔🍔

Rest. Phone #: (800) 746-2486
Location: Easy walk
Open: Daily: Breakfast, Lunch, and Dinner

PIREP If you dare, fly into Baraboo Wisconsin Dells. The Ho-Chunk Casino is little more than a stone's throw from the airport. The casino will pick you up in their courtesy car so you can, if you're lucky, enjoy an inexpensive and expansive buffet. Rental cars are usually available at the airport. Nearby Wisconsin Dells is filled with large motels/hotels complete with indoor and outdoor water parks and a multitude of restaurants. Baraboo is charming with the Circus World Museum.

Brookfield, WI (Capitol—02C)

Jake's 😋 😋 😋

Rest. Phone #: (262) 781-7995
Location: Across the street
Open: Daily: Breakfast, Lunch, and Dinner

PIREP Jake's is just across the street from Capital Airport and has excellent steak and potato fare. A very popular supper club just west of Milwaukee, be prepared to wait for a seat. They don't take reservations (even for pilots). Plan ahead to arrive early or to wait. Either way, it is worth a detour to have dinner there. Prices range from $10+ up for an entree, salad, and twice-baked potato.

Brodhead, WI (Brodhead—C37)

The Sandpiper 😋 😋 😋

Rest. Phone #: (920) 839-2528
Location: Close
Open: Daily: Breakfast and Lunch

PIREP The Sandpiper is a basic diner with, get this, a tea room attached! The diner has good burgers and very friendly service. They, too, have good corn. The tea room is a notch up in fancy. The French onion soup is good, as are the basic entrees. Neither the diner nor the tea room will set you back much ($5–$10 for a good lunch).

Note that The Sandpiper is popular but small. That is how we ended up in the tea room—the diner was full—the first time we ate there.

I give it a three out of five burgers. It is basic fare, but well done. The cafe would make 3.5 to 4, but for some reason, the folks who run the diner feel overwhelmed by the fly-in crowd, so the service isn't as friendly as it could be. Just a small quibble, though.

Until next time: happy flying, but remember; wipe your hands before grabbing the yoke!

Cable, WI (Cable Union—3CU)

Telemark Resort 😋 😋 😋

Rest. Phone #: (715) 798-3999
Location: Taxi to front door
Open: Daily: Breakfast, Lunch, and Dinner

PIREP Telemark is a great place to stop if you're in Northern, WI. A full service hotel with a ski resort and golf course. Food is nothing special, but being able to walk 50 feet from the ramp to the door is nice.

Chetek, WI (Chetek Muni-Southworth—Y23)

Pete's Landing 😋 😋 😋

Rest. Phone #: (715) 924-4400
Location: 1/4 mile
Open: Daily: Breakfast, Lunch, and Dinner

PIREP FBO name: EZ Heat
Crew car? Yes
FBO phone: (715) 924-4400
    FBO Comments: EZ Heat is not an FBO per se. They can, however, assist you with any questions you may have when visiting Chetek. Chetek Municipal has 100LL available 24/7, self-serve card-trol system. There is also a courtesy van available 24/7. Van information is available in the terminal building weather room, which is located next to the fuel pumps. There are many excellent restaurants within a short driving distance and there are two golf courses to test your skill.

PIREP Myself and a pilot friend and our wives were at Pete's Landing last Saturday evening (December 15). We had the prime rib and the ladies had shrimp. The food was very good and the drinks were great. Prices are quite attractive and service was very good.
    Pete's is under new ownership and the new owners are doing a nice job. Definitely worth the short walk from the airport.

PIREP The Chetek airport is right next to the lake so arrivals and departures are scenic. It's a short 1/4-mile walk to Pete's Landing. They have great, big burgers, fish fry on Fridays, and specials at night such as broasted chicken and prime rib. Staff is friendly. Packers sign over the bar. Clean kitchen. You can rent cabins and pontoon boats adjacent to the airport on the way to Pete's Landing.
    This is a favorite "fly out" place for the Faribault (MN) Area Pilots Association.

Cumberland, WI (Cumberland Muni—UBE)

Tower House 🍔🍔

Rest. Phone #: (715) 822-8457
Location: 3 miles, crew car available
Open: Daily: Lunch and Dinner

PIREP FBO Comments: Key for courtesy car located in computer room. Use NDB freq to unlock.

PIREP It is a bit of a drive into town (about three miles) but well worth it. Excellent Italian and American food, and their homemade dinner rolls are to die for. Very charming mansion built in the late 1800s that was converted into a restaurant and decorated with relics from the past. A must-see when in the Cumberland area.

Delavan, WI (Lake Lawn—C59)

Lake Lawn Lodge 🍔🍔🍔🍔

Rest. Phone #: (800) 338-5253
Location: On the field
Open: Daily: Breakfast, Lunch, and Dinner

PIREP FBO Comments: *Caution:* This airport closes at night! There are *no* lights or services of any kind on the field. Therefore, this is not a dinner destination.

PIREP The restaurant is a short walk (south) across the street. Very nice place. Just don't go when departure is after dusk.
    My rating: 4

PIREP   Try Lake Lawn Lodge in Delavan, WI about 30 miles west of Lake Michigan and 10 miles north of the WI/IL state line. The lodge owns the paved runway and provides transportation across the street and through the parking lot to the restaurant. The lodge is on the north end of Lake Delavan. When approaching from the east it is difficult to locate. The lodge itself has boating, an outdoor pool, horseback riding, golf, and so on. The food is good.
Great Sunday buffet.

PIREP   We would like to agree in saying that Delavan, WI is a great place to visit. We had a wonderful stay at Lake Lawn Lodge last spring break. The staff was wonderful, especially the waiters. They were polite and precise about service (especially Marvin). The room service people always had a smile on their faces and were always happy to help! Lake Lawn Lodge is by far the *best* place to visit!!!!

Eau Claire, WI (Chippewa Valley Regional—EAU)
Connell's II 🍔🍔🍔

Rest. Phone #: (715) 833-9400
Location: In the main terminal building
Open: Daily: Breakfast and Lunch

PIREP   FBO Comments: Heartland Aviation is a nice full-service FBO serving Phillips 66 fuel. A control tower is in the process of being built on the southwest side of the airport and should be completed by late summer 2005. A new terminal is also under construction, so there is a lot of ground activity to watch for during taxi.

PIREP   I have flown several hundred flights to EAU and approximately 80 percent of those times I have eaten at Connell's. Consistently great food with reasonable prices, and they offer a wide variety of choices for breakfast, lunch, and dinner. Great scallops!

PIREP   Following the recommendation of two fellow pilots, a friend and I flew to Eau Claire (from Lake Elmo) and had a very nice dinner at Connell's II restaurant. The food was excellent, as was the service, and the portions were quite ample. (Our waitress, Becky, was very friendly and took excellent care of us.)
     If you're looking for a place to fly to and eat, I highly recommend Connell's II. It's in the main terminal building at EAU, east of runway 4-22 and north of runway 14-32. Go check it out!

Ephraim, WI (Ephraim-Fish Creek—3D2)
Julie's Park Cafe and Motel 🍔🍔🍔

Rest. Phone #: (920) 868-2999
Location: Call for pickup
Open: Daily: Breakfast, Lunch, and Dinner

PIREP   FBO Comments: Not a real FBO, but several of the local pilots volunteer time at the little office at the airport, and fuel is available.

PIREP   If you fly in to Ephraim Airport in Door County, whoever is manning the little office at the airport will let you phone Julie's Park Cafe and Motel ((920) 868-2999—the phone number is on their counter if you forget it). Julie's is pilot-friendly and will immediately dispatch a car to pick you up. The food at their cafe is wonderful, with an assortment of delicious sandwiches, salads, and wraps for lunch, a full breakfast menu served all day, and an extensive dinner menu. The service is first rate and more than friendly, and the prices are extremely reasonable. After you eat you can wander off down the road and shop in town, and when you are finished, go back to the cafe and they'll happily drive you right back up to the airport. We've done this trip twice, and both times the drivers refused tips for their service. Julie's is only open April 30 through November 1; they are closed for the winter. This is the best hundred dollar hamburger trip we've found so far, and it's hard to imagine a better one.

Julie's Cafe has a very nice deck for eating outside when the weather is good. There were several bird feeders set up in the bushes and trees surrounding the deck, and it was a lot of fun watching the different species of birds stopping by for a snack.

Julie's is right next door to one entrance to Peninsula State Park. At that park entrance, there is a bicycle rental business, and the park has many great trails for biking. If after your meal you want to burn off some of your newly acquired calories, I can't think of a better way to do it than to rent a bike and ride through the beautiful park.

Fond Du Lac, WI (Fond Du Lac County—FLD)

Schreiner's Restaurant 😋 😋 😋 😋

Rest. Phone #: (920) 922-6000
Location: Call for pickup
Open: Daily: Lunch and Dinner

PIREP Fond du Lac, at the southern end of Lake Winnebago, is home to Schreiner's Restaurant, a 50+ year old eatery with some of the finest food you'll find—at an affordable price. Best of all, the owner is a pilot, and goes out of his way to bring us in!

Land at Fond du Lac County airport (runways 09/27 and 18/36, both paved), and the friendly folks at the FBO, Fond du Lac Skyport, will call Schreiner's for you. Within minutes they'll have a courtesy car there for you, free of charge.

They serve all types of food, from breakfast all day to steaks. We fed our family of two adults and two kids for less than 20 bucks, and walked away stuffed.

When you're done, simply tell the cashier that you need a ride back to the airport—and away you go! Terrific service, terrific food, and they bend over more than backward for pilots!

PIREP With the exception of the days before, during, and after EAA (approximately a three-week period) Schreiner's Restaurant in Fond du Lac can almost always respond to a pilot request for pickup at Fond du Lac Skyport. There is always a manager on duty, and we can usually respond to request for pickup within a 15–20 minute period. Pilot and crew can have a great meal and a ride back to Skyport. Shopping mall within walking distance of restaurant; we don't provide transportation—eat at Schreiner's and walk to mall and back to restaurant and we take you back to Skyport.

We have had small groups of planes (flying clubs) joy ride (joy fly?) to FDL and Schreiner's. We'd prefer not to deal with groups like that because of having to take the manager on duty away from his/her work to make multiple time-consuming trips. Because of insurance-related concerns we try to limit the "shuttle" drivers to members of management. More than four or five people to be picked up and we're in a jam and wouldn't be able to help.

Green Bay, WI (Austin Straubel International—GRB)

Tony Roma's 😋 😋 😋

Rest. Phone #: (920) 499-9070
Location: On the field
Open: Daily: Lunch and Dinner

PIREP Austin Straubel Field has to be the premier fly-in for lunch or dinner spot in WI!!

Stop at Titletown Jet Centre and they will see to it that you get to and from any number of great spots within visual site of the airport, including but not limited to Tony Roma's, a local sports bar, Subway, and Shenandohah Restaurant at the Radisson. Also at the Radisson is the largest casino in WI.

The terminal restaurant is an easy stop, walking distance if you go to Executive Air. Also walking distance from Executive Air (a *great* FBO for piston planes) is the Radisson and the restaurants they offer. Hey, it is Green Bay—jeans are dressy.

If you want to avoid the walk, the folks at Executive Air will arrange for a quick pickup to the Radisson and other local spots.

Hillsboro, WI (Joshua Sanford Field—HBW)
Country Style Cookin' 😋😋😋

Rest. Phone #: (608) 489-3539
Location: 5-minute walk
Open: Daily: Breakfast, Lunch, and Dinner

PIREP FBO Comments: Ramp parking at HBW is scarce, one tie-down is on ramp between front of two hangars. HBW is on the WI state aeronautical chart but *not* on the CHI sectional!

PIREP Country Style Cookin' Restaurant is one of those little small town, nonchain eateries with no pretensions but good simple food. Service is buffet style and fast and friendly, lines are very rare. The menu never seems to vary, because it doesn't need to; everything is good. (I don't think they actually have burgers, but the roast beef is delicious!) Prices are very reasonable. Hours are seven days a week, 6 AM to 6 PM (shorter on weekends).

Country Style Cookin' Restaurant is about a five-minute walk into town from the airport, which is tight up against the east side of Hillsboro. Come out the west end of the airport ramp and turn right (northwest) on the first street (Hwy FF), and then turn left on Water Ave. (Hwy 33), which is the main drag of town, and walk two short blocks up hill (southwest), and Country Style Cookin' Restaurant will be on the left (look for the sign overhead).

P.S. There is a cheese factory store directly across the street from the airport which has a very nice eight-year-old cheddar.

Iron River, WI (Bayfield County—Y77)
Trout Haus 😋😋😋

Rest. Phone #: (715) 372-4219
Location: 1 mile
Open: Daily: Lunch and Dinner

PIREP Ron and Cindy Johnson have created a beautiful B&B just one mile east of the Bayfield County Airport (Y77) in Iron River, WI. Enjoy an outstanding breakfast overlooking their trout ponds, go biking on the Tri-County Corridor that adjoins their property, drop in at Cindy's Java Trout espresso bar/gift shop in town, or ask Ron to take you on the fishing trip of a lifetime on the world-famous Brule River. Their retreat is a great stepping-off point for adventures in Northern Wisconsin.

Janesville, WI (Southern Wisconsin Regional—JVL)
CAVU Cafe 😋😋😋

Rest. Phone #: (608) 741-1100
Location: Short walk
Open: Daily: Breakfast, Lunch, and Dinner

PIREP Flew into Janesville and had lunch at Cavu Cafe. The service was excellent and friendly. I was glad to see them still in business. I will fly in again for this restaurant.

Glen Erin Golf Club 😋😋😋😋

Rest. Phone #: (608) 741-1100
Location: Short walk
Open: Daily: Breakfast, Lunch, and Dinner

PIREP   Great food at the Cursing Stone Pub. It is a traditional Irish-style pub at the Glen Erin golf course. Located at the southwest side of the airport property. They will pick you up. It's only a short walk. Check for golf off-season hours.

Lake Geneva, WI (Grand Geneva Resort—C02)

Grand Geneva Resort 🍔🍔🍔🍔🍔

Rest. Phone #: (414) 248-8811
Location: On the field
Open: Daily: Breakfast, Lunch, and Dinner

PIREP   Grand Geneva Resort is the former Playboy Club, located in Lake Geneva. They have one paved runway, and a nice FBO facility. This resort has a working ski hill, golf courses, and a beautiful restaurant overlooking the very scenic Lake Geneva area. We visited for their breakfast buffet, which was very reasonable for what we got. We're not talking scrambled eggs here; nope, they had eggs Benedict, salmon, the whole nine yards. Served on real china with cloth napkins, no less. This is not your greasy spoon FBO.

The FBO shuttles you to and from the restaurant, and will gladly fill you in on what else there is to do in the area.

PIREP   Took my wife to Grand Geneva Resort and Spa (the former Playboy Club) for Valentine's weekend. The strip (Grand Geneva C02, Unicom 122.8, runway 2-22) is ~4100'. The FBO offers free tie-down and shuttle to the resort (only about 300 yards away though). The facility has everything for a great getaway weekend year-round.

During the summer, there are two golf courses: The Brute, which is 7200 yds. long, and the Highlands, a links-style course. They look great but are pricey ($115 and $100, respectively, with cart). The resort also offers horseback riding, hiking, and mountain bike rental. Winter offers a tiny downhill ski hill (hey, it's the Midwest!) and cross-country skiing.

There's always a short trip into the town of Lake Geneva, with its shops and restaurants. Rooms at the resort start around $150 off season. Suites are nice but considerably more. There are four good restaurants in the resort including Italian, steaks, seafood, and an American cafe. The Red Geranium is within a mile for the resort and is good, too.

They have a great spa facility with swimming, racquetball, tennis, basketball, weights and cardio, sauna and whirlpool, as well as massage and aerobics. Daycare and babysitting are available if you need to get away from the kids.

We had a great (but expensive) time.

Land O' Lakes, WI (Kings Land O' Lakes—LNL)

Gateway Lodge 🍔🍔🍔🍔

Rest. Phone #: (715) 547-3321
Location: Next to the airport
Open: Daily: Breakfast, Lunch, and Dinner

PIREP   Land O' Lakes, WI has the Gateway Lodge immediately adjacent to the airport. We're talking 500 feet. This is a classic WI lodge with rooms, restaurant, bar. Very reasonable rates and a nine-hole golf course (not affiliated with the lodge) across the street. Because the lodge is on the border with MI, the third hole tee is in WI and ball drops in MI. The town of Land O' Lakes is a short walk. Has bowling alley, other restaurants, bar, and so on.

What's not to like about this setup?

La Pointe, WI (Madeline Island—4R5)

The Oasis 🍔🍔🍔

Rest. Phone #: (715) 747-3192
Location: 500 feet
Open: Daily: Breakfast, Lunch, and Dinner

PIREP   If you don't mind a $100 slice of pizza instead of the $100 hamburger, visit Bud's place, The Oasis. Located just 500' off the Madeline Island airstrip on the road to La Pointe, you will find a small cozy cabin nestled between pine trees.

Sit around the open fireplace sipping on a drink of your choice (beer, pop, coffee). Instead of pizza a full meal can be prearranged. Watch deer through a large window, and with a little luck you might see a bear. Alternatively sit around the bar and view one of Bud's videos about the Apostle Islands, which will form the material for a documentary he is currently working on.

Transportation to La Pointe can usually be arranged, and in the worst of cases it's just an enjoyable 20-minute walk to town. There is regular ferry service to Bayfield during the summer and windsled service in the winter.

The bar is open whenever Bud is there, which is usually during the summer months. Call ahead to confirm.

Plans are in the works for unicom services on a part-time basis starting this summer. The 3000' paved runway (4-22) is in good shape and usually plowed during the winter. Fuel is available in Ashland (ASX).

Lone Rock, WI (Tri-County Regional—LNR)

Picadilly Lilly Airport Diner 🍔 🍔 🍔

Rest. Phone #: (608) 583-3318
Location: On the field
Open: Daily: Breakfast and Lunch

PIREP   Flew in for a late breakfast and had very good biscuits and gravy. My wife had a pancake that hung over the sides of her plate. We paid less than $10 for good food and good service and parked the plane less than 100 yards from the diner. We'll definitely be back.

PIREP   I stopped in this afternoon for some apple cobbler and ice cream, which was great. If you land runway 27 you can taxi straight off the runway and into their parking lot. Very convenient.

Madison, WI (Dane County Regional-Truax Field—MSN)

The Jet Room 🍔 🍔 🍔

Rest. Phone #: (608) 268-5010
Location: In the terminal
Open: Daily: Breakfast and Lunch

PIREP   Pat O'Malley has made a great restaurant inside the new Wisconsin Aviation FBO! Luckily I can just drive over (I live five minutes away), but people come from all over to enjoy the delicious food while watching activity on the ramp. The food is moderately priced and is well worth the trip. The only negative is the hours. Last I checked, the Jet Room was open from 6 AM to 2 PM Monday–Saturday and only open until noon on Sunday. You probably should call before you go just to make sure that I have the hours right. The Jet Room is right inside the Wisconsin Aviation FBO on the *east* ramp.

PIREP   The Jet Room offers really good food at reasonable prices with an outstanding view of the airport. The new Wisconsin Aviation building gives the Jet Room first crack at the ramp windows for an unobstructed view of the whole airport.

My menu favorite is the Irish stew, followed by the veggie omelet (the Navigator) which is made with real cheese (this *is* Wisconsin!). The coffee is good, the seating is fast, and the ambiance is all airport. They have some very interesting old airport pictures hanging on the walls.

PIREP FBO Comments: I found that the Wisconsin Aviation crew to be friendly and efficient, both on the line and behind the front desk. The lobby and restrooms are clean, nicely furnished, and quite nice. The most amazing thing is that I didn't have to pay for this ambiance at the pump, since the fuel prices were competitive. They have a slogan near one of the doors that states, "We don't charge fees, we want to earn your business," and based on my experience, they've earned mine.

PIREP On a Monday morning, I decided to take play hooky from work and visit the Jet Room located in the Wisconsin Aviation (FBO) building.

After looking at their ample menu, I decided on the Aviation Burger, which is a third-pound burger with Swiss, American, *and* cheddar cheese melted over the top. This, along with a side of their hot, freshly made potato chips, made the meal very enjoyable.

The prices were very reasonable and the dining area has a really great view of the airfield. I'm glad I got there early for lunch, because the place filled up rather quickly, which is another testament to the quality of the food there.

The only downside that I can see is that the restaurant is only open until 2 PM, which means either I visit them for lunch on the weekends or miss another day of work. I have yet to decide on which I will do.

Manitowoc, WI (Manitowoc County—MTW)

Joe's 🍔🍔🍔

Rest. Phone #: (920) 793-1036
Location: On the field
Open: Daily: Lunch and Dinner

PIREP We flew up to Manitowoc on CAVU day. Got there at 1. We all ordered Joe's specialty sandwich: a hamburger patty and a brat patty on a toasted bun. Great service, great food.

We'll be up there again.

They close at 2. However, they are open until 7 PM on Fridays for their fish fry.

PIREP My wife and I dropped in on Joe's today only to find that they're only open until 2 PM on Saturdays. Folks at the FBO were surprised.

Hours of operation: didn't get the full rundown, but they close at 2 on Saturdays.

Middleton, WI (Middleton Muni–Morey Field—C29)

Scott's Bakery 🍔🍔🍔

Rest. Phone #: (608) 836-7333
Location: Inside the FBO
Open: Daily: Breakfast and Lunch

PIREP FBO Comments: New facility is now open with new runways (10/28 and 1/19 turf) and parallel taxiways. New terminal building is also open.

PIREP Scott's Bakery is an airport coffee shop and bakery. They offer baked goods and soups and sandwiches. They also do breakfast sandwiches, as well as coffee, tea, and hot chocolate.

They are open Tuesday through Saturday (hours are different on the weekends). I hope they will have Sunday hours when the warm weather returns.

PIREP Watch for *high towers* south and east.

Stopped in there yesterday and found the new FBO building is up and running, including Scott's Bakery inside the FBO serving pastries and breakfast sandwiches in the AM and sandwiches and soups for lunch.

Tallard's 🍔🍔🍔🍔

Rest. Phone #: (608) 664-9393
Location: Across the street
Open: Daily: Breakfast, Lunch, and Dinner

PIREP Yesterday I decided to fly up to Madison and get some lunch with a friend of mine. The restaurant in the old terminal building was very good. At any rate when we got there we found the building torn down. The FBO has moved to the south ramp and as of now there is no restaurant.

We decided to try someplace else. I recalled that there was a new restaurant (one year old) near the Middleton Airport (Morey Field). I saw it one day while visiting friends there by car. While on final for RW 30 we passed right over the new place. It's a gray building and the name is Tallard's.

The menu was large with standard lunches, deli sandwiches, and full dinners. Prices okay and food is good. The theme is a '40s style diner inside, and the place is very large.

If the old Jet Room rated four burgers, then I would rate Tallard's four burgers also. It's only a one-block walk from the airport parking lot. I don't know the hours or if there is a breakfast menu. I will be going back again and will get those answers then.

PIREP We went across the street from where the old FBO had been located. There is a restaurant there called Tallard's. It is large spacious sports bar, with ample seating.

We were making an evening flight and had to return before sundown but had no trouble getting waited on and served in a timely manner. I would certainly recommend this stop to anyone passing through the area.

Quaker Steak & Lube 🍔🍔🍔

Rest. Phone #: (608) 831-5823
Location: 2.5 miles
Open: Daily: Lunch and Dinner

PIREP FBO Comments: Fuel prices are for self serve, and it is available 24 hours. There is a heated hangar to store your aircraft in the winter for $50 a night, and they can also plug you in or add heat in the winter for a fee. Tie-down fee overnight is $5. Staff is extremely professional and friendly. Mechanics are highly skilled and members of EAA. Visit website or call for more info.

PIREP Morey's in Middleton offers Scott's Pastry Shoppe in the FBO. Tallard's is a short walk away. The Quaker Steak and Lube is about 2.5 miles away, and I *highly* recommend it. It has the most awesome hot wings I've *ever* had (about 25 different flavors—their Atomic wings require you sign a waiver). It has two bars, dining area, and an outside patio. Its atmosphere is set up like an old vehicle maintenance garage and there are bikes (Rockets and Cruisers) and cars (Corvettes) hanging from the walls. The hours are 11 AM–11 PM Sunday through Thursday, 11 AM–2 PM Friday and Saturday. Tuesday nights are all-you-can-eat wings for about $15, and Wednesdays are bike night and the place is packed with bikers. Service is great. Facility is brand new. They also have a menu of burgers, and Biker Bill's Bacon Burger is awesome. I've recommended the place to all my friends and no one has been disappointed.

There are many golf courses in the Madison Area. I recommend Pleasant View, and it is fairly close to the airport. Rental cars available at the FBO—not sure about crew cars.

Milwaukee, WI (Lawrence J Timmerman—MWC)

El Greco 🍔🍔🍔

Rest. Phone #: (414) 462-0541
Location: 1/2 mile
Open: Daily: Lunch and Dinner

PIREP   The FBO was very helpful at Timmerman and set us up with a very nice crew car for no charge! The El Greco was about a two-minute drive but could have been walked to. El Greco was a very warm, family-owned Greek restaurant with cheap but average food. The only bad part was the $5 ramp fee the FBO handed out but it was justified with their great service.

Hours of operation: Open early for breakfast and late for dinner.

PIREP   There is a good Greek restaurant, El Greco, just off the field at Timmerman MWC. It is open all day. The food is good and generally more that you can eat and the prices are inexpensive.

Milwaukee, WI (General Mitchell International—MKE)
Packing House Restaurant 🍔🍔🍔

Rest. Phone #: (414) 483-5054
Location: Across the street from Signature
Open: Daily: Lunch and Dinner

PIREP   Directly across the street from Signature FBO is the Packing House. Excellent food and service. Large selection of beef, fish, pork, and poultry. Reservations may be required on Friday and Saturday evenings: (414) 483-5054.

Signature FBO was also very friendly—helped us with a motel and so on.

Monroe, WI (Monroe Muni—EFT)
Baumgartner's 🍔🍔

Rest. Phone #: (608) 325-6157
Location: Call a cab
Open: Daily: Lunch and Dinner

PIREP   I would like to recommend the city of Monroe, WI. Baumgartner's restaurant is known across the country. Getting to the restaurant is a simple matter of calling for a taxi (cost was $4 for three people). Baumgartner's is representative of German restaurants, with lots of beer and sausage. (They have coffee for the pilots!)

The town is friendly, the food is good, and the price is quite reasonable. Going on a Sunday, however, one will find most of the town closed up. This is best visited on a VFR Saturday.

By the way, the airport is about 15 nm SW of the Janesville VOR in southern WI.

Oshkosh, WI (Wittman Regional—OSH)
EAA AirVenture Museum 🍔🍔🍔🍔🍔

Rest. Phone #: (920) 426-4818
Location: On the field
Open: Daily

PIREP   The EAA AirVenture Museum has become one of the world's most extensive aviation attractions, and a year-round family destination. It is located on the site of the world's largest aviation event, EAA AirVenture.

The collection of historic artifacts began in 1962 when Steve Wittman donated his famous air racer Bonzo. It now comprises more than 20,000 aviation objects of historic importance. Included are 250 historic airplanes, and the count grows almost weekly as exhibits are constantly added. Everything from a powered parachute to a B-17 Flying Fortress is maintained in airworthy condition! Some are available to give ordinary people the chance to fly in historic aircraft. Most are used to provide flight demonstrations or support special activities such as the EAA Young Eagles program.

The EAA AirVenture Museum's library contains almost 9,000 volumes. The collection covers a variety of topics including biographies, aerodynamics, history, fiction, aeronautics, air racing, and home building.

The library's photographic collection archives more than 100,000 images of aircraft, spacecraft, and the people who made and flew them. Many photos chronicle the home-building movement of the early 1950s. They tell the story of an emerging group of people determined to design, build, and fly their own aircraft. Important photo archives donated by private collectors are curated here, including:

The Radtke Collection: One thousand negatives of military aircraft, civilian aircraft, and famous aviators from the '30s.

The Worthington Collection: 125+ glass negatives donated, taken by an unknown photographer.

The Zeigler Collection: Over 200 glass negatives of early German aviators of post-WWI era.

The Norman Collection: Hundreds of 8×10 black and white photographs covering the golden years of aviation.

This is truly one of the great aviation museums. Do not pass up an opportunity to visit. Be warned! Timing your visit to coincide with the EEA's annual air show is a great mistake. It becomes very crowded and is hardly user friendly at those times. Plan your trip for a nice spring or summer weekend. Combine your trip with a visit to the Pioneer Airport; it is right next door. Here, you'll want to spend thoughtful time and soak up the history that is all around you. Plan to spend time here.

Food? Yes, they have food, but that's not the reason to come!

Phillips, WI (Price County—PBH)
Harbor View 😋😋😋

Rest. Phone #: (715) 339-2626
Location: Just across the street
Open: Daily: Lunch and Dinner

PIREP Nicely done decor with a great view of the lake. Been there twice in the last two weeks with the Cherokee. Food is very good. Good variety of burgers, chicken sandwiches, a few Mexican items, among others.

Friday night fish fry is also good, but quite busy. (Tough to wait more than 10 or 15 minutes when you can't enjoy a few cocktails.)

Cheap gas (well, that's relative), too at the FBO, which is about 500 yards from the restaurant. Sandy beach at the restaurant would accommodate float planes nicely.

Prairie Du Chien, WI (Prairie Du Chien Muni—PDC)
The Black Angus 😋😋😋

Rest. Phone #: (608) 326-2222
Location: Close by, short walk
Open: Daily: Lunch and Dinner

PIREP There's a place up in Prairie du Chein, WI called the Black Angus. It's a few hundred yards down the road from the airport, but they had a mean fish fry on Friday nights the last time I was up there.

PDC is also the location for Cabela's outdoor sports store. So if you are a sportsman, you don't want to miss them. They will be more than happy to come and shuttle you from the airport.

PIREP Located a quarter mile from the airport, the Black Angus restaurant is a delightful stop on any night. They serve an excellent steak and are always packed. This is my favorite stop with friends, and the return flight to Chicago on a clear night is to die for!

Racine, WI (John H Batten—RAC)
River Run 😋😋😋

Rest. Phone #: (262) 633-4019
Location: Short walk
Open: Daily: Lunch and Dinner

PIREP FBO Comments: Racine might have cold winters but these guys are the warmest in any FBO in the country. Friendly, hospitable, they can't do enough for you. A great group to hang around waiting for the weather for improve. First time I have ever been given the keys to a crew car (van) that had only 300 miles on it.

PIREP River Run: probably a 3/4-mile walk if the crew car isn't available. Family dinner menu. Top quality at a reasonable price.

If you have time, drive downtown (three miles) to the full range of dining.

Reedsburg, WI (Reedsburg Muni—C35)
Longley's Restaurant 🍔🍔🍔

Rest. Phone #: (608) 524-6497
Location: Easy walk
Open: Daily: Breakfast, Lunch, and Dinner

PIREP We've gone to Longley's quite a bit, usually for breakfast. If you make your way to the back room (to the right of the bathrooms and farther back) there are usually a few pilots.

One time we came and asked at the front desk it there were any pilots eating in the back and the waitress sharply said, "*No*, there are *no* pilots here." And when we looked in the back, sure enough, about eight pilots were ordering breakfast.

Ha! Guess our reputation precedes us.

Rice Lake, WI (Rice Lake Regional—Carl's Field—RPD)
Crossroads Cafe 🍔🍔

Rest. Phone #: (715) 458-2511
Location: About a mile
Open: Daily: Breakfast, Lunch, and Dinner

PIREP Jerry (or Gerry) from Rice Lake Regional airport recommended that we visit the Crossroads Cafe, about a mile from Carl's Field, Rice Lake, WI (RPD) Rice Lake Air Center (715) 458-4400.

It's close by, great service, I recommend the hot beef.

Richland Center, WI (Richland—93C)
Peaches Restaurant 🍔🍔🍔

Rest. Phone #: (608) 647-8886
Location: 1/4 mile
Open: Daily: Lunch and Dinner

PIREP Peaches Restaurant was the April stop on our airport's (Wautoma-Y50) monthly fly-out. Peaches is approximately a quarter mile (walking distance) from the airport ramp. (Walk down main entrance, turn left. Take a right past the bridge to the stop sign. You'll see Peaches across the road on Hwy 14 on your right.) *Fabulous* buffet offered Saturday nights and Sunday (11 AM–2 PM). Five of us had the Sunday brunch recently—less than $50 for all of us. All fresh, homemade food and lots of it: prime rib, BBQ ribs, crab legs, several types of fish, deep-fried large shrimp, buttered noodles, seafood dishes, potatoes, fried and baked chicken—too many dishes, too numerous to mention. Homemade breads, rolls, salad bar, homemade pies, and out-of-this-world bread pudding with hot caramel sauce. All included in the buffet. Highly recommend this stop! Reservations not needed unless a group. Call ahead for group reservations at (608) 647-8886.

Shawano, WI (Shawano Muni—3WO)

Launching Pad Bar/Restaurant 🍔🍔

Rest. Phone #: (715) 524-4098
Location: Close by
Open: Closed Sunday

PIREP We ate at The Launching Pad, just across the street from the airport transient parking ramp. We arrived around 5 PM, and it being Friday Fish Fry night, there was a pretty good crowd. But there still were some empty tables. We were seated quickly, and had our drink order taken. When my wife ordered a glass of water, the waiter "tsked" and rolled his eyes. I thought that was rude but didn't say anything. Our drinks came quickly, but after an hour, no exaggeration, our food still hadn't come and the wait-staff had not checked on us again.

Several groups of people who'd come in after us and were obviously locals known to the wait-staff, received their food while we waited. I finally went up to the counter and told the waiter that we either had to have our food that minute, but "to go," or they could cancel the order, since I had to get our rental airplane back. He apologized and boxed our food, explaining that it was "just coming off the grill!" So we carried it back home in the baggage hold and ate it later. The food was so-so, and the trip was a disappointment.

There is another establishment at Shawano Airport, over behind the T-hangars right alongside the lake, called The Lighthouse. We've been there twice. The service was slow both times, though not as slow as The Launching Pad's, but the hamburgers are some of the best I've ever had.

PIREP The Launching Pad Bar/Restaurant directly across the street from the airport office in Shawano, WI has *great* burgers, fries, fish, and so on. You will get change back from a $10 bill every time. One paved and two grass runways.

The Light House 🍔🍔🍔

Rest. Phone #: (715) 526-5600
Location: Just behind the hangers
Open: Daily: Lunch and Dinner

PIREP The Lighthouse is at Shawano Airport just behind the hangars, north of the west end of runway 11. It is right on the lake and has a boat launch on the side of the building. There is a deck where you can sit outside and watch the boats coming and going. Their hamburgers are outstanding and reasonably priced, but they need to hire another cook during the summer. Both times we visited, the service was quite slow. But the hamburgers are worth the wait. If the weather is nice, eating at a table out on the deck is very nice.

Shell Lake, WI (Shell Lake Muni—SSQ)

Through the Woods Cafe 🍔🍔

Rest. Phone #: (715) 468-2969
Location: Short walk
Open: Daily: Breakfast and Lunch

PIREP A friend of mine who is also a pilot and I fly almost weekly to a little cafe approximately .7 of a mile from Shell Lake, WI municipal airport. The name of the cafe is Through the Woods.

The 7/10 of a mile walk takes you along the shore of Shell Lake. The breakfasts are fantastic, real down-home cooking with fair prices. The airport is simple. The traffic pattern either has you flying over the lake on down wind, or has you over the water on final.

The cafe closes on Sundays at 2 PM, so come early.

Tomah, WI (Bloyer Field—Y72)
Burnstad's European Village 🍔🍔🍔

Rest. Phone #: (608) 372-3277
Location: Just off the field
Open: Daily: Breakfast, Lunch, and Dinner

PIREP There is a great restaurant around a quarter mile from the field called Burnstad's, which is part of a shopping area that includes a food store along with a very large selection of crafts.

The food and service were excellent, as well as inexpensive. The breads were all homemade, as were the soups. The desserts are also huge.

We called ahead and were met at the airport by Mr. Burnstad himself, who drove us back and forth to the restaurant.

Very highly recommended as a great experience.

PIREP Short walk to Burnstad's European Village with a nice restaurant and great atmosphere.

Watertown, WI (Watertown Muni—RYV)
Steak Frye 🍔🍔🍔🍔

Rest. Phone #: (920) 262-2222
Location: 1 block
Open: Daily: Lunch and Dinner

PIREP FBO Comments: Very friendly airport staff. We were welcomed into the pattern with current weather info. We were offered the courtesy car but opted to walk the short distance to the restaurant.

PIREP Flying into Watertown Municipal was a very pleasant experience. We were offered the courtesy car but opted to walk the block or so to Steak Frye restaurant for lunch. The food (a burger and Philly cheesesteak) and service were very good, and we left stuffed for about $15. I definitely want to go back for the do-it-yourself steaks.

PIREP We stopped at Watertown overnight on our way to the big show at OSH. We were allowed to camp out overnight on the field. There are three places to eat within about a 300-yard walk of the airport. Two of the restaurants are fast food type. The people at the FBO recommended that we try the Steak Frye restaurant. We had an excellent meal at a very reasonable price. You can choose your steak and cook it yourself or they will cook it for you. This is a very good stop prior to joining the line of traffic going into OSH.

Waukesha, WI (Waukesha County—UES)
Brisco County Grill Restaurant 🍔🍔

Rest. Phone #: (262) 251-8330
Location: About a mile
Open: Daily: Lunch and Dinner

PIREP Brisco County Grill Restaurant recently opened (a national chain?). I had lunch there yesterday with a pilot buddy—reasonable prices, burger, fries, chicken dumpling soup, garlic bread—very good food and service.

AMF Waukesha Lanes 🍔🍔

Rest. Phone #: (262) 544-9600
Location: Down the street
Open: Daily: Lunch and Dinner

PIREP Waukesha has a great bowling alley right down the street from the FBO. AMF Waukesha Lanes has a fair menu of pub grub. It is a great stop in the afternoon for a quick bite and some people watching.

The Hundred Dollar Hamburger

Wyoming

Alpine, WY (Alpine—46U)

Red Barron Cafe 😋😋

Rest. Phone #: (307) 654-7507
Location: 1 mile from airport
Open: Daily: Breakfast, Lunch, and Dinner

PIREP In Alpine Way (sometimes called Alpine Junction) (46U), there is a small restaurant owned by Ed Browning, of the 1970s Red Barron unlimited racer. It is a very nice little restaurant. Ed was very friendly. There are several photographs of his airplanes on the walls. The airport is about one mile from the airport. I carried bicycles with me for ground transportation.

Buffalo, WY (Johnson County—BYG)

Virginian Restaurant and Occidental Hotel 😋😋😋

Rest. Phone #: (307) 684-0451
Location: A few miles away, call for pickup
Open: Daily: Breakfast, Lunch, and Dinner

PIREP We met an incredible woman at Buffalo, WY, Dawn Wexo. She is the owner of the historic Occidental Hotel and the Virginian Restaurant. When we arrived at Buffalo, WY, Johnson County Airport, there wasn't a soul in sight. I used the pay phone and picked the hotel with the unusual name of Occidental Hotel. Dawn answered the phone and was so kind, she told me that she had just rented her last room but she would give me the number to another hotel in the town. Dawn even told me that she would come and pick us up at the airport and take us to the other hotel. As fate would have it, the History Mansion House Hotel did not have anyone that could give us a ride. So I called Dawn and she came within 10 minutes. Dawn took us to our hotel and then drove us to the Virginian Restaurant that was located in the Occidental Hotel.

The service was excellent, the decor was beautiful, and the food was fantastic. We chose to walk back to the hotel because we had stuffed ourselves and needed the exercise. The next morning I received a call from Dawn; she told me that she was ready to drive us back to the airport whenever we were ready.

We will return to Buffalo, WY with prior reservations for a stay in the beautiful Occidental Hotel. This is truly a getaway retreat.

Cheyenne, WY (Cheyenne Regional / Jerry Olson Field—CYS)

Sanford's Grub 'n Pub 🍔🍔

Rest. Phone #: (307) 634-3381
Location: 3 minutes by crew car
Open: Daily: Lunch and Dinner

PIREP   How about an Atom Bomb or a Sloppy Jaloppy? Those are just two of the amazing "junkyard burgers" at Sanford's. You can even get them mild, medium, or hot if you ask nicely. Sanford's is fun. It has got the largest menu (literally) I've ever seen and tons of choices of food. The burgers were just one page of the menu. The atmosphere was entertaining—lots to look at. Memorabilia, photographs, walls lined with bats, and the ceiling covered with sports team flags. The one thing that stood out was the roll of paper towel next to every table. It was on a pipe contraption that was connected to a hubcap as the base.

We borrowed a car at the airport and it takes about three minutes to get there.

PIREP   Sanford's Grub 'n Pub also has a great chicken fried steak too, but we are rarely in that area.

PIREP   There's a great restaurant downtown called Sanford's Grub 'n Pub, 1.3 miles from the airport. Sky Harbor FBO has a courtesy car. However, since this was a multiplane $100 hamburger expedition (5 planes, 11 people), we rented a limo to get us all there and back. I unabashedly plug Easy Street Limousine Service as a way to get to dinner if the courtesy car is already occupied.

Sanford's has a huge menu: lots of appetizers, salads, and main courses from the chicken, steak, sandwich, and Cajun food groups. And then, of course, there are hamburgers. About 25 varieties. Names like The Pile, Thermostat, and my favorite, the Atom Bomb. Don't tell 'em you want it hot unless you really mean it.

Riverton, WY (Riverton Regional—RIW)

Airport Cafe 🍔🍔🍔🍔

Rest. Phone #: (307) 856-2838
Location: On the airport
Open: Daily: Breakfast, Lunch, and Dinner

PIREP   Riverton, WY, on the airport, at the commercial terminal building (not the FBO) is the Airport Cafe. It is so good and reasonably priced, the locals drive out from town for Sunday dinner! "Mom" bakes the pies, and there were 30 of them available!

Unfortunately, on weekends, the FBO is not manned (different operation than the Airport Cafe) and they charge a $20 call-out fee for fueling an aircraft. However, the good news is if they fuel two or more aircraft on the same run, there is no call-out fee.

For good fuel prices in the Riverton area, fly north to Thermopolis!

Saratoga, WY (Shively Field—SAA)

The River Street Deli 🍔🍔🍔🍔

Rest. Phone #: (307) 326-8683
Location: 1/2 mile, downhill
Open: Daily: Lunch and Dinner

PIREP   Come to Saratoga, south-central Wyoming (SAA):

- Long runway, good services
- Walk to town, stay at a Victorian hotel

- Run the North Platte River, whitewater's up now, fishing and canoeing by early summer
- Fish Saratoga Lake
- Beautiful valley scenery, Continental Divide (7000' level here) views
- Free 24-hour mineral hot pool
- Golf, golf, golf
- Tennis
- Galleries
- More cowboys than you care to look at

Eat at the River Street Deli, best home-cookin' on the planet and they're pilot-friendly!

The deli is about half mile downhill from the airport, in the center of town, near the bridge over the mighty North Platte River. It's a small town, hard to miss the deli.

The deli specializes in home-cooked sandwiches, all premium quality meats, all grilled, on custom-baked breads. This should give you some insight: the favorite is the meatloaf sandwich, grilled with smoky barbecue sauce (this is *kinda* like a burger, right?). Last year's tourist customers are just now hitting town and saying, "Thank God you're still open—I've been wanting a *insert sandwich name here* since we left *insert hometown here*."

Other sure bets are the Papa Pastrami (cream cheese and pastrami) and the Buster's Roast Beef (5 ounces of marinated roast beef, top-secret mild horseradish sauce). Special attention paid to pilots (sympathy, understanding, tolerance for lies, and so on but unmitigated scorn for all taildragger pilots). In other words, pilots will get the straight story on what's up in town, what to see, cheapest drinks, and so on. Rumor is that you can even get a ride back to the airport if you ask *real* nice (but it's much less than a mile anyway)!

Best yet: the deli's now hiring, so taildragger pilots only have to worry about the cost of gas it takes to *get* here.

Lazy River Cantina 🥪🥪🥪

Rest. Phone #: (307) 326-8472
Location: 1/2 mile
Open: Daily: Lunch and Dinner

PIREP We flew in on a Sunday morning from Boulder. The FBO wins the "grim" award for unfriendly, but they did efficiently fuel the plane and provided a tie-down. Also had a courtesy car available, although the walk into town doesn't really require it.

The deli is closed on Sundays and we ate breakfast at the Lazy River Cantina. Good, solid, hometown type of food with friendly service. Surrounded by townsfolk as well as local ranchers, a nice atmosphere.

Shoshoni, WY (Shoshoni Muni—49U)

The Walleye Cafe 🥪🥪🥪

Rest. Phone #: (307) 876-2481
Location: 3/4 mile
Open: Daily: Lunch

PIREP I just returned from Shoshoni, WY (49U). The Walleye Cafe, approximately three-quarters of a mile from the airport, has what is probably the best biscuits and gravy I have ever eaten. (Sorry, Mom.) They have very good service and a complete menu; however, I can't remember the prices (reasonable) exactly.

Thermopolis, WY (Hot Springs Co-Thermopolis Muni—THP)

Legion Supper Club 🥪🥪🥪🥪

Rest. Phone #: (307) 864-3918
Location: Across the parking lot
Open: Daily: Breakfast, Lunch, and Dinner

PIREP   At the Thermopolis Airport, yes, exactly across the parking lot, is the Legion Supper Club. Great food, prices not cheap, but reasonable for the value received. I had the NY strip steak and it was *great*!

Thermopolis, WY airport is up on the ridge, northwest of town. The airport managers are wonderful, local folks, Walt and Gwenda Urbigkit. You know you're at a real airport when the airport dog is as friendly as Gin (whose last known cohort was Tonic, named for a previously favored drink of the owners!).

Runway slopes uphill to the south (19), so until winds are > 15 knots, land 19, takeoff 01. If in doubt, always request the local practice discussion from the airport operator, or the local instructor(s).

G & W Aviation

Hot Springs County Airport (THP)

PO Box 1368

Thermopolis, WY

(307) 864-2488 FBO; (307) 864-2831 home/after hours fuel

PIREP   Legion Supper Club winter hours are as follows:

Closed Monday; closed Saturday mornings; Lunch 11 AM–2 PM; Dinner Sunday, Tuesday, Wednesday, and Thursday 5 PM–9 PM; Dinner Friday and Saturday 5 PM–10 PM; Sunday breakfast buffet served 9 AM–1 PM.

DATE DUE

NOV 26 2008			
ILL · 6/25/18			

HIGHSMITH 45230